METAL TOXICITY IN MAMMALS · 2

Chemical Toxicity of Metals and Metalloids

METAL TOXICITY IN MAMMALS

Volume 1 • Physiologic and Chemical Basis for Metal Toxicity

Volume 2 • Chemical Toxicity of Metals and Metalloids

METAL TOXICITY IN MAMMALS · 2

Chemical Toxicity of Metals and Metalloids

B. VENUGOPAL

AND

T. D. LUCKEY

Department of Biochemistry
University of Missouri, Columbia

PLENUM PRESS · NEW YORK AND LONDON

Library of Congress Cataloging in Publication Data

Venugopal, B
    Metal toxicity in mammals.

    Bibliography: v. 1, p.
    Includes indexes.
    CONTENTS: v. 1. Physiologic and chemical basis for metal toxicity.–v. 2. Chemical toxicity of metals and metalloids.
    1. Metals–Toxicology. 2. Semimetals–Toxicology. 3. Mammals–Diseases. I. Luckey, Thomas D., joint author. II. Title.
RA1231.M52L82   599'.02'4   76-44859
ISBN 0-306-37177-4 (v. 2)

© 1978 Plenum Press, New York
A Division of Plenum Publishing Corporation
227 West 17th Street, New York, N.Y. 10011

Printed in the United States of America

PREFACE

Chemical Toxicity of Metals and Metalloids is presented as the second volume of *METAL TOXICITY IN MAMMALS. Physiologic and Chemical Basis for Metal Toxicity* was presented as Volume 1.

The general principles of methods involved in the analysis of trace amounts of metals present in biologic material and the physicochemical properties of metals from the toxicological viewpoint were discussed in Volume 1. An elementary basis for the chemical and physiologic phenomena involved in the absorption of metals and their inorganic salts from the skin and gastrointestinal and respiratory tracts of mammals was reviewed, including metal involvement in carcinogenesis and teratogenesis, along with a short summary of the chemical toxicity of all the metals of the periodic chart.

In this volume, a systematic presentation of the chemical toxicity of metals is offered. All available data on metal toxicity were collected and reviewed. The toxicity of metals and their inorganic salts form the main focus; organometallics and radioactive metals are excluded, but some attempt was made to discuss the chemical toxicity and metabolism of radioactive metals and their salts. The chemical toxicity of each nonradioactive metal is discussed in three parts: (1) geographic occurrence of the metal and its mineral ores, uses of the metal and its compounds, its chemistry, and its nutrient status in mammals, (2) metabolism of the metal including absorption from different routes, distribution to various tissues, storage, excretion, and homeostasis, and (3) toxicity, i.e., its specific adverse effects in humans and animals at the whole organism, cellular, and molecular levels, carcinogenesis, teratogenesis, and other pathologic conditions. Detoxication mechanisms and interrelationships among metals or between metals and other compounds such as proteins and vitamins in influencing metal toxicity are discussed. The toxicity data for every metal are provided in tables, both in terms of MLD (minimum lethal dose), LD_{50}, and LD_{100} (expressed as milligrams of compound or metal or millimoles per kilogram body weight of the organism) and in our new pT scale. These data were compiled from original articles and the *Toxic Substances List* published by the National Institute for Occupational Safety and Health; U.S. Department of Health, Education and Welfare. Data published in the handbooks and review articles were checked with original articles, since most of the review articles report the toxicity data of the metal compounds as the toxicity of metal. Specific references are provided wherever possible. Most of the literature survey extended up to June 1976.

In the final chapter, the perspective of metal toxicity compared with industrial and environmental poisons is discussed. Interspecies comparison and comparison on the basis of different modes of administration are presented. The toxicity of the metals is compared in terms of groups and periods of the chart and in terms of simple and complex salts.

The authors are deeply indebted to proofreader Pauline Luckey, secretary Emily Deuser, general readers Drs. Douglas Frost and Dave Hutcheson, and to the National Aeronautic and Space Administration.

CONTENTS

Chapter 6
TOXICITY OF GROUP VI METALS AND METALLOIDS 233

Chapter 7
TOXICITY OF GROUP VII METALS 261

Chapter 8
TOXICITY OF GROUP VIII METALS 273

Chapter 9
CHEMICAL TOXICITY OF METALS IN MAMMALS 307

TOXICITY OF GROUP I METALS

1

Lithium (Li), sodium (Na), potassium (K), rubidium (Rb), cesium (Cs), and the radioactive francium (Fr) constitute subgroup A, and copper (Cu), silver (Ag), and gold (Au) constitute subgroup B of Group I elements in the periodic chart. All these metals exhibit monovalence; in addition Cu exhibits divalence and Au exhibits di- and trivalence; Na and K are essential macroelements in living tissues; Cu is an essential trace metal, and evidence is accumulating to indicate the possibly essential nature of Li. The other metals are not essential; Au is found to be stimulatory; Na and K are considered to be harmless and generally nontoxic; the order of toxicity among the other alkali metals is Fr > Li > Cs > Rb. Francium is highly toxic owing to its high β-ray radioactivity and high electropositivity, but will not be considered further, since it exists only in the form of unstable radioactive isotopes. The elements Rb and Cs at low dosage levels can partially substitute for either Na or K but are of low to moderate toxicity. The subgroup B metals are moderately toxic. Metals of subgroup A differ considerably from those of subgroup B in physicochemical properties and general metabolism; the latter have lower electropositivity, lower aqueous solubility of their salts, the ability to form covalent coordination compounds, and higher toxicity.

SUBGROUP IA

Metals of subgroup IA, the alkali metals, possess a single valence electron in their outermost orbits, with the next inner orbit possessing a stable octet of electrons. Their ionization energies are low compared with other metals; the alkali metals lose valence electrons more readily than do other metals, and hence they are the most highly electropositive metals. Their electropositivity increases with increased atomic weight; Fr is the most electropositive of all metals.

The alkali metals form stable, water-soluble electrovalent compounds, and their salts ionize strongly in aqueous media. These metals do not tend to form complex ions or coordinate bonds because their ions are large and carry a single small charge; the heavier K, Rb, and Cs ions can hold little water of hydration to form strong complexes, such as those formed by ferrous ions. In biologic fluids and tissues the alkali metals exist as ions and are rarely bound directly to proteins and other macromolecules.

Group I Metals

Atomic number Atomic weight	Name (Symbol)		(Core) Active electrons Usual valences	
Subgroup A			Subgroup B	
3	Lithium (Li)	(He) 1 + 1		
6.941				
11	Sodium (Na)	(Ne) 1 +1		
22.99				
19	Potassium (K)	(Ar) 1 +1	29 63.55	Copper (Cu)	(Ar) 8, 1 +1, +2
39.10				
37	Rubidium (Rb)	(Kr) 1 +1	47 107.87	Silver (Ag)	(Kr) 8, 1 +1
85.47				
55	Cesium (Cs)	(Xe) 1 +1	79 196.97	Gold (Au)	(Xe) 8, 1 +1, +2, +3
132.91				
87	Francium (Fr)	(Rn) 1 +1		
(223)				

The metabolic behavior of the alkali metals is characterized by rapid and almost complete absorption into the blood irrespective of the mode of administration, fast distribution to almost all tissues from the blood, and effective urinary excretion. Lithium ions move across biologic membranes by passive diffusion, sodium ions by diffusion and active transport, potassium ions by active transport, and rubidium and cesium ions by facilitated diffusion. The major transient deposit sites for the alkali metals are the skeletal muscles and bones; these metals are in equilibrium with plasma stores and are rapidly depleted by excessive excretion. Potassium and sodium ions are involved in the maintenance of ionic, electric, and osmotic equilibrium conditions in living tissues. The other alkali metals substitute for Na^+ or K^+ or both and disturb the cellular environment, especially the pH, osmolality, and ionic equilibrium and thereby change the physico-chemical properties of the macromolecules. Due to its greater permeability, Li^+ is more effective than Rb^+ and Ca^+ in this substitution.

Lithium (Li)

Lithium occurs in nature in the form of complex silicates, spodumene and pentalite (lithium aluminum silicates), lepidolite, and eucryptite (potassium lithium alumium silicate). The earth's crust contains about 30 ppm Li, and seawater contains 100 ppb. Lithium occurs in traces in many plants grown in soils which contain as much as 100 ppm Li. Lithium is present in animal tissues. Humans ingest about 2 mg Li daily; about 2.2 mg is present in adults. Lithium is stimulatory in some animal systems, but it is not considered an essential nutrient.

Lithium and its salts are used extensively in industry and medicine. Lithium metal is used in alloys such as lithium-hardened bearing metals; it is used as a coolant or heat exchanger in reactors, as fuel in missiles, and as a catalyst in the manufacture of synthetic rubber and multipurpose lubricants. Lithium salts are used in making heat-resistant glass; in ceramics lithium salts are used for fluxing, vitrification, and bonding; LiCl is used as a dehumidifying agent in air conditioning; and alkaline batteries contain LiOH as an electrolyte. Prisms of LiF are used in infrared spectrophotometers. In medicine lithium benzoate is used as a lubricant for compressing tablets and LiBr as a hypnotic and a sedative. Both $LiCO_3$ and lithium citrate are used extensively to treat manic psychosis. Lithium and its salts are not industrial health hazards, but Li poisoning can occur as a result of therapeutic overdose. Lithium poisoning through dietary sources is rare, but cases resulting from excessive use of LiCl as a NaCl substitute used to occur frequently. The use of lithium for predation aversion in coyotes has been suggested, but apparently the concentration required to be effective approximates the lethal dose (Gustavson *et al.*, 1974).

Chemistry

Lithium is the lightest metal known; it exhibits monovalency and forms water-soluble, electrovalent compounds. Lithium is highly electropositive, has high ionization potential, and does not generally form coordinate covalent compounds. It resembles the alkaline earth metals more than the Group I alkali metals in the relative insolubility of its carbonate and hydroxide in water and in the greater solubility of its bicarbonate. The higher solubility of lithium urate was utilized by Garrod to treat gout in the 19th century. Within the group, Li resembles Na more than K in its chemical properties. Lithium readily forms a stable hydride, LiH. Lithium invariably occurs in the ionic form in tissues. There are no specific reports of Li complexing directly with biologically active macromolecules and polymers, except as indicated below.

Metabolism

Studies on Li metabolism became extensive after Cade (1949) found Li to be very effective against the manic phase of manic-depressive psychosis. Lithium metabolism has been reviewed extensively (Schou, 1957, 1958; Stokinger, 1963; Diding *et al.*, 1969; Noyes, 1969; Shopsin, 1970; Maletsky and Blachly, 1970; Davis and Fann, 1971; Shrader, 1972; de Feudis, 1973; Singer and Rotenberg, 1973; Angino *et al.*, 1972; and Johnson, 1975). The stimulatory character of Li was reviewed by Luckey (1975b).

In mammals Li^+ shares some of the properties of extracellular Na^+ and intracellular K^+ but is not metabolized precisely like Na^+ or K^+. Lithium displaces potassium intracellularly and accumulates at the expense of potassium. Evidence is accumulating to suggest that Li is essential for animals; subsituting LiCl for NaCl reduces hypertension in hypertensive patients; Li^+ is found to be specific and essential for the slow respiration in nuclear membranes (Abruden-Bodea and Bodea, 1973); and lithium resembles insulin in stimulating the uptake of glucose in isolated diaphragm (Bhattacharya, 1964).

Although no deficiency symptoms were found in rats fed low-Li diets (< 0.01 ppm), Patt (1976) noted that the pituitary maintained a constant high Li level (0.14 ppm). All other tissues reflected the dietary level of Li. Such effective homeostasis in the pituitary raises intriguing questions: Could hypothalamic releasing factors or pituitary peptide hormones be Li-specific ionophores? And might Li be involved in their storage, release, or biologic function?

The absorption of Li^+ from the digestive tract and from sites of parenteral administration is rapid and complete. Inhaled dusts of Li salts are also readily absorbed from all parts of the respiratory tract. Lithium is not bound to any specific plasma protein, but moves freely across semi-permeable membranes. Rat pituitary and bone contain much more Li than do brain, kidney, adrenals, or lung; erythrocytes contain more than plasma or large organs. The rate of Li^+ movement across the cell membrane is relatively slow compared with that of Na^+, and much slower than that of K^+. The reduction of plasma volume noted after Li administration is due primarily to water accompanying Li into the cells; each meq of Li^+ (6.94 mg) carries about 10–12 ml of water into cells. Lithium is distributed to all soft tissues, including thyroid, adrenals, uterus, ovary, and placenta. Lithium moves rapidly into the kidney, slowly into liver, muscles, and bone, and at a still slower rate into the brain and erythrocytes. Representative values found by Patt (1976) in rats fed commercial and laboratory diets are (ppb, dry weight): bone, 187; pituitary, 97; adrenal, 58; heart, brain, lung and kidney, 32; thymus, erythrocytes, and spleen, 20; and liver, blood, and plasma, 10. These values are somewhat lower than those reported by Echner and Opitz (1974). Bone reflects Li intake more than other tissues. In rats fed K- or Na-deficient diets, Li accumulates in the tissues. Under normal conditions, and if the animals are not exposed to excessive dietary Li, it does not accumulate nor do Li levels increase with age. Absorbed Li is excreted predominantly in urine, and to a lesser extent in feces, sweat, saliva, tears, and semen. About 50% of ingested Li is rapidly excreted in the urine within 6–8 hr; the rest is excreted slowly over a period of 2 weeks. The fractional urinary excretion of Li varies directly with Na and inversely with Li load. The fractional absorption of Li in the kidney is about 0.8, the same as for Na, indication Li reabsorption in the proximal tubules (Schou, 1957). However, large doses of Li will disturb the Na balance and cause serious Na depletion (Trautner and Morris, 1955).

Toxicity

Lithium intoxication develops slowly; chronic toxicity symptoms following ingestion of nonlethal doses of LiCl with low-NaCl diets are gastrointestinal irritation, tremor, thirst, and polyuria. Acute toxicity symptoms are predominantly central nervous system effects. The overall Li toxicity is a general effect on K-sensitive metabolism of cells leading to the breakdown of cellular metabolism.

Lithium toxicity studies were conducted to establish the lethal dose in

TABLE 1-1. Lithium Toxicity

Compound	Animal	Route[a]	Toxicity	Compound mg	Metal mg	mM	pT
Lithium hydride	Rat	inhal	MLD	22[b]	19.2	2.76	2.56
LiH							
Lithium fluoride	Guinea pig	oral	LD_{100}	200	53.5	7.71	2.11
LiF	Guinea pig	sc	MLD	2000	535	77.1	1.11
Lithium chloride	Mouse	ip	LD_{50}	604	99	14.2	1.85
LiCl	Mouse	ip	LD_{50}	1100	180	25.9	1.59
	Mouse	ip	LD_{50}	1060	172	24.8	1.61
	Rat	oral	LD_{50}	757	124	17.9	1.66
	Guinea pig	sc	LD_{100}	620	101	14.6	1.84
	Rabbit	sc	LD_{100}	531	87	12.5	1.90
	Cat	sc	LD_{100}	400	65	9.37	2.03
Lithium perchlorate	Mouse	ip	LD_{50}	1160	75.6	10.9	1.96
$LiClO_4$							
Lithium carbonate	Rat	oral	MLD	710	131	18.9	1.72
Li_2CO_3	Dog	oral	LD_{50}	500	94	13.5	1.87
	Man	oral	Txc[c]	7	1.31	0.19	3.72
Lithium acetate	Man	oral	Txc	24	1.63	0.23	3.63
$Li(C_2H_3O_2) \cdot 2H_2O$							

[a]inhal = inhaled; sc = subcutaneous; ip = intraperitoneal.
[b]Exposure of 4 hr; number of mg/m^3.
[c]Txc = dose at which toxic symptoms are manifested.

experimental animals; those data are summarized in Table 1-1. Cats appear to be more susceptible than other animals to Li toxicity, and LiF is the most toxic of the Li salts studied. The lethal plasma Li level is about 8 mM (5.5 mg Li/100 ml) for rats. Lithium toxicity affects the following systems in mammals: (1) gastrointestinal, (2) renal, (3) neuromuscular, (4) central nervous, (5) cardiovascular, and (6) endocrine. Anorexia, nausea, and diarrhea are due to gastroenteritis resulting from osmotic disturbance in the digestive tract. Tremor, ataxia, weakness, clonus, and hyperactive reflexes are caused by the derangement of neuromuscular activity. Clinical symptoms of Li toxicity in the central nervous system are slurred speech, blurred vision, dizziness, sensory loss, convulsions, and stupor. Cerebellar and basal gangliar mechanisms are involved in brain toxicity. Polyuria, glycosuria, and weight gain due to water and salt retention are caused by renal toxicity. Decreased thyroid function and goiter are the main endocrinologic toxicity symptoms.

Lithium toxicity is more prevalent and serious in males than in females as shown by retarded growth and decreased survival in rats fed Li_2CO_3

(Andreoli, 1968). Lithium chloride proves toxic to man when used as a NaCl substitute or when the serum Li level reaches 3.0 meq (2 mg Li/100 ml) (Corcoran *et al.*, 1949; Hanlon *et al.*, 1949; Greenfield *et al.*, 1950). In rats Li inhibits osmotic gradient formation and interferes with both solute and water absorption; it decreases the intestinal absorption of glucose and water. The cation pump mechanism is not able to extrude Li from the epithelial cells. For the animal, the overall result is nausea and diarrhea (Rosenweig and Hendrix, 1967). Lithium excretion lags behind intake, if high doses of Li^+ are taken with a Na^+-deficient diet, Li^+ accumulation is accentuated. Acute Li toxicity causes excessive renal excretion of Na and K. The lowering of the K level in heart muscle causes pulse irregularities, hypotension, circulatory failure, and collapse. Renal function collapses with the gradual development of polyuria, which, in turn, develops into oliguria and azotemia. High Na intake prevents toxic injury because Na competes with Li for reabsorption in the renal tubules and induces Li excretion.

Chronic toxicity impairs the concentrating and acidifying abilities of the kidney. Lithium toxicity may cause diabetes *insipidus* by interfering with water metabolism (Singer *et al.*, 1972). Lithium inhibits vasopressin stimulation of renal adenyl cyclase and induces polyuria (Dousa and Hector, 1970). This inhibition is reversible, indicating that Li interferes with hormone (vasopressin action) (Geisler *et al.*, 1972; Wraae *et al.*, 1972). In the kidney, adenyl cyclase and cAMP mediate ADH and PTH hormone responses which are responsible for water reabsorption and PO_4^{3-} and Ca^{2+} reabsorption, respectively. Singer and Rotenburg (1973) postulated that Li alters the cellular microenvironment necessary for hormone action; other hormones whose responses are mediated by adenyl cyclase and cAMP and which might be susceptible to Li^+ inhibition are TSH, LH, epinephrine, and glucagon.

The neuromuscular activity is affected more by transient Na or fluid intake and renal damage than by Li. In the brain Li^+ influences the activities of epinephrine and norepinephrine by accelerating their presynaptic destruction, inhibiting the neuronal release of norepinephrine and serotonin, and increasing the neuronal uptake of norepinephrine. Lithium also inhibits catecholamine metabolism and decreases the cerebral glutamate content, which may be directly related to lithium inhibition of glutamate dehydrogenase activity (Schoffeniels, 1966; De Feudis and Delgado, 1970). A decrease in glutamate may be responsible for the sedative effect, and, with high Li intake, glutamate depletion causes clinical symptoms similar to those of Li toxicity in the central nervous system. Release of γ-amino butyric acid (GABA) in the brain produces a sedative effect, and GABA is shown to be released dramatically *in vivo* from rat cerebral cortex following Li administration (De Feudis and Mitchell, 1970). Lithium blocks the sodium-dependent uptake of GABA by the cerebral cortex (Iverson and Neal, 1968) and the sodium-dependent binding of GABA (Elliot and Van Gelder, 1970). Lithium-mediated inhibition of acetylcholine release from the cerebral cortex also can cause heavy sedation; this inhibition has been demonstrated in cats *in vivo* (Bjegovic and Randic, 1971).

The toxic action of Li on heart muscle and its functioning is complex. Cardiac conduction depends upon specific cationic distributions and flux. Lithium ions enter cardiac cells, and the intracellular K^+ diffuses into the surrounding medium; this substitution disturbs the ionic environment and equilibrium. Lithium does not exit the cell during the nerve impulse as potassium does. Thus Li decreases spontaneous depolarization and nerve conduction for cardiac contraction. Lithium alters myocardial carbohydrate utilization and interferes with catecholamine metabolism. Decreased K^+ level, inhibition of adenyl cyclase with the consequent decrease in heart rate and metabolism, and the changed ionic environment of the contractile proteins contribute to circulatory failure and cardiovascular collapse in acute Li toxicity.

Thyroid depression and goiter development have been reported in patients after Li therapy (Schou *et al.*, 1958; Lazarus and Bennie, 1972) and can be reversed by simultaneous thyroxine treatment (Myers,1972). Shopsin (1970) reviewed Li toxicity and hypothyroidism. Acute Li toxicity causes Li accumulation in thyroid and pituitary glands, reduction in serum thyroxine levels, and an increase in serum TSH levels; Li inhibits TSH-stimulated adenyl cyclase. The elevated TSH stimulates thyroid growth and produces goiter in both humans and experimental animals. Inhibition of adenyl cyclase and the subsequent reduced cAMP level appear to block some step in the biosynthesis of thyroxine. In pregnant goats Li^+ causes abortion with complete degeneration of fetal liver; newborn kids have liver necrosis, and there is a high concentration of Li in the amniotic fluid (Boulos *et al.*, 1973).

In chronic toxicity Li interferes with carbohydrate metabolism in the brain and with both carbohydrate and lipid metabolism in liver and adipose tissues by disturbing hormone-regulated, cAMP-mediated biochemical processes. Liver glycogen levels decrease with corresponding increases in blood and brain glucose levels, caused by increased glucagon secretion (Plenge *et al.*, 1969). The increased carbohydrate metabolism results in excessive weight gain in patients undergoing Li therapy. Chronic Li intoxication inhibits the development of the follicle and ovum in rats, shown by a reduction in size and number of corpora lutea (Trautner *et al.*, 1958). Lithium depresses the metabolism of human sperm and renders human erythrocytes fragile.

Reports on the teratogenicity of Li in both mice and rats include skeletal abnormalities, incidence of resorption, and decreases in litter size and fetal survival. The incidence of cleft palate in young born to pregnant mice fed daily doses of 300 mg Li_2CO_3/kg (56 mg Li/kg) on days 6–15 of gestation was reported by Szabo *et al.* (1970). They established a dose of 200 mg Li_2CO_3 or 37.5 mg Li/kg per day as a possible threshold for teratogenicity. Similar teratogenic and more specific embryotoxic effects were observed in rats (Szabo, 1970; Wright *et al.*, 1970, 1971). Intraperitoneal injections of 50 mg LiCl per day were given to rats weighing 200 g on days 1, 4, 7, and 9 of gestation and 20 mg per day from days 10 to 17. Cleft palate, and eye and external ear defects occurred in 40% of the fetuses. The resorption of fetuses was high. The average dose per rat per day was about 8 mg Li per 200-g rat; that used in the mouse was 1.4 mg Li per 25-g mouse per day; but in terms of dosage/kg body weight, mice had the higher dosage. These levels given to the experimental animals are comparable to, or a little less than, oral therapeutic doses (400 mg of LiCl/kg per day or 65.5 mg Li/kg) given to patients for treatment of manic psychosis; the plasma Li level in rats under these conditions ranges from 1.0 to 1.7 meq (0.7–1.1 mg Li/100 ml). When lower oral doses, ranging from 7 to 23 mg Li/ kg per day, were fed to pregnant rats for 20 days, no skeletal abnormalities were seen in the newborn (Johansen and Ulrich, 1969). Gralla and McIllhenny (1972) also report no embryotoxic effects in newborn rats, rabbits, and rhesus mondeys whose mothers were fed Li_2CO_3 in doses ranging from 0.675 to 4.0 meq (4.68–27.7 mg) Li/kg per day; these levels are considerably lower than the levels in Wright's and Szabo's studies.

The incidence of malformed babies born to mothers receiving Li treatment is not significant (Schou and Amdisen, 1970). Aoki and Ruedy (1971) reported a severely deformed "lithium baby" born to a mother being treated for manic psychosis with Li_2CO_3; however, it was impossible to attribute the severe deformity to Li teratogenicity. Newborn children of women under Li therapy during their pregnancy show characteristic symptoms of Li intoxication (Wilbanks *et al.*, 1970), but Li does not seem to have teratogenic effects in humans at the therapeutic levels used in treating manic psychoses. Lithium teratogenicity in experimental animals is clearly dose-related and differs from species to species.

There are no reports of the existence of detoxication mechanisms for Li toxicity. However, as noted previously, a high Na concentration alleviates much of the Li toxicity syndrome.

According to Singer and Rotenburg (1973) a potential mechanism of the toxic action for Li is its degree of substitution for normal extracellular and intracellular cations in physiologic processes. The resulting change affects the cellular microenvironment, especially pH, ionic equilibrium, and osmolality, and the physicochemical properties of the biologic macromolecules. Lithium toxicity could be due to its capacity to disturb the normal processes of ion distribution which produce and maintain the required optimal osmotic conditions and the electrochemical gradients in living tissues. Other mechanisms of toxicity include (1) the inhibition of adenyl cyclase and cAMP formation, (2) the derangement of catecholamine metabolism, and (3) a disturbance in cerebral carbohydrate metabolism.

Sodium (Na)

Sodium is the most abundant of the alkali metals and is too active chemically to occur free in nature. Sodium chloride is the most abundant compound of Na; other minerals are soda niter, cryolite, amphibole, sodalite, and zeolite. Sodium ranks sixth in abundance among elements and forms about 2.8% by weight of the earth's crust; seawater contains 1.05% Na as NaCl. Sodium is present in all living matter. The normal human adult has about 100 g of Na, one-third present in the inorganic portion of the skeleton and most of the rest found in extracellular fluids of the body.

Metallic sodium is used extensively in industry to improve the structure of alloys, descale metals, purify molten metals, serve as a heat transfer medium, and as a constituent in the manufacture of sodamide and a variety of sodium salts, which are used in the paper, glass, soap, textile, and petroleum industries. Sodium salts such as the azide, borate, and bromide are used therapeutically; sodium fluoride is used as an insecticide.

Sodium poisoning is mostly accidental; exposure of workers in the caustic soda and soda ash industries, mistaken substitution of salt for sugar in feeding babies, and therapeutic overdosages are recurrent problems.

Chemistry

Sodium is highly electropositive and forms higher oxides, but its stable valence is $+1$. The ionization energy associated with the alkali group metal atoms shows that they have little attraction for their own valence electrons; these metals can form stable electrovalent compounds and their salts are water-soluble. Sodium is no exception, although it can form stable coordination complexes such as sodium benzoyl acetone, which has the characteristic properties of a salt; salicylaldehyde also forms a coordination complex with Na. However, such compounds are not likely to occur in living tissues. Sodium and other alkali metal ions are generally present as free ions in biologic fluids and tissues, but some binding of Na is reported for chondroitin, sulfate, brain lipids, and DNA (Forbes, 1963).

Metabolism

Sodium is essential for all living organisms. Extracellular functions in mammals include osmotic pressure regulation, buffer systems, CO_2 transport (as $NaHCO_3$), hydration of proteins, cell permeability, and solubilization of organic acids, while intracellular functions include, in addition to the above, neuromuscular irritability and Na pump action to regulate the intake of many metabolites. Diarrhea, vomiting, excessive sweating, and renal inefficiency can cause Na^+ depletion; these and dietary deficiency may produce loss of appetite, nausea, muscle atrophy, retarded bone development, poor growth, weight loss, and death from Na deficiency. The metabolism of Na has been well characterized and reviewed (Forbes, 1963). The average daily intake of adult humans varies from 5 to 15 g; the Na content of the body as a whole declines during growth in man and other species. Although Na is an essential nutrient the exact amount required by man and most other mammals is not known.

Sodium is rapidly and fully absorbed from all parts of the alimentary tract and from parenteral injection sites. Jackson and Smyth (1971) indicate the trifaceted complexity of Na absorption. At the brush border of intestinal mucosal cells Na penetrates by simple diffusion down an electrochemical potential gradient (Schultz, 1974). Extrustion of Na into the serosal fluid occurs by active transport which requires energy from aerobic oxidation. The active transport component is linked to K–Na-ATPase, which is inhibited by millimolar concentrations of glucose (Sernka, 1974). Skin and the lungs also absorb Na rapidly, by simple diffusion and ion exchange. Excretion of Na is mainly urinary, with appreciable amounts excreted in feces, sweat, and tears. Mammalian renal excretion of Na is a two-phase process involving glomerular filtration and reabsorption in proximal tubules; of about 600 g Na involved in 24 hr glomerular filtration, approximately 99.5% is reabsorbed in human adults. The small Na^+ absorption in the colon is important to maintain positive dietary sodium balance (Jackson and Smyth, 1971).

A homeostatic mechanism for Na functions at the renal excretory level; this homeostasis is controlled by aldosterone, posterior hypophyseal antidiuretic hormone, body pH, and thirst which is stimulated by osmoreceptors in the hypothalamus.

Toxicity

Sodium salts are generally considered to be nontoxic, although some anions contribute greatly to the toxicity of such salts as sodium chromate, arsenate, and vanadate. The hydroxyl ions contribute to the caustic action of the alkali metal hydroxides. The toxicity of sodium salts is summarized in Table 1-2. It appears obvious that certain anions obscure any sodium toxicity; arsenite, fluoride, nitrite, and sulfides are particularly confusing. Other salts, such as metaphosphates, are toxic probably because of their excess alkalinity rather than from simple Na excess. Therefore, more work has been done with relatively nontoxic salts such as NaCl.

TABLE 1-2. Sodium Toxicity

| | | | | Dosage/kg body weight | | | |
| | | | | Compound | Metal | | |
Compound	Animal	Route[a]	Toxicity	mg	mg	mM	pT
Sodium hydroxide NaOH	Rabbit	oral	LD_{50}	500	287	12.48	1.90
Sodium fluoride	Mouse	oral	MLD	97	53	2.31	2.64
NaF	Rat	oral	LD_{50}	180	99	4.31	2.37
	Rat	ip	MLD	28	15.3	0.67	3.18
	Guinea pig	oral	MLD	250	137	5.96	2.22
	Rabbit	sc	MLD	400	219	9.53	2.02
	Rabbit	oral	MLD	100	55	2.39	2.62
	Man	oral	MLD	75	41.1	1.79	2.75
Sodium-acid fluoride	Guinea pig	oral	MLD	200	74.2	3.23	2.49
NaF·HF	Guinea pig	sc	MLD	250	92.7	4.03	2.39
Sodium chloride	Mouse	ip	LD_{50}	2600	1020	44.4	1.35
NaCl	Rat	oral	LD_{50}	3000	1180	51.33	1.29
	Rat	ip	LD_{100}	5000	1970	85.7	1.07
	Rat	sc	MLD	3500	1380	60.0	1.22
	Guinea pig	iv	MLD	2910	1150	50.0	1.30
	Man	oral	MLD	8200	3230	14	0.85
Sodium chlorate	Mouse	ip	LD_{50}	550	119	5.18	2.29
$NaClO_3$	Rat	oral	LD_{100}	12000	2590	113	0.95
Sodium bromide	Mouse	ip	LD_{50}	5000	1120	48.7	1.31
NaBr	Rat	oral	LD_{50}	3500	780	33.9	1.47
Sodium bromate $NaBrO_3$	Rabbit	oral	MLD	250	38.1	1.66	2.78
Sodium iodide	Rat	iv	MLD	1300	199	8.66	2.06
NaI	Rat	oral	LD_{50}	4340	666	29.0	1.54
Sodium nitrate $NaNO_3$	Rat	oral	MLD	200	54	2.35	2.63
Sodium nitrite	Rat	oral	LD_{50}	180	60	2.61	2.58
$NaNO_2$	Rat	sc	MLD	15	5		
Sodium sulfate	Mouse	ip	MLD	193	62.5	2.72	2.57
Na_2SO_4	Mouse	ip	LD_{50}	227	73.5	3.20	2.50
	Mouse	iv	LD_{50}	130	42.1	1.83	2.74
	Rat	ip	LD_{50}	650	210	9.13	2.04
	Rat	iv	LD_{50}	115	37.2	1.62	2.79
	Rabbit	ip	LD_{50}	300	97.1	4.22	2.37
	Rabbit	iv	MLD	4470	1450	63.1	1.20
	Rabbit	iv	LD_{50}	65	21.0	0.91	3.04
	Dog	ip	LD_{50}	244	79.0	3.44	2.46
	Hamster	iv	LD_{50}	95	30.8	1.34	2.87
Sodium sulfite	Mouse	iv	LD_{50}	175	63.9	2.78	2.56
Na_2SO_3	Rabbit	oral	MLD	1180	430	18.7	1.73
Sodium bisulfite $NaHSO_3$	Rat	iv	LD_{50}	115	25.3	1.10	2.96

(Cont'd)

TABLE 1-2. (Cont'd)

Compound	Animal	Route[a]	Toxicity	Compound mg	Metal mg	Metal mM	pT
Sodium persulfate $Na_2S_2O_8$	Rabbit	iv	MLD	178	34	1.48	2.83
Sodium sulfide	Mouse	ip	LD_{50}	53	31	1.35	2.87
Na_2S	Rabbit	iv	LD_{100}	6	3.5	0.15	3.82
Sodium monophosphate NaH_2PO_4	Rat	im	LD_{50}	250	41.7	1.81	2.74
Sodium diphosphate Na_2HPO_4	Rat	ip	MLD	2000	648	28.2	1.55
Sodium triphosphate Na_3PO_4	Rat	ip	LD_{50}	326	137	5.96	2.22
Sodium metaphosphate $(NaPO_3)_6$	Rat	iv	MLD	130	29.3	1.27	2.89
	Rabbit	iv	MLD	130	29.3	1.27	2.89
	Dog	iv	MLD	140	31.5	1.37	2.86
Sodium borate $Na_2B_4O_7$	Rat	oral	LD_{50}	2600	295	12.8	1.89
Sodium carbonate Na_2CO_3	Rat	oral	MLD	4000	1740	75.7	1.12
Sodium cyanide	Mouse	sc	MLD	10	4.7	0.20	3.69
NaCN	Rat	oral	LD_{50}	6.44	3.0	0.13	3.88
	Rabbit	sc	MLD	2.2	1.03	0.045	4.35
	Dog	iv	MLD	1.3	0.61	0.026	4.58
	Dog	sc	MLD	6	2.8	0.12	3.91
Sodium acetate	Mouse	iv	MLD	240	67.3	2.93	2.53
$Na(C_2H_3O_2)$	Mouse	iv	LD_{50}	335	94.0	4.09	2.39
	Rat	iv	LD_{50}	380	106	4.61	2.34
	Dog	iv	MLD	3000	842	36.6	1.44
Sodium tricitrate $Na_3(C_6H_5O_7)$	Rat	ip	LD_{50}	1760	457	19.9	1.70
Sodium monoarsenate	Mouse	ip	LD_{50}	9	2.20	0.10	4.02
NaH_2AsO_4	Rat	ip	MLD	35	8.48	0.37	3.43
Sodium diarsenate	Rat	ip	MLD	30	7.24	0.31	3.50
Na_2HAsO_4	Rat	ip	LD_{50}	50	12.4	0.54	3.27
Sodium metaarsenite	Mouse	sc	LD_{50}	10	1.77	0.084	4.11
$NaAsO_2$	Mouse	ip	MLD	10	1.77	0.084	4.11
	Rat	ip	MLD	4	0.69	0.030	4.52
	Rat	oral	LD_{50}	41	7.10	0.31	3.51
Sodium selenide Na_2Se	Mouse	ip	50	3.4	1.25	0.055	4.26
Sodium molybdate $Na_2M_0O_4 \cdot 2H_2O$	Rat	ip	M	290	55.1	2.40	2.62

[a] iv = intravenous; im = intramuscular.

Sodium toxicity is best assessed by using NaCl. Excessive intake of NaCl in mammals causes a tendency to drink large amounts of water, diarrhea caused by osmotic changes in the digestive tract, stiff gait, salivation, muscular fibrillation, exhaustion, and death. The pathological signs at autopsy include a violent local inflammatory reaction in the gastrointestinal tract caused by capillary and venous congestion of the lamina propria and submucosa of the pyloric stomach, small bowel, cecum, and colon, and dehydration in other body organs such as the adrenals, brain, liver, and spleen. Death is attributed to respiratory failure associated with acute encephalopathy and congestion of body organs. Chronic toxicity symptoms include inhibition of growth, increased water intake and urinary volume, and osteosclerosis. Sublethal doses cause nausea, vomiting, abdominal stress, convulsive seizures, and diarrhea.

The toxic effects of chronic ingestion by rats of large amounts of dietary NaCl ranging from 2.8 to 9.8% of the diet were studied by Meneely and Ball (1958). Animals with diets containing 7% or more NaCl suffered from a syndrome resembling nephrosis. This syndrome is characterized by a sudden onset of massive edema, and by hypertension, anemia, lipemia, severe hypoproteinemia, azotemia, and finally death. Severe arterial disease was also observed. Among the animals eating less than 7% NaCl in the diet, the increase in blood pressure was related to the NaCl intake. A shortened life span was reported for rats fed a diet with 2.8–5.6% NaCl (Boyd and Shanas, 1963). (This amount is equivalent to 14–28 g/day per human adult.) A chronic oral toxic dose (LD_{100} in 100 days) in albino rats was 2.06–3.23 g NaCl/kg per 24 hr. In addition to the histopathological effects associated with NaCl toxicity, inhibition of spermatogenesis was observed. The toxic effects of NaCl, including congestive heart failure and hypertension in man and animals, have been known for many years (Ambard and Beaujard, 1904). The effects are due to increased blood volume, presumably from the entry of tissue water into blood to maintain the isotonicity of the blood. It is possible to produce NaCl lethality in pigs fed 0.5–1% NaCl in the diet if water is withheld; sodium concentration in the cerebrospinal fluid increases and histopathologic lesions include eosinophilic infiltration of the cerebral cortex. Excessive dietary Na and K intake in rats leads to a moderate increase in insulin activity (Lewis *et al.*, 1944). The lethal parenteral dose of NaCl in mice, rats, and guinea pigs ranges from 2.5 to 5.0 g/kg; the mean lethal dose is approximately equal to the total amount of NaCl reported to be present in the body of these animals.

Human infants develop acute toxicity symptoms after an oral intake of 4 g NaCl/kg body weight per day for five days. Adult humans can tolerate an intragastric drip of 0.8 g NaCl/kg for over 21 hr, but an intravenous infusion of 20% NaCl at a dose of 1 g NaCl/kg causes epileptiform convulsions, somnolence, and stupor. Sodium chloride is drastically hypertensive in humans whose NaCl intake is above 30 g/day for a long period of time (Meneely *et al.*, 1957; Meneely, 1973). Potassium chloride counteracts the toxic effect of chronic excess of NaCl; it ameliorates the hypertensive effect of NaCl in both humans and animals. Serum hypernatremia occurs owing to hyperactivity of the adrenal cortex, or to dehydration associated with diabetes insipidus or excessive sweating; these conditions are different from dietary NaCl toxicity.

Potassium (K)

Potassium is too active chemically to occur in the free state; its most common minerals are sylvite (KCl), carnallite (KCl · MgCl$_2$), and polyhalite (K$_2$SO$_4$ · MgSO$_4$ · 2CaSO$_4$). Potassium is the seventh most abundant element on the earth, forming 2.5% of the earth's crust; seawater contains about 0.038% K. Plants contain more K than Na, while the opposite is true of animals. Fresh fruits and nuts such as almonds have a very high K content. The adult human body contains about 140 g K, 98% of which is held intracellularly. Potassium is an essential nutrient.

An alloy of K and Na is used as a heat transfer fluid or coolant in internal combustion engines and nuclear reactors. Diverse potassium salts are used in the manufacture of such products as textiles, paper, glass, ceramics, photographic plates and papers, insecticides, and fertilizers. Potassium salts are used therapeutically in human and veterinary medicine. Poisoning caused by K compounds is rare and mostly accidental, from the rapid intravenous injection of K salts, the substitution of KCl for sugar in baby foods, or accidental splashing in the caustic potash industry. Therapeutic overdoses also cause potassium poisoning.

Chemistry

Potassium exhibits a characteristic oxidation state of +1 and forms electrovalent compounds; due to its small electric charge it does not tend to form complex ions or coordinate bonds. Potassium forms electrovalent bonds with complex anions such as chloroplatinates and chloropalladates. Potassium salts are higher water-soluble and readily undergo ionization. In tissues K exists as free, positively charged ions.

Metabolism

Potassium is essential for muscle contraction, nerve function, cell permeability, and intracellular osmotic pressure and buffering. Diarrhea, emesis, diuresis, starvation, prolonged saline infusion, or dietary deficiency may lead to K deficiency, which is characterized by muscle weakness, cardiac arrhythmia, paralysis, bone fragility, sterility, adrenal hypertrophy, decreased growth rate, loss of weight, and death. The average daily intake of potassium salts for human adults ranges from 2 to 6 g; vegetarian diets contain more potassium salts than do meat diets, but the quantity required has not been established for man.

The metabolism and excretion of K in mammals is well reviewed in standard textbooks on pharmacology and physiology; Kones (1975) reviewed the metabolism of K in the heart. Potassium is readily and rapidly absorbed by passive diffusion in the high-conductance membrane of the upper intestine and may be excreted in the colon (Jackson and Smyth, 1971). About 90% of the ingested dose is absorbed. It is distributed to all tissues where it is the principal intracellular cation. About 15% of the total excreted is found in feces; the major portion of the remainder is excreted through the kidney in urine, and traces are excreted in sweat and tears. There is an unstable equilibrium between the high intracellular and the low extracellular concentration of K. The renal excretory mechanism is designed for efficient removal of excess K, rather than for its conservation during deficiency. Even with no intake of K, man loses a minimum of 15–30 meq (585–1170 mg) K per day. Renal K losses are exaggerated by diuresis, acidosis, and conditions which produce excessive secretion of adrenocortical hormone.

Toxicity

Potassium salts are not generally considered to be toxic; studies on the acute toxicity of K are rare. Available data on toxic doses of potassium salts in experimental animals are given in Table 1-3. The most toxic salts reflect the toxicity of the anions, such as chromate, arsenite, and fluoride. Chlorate ions cause irritation of the gastrointestinal tract and kidney tissue hemolysis of erythrocytes and methemoglobinemia. Potassium chloride, bicarbonate, and phosphate cause severe ulceration in the gastrointestinal tract at a level equivalent to 1000 mg K^+ ion when given to monkeys in the form of a pill (Earl *et al.*, 1966). Excessive dietary K intake in rats causes a moderate increase in insulin activity (Lewis *et al.*, 1944) and growth depression (Grunert *et al.*, 1950). A maximal nontoxic oral dose of KCl in man varies from 0.2 to 1.0 g K/kg per day, depending upon the efficiency of the individual renal excretory mechanism; lower doses sometimes cause impairment of renal function as shown by reduced inulin and urea clearance. If the serum K level reaches 30 mg/100 ml, paresthesia of extremities and general flaccid paralysis occurs. A serum K level of 40 mg/100 ml is fatal in man.

Clinical symptoms of acute oral potassium toxicity induced in experimental animals include tonoclonic convulsions, diarrhea, polydipsia diuresis, fever, prostration, and respiratory failure. Other toxic effects are dilation of the heart, collapse of the alveoli of the lungs, acute gastroenteritis, dehydration, loss in wet weight of body organs, and necrosis of renal epithelium (Boyd and Shanas, 1961). The central nervous system is paralyzed by excess K; the reflexes suffer, and then a medullary depression of the respiratory center leads to asphyxial convulsions. The toxic effect on the heart is attributed to an intraventricular conduction defect, which can be described from deviations on an electrocardiogram. Similar symptoms were induced by rapidly injecting KCl either intravenously or intramuscularly. Cardiac arrhythmias result from K excess or deficiency.

The toxic elevation of serum K (hyperkalemia) is restricted to human patients suffering from renal failure, advanced dehydration caused by diabetes insipidus, dehydration, shock, or adrenal insufficiency. This condition is also produced by injecting KCl solution intravenously, or by ingestion of KCl tablets (Baker *et al.*, 1964; Jefferson and Aukland, 1974). Hyperkalemia causes cardiac and central nervous system depression. Clinical symptoms of hyperkalemia in humans include mental confusion, weakness, vomiting, numbness, tingling, and a flaccid paralysis of the extremities (Bacon, 1974). Thus overdose of K pills or sudden infusion with high K concentrations are dangerous.

In mammals K is not stored in any specific tissues; there is no evidence for the presence of any detoxication mechanism. The efficient renal excretion is not considered a part of any homeostatic mechanism.

Rubidium (Rb)

Rubidium is present in small quantities, as RbCl, in minerals such as carnallite and beryl. There are no specific ores or minerals for Rb, although it is present in iron ores. Rubidium occurs in the earth's crust at about 120 ppm and in seawater at 120 ppb; RbCl is found in plant tissues and in most animal tissues; Rb is present in living tissues in higher concentrations than in the earth's crust or the environment, up to 1/2500 those of either K or Na. Rubidium may stimulate growth, but it is not essential for mammals.

Normal human adults contain about 320 mg Rb, which is mostly associated with potassium in intracellular fluids.

Industrial uses of rubidium salts are not extensive except in the manufacture of photoelectric cells; formerly RbI was used in iodide therapy.

TABLE 1-3. Potassium Toxicity

| Compound | Animal | Route | Toxicity | Dosage/kg body weight | | | |
| | | | | Compound | Metal | | |
				mg	mg	mM	pT
Potassium hydroxide KOH	Rat	oral	LD_{50}	365	254	6.51	2.19
Potassium fluoride	Rat	oral	LD_{50}	245	164	4.19	2.38
KF	Guinea pig	oral	MLD	250	160	4.09	2.39
	Guinea pig	sc	MLD	350	236	6.05	2.22
Potassium-acid fluoride	Guinea pig	oral	MLD	150	75	1.92	2.72
KHF_2	Guinea pig	sc	MLD	250	125	3.2	2.50
Potassium chloride	Mouse	ip	LD_{50}	552	289	7.4	2.13
KCl	Rat	oral	LD_{100}	2430	1270	32.5	1.49
	Rat	sc	MLD	1200	628	16.1	1.79
	Rat	ip	LD_{100}	825	432	11.1	1.96
	Rat	ip	LD_{50}	660	346	8.85	2.05
	Rat	iv	MLD	90	47	1.2	2.92
	Guinea pig	oral	LD_{50}	2500	1310	33.6	1.47
	Guinea pig	sc	LD_{100}	1140	597	15.3	1.82
	Guinea pig	iv	MLD	85	44.5	1.14	2.94
	Guinea pig	ip	LD_{100}	900	471	12.1	1.92
Potassium chlorate	Rat	oral	MLD	1500	478	12.3	1.91
$KClO_3$	Rat	ip	MLD	1500	478	12.3	1.91
Potassium sulfate K_2SO_4	Guinea pig	sc	MLD	3000	1350	34.5	1.46
Potassium pyrophosphate $K_4P_2O_7 \cdot 3H_2O$	Mouse	oral	LD_{50}	1600	650	16.7	1.78
Potassium triphosphate $K_3PO_4 \cdot 8H_2O$	Rat	oral	LD_{50}	1400	153	3.92	2.41
Potassium acetate $K(CH_3COO)$	Rat	oral	LD_{50}	3250	1290	33.0	1.48
Potassium carbonate K_2CO_3	Rat	oral	LD_{50}	1870	1060	17.1	1.57
Potassium cyanide	Rat	oral	LD_{50}	10	6	0.15	3.81
KCN	Mouse	sc	LD_{50}	6	3.6	0.09	4.04
	Dog	oral	MLD	3.8	2.28	0.058	4.23
Potassium thiocyanate	Rat	oral	LD_{50}	854	343	8.8	2.06
KCNS	Human	oral	MLD	80	32	0.82	3.08
Potassium antimonyl tartrate	Rat	oral	LD_{50}	115	13.8	0.344	3.45
$KSbC_4H_5O_7$	Mouse	oral	LD_{50}	600	72	1.84	2.73
	Mouse	ip	LD_{50}	50	6	0.15	3.81
	Mouse	sc	LD_{50}	55	6.6	0.16	3.77
	Mouse	iv	LD_{50}	65	7.8	0.20	3.70
	Human	oral	MLD	2	0.24	0.006	5.21

(Cont'd)

TABLE 1-3. (Cont'd)

| Compound | Animal | Route | Toxicity | Dosage/kg body weight | | | |
| | | | | Compound | Metal | | |
				mg	mg	mM	pT
Potassium arsenite $KASO_2 \cdot HASO_2$	Rat	oral	LD_{50}	14	2.15	0.055	4.25
Potassium borohydride KBH_3	Rat	oral	MLD	160	118	3.02	2.52
Potassium borofluoride KBF_4	Rat	ip	LD_{50}	240	74	1.89	2.72
Potassium chromate K_2CrO_4	Rabbit	sc	MLD	12	0.08	0.0021	5.68
	Human	oral	MLD	430	17.2	0.44	3.36
Potassium hexafluoro-titanate	Guinea pig	oral	MLD	200	65	1.66	2.78
K_2TiF_6		sc	MLD	450	146	3.74	2.43
Potassium niobate	Rat	oral	LD_{50}	3000	1280	32.8	1.48
K_3NbO_4	Rat	ip	LD_{50}	225	96	2.46	2.61
Potassium oxopentafluoro-niobate K_2NbOF_5	Mouse	oral	LD_{50}	130	33.8	0.87	3.06
Potassium permanganate $KMnO_4$	Rat	oral	LD_{50}	1090	269	6.89	2.16
Potassium perrhenate $KReO_4$	Mouse	ip	MLD	692	93	2.38	2.62
Potassium selenocyanate $KSeCN$	Rat	Ip	MLD	200	54	1.38	2.86
Potassium silicofluoride	Guinea pig	oral	MLD	500	177	4.54	2.34
K_2SiF_6	Guinea pig	sc	MLD	250	88.5	2.27	2.65
Potassium silver cyanide $KAg(CN)_2$	Rat	oral	LD_{50}	21	4.1	0.105	3.78
Potassium tantalum heptafluoride	Mouse	oral	LD_{50}	110	22	0.56	3.25
$KTaF_7$	Mouse	ip	LD_{50}	97	19.5	0.50	3.30
	Rat	oral	LD_{50}	2500	503	12.9	1.89
	Rat	ip	LD_{50}	375	75.5	1.93	2.71
Potassium tellurite K_2TeO_3	Dog	iv	MLD	35	5.4	0.138	3.86

Chemistry

Rubidium is similar to potassium and cesium in its chemical properties; it is monovalent and more electropositive than potassium. Like other alkali metal ions, Rb occurs mostly as free ions in tissues and biologic fluids. There are few reports of the binding of Rb to biologically active macromolecules or of the existence of Rb complexes in tissues. Rubidium poisoning in man and animals is rare.

Metabolism

Trace quantities of Rb do not appear to be a dietary essential for rats (Glendening *et al.*, 1956), other animals, or man, but Rb can replace K as an essential mineral for yeast (Laznitski and Szorenyi, 1934) and for certain bacteria (Macleod and Snell, 1950). Rubidium in small doses can partially substitute for K in the diet of experimental animals; this led to the speculation that Rb may have stimulatory effects (Luckey, 1975b). Dietary Rb prevented experimental dental caries in rats fed a cariogenic diet (Bobyleva, 1968). The metabolism of Rb was reviewed by Relman (1957) and Underwood (1971).

Studies on Rb metabolism reveal the general relationship between Rb and K. The physiological similarity between K and Rb is noticed in the pattern of their distribution, their excretion, and their capacity to affect the contraction of isolated frog hearts, the motility of spermatozoa, and yeast fermentation. In potassium-depleted animals the acid–base metabolism is disturbed due to increased intracellular and extracellular acidosis. Either K or Rb can restore this balance by displacing intracellular H^+ ions; the neutralizing effect of Rb is more pronounced than that of K. Rubidium and potassium both stimulate aldosterone production in rats (Bach *et al.*, 1967). Both Rb and K are present in relatively high concentrations in erythrocytes and muscle tissue and in lower concentrations in bone and plasma. The distribution pattern of Rb at the cellular level is different from that of K (Zorlein *et al.*, 1969). In mammals rubidium salts are rapidly and fully absorbed from the digestive tract and from injection sites of parenteral administration. There are no reports of Rb absorption from the lung after inhalation; however, the absorption from the lung should also be rapid, since cell membranes are freely permeable to Rb ions. After absorption Rb is distributed to all soft tissues, where Rb levels are high compared to other trace metals and range from 20 to 60 ppm on a dry basis (Sheldon and Ramage, 1931). Rubidium levels in muscle, liver, kidney, and brain may reach 100–200 ppm, depending upon the diet. Soybeans contain 160–225 ppm Rb; beef contains about 140 ppm. Glendening *et al.* (1956) found Rb levels in soft tissues up to 8000–10,000 ppm when rats were fed high and toxic levels of Rb. Monkeys accumulated orally ingested Rb in the brain, erythrocytes, and the pituitary (Spirtes and Gary, 1973). There is no other specific Rb accumulation in any individual organ or tissue (Stich, 1957), but there is a transient Rb accumulation in soft tissues during K depletion.

The excretion of Rb is rapid and mainly urinary, although some Rb is lost in sweat and tears. The average American adult consumes about 1.5 mg Rb daily, with urinary excretion accounting for 75% of the intake. Rubidium does not accumulate in humans with aging. This lack of accumulation suggests a poorly defined homeostatic mechanism for Rb.

Toxicity

Toxicity studies on Rb are few; the data are summarized in Table 1-4. Acute toxic symptoms are violent tetanic spasms and convulsion which precede death. Rubidium is not toxic to rats at 100 ppm of the diet, but at 1000 ppm growth retardation, impaired reproduction, and decreased longevity were observed. Young are born but do not survive beyond weaning (Glendening *et al.*, 1956). Rubidium substituting for potassium in a relatively low-potassium diet is more toxic than in a high-potassium diet. Addition of rubidium to a potassium-free diet permits almost normal growth for two weeks; then the animals develop toxic symptoms and die.

The possible substitution of Rb for K in the chick diet is limited, since Rb at the 0.4% level proved to be toxic, causing neuromuscular incoordination and irritability (Sasser *et al.*, 1969). The Rb:K molar ingestion ratio influences rubidium toxicity in rats (Meltzer and Liebermann, 1971). Subtoxic levels of Rb show toxicity when this ratio is 0.1 or greater; to a level equivalent to 10% of the dietary K, Rb substitutes for K, and beyond that level accumulated Rb disrupts the cell function by unknown mechanisms. Although the levels reported are high, intravenous administration of Rb is very toxic and should be avoided (Meltzer *et al.*, 1969). Rubidium is known to suppress NH_4^+ excretion in the urine. *In vitro*, Rb causes erythrocyte agglutination. Rubidium has a depolarizing effect on cell membranes which changes the membrane potentials; heart muscles can be affected by this action leading to myocardial irritability. The neuromuscular effect may cause convulsions due to acidosis, which in turn may be caused by profound changes in intracellular cation composition.

Cesium (Cs)

Cesium occurs in nature in lepidolite and pollucite (hydrated cesium aluminum silicate) which contain about 28% Cs. Carnallite and other alkali metal ores contain small amounts of Cs. The earth's crust contains 1 ppm

Cs, and seawater contains about 0.3 ppb; Cs is present in high concentration in radioactive fallout. Cesium is also present in trace amounts in vegetable and animal tissues. The human adult body contains about 1.5 mg Cs, which is mainly distributed in soft tissues.

Cesium is used in photoelectric cells and as a "getter" in vacuum tubes; the cesium atomic clock is based on the vibrational frequency of the cesium atom. Cesium has been proposed for ion propulsion systems in outer space and as an encapsulated energy source in place of ^{60}Co. Alkaline storage batteries contain CsOH; CsBr is used in X-ray fluorescent screens, in spectrometer prisms, and in absorption cell windows. In the past CsI was used in iodine therapy. Accidental cesium poisonings are rare. The high concentration of ^{137}Cs in radioactive fallout poses a health hazard, because ^{137}Cs disintegrates into radioactive Ba with powerful β and γ rays.

Chemistry

Cesium forms monoelectrovalent salts which are water soluble; Cs is the most electropositive nonradioactive metal. Double salts of Cs with mercuric chloride and bromide or with rhodium and vanadium sulfates are stable; Cs also forms complex salts with such anions as chloroplatinate and chloroaurate. Cesium complexes of biologically active macromolecules are not known, but Mraz *et al.* (1957) reported that muscle and kidney cells retain Cs in a bound form.

Metabolism

Cesium is not essential for mammals. Most studies on Cs in mammals involve radioisotopes of Cs (Hood and Comar, 1953; Ballou and Thompson, 1958; Ekman, 1961; Wassermann, 1962; Rosoff, 1963; Rundo, 1964). There are no reports about Cs involvement in metabolism or enzyme systems. Cesium can replace potassium to some extent in mammals (Relman, 1957). Small amounts of Cs prevent characteristic lesions in kidney and muscle of K-depleted rats (Follis, 1943). Data about the daily intake of Cs by man are not available.

The absorption of Cs from the digestive tract and from parenteral injection sites is rapid in mammals; absorption is about 90% in monogastric animals and about 80% in ruminants (Stara *et al.,* 1971). The rate of Cs absorption is highest in the small intestine. The distribution pattern of Cs to soft tissues is similar to that of K and Rb. Cesium occurs in traces in various soft tissues of mammals including the retinal layers of eyes in pigs, sheep, and oxen (Scot and Canaga, 1939).

Cesium differs from other alkali metals in its urinary and fecal excre-

TABLE 1-4. Rubidium and Cesium Toxicity

Compound	Animal	Route	Toxicity	Compound mg	Metal mg	Metal mM	pT
Rubidium hydroxide	Mouse	oral	50	900	747	8.74	2.06
RbOH	Rat	oral	50	586	488	5.79	2.24
Rubidium chloride	Mouse	ip	50	1160	825	9.65	2.02
RbCl	Rat	ip	50	1200	854	9.99	2.00
Rubidium iodide RbI	Rat	oral	50	4710	1900	22.2	1.65
Cesium hydroxide	Rat	ip	50	100	84.6	0.64	3.19
CsOH	Rat	oral	50	1026	869	6.54	2.18
Cesium chloride	Mouse	ip	50	1550	1330	9.99	2.00
CsCl	Rat	ip	50	1500	1120	8.41	2.08
Cesium bromide CsBr	Rat	ip	50	1400	874	6.58	2.18
Cesium iodide CsI	Rat	ip	50	1400	716	5.39	2.27
Cesium nitrate $CsNO_3$	Rat	ip	50	1200	818	6.15	2.21
Dicesium sulfate Cs_2SO_4	Guinea pig	ip	50	1270	931	7.00	2.15
Cesium carbonate Cs_2CO_3	Mouse	iv	50	940	767	5.77	2.24

Header note: Dosage/kg body weight spans the Compound mg, Metal mg, Metal mM, and pT columns.

tion; the ratio of urinary to total excretion of cesium varies from 0.71 to 0.91 in different mammalian species. Ruminants excrete part of their dietary Cs in the feces. Rats fed a K-deficient diet retain Cs in their soft tissues temporarily before excreting it; skeletal muscle accumulates Cs in preference to K (Relman *et al.*, 1957). The lack of accumulation of Cs in humans with age suggests a poorly defined homeostasis for Cs.

Toxicity

Cesium salts resemble those of rubidium and potassium in their metabolism and toxicity, but CsOH is more toxic than RbOH or KOH; the greater toxicity of CsOH could be due to its caustic action (Cochran *et al.*, 1950). Data on Cs toxicity are summarized in Table 1-4. Liver disorder, neuroendocrine disturbance, and neuromuscular toxicity leading to irritability and convulsions are clincal symptoms of cesium toxicity (Follis, 1943; Relman, 1957; Kiryushkin *et al.*, 1963). The mechanism of Cs toxicity seems to be similar to that of Rb toxicity.

SUBGROUP IB

Metals of subgroup IB also possess a single valence electron in the outermost orbit, but the next inner orbit does not possess a stable octet of electrons. Their ionization energy is double that of alkali metals, and their ionic radius is smaller (Au = 1.4 Å, Li = 1.56 Å), which indicates that the electrons are more densely packed in the case of the subgroup IB metals; their elecrons are less readily lost. The most stable oxidation states for these metals in aqueous solution are Ag^+, Cu^{2+} and Au^{3+}.

Within subgroup IB, Cu forms more water-soluble salts than either Au or Ag. Unlike the alkali metals, Cu, Ag, and Au form coordination complexes; the empty d orbitals in their atoms are available for hybridization. Copper forms both tetrahedral and square planar configurations, while silver and gold form only square planar configurations. The coordination number varies from 2 to 6 for each. The three metals have great affinity for S- and N-containing ligands. Copper is a component of a number of functional proteins involved in oxidation and reduction. Silver forms stable insoluble complexes with proteins and amino acids. Gold compounds, however, are not stable in living tissues, but tend to decompose into the elemental form.

The metabolic behavior of these metals is quite different from that of alkali metals. Their absorption from the digestive tract is poor, and their excretion is mainly fecal. Copper has specific functions in living tissues, whereas silver and gold have no essential functions.

Copper (Cu)

Copper occurs as the native metal and in sulfide ores such as malachite, cuprite, and chalcopyrite. Copper is present in the earth's crust at about 4.5 ppm and in seawater at 1–25 ppb. Copper is an essential nutrient which is widely distributed in animal and plant tissues; the adult human body contains about 100 mg Cu, one-third in muscle tissues. Liver and brain are rich in Cu, and the concentration of Cu in the human fetus is ten times higher than in the liver of adults (Widdowson *et al.*, 1972). Oysters contain 137 ppm Cu and black pepper contains 53 ppm.

Copper has been utilized in tools for 6000 years and is used extensively in the manufacture of electrical equipment, a large number of alloys, coinage, and chemical apparatus. Copper salts are used in antifouling paints, in insecticides and fungicides, and as algicides. Copper arsenite is used as a pigment and rodenticide; copper-chrome arsenate is the best-known wood preservative; cupric bromide is used as an intensifier in photography and as a humidity indicator; cupric chloride, carbonate, and sulfate are additives in animal feeds and seed fungicides, and cupric sulfate is also an emetic in veterinary medicine. Copper is an industrial health hazard, copper poisoning occurring most frequently following accidental overexposure to insecticides or other toxic copper salts, inhalation of metal dust, suicidal or homicidal ingestion of copper-containing solutions, or consumption of acidic beverages stored in containers or transferred through pipes which contain copper. About one outbreak per year is reported from ingestion of beverages stored in copper vessels (according to the Center for Disease Control, U.S.A.). Copper is a potential toxicant in agricultural sprays and their residues in food. Several snake venoms contain copper which is associated with the toxic proteins (Tu, 1977).

Chemistry

Copper exhibits mono- and divalence and forms water-soluble cationic simple salts (Cu^+ can only exist in aqueous solution); Cu also forms coordination complexes by association with sulfur- and nitrogen-containing ligands with a coordination number of 4. Complexes of Cu^+ have a tetrahedral configuration and can coordinate with 2 or 6 ligands. Cuprous ions can also form coordination compounds with coordination number 2 or 3. Cupric (Cu^{2+}) coordination complexes are of square planar configuration and can coordinate with 6 ligands involving four short and two long metal–ligand bonds. Cuprous ions, with the three d orbitals filled, are involved in sp^3 hybridization while Cu^{2+} is involved in dsp^2 hybridization; these form more stable complexes with most ligands than do other transition metals. Owing to this Cu is present in a number of metalloenzymes and other proteins. The protein-binding capacity of Cu is high; it confers stability and/ or maintains the conformation of proteins. Copper metalloenzymes are involved in oxidation–reduction reactions with O_2 as the electron acceptor when copper participates directly in the electron transfer. The cupric ion is present in phenolases and other oxidases, and predominates in oxygen carriers and enzymes reacting directly with molecular oxygen. Cuprous ions are oxidized to the stable cupric state by peroxide accumulation in the tissues; however, Cu^{2+} can bind with —SH-containing compounds and reversibly react to form Cu^+ and disulfide. Copper oxidase functions through a cyclic shuttling between the valence states of copper, with a continuous uptake of molecular oxygen.

Metabolism

Copper is an essential nutrient; it also stimulates growth when moderately high levels, about 100 times the dietary allowances, are fed to animals.

Copper is essential to iron utilization and functions in enzymes for energy production, connective tissue formation, and pigmentation. Copper deficiency results in anemia, abnormal bones, poor growth, defective connective tissue, cardiovascular failure, and death. Copper metabolism in man and animals has been extensively reviewed (Scheinberg and Sternleib, 1960; Adelstein and Vallee, 1961; Peisach *et al.*, 1966; Schroeder *et al.*, 1966b; Underwood, 1971; O'Dell and Campbell, 1971; Linder and Munro, 1973). The copper content of a 70-kg human adult is between 80 and 150 mg, and the average daily dietary intake by a normal adult is about 4–5 mg (Butler and Daniel, 1973).

Ionic copper is absorbed from the stomach, duodenum, and jejunum (Van Campen, 1971; Decker *et al.*, 1972). The initial absorption is about 30%, but the effective net absorption is only about 5% due to excretion of Cu into the bile; biliary copper is bound to protein, and this complex is not reabsorbed. Absorption is influenced by a number of factors including the chemical forms of Cu: oxides, hydroxides, iodides, glutamates, citrates, and pyrophosphates of copper are readily absorbed, but copper sulfides and other water-insoluble salts are poorly absorbed. Copper complexes of some amino acids are easily absorbed, whereas copper porphyrins present in meat are very poorly absorbed. The presence of molybdate, sulfate, phytate, Zn^{2+}, Fe^{2+}, Cd^{2+}, or ferrous sulfide in the diet decreases copper absorption. Leucine enhances the absorption of dietary Cu in rats and humans (Krishnamachari, 1974). A duodenal protein which binds Cu is involved in Cu absorption. Younger rats retain more dietary Cu than do older animals.

The absorption of Cu salts from sites of parenteral injection is gradual, depending upon the solubility of the Cu compounds injected. Absorption through the skin is minimal. The absorption of inhaled Cu-containing dust from the lung is gradual, part of it remaining in the lung tissue. Intravenously injected Cu^{2+} is sequestered by erythrocytes, and in serum Cu is present as an exchangeable and loose complex with serum albumin as copper amino acid complexes and as the copper metalloprotein, ceruloplasmin, which contains firmly bound Cu. Ceruloplasmin contains about 95% of the Cu found in adult human plasma. The Cu–albumin complex helps to transport Cu across membranes, and distribute it to soft tissues. Liver, brain, and kidneys retain more Cu than do other soft tissues, although muscle tissues contain about 35% of the total body Cu. Within the liver cell about 65% of the Cu is present in the soluble fraction, and 8% in the mitochondria. Metallothioneine can retain Cu. Copper storage proteins in the tissues include erythrocupreine, hepatocupreine, cerebrocupreine, and mitochondrocupreine; the first three were formerly known as cytocupreine, and are now identified as superoxide dismutase (Evans, 1973).

The major pathway of copper excretion in man, animals, and birds is the biliary system. The total fecal Cu excretion represents unabsorbed dietary Cu and 90% of the absorbed Cu including that from sloughed cells; only 2–4% of the absorbed Cu is excreted in the urine, and a small amount is excreted in sweat. About 80% of the absorbed Cu is excreted in the bile, while about 15% is passed directly into the bowel. Intravenously injected Cu does not increase the urinary output.

An efficient homeostatic mechanism for Cu exists in man. In addition to the liver, the primary organ regulating Cu metabolism, the intestinal mucosa acts as a regulatory barrier to the absorption of excessive Cu and for the release of Cu in the succus entericus. Absorbed Cu is freely exchangeable with the Cu loosely bound to serum albumin; out of eight Cu atoms present in ceruloplasmin, four are exchangeable. This exchange and rapid turnover of Cu are responsible for maintaining the normal level of mobile Cu in man.

Toxicity

The range between deficiency and toxicity of Cu is wide for mammals, although it is narrow for bacteria and fungi, and Cu is highly toxic to aquatic organisms. The relative toxicity appears to be related to the efficiency of the absorptive and excretory mechanisms. Among mammals, ruminants are more susceptible to Cu toxicity than are monogastric animals; Salvidio *et al.* (1963) suggest that this is due to their low glucose-6-phosphate dehydrogenase levels. Young calves, whose rumens are not fully developed, are more susceptible to copper toxicity than are older ruminants (Underwood, 1971). Among monogastrics, guinea pigs and rabbits are especially susceptible. Symptoms of chronic copper poisoning in mammals are nausea, vomiting, epigastric pain, yellow watery diarrhea, dizziness, general debility, jaundice, and green stools, saliva, and vomitus. Symptoms of acute copper toxicity are sporadic fever, tachycradia, hypotension, hemolytic anemia with intravascular hemolysis, oliguria, uremia, coma, cardiovascular collapse, and death. The prompt emetic effect of Cu limits its oral toxicity; Cu irritates the nerve endings in the stomach and initiates the vomiting reflex in higher animals. Inhalation of dusts and fumes of metallic Cu and its salts causes congestion of nasal mucous membrane, ulceration and perforation of the nasal septum, and pharyngeal congestion. The toxicity of Cu salts is summarized in Table 1-5. Highly water-soluble Cu salts are more toxic than sparingly soluble salts; anions such as arsenite and chromate enhance apparent Cu toxicity.

Excessive ingestion of Cu (about 300 to 500 times the normal intake) by mammals leads to its accumulation in the tissues; the extent of accumu-

TABLE 1-5. Copper Toxicity

Compound	Animal	Route	Toxicity	Dosage/kg body weight			
				Compound	Metal		
				mg	mg	mM	pT
Cuprous oxide Cu_2O	Rat	oral	LD_{50}	470	417	6.56	2.18
Cupric oxide CuO	Rat	oral	LD_{50}	710	567	8.92	2.05
Cupric chloride $CuCl_2$	Mouse	ip	LD_{50}	41	19.4	0.31	3.52
	Rat	oral	LD_{50}	140	66.1	1.04	2.98
	Guinea pig	sc	LD_{50}	100	47.2	0.74	3.13
Copper oxychloride $CuO \cdot CuCl_2$	Rat	oral	LD_{50}	700	415	6.53	2.19
Cupric perchlorate $Cu(ClO_4)_2 \cdot 2H_2O$	Rat	ip	LD_{50}	29	6.17	0.09	4.01
Cupric nitrate $Cu(NO_3)_2 \cdot 3H_2O$	Rat	oral	LD_{50}	940	247	3.89	2.41
Cupric sulfate $CuSO_4$	Mouse	oral	LD_{100}	50	20	0.31	3.50
	Mouse	ip	MLD	7	2.8	0.044	4.36
	Rat	oral	MLD	300	120	1.88	2.72
	Guinea pig	iv	MLD	2	0.8	0.012	4.90
	Rabbit	oral	LD_{100}	50	20	0.31	3.50
	Rabbit	iv	LD_{100}	4.5	1.8	0.028	4.55
Cupric sulfate $CuSO_4 \cdot 5H_2O$	Mouse	ip	LD_{50}	33	8.4	0.13	3.88
	Rat	oral	LD_{50}	960	244	3.84	2.42
Cupric carbonate $CuCO_3$	Rat	oral	LD_{50}	159	81.7	1.29	2.89
	Goat	oral	LD_{100}	420	216	3.40	2.47
Cuprous cyanide $CuCN$	Rat	ip	MLD	50	35.4	0.56	3.25
Cupric acetate $Cu(C_2H_3O_2)_2 \cdot H_2O$	Rat	oral	LD_{50}	710	226	3.55	2.45
Cupric acetoarsenite $C_4H_6As_6Cu_4O_{16}$	Rat	oral	LD_{50}	22	5.5	0.086	4.06

lation and subsequent toxicity depends upon the species, the dietary levels of zinc, iron, molybdate, and sulfate, and the efficiency of the animals' excretory mechanisms; copper toxicity is greater with low dietary intake of molybdate, sulfate, zinc, and iron.

Rats and pigs can tolerate about 200 ppm Cu in their diet. Pigs reveal no signs of Cu toxicity when fed diets containing 250 ppm Cu, and show growth stimulation when adequate amounts of Zn (100 ppm) and Fe (750 ppm) are added to the diet (Barber *et al.*, 1955, 1956; Ritchie *et al.*, 1963; Suttle and Mills, 1966), while, without substantial Zn and Fe in the diet, 250 ppm of dietary Cu may be harmful to pigs (Gipp *et al.*, 1974). Symptoms of mild Cu toxicity (500 ppm) in pigs are reduced feed intake, reduced growth, and low hemoglobin, and continued ingestion of this amount of Cu results in degenerative changes in liver, kidney, and spleen. Pigs fed 600–750 ppm dietary Cu have high serum aspartate amino transferase and ornithine carbamyl transferase activities indicative of general tissue damage. Zinc supplements at 750 ppm relieve the Cu-induced anemia and jaundice, and restore the serum copper and aspartate transaminase levels. Laboratory animals can tolerate about 100 times the normal Cu intake; however, the Cu content of the liver increases about 14-fold. Above this level, the Cu-binding sites in liver become fully saturated; necrotic hepatitis ensues and Cu is suddenly released into the blood causing hemolytic anemia with methemoglobinemia, jaundice, and hemoglobinuria.

In sheep a sustained or chronic Cu dietary level of 10–15 ppm can lead to hepatic Cu accumulation and sudden hemolytic crisis. Chronic Cu poisoning of sheep can occur in two ways: by feeding on copper-containing subterranean clover (20 ppm dry weight), or by feeding on weeds containing hepatotoxic alkaloids. (These alkaloids cause hepatic accumulation of Cu.) Small quantities of Cu (50 mg $CuSO_4$) given intravenously cause hemolytic crisis (Thompson and Todd, 1970; Todd and Thompson, 1963), as do subcutaneous injections of Cu–EDTA (2–5 mg Cu/kg). During chronic ingestion of copper, passive accumulation of copper may occur over a period ranging from a few weeks to more than a year with no toxic symptoms; suddenly the animals develop jaundice, methemoglobinemia, and hemoglobinuria, and die within a few days.

The toxic syndromes of chronic copper poisoning in sheep and calves are comparable. Calves fed 20 to 125 ppm Cu develop jaundice after about 20 weeks; hemolysis occurs following liver injury (Doherty *et al.,* 1969). Serum lactic dehydrogenase, glutamic oxalacetic transaminase, and glutamic pyruvic transaminase activities increase while reduced glutathione (GSH) and hemoglobin levels decrease prior to the hemolytic crisis (Todd and Thompson, 1965). Pulmonary edema, subendocardial hemorrhage, and congested kidneys with tubular epithelial necrosis were observed in sheep that died from copper intoxication.

Excess dietary Cu produces pathological changes in brain tissue. A spongy transformation of the white matter of midbrain, pons, and cerebellum in lambs has been reported (Doherty *et al.,* 1969). Similar observations were reported in acute and chronic copper poisoning in rats (Wiederanders *et al.,* 1968). Molybdate in the presence of sulfate inhibits copper retention, and so providing supplemental Mo to maintain a more balanced dietary Cu/Mo ratio is very effective in reducing copper levels in the liver (Dick, 1956). Selenium is reported to decrease copper toxicity (Hill, 1974); a high mortality and a high incidence of exudative diathesis and muscular dystrophy were noted in chicks fed 800 or 1600 ppm copper (Jensen, 1975), but the addition of 0.5 ppm selenium in the diet prevented copper toxicity.

Exact data on oral copper toxicity in humans are not available. From comparable data in other monogastric animals such as pigs, it could be inferred that with adequate Fe, Zn, Mo, and Se, a dietary level of 100 ppm Cu, on a dry-weight basis, could be tolerated by human adults for prolonged periods; this level is equivalent to 10–15 times the level of Cu normally present in the human diet. Drinking soft water may increase the daily intake by about 1.4 mg; kidney, liver, oysters, and shell fish in the diet increase the Cu intake, since these foods contain 200–400 ppm Cu on a dry-weight basis.

Acute Cu poisoning in humans occurs typically 10–90 min following ingestion of acidic beverages which have been stored in or drawn through Cu containers. Abdominal cramps, vomiting, and diarrhea last less than 24 hours (Shreir *et al.*, 1974). Copper is more toxic in drink than in food. In infants, 7 ppm Cu is fatal (Walker-Smith and Blomfield, 1973). Copper sulfate taken as an emetic is toxic if vomiting does not occur (Decker *et al.*, 1972); ingestion of 175–250 mg Cu in the form of its sulfate is sometimes fatal in adults (Ceresa, 1947; Chuttani *et al.*, 1965). Individuals deficient in glucose-6-phosphate dehydogenase may be more sensitive to copper excess (Bremner, 1974*a,b*); some human patients who recovered from acute poisoning have died later from necrosis of the kidney.

Contact dermatitis and eczematous dermatitis caused by internal exposure to Cu in the form of intrauterine devices have been reported (Saltzer and Wilson, 1968; Barranco, 1972).

Wilson's disease, a hepatolenticular degenerative disorder of Cu metabolism in humans, is inherited as an autosomal recessive trait. The clinical symptoms of this disease are tremor, ascites, psychosis, slurring of speech, and other manifestations; the disease is reviewed by Scheinburg and Sternlieb (1964). Excess deposition of Cu may be due to excessive absorption, poor excretion, or the inability to store Cu in ceruloplasmin. Plasma ceruloplasmin decreases and the level of unbound plasma Cu is elevated. Copper accumulates to an extent ten times the normal level in the liver, kidney, and cornea, and later in the brain, owing to increased absorption of Cu from the digestive tract and biliary obstruction to Cu excretion. This accumulation causes hepatic cirrhosis, necrosis and sclerosis of the corpus striatum, and trauma in the brain, which lead to death. Wilson's disease can be arrested by administering a chelating agent, penicillamine. Although this disease in humans is not related to the toxicity caused by unusually high Cu intakes, the mishandling of Cu following accumulation is similar. In acute copper intoxication, high levels of Cu are found in the intestine, bile, skin, nails, and hair, while copper levels are normal in these tissues in Wilson's disease. Ionic Cu can accumulate in tissues such as the liver and brain in both Wilson's disease and in acute copper intoxication; these tissues incorporate Cu into ceruloplasmin and thus store it; hence these tissues are susceptible to Cu overload. Organs which are incapable of making ceruloplasmin but require ceruloplasmin for Cu storage do not accumulate ionic Cu and hence are not susceptible to Cu overload.

Hemolytic anemia, a recognized complication of Wilson's disease, is due to oxidative stress on the erythrocytes as a consequence of Cu accumulation within the cells (Deiss *et al.*, 1970). There is also evidence that hemolytic anemia is caused by either acute or chronic Cu intoxication in animals (Todd, 1969). High plasma levels of Cu are recorded in Fe-deficiency anemias such as those of pregnancy, rheumatoid arthritis, or infection, and high hepatic level of Cu are observed in anemias of premature birth, cirrhosis, or hemochromatosis (Seelig, 1972).

Administration of synthetic chelating agents to overcome the toxic effects of excess Cu results in tissue redistribution of Cu rather than its elimination; the excess is transferred from one organ to another. Parenteral administration of sodium diethyl dithiocarbamate (DDC) to rats results in the accumulation of Cu in the cerebral cortex, stem, and spinal cord (Iwata *et al.*, 1973); the ready penetration of the central nervous system is attributed to the lipid solubility of the Cu–DDC chelate. Diethyl dithiocarbamate increases Cu toxicity when it is administered to rats parenterally with the Cu salt being tested; the retention of Cu in the body is also increased and the Cu content of blood, brain, heart, and kidneys is higher than in controls, although the hepatic Cu level is low (Koutensky *et al.*, 1971).

The mechanism of copper toxicity in mammals is complex. It involves increased cellular permeability in erythrocytes with consequent lysis, inhibition of glutathione reductase, and loss of intracellular reduced glutathione (Mital *et al.*, 1966), agglutination (Todd and Thompson, 1963, 1965), and an excessive stimulation of the hexose monophosphate shunt. These lead to oxidative stress in erythrocytes, and to accelerated loss of intracellular reduced glutathione (Metz and Sagone, 1972), since the nonenzymatic regeneration of glutathione within the cell is restricted in Cu toxicity (Kosower *et al.*, 1967). Copper ions induce mitochondrial swelling and inhibit oxygen consumption (Bowler and Duncan, 1970). The affinity of Cu^{2+} to —SH groups of hemoglobin, erythrocyte, and other membranes increases the permeability and lysis of erythrocytes. Copper increases the fragility of lysosomal membranes and releases acid hydrolases causing cell degeneration (Chvapil *et al.*, 1972); subcutaneous bleeding and anemia in animals is caused by failure of collagen formation in the walls of arterioles in copper-deficient animals.

A high incidence of cancer among coppersmiths, and of stomach cancer in humans living in regions with a high Zn/Cu ratio in the soil, are suggestive of a carcinogenic capacity of Cu. This epidemiologic evidence is not supported by experimental studies with rats. However, malignant tumors that are induced by organic carcinogens contain appreciable levels of Cu, Zn, and Fe.

The interrelationships between Cu and other minerals influences the toxicity of Cu and renders the study of the mechanisms of its intoxication more difficult. Tissue retention of Cu is considerably influenced by this interrelationship. High dietary levels of zinc, nickel, and manganese limit copper storage in the liver, and molybdenum and sulfate reduce copper absorption and enhance its excretion. Dietary Cu and sulfate counteract the harmful effects of excess dietary Mo.

The livers of newborn mammals contain a Cu-binding protein rich in cystine; this protein is believed to have either a detoxicating or storage function analogous to that of ferritin (Porter, 1971); it is localized in the mitochondrial fraction or in lysosomes. This Cu-containing protein differs from metallothioneine of adult liver since it contains no methionine and only one-third as much aspartate as metallothionein.

Silver (Ag)

Silver occurs as the free metal, in ores such as argentite (Ag_2S) and horn silver (AgCl), and in ores of other metals; the earth's crust contains 0.1 ppm and seawater about 0.15 ppb of Ag. Humus from plant remains may contain as much as 5 ppm of Ag. The normal adult human body contains about 1 mg silver, varying according to the exposure of the individual to silver compounds, but silver is not essential for mammals.

Silver has been known to man since about 3000 B.C. Metallic Ag is used in industry and in jewelry, and silver salts are much used in therapeutics. Industrial uses of Ag are: (1) manufacture of scientific instruments, (2) bearing linings in air-cooled aircraft engines, (3) storage batteries, (4) different alloys with Cu, Cd, and Pb, (5) special pasteurizing coils, nozzles, and valves for the dairy and brewing industries, and (6) electroplating. Silver salts (Ag_3PO_4 and AgBr) are used in photographic plates and as bactericides for sterilizing water and fruit juices. In medicine silver salts such as $AgNO_3$ and AgBr are used as external antiseptics and astringents; colloidal silver salts are good antiseptics since they coagulate and inactivate viral and bacterial proteins. In the past silver salts were used in nervous disorders with no proven effects. Silver and its salts are not industrial hazards; cases of poioning are rare, and usually occur following indiscriminate therapeutic use of silver salts; poisoning does not occur through food, although food products such as wheat bran contain about 0.9 ppm Ag and some mushrooms have 300 ppm Ag.

Chemistry

Silver exhibits mostly monovalence and forms stable electrovalent compounds; most inorganic silver salts, with the exception of halides, are water-soluble. Silver halocomplexes of the type $[AgX_2]$ and $[AgX_3]^{2-}$ are known. Because of the $4d$ empty electronic orbitals, which are available for SP^3 or linear hybridization, Ag ions form stable coordination complexes with organic groups; under these conditions Ag exhibits valences of +1, +2, and even +3. In both the +1 and +2 oxidation states, Ag exhibits a coordination number of 4. In the monovalent state Ag forms tetrahedral complexes with SP^3 hybridization, while Ag^{2+} forms square planar complexes with dSp^2 hybridization and promotion of one electron to the p orbital. Silver complexes with ethylene thiourea to form $Ag[S—C(NH_2)_2(CH_2)_2]_3Cl$ and $Ag_2S—C(NH_2)_2(CH_2)_2Cl$. Stable chelates are formed with ethylene diamine, e.g., $[Ag_2en_2]^2$. Ethylenebiguanidine forms a stable quadridentate complex with trivalent Ag:

$$
\begin{array}{ccc}
 & \overset{\displaystyle NH}{\underset{\parallel}{}} & \overset{\displaystyle NH}{\underset{\parallel}{}} \\
CH_2\!-\!NH\!-\!C\!-\!NH\!-\!C\!-\!NH_2 \\
 & \diagdown\ \diagup \\
 & Ag \\
 & \diagup\ \diagdown \\
CH_2\!-\!NH\!-\!C\!-\!NH\!-\!C\!-\!NH_2 \\
 & \underset{\displaystyle NH}{\overset{\parallel}{}} & \underset{\displaystyle NH}{\overset{\parallel}{}}
\end{array}
$$

Silver has great affinity for thiol, sulfide, and selenosulfide; it also reacts with amino, imidazole, carboxyl, and phosphate groups present in biologically active macromolecules. The reaction of ionic silver with proteins results in insoluble and complex silver proteinates with unknown structures which may be similar to the quadridentate structure mentioned above. (Although there are no reports of the existence or isolation of Ag macromolecular complexes, there is a strong possibility of their existence.) These complexes are capable of slowly liberating Ag ions, contributing to the antiseptic activity of Ag.

Metabolism

Silver metabolism in man and animals has not been thoroughly studied due to the poor absorption of ingested silver salts and the insolubility of silver complexes formed when silver salts are injected parenterally. Radioactive Ag was used in the few studies reported. Ingested silver salts, with the exception of $AgNO_3$, are not highly toxic. The average daily dietary intake of Ag by humans is less than 0.1 mg. Silver ions are known to inhibit the winding and rewinding of DNA *in vitro*.

The gastrointestinal absorption of monovalent silver salts by mammals is poor due to the sparingly soluble nature of the salts; the soluble sulfate and nitrate are converted into insoluble chlorides in the stomach. These insoluble salts can cause corrosion of the mucosa of the digestive tract and remain in the epithelial cells until the cells are sloughed off. A small amount of Ag absorption takes place at the mucous membrane of the upper intestine and respiratory tract; inhaled Ag dusts accumulate in the subepithelial area of the mucous membrane of the lung. Parenterally injected silver salts remain at the site of injection bound to tissue proteins; only a fraction is absorbed. Large doses of intravenously injected silver salts cause hemolysis and agglutination of erythrocytes; in low doses Ag complexes with serum albumin and is transported to the tissues. In blood the colloidal or particulate AgCl or silver proteinate is present in plasma or leukocytes, not in erythrocytes. The pattern of distribution of Ag to the soft tissues after an intravenous administration of radioactive Ag colloids to rats was: spleen > liver > bone > marrow > lungs > muscle > skin (Gammil *et al.*, 1950). The macrophages of the reticuloendothelial system apparently retain the Ag colloids.

Excretion of orally ingested Ag^+ is mainly fecal, representing the unabsorbed portion. The absence of Ag in the urine indicates its enterohepatic circulation (Kent and McCance, 1941*a*). However, about 0.42 to 3.8 ppb Ag^+ has been found in human urine by recent sensitive analytical methods. Silver appears to accumulate in humans with age; but the nature, site, and extent of accumulation is not definitely known. It is difficult to speculate about homeostasis for Ag in mammals.

Toxicity

The meager toxicity data for Ag are summarized in Table 1-6. The free metal is quite toxic, particularly in the colloidal state. Acute toxicity symptoms following the ingestion of caustic $AgNO_3$ are severe gastroenteritis, diarrhea, fall in blood pressure, decreased respiration, spasms, and paralysis leading to death; the heart is little affected, while paralysis first affects the diaphragm muscle; the central nervous system also is affected in Ag toxicity. Chronic toxicity symptoms from prolonged intake of low doses of silver salts are fatty degeneration of liver and kidney, changes in blood cells, and argyria, a black pigmentation of the tissue, which is caused by the deposition of a silver–protein complex or its metabolized product, Ag_2S or

TABLE 1-6. Silver and Gold Toxicity

Compound	Animal	Route	Toxicity	Compound mg	Metal mg	mM	pT
Silver metal powder Ag	Human	inhal	Txc[a]		1[b]	0.009	5.03
Silver metal colloidal	Mouse	oral	LD_{50}		100	0.927	3.03
Ag	Human	iv	MLD		0.7	0.0064	5.19
Silver oxide Ag_2O	Rat	oral	MLD	2820	2630	24.3	1.61
Silver fluoride	Guinea pig	oral	MLD	300	255	2.36	2.63
AgF	Guinea pig	sc	MLD	800	680	6.3	2.20
Silver nitrate	Mouse	oral	LD_{50}	129	82	0.76	3.12
$AgNO_3$	Mouse	ip	LD_{50}	50	31.7	0.29	3.53
	Rabbit	iv	LD_{100}	8.8	5.6	0.052	4.28
	Human	oral	LD_{100}	140	88.9	0.82	3.08
Silver cyanide AgCN	Rat	oral	LD_{50}	123	99	0.92	3.04
Potassium silver cyanide $KAg(CN)_2$	Rat	oral	LD_{50}	21	11.4	0.105	3.98
Gold trichloride $AuCl_3 \cdot 2H_2O$	Mouse	sc	LD_{100}	1500	870	4.42	2.35
Gold sodium thiosulfate	Rat	im	MLD	35	13	0.066	4.19
$Na_3AuS_4O_6 \cdot 2H_2O$	Rat	iv	MLD	80	30	0.15	3.81
	Rat	iv	LD_{50}	100	37.4	0.19	3.72
	Rabbit	iv	LD_{100}	100	37.4	0.19	3.72
	Human	im	Txc[a]	1.0	0.375	.0020	5.70
Gold thioglucose $AuC_6H_{12}O_5S$	Mouse	sc	MLD	400	200	1.01	2.99
Gold sodium thiomalate $AuNa_2C_4H_3O_4S$	Human	im	Txc[a]	0.14	0.07	0.0003	6.45

[a]Txc = dose at which toxic symptoms are manifested.
[b]Exposure of 4 hr; number of mg/m^3.

Ag_2Se, as opaque granules in the elastic fibers of tissues, sometimes bound to cell nuclei. This black pigment can be deposited in all tissues except the nerves; it first appears in retinculoendothelial cells and leukocytes, and is then distributed throughout the body, starting with the conjuctiva in the eye and eyelid. The maximum concentration of this pigment is in the liver and kidney (Sollman, 1963; Goodman and Gilman, 1970). Presence of the black pigment in mucous membranes, conjunctiva, or lunulae of the nails is a warning of serious poisoning.

Ingestion of colloidal Ag or $AgNO_3$ produces liver necrosis in rats fed a vitamin-E-deficient diet (Grasso, 1969); Ag reduces the availability of Se to nonheme Fe proteins and impairs the ability of these proteins to function as electron carriers. Ingestion of subtoxic levels of $AgNO_3$ induces tocopherol and Se deficiencies which are preventable by an increase of Se or tocopherol in the diet (Frost, 1972).

Tumors have been induced in rat liver and spleen by intravenous injection of colloidal Ag (Schnahl and Steinhoff, 1960). Silver foil was found to be highly toxic when implanted in brain tissues of animals (Dymond *et al.*, 1970), while subcutaneous implantation of thin silver foil induces fibrosarcomas in rats (Oppenheimer *et al.*, 1956). Argyria can be considered a mechanism to detoxify Ag by sequestering it in the tissues as harmless Ag–protein complex or Ag_2S. The toxicity of Ag ions can be attributed to their ability to form stable complexes with structural and functional proteins, and thereby disrupt life processes.

Gold (Au)

Gold occurs free in nature and in the minerals calaverite, sylvanite, petzite, and nagyagite, and also in minute quantities in the ores of silver, copper, lead, bismuth, etc. Seawater contains 0.004 ppb Au and the earth's crust contains about 0.005 ppm Au. The adult human body contains < 10 mg Au, about 50% of which is concentrated in the bones. Gold is not normally present in plant tissues, and it is not an essential nutrient.

Gold, used in jewelry and coinage for 5000 years, now has new uses in the coating of space satellites and space suits, where it serves as a reflector of radiation. Gold salts are used therapeutically in rheumatoid arthritis and syphilis, and gold alloys in dentistry. Industry utilizes gold stannate $[AuSn(OH)_6]$ for making glazes and ruby glass. Gold poisoning is very rare, and the few known cases are the result of therapeutic overdose and other side effects during gold therapy in rheumatoid arthritis (Freyberg, 1960). Because of the malleability of the Au alloys used in dental prostheses, there is only negligible loss from wear in gold fillings over many years.

Chemistry

Gold exhibits valences of $+1$ and $+3$, and forms both cationic and complex anionic salts such as chloroaurates. Simple Au^+ and Au^{3+} ions do not exist in solution, but double salts such as gold-sodium thiosulfate $[AuNa_3(S_2O_3)_2]$ are both stable and water-soluble.

Trivalent Au forms square planar compounds with a coordination number of 2; Au has strong affinity for S, weak affinities for C and N, and no affinity for O; only in unusual types of chelate formation does Au complex with O. Gold readily complexes with sulfur-containing ligands, e.g., those of α-thiol fatty acids and gold thiourea. In living tissues, gold compounds tend to decompose into the elemental form because of their low chemical reactivity.

Metabolism

The available information on Au metabolism in mammals is mostly from studies involving the therapeutic use of Au compounds in arthritis; involvement of Au in biochemical systems of living tissues has been little studied. Gold salts inhibit *in vitro* some of the lysosomal proteases such as collagenase and elastase (Persellin and Ziff, 1966; Janoff, 1970). Gold is not essential to mammals, but may be stimulatory (Luckey, 1975*b*). There are no reports of the dietary intake of Au by man; fecal analyses indicate no detectable Au except in persons exposed to it in their work, e.g., astronauts.

Absorption of gold compounds from the mammalian digestive tract is poor and erratic; the metal, its oxide, and simple cationic salts are poorly absorbed, while water-soluble gold compounds such as $Na_3Au(S_2O_3)_2$ or gold sodium thiomalate are absorbed rapidly (Orestano, 1933). At tissue pH most soluble trivalent gold salts are hydrolyzed to colloidal compound salts or oxides of the metal. Gold sodium thiosulfate and gold thiomalate are rapidly absorbed into the blood from the sites of parenteral injection. Colloidal forms of gold compounds are more slowly absorbed than are crystalline compounds from intramuscular injections (Block *et al.,* 1942); these gold compounds are present mostly in the blood plasma and are loosely bound to α globulin. Intravenous administration of microquantities of colloidal Au is harmless. However, gold salts may be toxic, because *in vitro* studies had shown that 15–50 μmol chloroauric acid ($HAuCl_4$) causes agglutination and hemolysis of erythrocytes (Jandl and Simmons, 1957). The site and nature of distribution of Au salts depend on their solubility; the skeleton permanently retains about 50% of the absorbed Au. Colloidal compounds are phagocytized by macrophages and are distributed mainly as follows: liver > spleen > kidney, but the distribution of soluble gold compounds is: kidney > liver > spleen. About 80% of the absorbed Au is retained. Gold salts are poorly absorbed from the skin. There are no reports on inhalation of gold compounds, but it is presumed that clearance from the lungs would be similar to that from parenteral administration.

Excretion of absorbed Au is both fecal and urinary in a ratio of 1.4; unabsorbed dietary Au compounds are excreted predominantly in the feces. The retention of Au in the tissues is called chrysiasis, similar to argyria but less pronounced; the mechanism is probably similar. In mice organic gold compounds such as gold thioglucose are rapidly absorbed and metabolized, and gold is deposited in the diencephalon, especially the hypothalamus (Debons *et al.,* 1970), the reticuloendothelial system, and sites of inflammation (Swartz *et al.,* 1960).

Toxicity

Owing to its poor absorption, gold dust and insoluble salts of gold are not toxic when given orally. However, soluble gold compounds are toxic. The available toxicity data is summarized in Table 1-6. The toxicity of simple cationic Au is less than that of the more soluble double salts or organic salts. Acute Au toxicity symptoms are similar to arsenic toxicity symptoms, i.e., violent diarrhea, gastritis, colitis, simple erythema to severe exfoliative dermatitis of mucous membranes, and degeneration and necrosis of renal epithelium. Other toxic symptoms in mammals include blood dyscrasias: leukopenia, agranulocytosis, and aplastic anemia. Auro-thioglucose induces obesity and sterility in mice, while food restriction and exercise restore fertility (Einer-Jensen *et al.,* 1970); this compound was highly toxic to chicks when given intramuscularly at 0.1% body weight (Simkins and Pensack, 1970). Proteinuria developed in some rheumatoid arthritis patients during gold therapy (Silverberg *et al.,* 1970). Lethal hemorrhagic diathesis has been noted in experimental animals after gold therapy (Waechter *et al.,* 1969). Allergic contact dermatitis from gold and gold-induced thrombocytopenia in humans have been reported. Gold sodium thiomalate is quite toxic in humans; when given to rats subcutaneously it was found to inhibit transaminase activity and the formation of glucosamine-6-phosphate in connective tissue but not in liver (Bollet and Shuster, 1960). Experimental animals can develop tolerance when they are given repeated injections of gold compounds at sublethal doses (Cortel and Richards, 1942). Arena (1974) emphasizes the great variation in tolerance to Au salts. The major symptoms of Au toxicity in man are fever, nausea with vomiting, photosensitivity, metallic taste, nephritis and nephrosis, uremia, aplastic anemia, and a variety of skin disorders.

Some Au compounds can suppress or prevent experimental arthritis (Freyberg, 1960), and soluble Au salts are effective in the therapy of rheumatoid arthritis when given intraarticularly, by inhibition of the collagenase and elastase implicated in cartilage breakdown. The macrophages phagocytize Au salts and accumulate them in lysosomes (Cohn and Fedorko, 1969). Gold salts reversibly inhibit intralysosomal proteases by binding sulfhydryl groups (Persellin and Ziff, 1966; Janoff, 1970; Davies *et al.,* 1971), and alter the properties of collagen in rats, presumably by increasing the cross-linkages (Adam and Kuhn, 1968). The significance or involvement of this change in collagen properties during therapy for arthritis is not clear.

The mechanism of Au ion toxicity is its affinity for thiol groups and the resulting inhibition of thiol enzymes (Libenston, 1945). *In vitro* studies show that Au readily binds to deoxynucleotides; the binding is rapid with dAMP and dGMP, and slow with pyrimidine derivatives. Gold salts can bind to DNA, and this binding is involved in gold toxicity.

There are no specific biochemical reactions involved in Au detoxication; immobilization of Au in the bone and in chrysiasis could be considered detoxication mechanisms.

TOXICITY OF GROUP II METALS

2

Beryllium (Be), magnesium (Mg), calcium (Ca), strontium (Sr), barium (Ba), and radium (Ra) form subgroup A, and zinc (Zn), cadmium (Cd), and mercury (Hg) form subgroup B of Group II in the periodic table of elements. The elements Mg and Ca are essential macrometals for mammals, and Zn is an essential trace metal; Sr and Ba are not essential, although Ba is stimulatory. The elements Be, Cd, and Hg are stimulatory at very low doses but become highly toxic at relatively low levels. Magnesium, calcium, strontium, and barium are considered relatively nontoxic metals at physiologic levels but prove to be definitely toxic at higher levels.

All of the Group II metals exhibit a valence of two; Hg exhibits both mono- and divalence. Group II metals form simple cationic salts which are water soluble, exceptions being the sulfides of the metals and sulfates of Sr and Ba. The electropositivity of these metals increases in direct proportion to their atomic weights. While subgroup IIA metals form mostly electrovalent salts and remain as ions in living tissues, subgroup IIB metals readily form coordinated covalent compounds in biologic fluids and tissues.

Group II metals coordinate better with organic ligands than do subgroup IA metals; the stability of their complexes with albumin increases with the metal-ion radius. The affinity of subgroup IIB metals toward thiol ligands is greater than that of subgroup IIA metals, while the affinity toward phosphate ligands is the opposite. Subgroup IIA metals are involved in maintaining ionic equilibria, and performing essential functions such as the transmission of nervous impulses, muscular contractions, and activation of enzyme systems; all of these metals are bone-seekers and settle in the bone mineral matrix. Zinc, of subgroup IIB, is involved in a number of metal-activated and metalloenzymes. The subgroup IIB metals also accumulate in soft tissue. Subgroup IIA metals are excreted in the urine, while subgroup IIB metals are excreted in the feces. Subgroup IIB metals Cd and Hg are more toxic than the corresponding subgroup IIA metals. Beryllium, the lightest, and mercury, the heaviest of Group II metals, are more toxic than the others. Homeostatic mechanisms maintain normal levels of Mg, Ca, Zn, and to a certain extent Sr and Ba; apparently the other metals are not controlled by homeostasis; Be, Ba, Cd, and Hg are retained in the tissues, and the body levels of these metals increase with age. Radium, the radioactive metal, should be toxic chemically; however, its radiation damage masks clear evaluation of chemical toxicity.

Group II Elements

Atomic number Atomic weight	Name (Symbol)		(Core) Active electrons Usual valences
Subgroup A		Subgroup B	
4 9.01	Beryllium (Be)	(He) 2 +2	
12 24.3	Magnesium (Mg)	(Ne) 2 +2	
20 40.08	Calcium (Ca)	(Ar) 2 +2	30 65.38	Zinc (Zn)	(Ar) 8, 2 +2
38 87.62	Strontium (Sr)	(Kr) 2 +2	48 112.40	Cadmium (Cd)	(Kr) 8, 2 +2
56 137.34	Barium (Ba)	(Xe) 2 +2	80 200.59	Mercury (Hg)	(Xe) 8, 2 +1, +2
88 226.03	Radium (Ra)	(Rn) 2	

SUBGROUP IIA

Metals of subgroup IIA are similar in their distribution of valence electrons in the outermost orbits. Calcium, strontium, barium, and radium have empty inner orbitals, while beryllium and magnesium do not. The electropositivity of these metals increases progressively with increase in atomic weight. The metabolism of Ca, Sr, Ba, and Ra is similar. Calcium, strontium, and barium are the least toxic. Beryllium, with the highest charge-to-radius ratio among these metals, differs from the rest of subgroup IIA metals in its metabolism, poor gastrointestinal absorption, high retention in the skeleton and soft tissues, and high toxicity; it causes cellular death in all tissues where it accumulates. Magnesium resembles zinc more than the other metals of subgroup IIA in its metabolic and biochemical activities, and has a gastrointestinal absorption pathway similar to that of calcium. The binding energy of Mg with other ligands is high among the essential metals due to its small ionic radius. Hence, Mg forms a variety of complexes in biologic tissues, and when given intravenously in medium doses it is fatal.

Beryllium (Be)

Beryllium occurs as the oxide, bromelite (BeO); as complex alumino-silicates, beryl ($3BeOAl_2O_36SO_2$), betrandite [$Be_4Si_2O_{10}(OH)_2$], euclase ($BeHAlSiO_5$), and phenacite (Be_2SiO_4); and as the aluminate [$Be(AlO_2)_2$]. These minerals are widely distributed but are present in small quantities. The earth's crust contains about 2.8 ppm Be, and the Be level in seawater is about 0.0006 ppb. Beryllium might be expected to be present in larger quantities on the earth since it is a light and stable metal, but the Be nucleus is easily destroyed by collision with high-energy protons and hence could not be stable in the sun's atmosphere. No reports of its presence in plant tissues have been found; however, Be is present in coal at 1.5 to 2.5 ppm. The normal adult human body contains traces of Be, 75% of which is in the bone. Beryllium is not essential for mammals, and is oncogenic.

There is extensive use in industry of Be metal as a hardening agent in alloys, especially with Cu; Be–Cu alloys, which can be of high tensile strength and high conductivity, are used in parts exposed to abnormal wear, electric contact switches, and nonsparking tools; Be–Ni alloy is used in diamond drill matrixes and in the structural material of spaceships. In atomic reactors Be acts as a good moderator to retard neutrons, due to its low neutron absorption properties and its high-scattering cross section. Finely powdered Be is an additive in rocket fuels. Formerly Be was used in fluorescent lamps (discontinued because of the danger from the dust of broken lamps), as electrodes in neon lamps, and in X-ray diffraction tubes. Fluoroberyllates and BeO_2 are used in the ceramics and glass industries. Beryllium potassium sulfate is used in chromium and silver plating, and $Be(NO_3)_2$ is used to stiffen mantles in gas and acetylene lamps.

Beryllium is an industrial health hazard. Acute and chronic Be poisoning occurs in workers in Be metal extraction and refining plants, and Be is an environmental hazard near rocket bases, nuclear reactor plants, and in areas where coal is burned extensively. Poisoning occurs mainly by the inhalation of industrial gases and dust; Be uptake from food sources is negligible. Beryllium is an insidious poison and exhibits latent toxicity.

Chemistry

Beryllium is highly electropositive and forms stable water-soluble divalent cationic salts; it has the highest charge-to-radius ratio ($Z/\gamma = 6.45$) among the subgroup IIA metals; the Z/γ ratio of the other members range from 0.65 to 1.5. Owing to this high ratio Be^{2+} forms a series of charged and hydrated hydroxides [e.g., $Be_3(OH)_3^{3+}$] which vary according to pH and are heavily hydrated in aqueous solutions. Beryllium forms linear molecules only in gaseous condition. Owing tc its small size the Be atom assumes tetrahedral geometry in coordination compounds and cannot expand its coordination number above 4. Soluble Be compounds tend to become colloidal solutions (in the absence of suitable chelating agents) in concentrations exceeding micromolar range. Beryllium is ligated through an oxygen ligand for stable Be-chelate formation. If the chelating agent has two function groups with ionizable protons, Be forms stable 5- or 6-membered chelate rings with compounds such as aluminum or gold tricarboxylic acid. The beryllium salts of citric, aspartic, succinic, maleic, and sulfosalicylic acids are stable. The affinity of Be^{2+} is restricted to selective ligands which include N ligands, and hence Be^{2+} binding to proteins is highly specific; beryllium does not bind indiscriminately to biological macromolecules; beryllium phosphate binds loosely to plasma proteins such as α globulins.

Metabolism

Beryllium is not essential for the nutrition of animals, nor is it stimulatory for any known physiologic function. It is toxic, and absorbed Be accumulates permanently in mammalian bones and liver, with no known excretory mechanism. Studies on Be metabolism are few, and most use radioactive Be. Interest in Be metabolism is mainly due to its toxicity, and thus reviews on Be relate more to its toxicity than to its metabolism (Aldridge *et al.*, 1949; Schubert, 1958; Tepper *et al.*, 1961; Hardy, 1965; Stokinger, 1966; Browning, 1969, Durocher, 1969; Tepper, 1971).

The gastrointestinal absorption of soluble Be salts in mammals is poor and depends upon the ingested dose (Reeves, 1965). Soluble Be salts form insoluble $Be_3(PO_4)_2$ or bind to proteins in the epithelial cells and are not further utilized. Dietary beryllium phosphates are not available to the organism. A small amount (less than 0.01%) is absorbed when a large dose of soluble Be citrate or other organic acid complex is ingested. Absorption of small doses of peritoneally or intramuscularly injected soluble Be salts is slow; Be is retained at the site, causing cellular damage and dermatitis until is is cleared from the site and finally deposited in the bone. Upon intravenous administration to rats, soluble $BeSO_4$ is distributed in three ways: part of Be^{2+} binds to the surface of erythrocytes, sensitizing the cells; another part forms diffusible small particles of beryllium citrate or other plasma organic acids and is distributed to bone, liver, and other soft tissues; and the rest forms a nondiffusible colloidal fraction of $Be_3(PO_4)_2$ loosely bound to α globulins which gradually accumulate in the reticuloendothelial system (Vacher and Stoner, 1968*a,b*). The clearance of these aggregates from the blood follows an exponential curve, with the rate of clearance being inversely proportional to the dose. When insoluble $Be_3(PO_4)_2$ is injected into the blood, it forms a colloidal aggregate which accumulates in Kupffer cells and effectively blocks the function of the reticuloendothelial system by saturating absorption sites (Vacher *et al.*, 1973).

Inhalation studies in dogs reveal that soluble Be salts are transiently retained in the lung, and a small part is absorbed into the blood stream for distribution to bone, kidney, liver, and other tissues. Insoluble salts are retained in the lung for long periods extending to many months; pulmonary lymph nodes constitute the other major retention site. Beryllium absorbed from the digestive tract is stored mainly in bones following a transient retention in the liver and kidneys. Intratracheally deposited Be also is retained in the lung and part of it is deposited in the bone. The retention of Be in tissues depends upon the dosage and nature of the Be salt; larger doses result in greater retention in soft tissues. With sulfates the chief sites of absorbed Be are the liver and bones; with beryllium citrate the chief site is the bones (Crowley *et al.*, 1949; Klemperer and Liddey, 1952; Van Cleave and Kaylor, 1955). Inhalation of BeO results in 99% retention in the lungs, whereas bones absorb 50–80% of the inhaled soluble $BeSO_4$ (Stokinger *et al.*, 1951). Parenterally administered Be is stored ultimately in the bone.

The excretion of unabsorbed dietary Be is fecal in rats and other animals; 96% of a single dietary dose is excreted within 24 hr and the rest slowly over a period of 4 to 5 weeks. Parenterally injected and inhaled Be salts, as well as the Be absorbed from the digestive tract, are excreted very slowly if at all; less than 1% is excreted in the urine and feces within the first 24 hr, but fecal excretion is persistent in barely detectable amounts and is extended over a long period. In humans exposed to about 3 μg Be/m^3 inhaled air, excretion is mainly urinary and the amount excreted is determined by the solubility of the Be salt. There is prolonged excretion in the urine, with Be detected in the urine for 10 years following exposure to Be. There seems to be no homeostatic regulatory mechanism for Be; Be accumulates in mammalian tissue with age.

Toxicity

Beryllium has latent toxicity, and symptoms may be observed years after initial exposure. Symptoms of acute Be toxicity in man and animals include skin and eye ulcers, and inflammation of lungs, nose, throat, and the mucous membrane of both trachea and bronchi. These disorders are partially reversible if exposure to Be ceases in time. Other acute responses are hyperemia, edema, hemorrhage with mild edema of the brain, inflammation of the liver, and focal hemorrhage of the spleen. Most of the symptoms of Be toxicity reported in the literature result from inhalation of dusts of Be metal and its salts. Chronic Be toxicity in man, caused by exposure to microgram quantities of airborne dusts of Be or its salts, develops into a granulomatous, crippling, and incurable lung disease. The effects of chronic Be toxicity both in man and animals, noticed in some cases years after chronic inhalation exposure to Be, are pneumonitis, pulmonary dysfunction, heart enlargement and congestive heart failure, cyanosis, enlargement of liver and spleen, digital clubbing, formation of kidney stones, malignant tumors, cellular infiltration in the interstices of various organs and tissues, and calcific inclusions in cells and tissues. Although the main organ affected through Be inhalation is the lung, toxic effects are observed in all other tissues and organs; on penetration into living tissues, Be produces peculiar and harmful effects and causes cellular death in all tissues where it accumulates.

Very few reports are available about Be toxicity following oral ingestion or parenteral administration. The available data on the toxicity of Be are summarized in Table 2-1. Chronic forms of beryllosis are associated with insoluble Be salts, and acute forms with soluble Be salts. Soluble Be salts are highly toxic by all routes of administration; death in experimental animals seven weeks after inhalation exposure illustrates the severity of chronic Be toxicity. The cat and mouse appear to be slightly less susceptible to Be toxicity than are other experimental animals.

Sparingly soluble Be or BeO is not toxic when fed to rats as 5% of the diet, but $BeCO_3$ inhibits growth at that dietary level. Beryllium sulfate and oxyfluoride inhibit growth at 1.4% and 0.1% of the diet, respectively; both induce macrocytic anemia, hemorrhagic necrosis of gastric mucosa, and rachitic conditions.

TABLE 2-1. Beryllium Toxicity

				Dosage/kg body weight			
				Compound	Metal		
Compound	Animal	Route	Toxicity	mg	mg	mM	pT
Beryllium	Human	inhal	Txc		0.1^b	0.01	4.95
Be	Pig	sc	MLD		7	0.77	3.11
Beryllium oxide	Rat	inhal	MLD	54	19.4	2.15	2.67
BeO	Rabbit	im	MLD	10	3.59	.398	3.40
	Pig	sc	MLD	7	2.5	.27	3.56
Beryllium hydroxide, α $Be(OH)_2$	Rat	iv	LD_{50}	3.8	0.8	.09	4.05
Beryllium hydroxide, β	Rat	iv	LD_{50}	12	2.5	.28	3.56
Beryllium hydroxide, α and citrate	Rat	iv	LD_{50}	7.7	0.35	.039	4.41
Beryllium fluoride	Mouse	oral	LD_{50}	100	19.1	2.12	2.67
BeF_3	Mouse	sc	LD_{50}	20	3.8	0.42	3.37
	Mouse	iv	LD_{50}	1.8	0.34	.038	4.42
	Rat	oral	LD_{50}	98	18.8	2.08	2.68
Beryllium oxyfluoride $BeO \cdot BeF_2$	Rat	oral	LD_{50}	146	18.3	2.02	2.69
Beryllium chloride	Mouse	im	LD_{50}	12	1.3	.144	3.84
$BeCl_2$	Rat	oral	LD_{50}	86	9.8	1.09	2.96
	Rat	ip	LD_{50}	4.4	0.5	.056	4.26
	Guinea pig	ip	LD_{50}	50	5.6	.62	3.21
	Guinea pig	ip	LD_{100}	56	6.3	0.70	3.15
Beryllium nitrate $Be(NO_3)_2$	Guinea pig	ip	LD_{50}	50	3.48	.386	3.41
Beryllium nitrate trihydrate $Be(NO_3)_2 \cdot 3H_2O$	Rat	ip	LD_{50}	501	24	2.66	2.57
Beryllium sulfate	Mouse	oral	LD_{50}	80	6.95	0.81	3.08
$BeSO_4$	Mouse	sc	LD_{50}	1.5	0.13	0.015	4.82
	Mouse	iv	LD_{50}	0.5	0.043	0.0050	5.35
	Rat	oral	LD_{50}	82	7.02	0.818	3.09
	Rat	inhal	LD_{50}	1.8	0.15	0.0175	4.76
	Rat	ip	LD_{50}	18	1.54	0.179	3.74
	Rat	sc	LD_{50}	1.5	0.13	0.15	3.81
	Rat	iv	LD_{50}	7.2	0.62	0.072	4.16
	Dog	iv	LD_{50}	1.0	0.086	0.010	5.00
	Dog	it^a	LD_{50}	1.0	0.086	0.010	5.00
	Monkey	iv	LD_{50}	0.6	0.051	0.0059	5.23
	Monkey	it^a	LD_{50}	1.0	0.036	0.010	5.00
	Rabbit	sc	LD_{50}	1.5	0.15	0.0175	4.76

(Cont'd)

TABLE 2-1. (Cont'd)

Compound	Animal	Route	Toxicity	Compound mg	Metal mg	Metal mM	pT
Beryllium phosphate	Mouse	iv	LD_{50}	11.3	1.4	0.16	3.81
$Be_3(PO_4)_2$	Mouse	iv	LD_{50}	3.0	0.265	0.030	4.52
	Mouse	inhal	MLD	47^b	4.02	0.45	3.35
	Rat	oral	LD_{50}	82	6.5	0.72	3.14
	Rat	iv	LD_{50}	4.2	0.36	0.040	4.40
	Rat	inhal	LD_{50}	10^b	0.86	0.095	4.02
	Guinea pig	ip	LD_{50}	100	8.6	0.95	3.02
	Cat	inhal	MLD	10^b	0.86	0.10	4.00
	Cat	inhal	LD_{100}	47^b	4.02	0.45	3.35
	Guinea pig	inhal	LD_{50}	47^b	4.02	0.45	3.35
	Rabbit	inhal	MLD	10^a	0.86	0.095	4.02
Beryllium carbonate $BeCO_3$	Guinea pig	ip	LD_{50}	50	1.2	0.133	3.88
Beryllium acetate $Be(C_2H_3O_2)_2$	Rat	ip	LD_{50}	317	22.4	2.49	2.60
Beryllium lactate $Be(C_3H_5O_2)_2$	Rat	iv	LD_{50}	0.53	0.024	0.0026	5.59

[a] it = intratracheal.
[b] Exposure of 4 hr; number of mg/m^3.

Parenteral injection of soluble Be salts into rats causes necrosis of the liver, epithelial necrosis in kidney tubules, and degenerative changes in the hemopoietic system; hepatotoxicity is due to the destruction of Kupffer cells. Osteosarcoma is produced in rabbits, and macrocytic anemia develops in dogs following a slowdown in the synthesis of hemin and globin with the consequent decrease in the hemoglobin content of erythrocytes and a small decrease in erythrocytes in the blood (Stokinger *et al.*, 1953). Beryllium decreased the ratio of phospholipid to free cholesterol in erythrocytes and increased the ratio of uric acid to creatinine in urine (Spiegel *et al.*, 1953).

Experimental Be granulomata and necrosis can be produced in the lungs and skin; Be phosphor implantation caused Be granulomata in the skin of pigs, and intratracheal injection of amorphous BeO caused Be granulomata in the lungs of rats. In adult guinea pigs Be induces either a skin sensitization or an immunologic tolerance. The delayed hypersensitivity results only from contact of Be^{2+} with skin; however, parenteral Be^{2+} injection fails to sensitize but induces tolerance (Vacher, 1972). The immunological responses depend upon the route of administration and the nature of the Be salt. Beryllium disease or berylliosis is a complex body-wide systemic disease caused by the inhalation of Be compounds; acute beryllosis is manifested by chemical pneumonitis ranging from transient pharyngitis or tracheobronchitis to severe pulmonary reaction, whereas under chronic beryllosis lung lesions cause serious respiratory damage and progressive pulmonary fibrosis, leading to death.

Inhalation of Be salts by animals produces primarily a complex pneumonitis; fine particles with a large surface area cause greater damage to the lungs than do larger particles with a smaller surface area. Hydrogen fluoride gas potentiates the inhalation toxicity of soluble Be salts; BeF_2 is acutely more toxic than either the sulfate or the oxide. Prolonged inhalation of Be salts results in recognizable Be injury to many tissues and organs. Experimentally produced pneumonitis in animals is similar to human pneumonitis observed in workers in the Be industry. When the concentration of soluble Be has exceeded 1000 $\mu g/m^3$ in the atmosphere of beryllium refineries, acute beryllosis has occurred in all personnel (Hall *et al.*, 1950). Beryllium oxide calcined at 1600°C appears to be less toxic than the low-fired BeO (500°C). Experimental animals tolerated high exposure to the high-temperature BeO, but developed acute Be toxicity symptoms when exposed to similar concentrations of low-temperature oxides (Sterner and Eisenbud, 1951).

Certain aerosols of $BeSO_4$ are capable of developing an acute and often lethal inflammatory reaction, even in extremely small concentrations (Vorwald *et al.*, 1966). Epidemiologic and experimental evidence have established the carcinogenicity of Be in mammals; the prolonged retention of Be in the tissues enhances this involvement. Primary lung tumors develop in rats and monkeys following prolonged inhalation of Be compounds such as the sulfate or oxide (28 $\mu g/m^3$ for 10 months); these tumors are transplantable and also metastasize to other organs (Vorwald, 1959; Schepers *et al.* 1957; Vorwald *et al.*, 1966). Beryllium induces tumors when about 20 mg Be is deposited in the medullary cavity of long bone (Tapp, 1966); osteogenic sarcomas are induced in rabbits following intravenous injection of soluble Be salts. The incidence of lung cancer among workers in Be refineries and associated industries, about 2%, implicates Be as an environmental carcinogen. However, the few incidences of lung cancer and bone tumor in human beryllosis victims, about 11 in 856 Beryllium Registry cases, cannot justifiably incriminate Be as an occupational carcinogen in humans.

The pathogenesis of beryllium disease is still unknown, although a threshold nontoxic level of 2 $\mu g/m^3$ for Be has been set (Stokinger, 1966). The mechanism of action in Be toxicity is complex; beryllium inhibits a number of enzymes *in vitro*, such as alkaline phosphatase, phosphatidic acid phosphatase, pyrophosphatase, phosphoglucomutase, Ca^{2+}-activated ATPase, enolase, amylase, Mg^{2+}-activated RNA polymerase, chymotrypsin, and deoxythymidine kinase, but it has no effect on the activities of acid phosphatase, arginase, DNase, and RNase (Thomas, M., and Aldridge,

1966; Aldridge and Thomas, 1966). However, phosphatases of serum and adrenal cortex, hyaluronidase, and phosphoglucomutase of liver and muscle of rats poisoned with Be are all inhibited to varying degrees; the activity of these enzymes is considerably reduced in animals fed Be salts or suffering from beryllosis.

Beryllium inhibits the synthesis of nuclear proteins and influences the steps leading to DNA synthesis in animals; and it inhibits thymidine and thymidylate kinases and DNA polymerase; doses of Be too small to cause hepatic necrosis inhibit incorporation of thymidine ^{14}C into DNA (Witschi, 1968, 1970, 1971). Since these enzymes are peripherally involved in DNA synthesis, Be inhibits DNA synthesis. It can inhibit phosphorylation by binding strongly with high-energy phosphate esters (HSAB theory); Be probably combines with O or N ligands of the unphosphorylated enzymes and competes with Mg for the enzymes; the Be–enzyme complex is unable to chelate with ATP, thus preventing phosphorylation of substrate. Beryllium displays a high and selective affinity for cell nuclei but also accumulates in lysosomes before it reaches the nucleus. Beryllium has been shown to interfere *in vitro* with the differentiation of developing embryonic tissue. Ultrastructural changes in intact and regenerating liver induced by beryllium sulfate include vacuolization, loss of fibrils, and distortion of bile canaliculi (Goldglat *et al.*, 1973).

There are conflicting reports regarding the immunologic aspects of Be toxicity. Immunological abnormality or an inborn error of metabolism have been attributed to the (low) incidence of beryllosis among the industrial workers exposed to Be. Van Ordstrand and Deodhar (1974) attribute a delayed immune response to chronic Be disease and demonstrated a delayed hypersensitivity to Be in patients with chronic Be disease and in a small number of exposed workers who do not appear to have the disease; blood from both groups showed a positive response for the transformation of lymphoblasts into lymphocytes by Be ions. However, it should be pointed out that the lung lesions noticed in dogs after exposure to Be oxides resembled the more classic reactions to a foreign body than immunologic reactions (Robinson *et al.*, 1965). An altered adrenal function is suggested as an inducer of latent chronic Be disease (Clary *et al.*, 1972).

There are no reports about a mechanism for the detoxication of Be. Administration of chelating agents such as aurine tricarboxylic acid increases urinary excretion of Be.

Magnesium (Mg)

Magnesium occurs widely in nature as carbonates, dolomite and magnesite ($MgCO_3$); silicates, olivine, serpentine, talc, and asbestos; and as sulfate epsom salt ($MgSO_4$); and chloride ($MgCl_2$) in seawater, brine wells, and salt deposits. About 2% Mg is present in the earth's crust and about 0.135% in seawater. Magnesium is the fourth most abundant cation in the living human body. A normal adult body contains 21–28 g Mg, about 60% of it present in the skeleton, 29% in muscle tissues, 10% in soft tissues, and the rest in extracellular fluids. Like K, Mg is present inside the cell in higher concentrations than in extracellular fluids. Magnesium is essential to both plants and animals and is considered one of the major cations in the Precambrian seas, from which life probably first evolved.

Magnesium metal is used extensively as a lightweight structural material in parts of airplanes, automobiles, boats, machinery, and handling equipment; powdered magnesium metal is used in the manufacture of flares, incendiary bombs, tracer bullets, and flashlight powders. Magnesium is used with different metals in making alloys with specific functional properties such as reduced vibrational stress or corrosion resistance; magnesium trisilicate is the refractory material for furnaces. Magnesium salts find therapeutic use: magnesium benzoate for rheumatism, magnesium salicylate as an intestinal antiseptic, magnesium bromide as a sedative and an antiepileptic, magnesium chloride and sulfate as laxatives and cathartics, and magnesium borate as an antiseptic and a fungicide. Magnesium and its salts are not considered to be industrial health hazards; magnesium poisoning is mostly accidental from therapeutic overdoses given intravenously. Renal failure in humans due to other causes can induce Mg toxicity.

Chemistry

Magnesium is a reactive metal exhibiting divalence and forming stable electrovalent cationic compounds. Most Mg salts are water-soluble; double halides, such as $KMgF_3$, are not anionic complexes. Magnesium has the smallest ionic radius among the biologically important cations; the smaller the size of the ion, the greater its binding energy and its tendency to form complexes. Magnesium complexes with many molecules of dipolar character (e.g., water) and with ammonia, amino acids, heterocyclic nitrogen compounds, and oxygen-containing ligands; its affinity for phosphate is high. Normally Mg complexes assume tetrahedral structures with a coordination number of 4; however, Mg also forms complexes with coordination numbers 5, 6, and 8. Chlorophyll is usually associated with a tetrahedral configuration having Mg in the center with a coordination number of 4. However it may be more complex, since it has been found to have a more or less planar configuration with Mg outside the plane of the macromolecule, suggesting that the Mg coordination number is 5. Magnesium halides form complexes with alcohols, ethers, esters, and amides where Mg is hexacoordinated. Magnesium-containing porphyrins form stable dipyridine complexes similar to hemichromes in which Mg has an octahedral coordination. Magnesium reacts with flavins and flavoquinones to form monodentate complexes, and binds to phosphate groups in nucleotides. It is quite versatile, can react with a variety of compounds in biological systems, and can also bring reacting substances physically close to one another.

Metabolism

Magnesium is an essential nutrient for living organisms because it forms part of the structure of the body and plays a critical role in cell metabolism. Magnesium is an activator of many enzyme systems, especially those which utilize ATP; Mg metabolism is closely related to that of other elements such as Ca, Na, K, Mn, and P. Magnesium is essential for neuromuscular action and muscle contraction. High metabolic activity is shown by a high ratio of Mg to Ca inside cells of muscle, nerve, brain, and heart. Magnesium-deficiency syndromes in man and animals include poor growth, hyperirritability, convulsions, and tetany leading to death. Diets deficient in magnesium or protein, malabsorption of Mg, alcoholism, diarrhea, and renal failure can lead to Mg deficiency. The chemistry and metabolism of Mg, the role of Mg deficiency in chronic diseases of the cardiovascular and renal systems, and the involvement of Mg in biochemical systems have been well reviewed (Stokinger, 1963*a*; Wacker and Vallee, 1964; MacIntyre, 1967; Schroeder *et al.*, 1969; Aikawa, 1963, 1971; Wustenberg, 1972), and will not be presented here.

The gastrointestinal absorption of Mg in mammals depends upon the species, dose, and solubility of the Mg compounds. The absorption of dietary Mg is about 15% in dogs and rabbits, 40% in calves, and 60% in rats; young animals absorb more Mg than do mature animals. In man the absorption of Mg is about 75% with a daily dietary intake of 24 mg Mg, about 44% with 240 mg Mg, and 24% with 564 mg Mg. Absorption occurs more in the duodenum than in the ileum. Phytates in the diet drastically decrease Mg absorption. Calcium and magnesium share similar mechanisms by which they are transported across mucosal walls; the absorption of Mg is high in Ca deficiency. Vitamin D in pharmacologic doses increases the intestinal absorption and reduces urinary Mg loss (Aikawa, 1971). Absorption of physiologic doses of soluble Mg from the sites of parenteral injection is fairly rapid and complete; higher doses cause local necrosis and swelling. Intravenous injection of Mg salts is risky and may prove to be toxic depending upon the dose and the rate of injection; Mg also can induce general anesthesia. If the plasma Mg level does not rise above 5 mg/100 ml as a result of intravenous administration, two-thirds of the injected Mg is absorbed by the erythrocytes, and the rest remains in the plasma. Part of the plasma Mg is loosely bound to plasma protein, and this labile Mg–

protein complex is readily distributed to the soft tissues; the rest of the Mg is present in a dialyzable form in the plasma. Inhalation of soluble Mg salts results in transient retention of Mg in the lung before final absorption into the blood. Insoluble compounds such as MgO are retained for a longer period, causing local irritation and a temporary fall in body temperature. Magnesium is retained in the skeleton, muscle, and soft tissues.

Excretion of Mg is both fecal and urinary; fecal Mg consists of unabsorbed Mg and the Mg excreted into the gastrointestinal tract from the bile; fecal elimination of parenterally injected Mg is usually 15–25% of the dose. About 45% of the absorbed Mg is excreted in the urine within the first 24 hours, while the rest is excreted more slowly.

Hemostatic regulatory mechanisms for Mg control absorption in the gastrointestinal tract and renal excretion. Retention of Mg by the kidney occurs rapidly in response to any restriction in the dietary intake and absorption; this is accomplished by glomerular filtration and partial reabsorption of the filtered material by renal tubules. Only with impaired kidney function could large oral doses of Mg cause toxicity. The control of the plasma Mg level is influenced by parathyroid hormone through a negative feedback mechanism (MacIntyre *et al.*, 1963). A decrease in the fractional rate of cellular Mg influx occurs during hypermagnesemia, preventing accumulation of intracellular Mg (Wallach *et al.*, 1967).

Toxicity

Magnesium toxicity is usually acute; the symptoms are nausea, malaise, muscular weakness and paralysis, general depression, and paralysis of the respiratory, cardiovascular, and central nervous systems. Magnesium salts are not generally toxic when given orally.

Oral $MgSO_4$ acts as a purgative; due to its poor absorption, osmotic changes leading to water influx into the intestine cause diarrhea and dehydration. The distensive stimulus overcomes the depressant action of Mg on intestinal muscle. Prolonged $MgSO_4$ enema causes intoxication (Fawcett and Gens, 1943), and fatal poisoning has been reported in both adults and children (Arena, 1974). Weanling rats exhibited diarrhea, depressed growth rate, and intestinal distension when fed 1% $MgSO_4$ (Moinuddin and Lee, 1960). Magnesium inhalation by miners and industrial workers causes leukocytosis, granulomas, and fever (Sollman, 1963). The toxicity of Mg is summarized in Table 2-2. Magnesium acetate appears to be highly toxic when given intravenously. Renal failure can elevate the plasma Mg level; when this level exceeds 5 mg/100 ml, Mg becomes toxic and causes hypermagnesemia. Feeding high doses of Mg to cows (O'Kelley and Fontenot, 1973) and sheep (Ewer, 1951) increases Ca excretion.

Newborn infants may develop hypermagnesemia if their mothers have

TABLE 2-2. Magnesium Toxicity

Compound	Animal	Route	Toxicity	Dosage/kg body weight			
				Compound	Metal		
				mg	mg	mM	pT
Magnesium Mg	Dog	oral	MLD		230	9.46	2.02
Magnesium oxide MgO	Human	inhal	Txc	400[a]	240[a]	9.87	2.01
Magnesium fluoride MgF_2	Guinea pig	oral	MLD	1000	390	16.0	1.79
	Guinea pig	sc	MLD	3000	1170	48.1	1.32
Magnesium chloride $MgCl_2$	Mouse	ip	LD_{50}	342	87.3	3.59	2.44
	Mouse	iv	LD_{50}	14	3.5	0.14	3.84
	Rat	oral	LD_{50}	2800	715	29.4	1.53
	Rat	sc	MLD	900	230	9.46	2.02
	Rat	ip	MLD	225	57.5	2.37	2.63
Magnesium chloride $MgCl_2 \cdot 6H_2O$	Rat	iv	LD_{100}	175	21	0.86	3.06
Magnesium chlorate $Mg(ClO_3)_2 \cdot 6H_2O$	Mouse	oral	MLD	1000	81.2	3.43	2.48
	Mouse	it	MLD	500	40.6	1.67	2.78
	Rat	oral	MLD	5250	426	17.5	1.76
	Rat	ip	MLD	1100	89	3.66	2.44
	Rat	it	MLD	700	57	2.34	2.63
	Guinea pig	oral	MLD	1000	81.2	3.34	2.48
	Guinea pig	it	MLD	700	57	2.34	2.63
Magnesium perchlorate $Mg(ClO_4)_2$	Mouse	ip	LD_{50}	1500	163	6.71	2.17
Magnesium sulfate $MgSO_4 \cdot 7H_2O$	Dog	oral	LD_{100}	2800	276	11.4	1.94
	Dog	sc	MLD	1750	172	7.08	2.15
	Dog	iv	MLD	750	74	3.04	2.52
	Guinea pig	sc	MLD	1800	177	7.32	2.14
	Rabbit	sc	MLD	1750	172	7.08	2.15
	Cat	sc	MLD	1000	98.6	4.06	2.39
Magnesium thiosulfate $MgS_2O \cdot 6H_2O$	Rat	ip	LD_{50}	194	19.3	0.79	3.10
Magnesium phosphide Mg_3P_2	Rat	inhal	MLD	580[a]			
	Cat	inhal	MLD	173[a]			
	Guinea pig	inhal	MLD	288[a]			
Magnesium arsenate $Mg_3(AsO_4)_2$	Rat	oral	MLD	280	58.2	2.39	2.62
	Mouse	oral	LD_{50}	315	65.5	2.69	2.57
	Rabbit	oral	MLD	80	16.6	0.68	3.17
	Rabbit	sc	MLD	2100	436	17.93	1.75
Magnesium hexafluorosilicate	Guinea pig	oral	LD_{50}	200	17.7	0.728	3.14
$MgSiF_6 \cdot 6H_2O$	Guinea pig	sc	MLD	400	35.4	1.46	2.84

[a]Exposure of 4 hr; number of mg/m^3.

been treated with $MgSO_4$ (Lipsitz and English, 1967). Magnesium toxicity depresses the central nervous system and peripheral neuromuscular junctions resulting in ataxia, lethargy, listlessness, poor tendon reflexes, and coma. The toxic effects are attributed to a decreased liberation of acetylcholine at the neuromuscular junctions and sympathetic ganglia (Del Costillo and Engback, 1954). Magnesium–calcium imbalance impairs neuromuscular transmission; excess Mg ions block synaptic transmission through ganglia of the sympathetic nervous system. The toxic effects of Mg on the cardiovascular system include hypotension, cutaneous vasodilation, increased sensitivity of the carotid sinus, and cardiac arrest.

When injected intravenously, Mg ions act as a local intraspinal and general anesthetic; the mechanism is not clear, but it could be due to depression in cardiac muscle activity and to hypoxia that is secondary to paralysis of the respiratory muscle. The effective anesthetic dose is dangerously close to the fatal dose which causes respiratory paralysis, 17–20 mg Mg/100 ml blood compared with normal Mg content of 4–5 mg/100 ml. Intramuscular or intraperitoneal administration of Mg does not produce anesthesia.

Calcium (Ca)

Calcium occurs as limestone ($CaCO_3$), gypsum ($CaSO_4 \cdot 2H_2O$), and apatite (fluorophosphate and chlorophosphate of Ca), in addition to other less important minerals. Calcium is the fifth most abundant element, forming 3.64% of the earth's crust; seawater contains about 0.04% Ca. Calcium is present in all plant and animal tissues and is a major essential constituent of bones, teeth, and soft tissues. The human adult body contains about 1100 g Ca, about 1.5% of the body weight; the skeleton contains 99% of the total body Ca.

Calcium metal and its compounds are used extensively in industry, medicine, and veterinary medicine. Metallic Ca is used as a deoxidizer in the purification of metals such as Cu and Be, and alloys such as steel. With Ce, Ca is used as flints in gas lighters. Some uses of Ca salts are: $CaBr_2$ as a fixer in photography, calcium acetate in the dyeing, tanning, and curing of skins; calcium arsenate and aresenites as insecticides and pesticides; calcium chlorate as an insecticide and seed disinfectant; calcium cyanamide as fertilizer, as a defoliant for cotton plants, and in the manufacture of melamine resin used in making plastic products; and calcium borate in glycol antifreeze and in the manufacture of sterile porcelain insulators. Slaked lime, with countless uses, is a most inexpensive base used in the chemical industry.

Calcium salts used in medicine are: $CaCl_2$ as a diuretic, antiallergenic, and urinary acidifier; $CaBr_2$ as a sedative and anticonvulsant; $CaCO_3$ as a gastric antacid and antidiarrheal; and calcium lactate, gluconate, and ferrous citrate as dietary Ca supplements.

Calcium poisoning in man and animals is rare; Ca toxicity through dietary sources is negligible, except in overdosage of dietary vitamin D. Calcium is present in stoichiometric amounts in snake venoms (Tu, 1977), and appears to be the important divalent cation for proteinase B, phospholipase A_2, and hemolysis activities. Inhalation of CaO dusts produces only temporary irritation in the respiratory tract. Accidental overdoses by parenteral injections of Ca salts during therapy are mainly responsible for Ca poisoning.

Chemistry

Calcium forms stable salts which exhibit divalence; some of the cationic salts are water-soluble. Calcium forms coordination covalent bonding with mucopolysaccharides, mucoproteins, and plasma proteins. It chelates strongly with EDTA and similar chelating agents; other Ca coordination compounds are either rare or unstable. The affinity of Ca toward phosphoryl groups is high.

Metabolism

Calcium is an essential macroelement; it performs essential functions in bone structure, muscle contraction, nerve impulses, membrane permeability, blood clotting, intercellular cement, and enzyme catalysis. Calcium deficiency leads to stunted growth, bone demineralization, muscle weakness, decreased serum Ca^{2+} level, parathyroid hyperplasia, hyperirritability, tetany, and death. The recommended daily allowance of Ca for North American adults is 800 mg, and this amount should increase by about 50% during pregnancy and lactation. Requirements of Ca for humans vary in different geographic regions; the biochemistry and physiology of Ca has been extensively studied (Comar and Bronner, 1969; Thomas and Howard, 1962; Bronner, 1964; Irving, 1973).

The gastrointestinal absorption of Ca in mammals is about 40% of a normal physiologic dose. Within limits Ca absorption is directly related to the content of the diet; acute transport is directly linked to energy (DeGrazia, 1971). Low dietary protein or high fat, or the presence of anions such as phytate and oxalate, corticosteroids, and thyroxine decrease the absorption of Ca from the digestive tract; Ca absorption is greatly enhanced by high dietary protein, lysine, and arginine, by lactose, and particularly by vitamin D. Active vitamin D promotes the formation of a Ca-binding protein. Impaired fat absorption results from fatty acids reacting with Ca in the digestive tract, forming insoluble Ca salts which inhibit Ca absorption. Excessive dietary Zn and other divalent metals that utilize the same absorption sites as Ca also inhibit Ca absorption. Any change which elevates the pH of the intestinal contents will also decrease Ca absorption, since the solubility of Ca salts decreases with high pH. A Ca/P ratio outside the range 0.5–2.0 leads to poor absorption of both elements. Calcium absorption has been well reviewed (DeGrazia, 1971; Holdsworth and D'A Welling, 1971; Ogata *et al.,* 1971).

Absorption of Ca salts from the sites of parenteral injection is gradual. Intravenous injection of Ca must be done slowly and in physiological doses in order to avoid Ca toxicity. Plasma Ca exists in three forms: easily diffusible ionic Ca, nondiffusible protein-bound Ca, and Ca complexed to citrates and other compounds. There is rapid distribution of Ca from the blood to all tissues.

Excretion of Ca is both fecal and urinary; unabsorbed dietary Ca and endogenous excretion into the intestine account for fecal excretion. Although a large amount of Ca passes through the renal glomeruli, only a small amount of it is excreted into the urine. In human adults the average daily endogenous excretion of Ca is: urinary, 175 mg; fecal, 125 mg; and dermal, 25 mg.

An efficient homeostatic mechanism associated with energy-linked transport operates at the intestinal absorption level; it involves glycolysis and K^+, Na^+, Mg^{2+}, and H^+ exchange. Calcium accumulates in the aorta, kidneys, prostate, thyroid, and trachea with age.

Toxicity

Calcium salts are considered nontoxic except at very high doses; toxicity depends upon the nature of the calcium salt and the mode of administration. By oral ingestion most Ca salts except the arsenate and molybdate are practically nontoxic; $CaCl_2$ acts as a gastric irritant. The general symptoms of parenterally administered Ca salts at sublethal doses are apathy, dullness, stupor, and confusion. Subcutaneously, $CaCl_2$ is highly irritating to the tissues, causing painful sloughing of cells. Intramuscular injection also causes high local irritation; in mice 0.3–0.4 g Ca (as $CaCl_2$) per kg body weight causes death due to central nervous system paralysis and respiratory depression; at 60–90 mg/kg it impairs equilibrium and the postural reflexes of skeletal muscles. Small doses of $CaCl_2$ given intravenously raise blood pressure, while larger doses produce successively bradycardia, ventricular arrhythmias, ventricular fibrillation, and finally cardiac failure.

Inhalation of moderately caustic calcium oxide or hydroxide causes chemical pneumonia and severe irritation of the upper respiratory tract. In man, inhalation of calcium cyanamide causes transient vasomotor disturbances of the upper portion of the body; higher doses cause dermatitis, permanent vasomotor changes, and dyspnea (DeLarrad and Lazarini, 1954). The toxicity of Ca is summarized in Table 2-3.

Hypercalcemia in mammals is a pathologic condition which is unlikely to be caused by Ca toxicity or excess of dietary Ca; it may be caused by hyperthyroidism or excess vitamin D. Neoplasms with metastases to the bones also cause hypercalcemia; under this condition calcium is deposited in the kidneys, causing impaired renal function (Epstein, 1963). In man hypercalcemia causes coma and death quickly if the serum Ca level rises to 16 mg/100 ml; other symptoms include dehydration, lethargy, nausea, vomiting, and anorexia (Arena, 1974).

High dietary levels of Ca decrease the gastrointestinal absorption of Zn, Mg, Fe, Mn, and Cu to a critical level if the latter minerals are at marginal levels in the diet, leading to deficiency symptoms of these minerals. In cattle intravenous Ca^{2+}, either as $CaCl_2$ or calcium gluconate, produced toxic symptoms by lowering the serum phosphate level (Hapke, 1971); cattle tolerated twice as much calcium gluconate (40 mg/kg) as $CaCl_2$. The toxic effects were limited to cardiac malfunctioning; extra systoles and ventricular fibrillations occurred, leading eventually to death when the total amount of intravenous injection of $CaCl_2$ reached 100 g per animal. Excess calcium has a teratogenic action in chicks when about 0.1 mg of calcium gluconate or chloride is injected into the egg (Grabowski, 1966), but there are no reports of similar teratogenic effects in mammals.

The mechanism of action of Ca toxicity is not fully understood. Intravenously the amount of Ca^{2+} given per unit time rather than the total amount injected is considered to be responsible for cardiac failure; toxicity is attributed to the sudden change in osmotic pressure due to the formation of colloidal calcium phosphate (Plume, 1971). An excess of Ca ions depresses the functioning and increases the threshold of excitability of muscle and nervous tissues, leading to a state of stupor. Calcium can also affect the viscosity of mucous secretions by coordinate covalent bonding with mucopolysaccharides and mucoproteins.

Strontium (Sr)

Strontium occurs in nature as the carbonate, strontianite ($SrCO_3$), and as the sulfate, celestite ($SrSO_4$). The earth's crust contains 450 ppm Sr, and

TABLE 2-3. Calcium Toxicity

| | | | | Dosage/kg body weight | | | |
| | | | | Compound | Metal | | |
Compound	Animal	Route	Toxicity	mg	mg	mM	pT
Calcium chloride	Mouse	ip	LD_{50}	280	100	2.50	2.60
$CaCl_2$	Mouse	iv	LD_{50}	42	15.1	0.38	3.42
	Rat	oral	LD_{50}	1000	1440	35.9	1.44
	Rat	ip	MLD	500	180	4.49	2.35
	Rat	iv	MLD	161	60.5	1.51	2.82
	Rabbit	oral	LD_{100}	1380	49	12.2	1.91
	Rabbit	sc	LD_{100}	472	170	4.2	2.37
	Rabbit	iv	LD_{100}	274	98.6	2.46	2.61
	Cat	sc	LD_{100}	250	90	2.25	2.65
	Cat	iv	LD_{100}	250	90	2.25	2.65
	Dog	sc	LD_{100}	275	99	2.47	2.61
	Dog	iv	LD_{100}	275	99	2.47	2.61
Calcium chlorate	Rat	oral	LD_{50}	4500	871	21.73	1.66
$Ca(ClO_3)_2$	Rat	ip	LD_{50}	625	121	3.02	2.52
Calcium thiosulfate	Rat	ip	MLD	150	39.5	0.99	3.01
CaS_2O_3							
Calcium arsenate	Mouse	oral	LD_{50}	794	79.8	1.99	2.70
$Ca_3(AsO_4)_2$	Rat	oral	LD_{50}	298	29.9	0.75	3.13
Calcium molybdate	Rat	oral	LD_{50}	100	20	0.50	3.30
$CaMoO_4$							
Calcium cyanamide	Rabbit	oral	LD_{50}	1400	700	17.5	1.76
$CaCN_2$							
Calcium cyanide	Rat	oral	LD_{50}	39	16.9	0.42	3.38
$Ca(CN)_2$							
Calcium silicofluoride	Guinea pig	oral	MLD	250	54.9	1.37	2.86
$CaSiF_6$	Guinea pig	sc	MLD	450	98.9	2.47	2.61
Calcium acetate	Rat	oral	LD_{50}	4280	1085	27.1	1.57
$Ca(C_2H_3O_2)\cdot 2H_2O$							
Calcium acetate	Mouse	iv	LD_{50}	52	13.2	0.33	3.47
$Ca(CH_3COO)_2$	Rat	iv	MLD	147	37.3	0.92	3.12
	Rat	iv	LD_{50}	245	62	1.55	2.81
Calcium pantothenate	Rat	ip	LD_{50}	820	65.5	1.63	2.79

seawater contains about 8 ppm Sr. Strontium is the 15th most abundant element in nature and the most abundant trace element in seawater. Strontium is present in both plant and animal tissues; marine plants and animals are rich in Sr. The human adult body contains about 320 mg Sr, varying according to the geographic location. The skeleton contains 99% of the total body content of Sr, and the rest is distributed in soft tissues. The evidence that Sr might be essential for mammals is not substantial.

Strontium is used in pyrotechnical devices such as tracer bullets and distress signal rockets for its characteristic red flame. Strontium salts are used for rubber fillers, corrosion inhibitors, plastics, luminous paints, and greases and in tooth pastes and depilatories. Strontium carbonate and hydroxide are used in sugar refining. In medicine $SrBr_2$ is used as a sedative, an antiepileptic, and a therapeutic in urticaria and parathyroid tetany. Strontium and its salts are not industrial health hazards; strontium poisoning is rare and is mostly accidental due to inadvertent therapeutic use. Strontium-90, present in radioactive fallout from nuclear blasts, is a potent environmental health hazard.

Chemistry

Strontium exhibits divalence and forms electrovalent compounds; its common cationic salts are water-soluble; it forms chelates with compounds such as EDTA. Strontium coordination compounds are not common; it reacts with phosphate and carboxylate groups of biologically active macromolecules, e.g., strontium proteinates in blood serum. Its affinity for phosphate results in the deposition of Sr in bone. Strontium resembles calcium more than barium in its chemical and physiologic properties.

Metabolism

The metabolism of Sr in mammals has been studied in conjunction with Ca and has been reviewed (Comar and Wasserman, 1964; Lenihen *et al.*, 1967; Schroeder *et al.*, 1972). Strontium can replace calcium in the nutrition of some bacteria and fungi; Sr can substitute for Ca in some of the metabolic functions of Ca in mammals, but Ca is always preferred to Sr in absorption from the digestive tract, urinary excretion, and utilization in bone formation or milk secretion. Strontium metabolism is controlled by Ca levels in the tissues. Strontium is probably not essential to mammals, since no confirmed demonstrable effects on growth or other essential functions have been found. However, Sr increases the mineralization of bone and teeth in experimental animals and acts as an anticariogenic agent (Rygh, 1949). A caries survey showed a good correlation between the absence of caries and the presence of Sr in drinking water (Losee and Adkins, 1969). The low toxicity of Sr, its association with Ca in the food chain from soil to plants to animals and man, and its presence in the tissues of newborn infants suggest a possible essential role for this metal. Rygh (1949) reported growth failure in rats fed a Sr-free diet. The average daily intake of Sr by an adult man is about equal to the Sr level in soft tissues, suggesting a rapid turnover of Sr in the tissues and the existence of a homeostatic regulatory mechanism. Comar and Wasserman (1964) claim that the Sr level in the body is regulated by either Ca concentrations or the combined concentrations of Ca, Sr, and Ba, and not by Sr levels alone.

The gastrointestinal absorption of soluble Sr salts varies from 5% to 25% of the ingested dose in different animals under normal physiological conditions; absorption is high in man and low in dogs and goats. The major factors are discussed by Volf (1971). Insoluble Sr salts are poorly absorbed, about 5% in man. Young weanling animals absorb Sr more rapidly and completely than do mature animals (Forbes and Reina, 1972). Mature rats absorb 68% of a dietary dose ranging from 0.5 to 2.0 mg Sr^{2+}; in man the absorption is about 36% of an ingested dose ranging from 100 to 250 mg Sr^{2+} (Harrison *et al.*, 1955). The major site of Sr absorption in the rat's digestive tract is the ileum, 66% of the absorbed dose, and the least is the stomach, 2% (Cramer and Copp, 1959). Alginates and high-molecular-weight uronic acids reduce the intestinal absorption of Sr. In man, the absorption of Sr is high when there is a low dietary Ca intake (Spencer-Lazlo *et al.*, 1963).

Absorption of Sr^{2+} into the blood from sites of intraperitoneal and intramuscular injections is gradual; insoluble salts remain at the site for a long time and are cleared by macrophage phagocytosis. Intravenously administered ionic Sr forms colloidal or particulate strontium phosphate, or binds to plasma proteins to form partly diffusible complexes. This protein complex is distributed to the skeleton and soft tissues; the major organs of Sr deposition among the soft tissues are the kidneys, liver, spleen, brain, and aorta. The accumulation of orally ingested Sr in the skeleton is proportional to the dose, and averages about 3%; the bone retention factor or discrimination factor for Sr (number of Sr atoms/1000 Ca atoms divided by the same ratio in the diet) averages about 0.2 (Sowder and Stick, 1957). In rats, administration of fluoride decreases Sr retention (Muhler *et al.*, 1959), and administration of parathyroid hormones decreases the retention of Sr in bone, decreases its fecal excretion, and enhances its urinary excretion (Fay *et al.*, 1942). Thyroid hormone also decreases Sr retention in growing mice (Simmons and Failla, 1966). Strontium ions can pass through the placental and mammary barrier in rats (Pecher and Pecher, 1941); babies have Sr in their tissues; Sr retention increases with age (Kahn *et al.*, 1969).

The pattern of Sr excretion in mammals differs from species to species. In humans, the excretion of absorbed Sr is about 90% through urine; the small amounts found in feces come from intestinal secretions and desquamated cells. Unabsorbed dietary insoluble salts are excreted in the feces. In mice the fecal route is the main excretory pathway for Sr, and in rabbits about 30% of the absorbed dose is excreted in the feces. The ratio of Sr/Ca in the urine is higher than in plasma due to selective preference of Ca over Sr in renal tubular absorption. Homeostatic mechanisms for Sr are efficient in soft tissues but are dominated by Ca metabolism (Comar and Wasserman, 1964).

Toxicity

Strontium salts are considered relatively nontoxic or of very low toxicity; strontium toxicity is summarized in Table 2-4. Parenterally administered strontium acetate is considerably less toxic than calcium acetate; at

TABLE 2-4. Strontium Toxicity

| | | | | Dosage/kg body weight | | | |
| | | | | Compound | Metal | | |
Compound	Animal	Route	Toxicity	mg	mg	mM	pT
Strontium fluoride	Rat	iv	LD_{100}	625	436	4.98	2.30
SrF_2	Guinea pig	oral	MLD	5000	3490	39.8	1.40
	Guinea pig	sc	MLD	5000	3490	39.8	1.40
Strontium chloride	Mouse	iv	LD_{50}	148	82	0.94	3.03
$SrCl_2$	Mouse	ip	LD_{50}	908	502	5.73	2.24
	Rat	ip	LD_{50}	405	224	2.56	2.59
	Rat	iv	MLD	123	68	0.78	3.11
	Rabbit	iv	LD_{50}	1060	590	6.73	2.17
Strontium chloride $SrCl_2 \cdot 6H_2O$	Rat	iv	MLD	400	221	2.52	2.60
Strontium bromide	Rat	ip	LD_{50}	1000	246	2.81	2.55
$SrBr_2$	Rat	iv	MLD	500	177	2.02	2.69
Strontium iodide SrI_2	Rat	ip	LD_{50}	800	156	1.78	2.75
Strontium nitrate $Sr(NO_3)_2$	Rat	ip	LD_{50}	540	224	2.56	2.59
Strontium fluoroborate SrB_4F_8	Rat	oral	MLD	500	155	1.77	2.75
Strontium acetate	Mouse	iv	MLD	123	52.3	0.60	3.22
$Sr(C_2H_3O_2)$	Mouse	iv	LD_{100}	383	163	1.86	2.73
	Rat	iv	MLD	239	101	1.15	2.94
	Rat	iv	LD_{100}	238	101	1.15	2.94
		iv	LD_{50}	105	44.5	0.51	3.29
Strontium lactate $Sr(C_3H_5O_3)_2$	Rat	ip	LD_{50}	900	247	2.82	2.55
Strontium salicylate $Sr(C_7H_5O_3)_2 \cdot 2H_2O$	Rat	ip	LD_{50}	363	88	1.00	3.00

equal amounts (in millimoles) of compound/kg, the mortality in rats was about 5% from Sr acetate and about 70% from calcium acetate (Loeser and Konweiser, 1929). Strontium salicylate is the most toxic among the strontium salts, and $SrBr_2$ is the least toxic (Stokinger, 1963). Excessive salivation, vomiting, colic, and diarrhea are some symptoms of acute Sr poisoning. Acute Sr toxicity results in respiratory failure and death in rats, while acute Ca toxicity results in cardiac failure. In cats, injection of $SrCl_2$ raises the blood pressure owing to stimulation of cardiac muscle, but large doses produce apnea, inhibit heart functioning, and cause death by cardiac arrest. Poor growth is observed in mice fed 16,000 ppm strontium lactate in drinking water (Forbes and Mitchel, 1957), and strontium carbonate at dietary levels of 2% caused rickets in experimental animals (Follis, 1955). In Ca deficiency, Sr can activate catecholamine secretion from the adrenal medulla, and can be stimulatory (Douglas, 1964). Specific involvement of Sr toxicity in enzyme or biochemical systems is not known, although Sr involvement in the inhibition of bone calcification could be predicted.

Barium (Ba)

Barium occurs in nature as the sulfate, barite or heavy spar ($BaSO_4$), and as the carbonate, witherite ($BaCO_3$); it also occurs in zinc or iron ores. The earth's crust contains 450 ppm Ba, and seawater contains about 0.03 ppm. Air has a varied distribution of Ba compounds ranging from 0.005 to 1.5 $\mu g/m^3$. Barium is present in all living organisms, especially marine plants and animals. The human adult body contains about 22 mg Ba, 66% of it present in the bones. Mammalian eyes contain Ba in the pigmented part in concentrations varying from 206 to 1110 $\mu g/g$ wet tissue. Barium may be stimulatory, but it is probably not essential for mammals.

Barium and its salts are used extensively in the manufacture of alloys for such products as nickel–barium parts used in ignition equipment for automobiles and in the manufacture of lithopone, glass, ceramics, and television picture tubes. Barium salts are employed in brick and refractory tiles, vinyl stabilizers, lubricating oil additives, green signal flares, and fireworks. Barium hydroxide is used in sugar refining; barium carbonate and fluorosilicates are used as rodenticides and insecticides; barium sulfate is used as a lubricating agent in drilling oil wells. Barium organometallic compounds are employed in reducing black smoke emission from diesel engines. Man is exposing himself to increasing concentrations of Ba and its salts. Barium is not considered an industrial health hazard, although inhalation of fine dusts of $BaSO_4$ produce benign pneumoconiosis. Barium chloride was formerly used in medicine as a cardiac muscle stimulant, but barium salts are no longer used therapeutically for fear of toxic effects. Barium poisoning is mostly accidental, from the intake of rodenticides such as $BaCO_3$.

Chemistry

Barium is the most electropositive of the subgroup IIA metals; it forms electrovalent salts exhibiting a valence of $+2$. The high hydration enthalpy maintains Ba and other ions of this subgroup in the divalent state in aqueous solution. Most of the simple cationic salts of Ba are water-soluble; among the hydroxides of subgroup IIA metals, $Ba(OH)_2$ is the most soluble. The affinity between Ba^{2+} and sulfate and phosphate ligands is very high, while the affinity of Ba^{2+} toward NH_2, SH, and COOH groups is rather low.

Metabolism

Although Ba associates with Ca and Sr in the food chain from plants to animals, it is not considered essential for mammals for growth or any biologic function. Rygh (1949) showed poor growth rate in rats fed Ba- and Sr-deficient diets. The average daily Ba intake of the human adult is about 1.3 mg (0.65–1.7 mg); about $\frac{2}{3}$ of the Ba present is to be found in the soft tissues. There is no conclusive evidence for any enzymatic, osmotic, or electrochemical function for Ba in mammalian tissues, but it appears to have a structural function because of its irreversible lodgment and retention in bone. The presence of Ba^{2+} in the iris suggests a special function in vision.

The metabolism of Ba in mammals has been studied with radioactive isotopes and shown to be essentially similar to that of Ca and Sr (Bauer *et al.* 1956, 1957; Castagnow *et al.*, 1957). The biochemistry and physiology of Ba has been reviewed by Schroeder *et al.* (1972). The principal physiological activity of Ba is stimulation of all types of muscle, irrespective of their innervation (Cushney, 1941); the mechanism of muscle stimulation is not known.

Mammalian intestinal mucosa is highly permeable to Ba^{2+} ions and is involved in the rapid flow of soluble Ba salts into and out of the blood (Castagnow *et al.*, 1957). Hence, soluble Ba salts given to experimental animals are rapidly absorbed following ingestion. However, absorption of naturally occurring Ba in the food is only about 2% of the total dietary Ba content, because Ba occurs in bound or insoluble forms. The absorption of soluble Ba salts into blood from the sites of intramuscular injections and from the peritoneal cavity is rapid, while insoluble salts such as $BaSO_4$ are very poorly absorbed. Small quantities of intravenously administered water-soluble Ba salts are transported quickly to skeletal muscles and other soft tissues; larger amounts of Ba salts form insoluble barium phosphates in the plasma and are removed slowly by phagocytosis. In man, inhaled insoluble Ba salts are retained in the lung, and their accumulation in the lung tissues increases with age. Analysis of human tissues reveals the presence of absorbed Ba in the following tissues: adrenal, aorta, thyroid, lung, muscle, testes, ovary, uterus, and urinary bladder, indicating wide distribution of Ba in mammalian soft tissues (Schroeder *et al.*, 1972). Barium deposition in bone is similar to that of Ca, but Ba deposition is faster and irreversible (Bauer *et al.*, 1957). Excretion of Ba is both fecal and urinary, depending upon the mode of administration; within 24 hours, 20% of the ingested dose appears in the feces while 5–7% is excreted in the urine, but when an equal dose of soluble Ba salt is injected, Ba excretion via the urine is high. There is higher reabsorption of Ba than of Ca in the renal tubules. The fecal excretion of absorbed Ba indicates an enterohepatic circulation. There is no increased total accumulation of Ba with age in humans, except in the lungs and aorta. Since Ba is found in newborn young, Ba ions cross both the mammary and placental barriers; the concentration in infant tissues is higher than in adult tissue. Homeostatic regulatory mechanisms which control the tissue Ca and Sr levels may also be involved in regulating Ba levels.

Toxicity

The symptoms of acute Ba poisoning in humans and other mammals are excessive salivation, vomiting, colic, violent diarrhea, tremors, muscular paralysis, and paralysis of the central nervous system. Strong vasoconstriction, due to the direct Ba stimulation of arterial muscles, raises the blood pressure; high blood pressure causes hemorrhage in the stomach, intestines, and kidneys. Symptoms of chronic Ba poisoning are similar but of lesser severity. Barium toxicity is summarized in Table 2-5; insoluble salts such as $BaSO_4$, are not toxic and may be used as fecal or nutritional markers, or as opaque material in X-ray examinations. The only reported acute toxicity of $BaSO_4$ in rats occurred, following rupture due to bowel obstruction, with a dose reaching 40% of the body weight (Boyd and Abel, 1966). Guinea pigs, dogs, and mice are more susceptible to Ba toxicity than are chickens, rats, and rabbits; humans also are susceptible to Ba toxicity. The toxicity of soluble Ba salts is higher than that of corresponding salts of Ca, Mg, and Sr (Syed and Hosain, 1972). Acute Ba toxicity results in low serum K levels and leukocytosis in both experimental animals (Lydtin,

TABLE 2-5. Barium Toxicity

| Compound | Animal | Route | Toxicity | Dosage/kg body weight | | | |
| | | | | Compound | Metal | | |
				mg	mg	mM	pT
Barium oxide BaO	Mouse	sc	MLD	29	17.9	0.13	3.88
Barium peroxide BaO_2	Mouse	sc	MLD	100	81.1	0.59	3.23
Barium fluoride BaF_2	Guinea pig	oral	LD_{100}	350	274	2.00	2.70
Barium chloride $BaCl_2$	Mouse	iv	LD_{50}	12	7.91	0.058	4.24
	Mouse	sc	MLD	10	6.59	0.048	4.32
	Mouse	oral	LD_{100}	105	69.2	0.048	3.32
	Mouse	ip	LD_{50}	54	35.6	0.26	3.59
	Rat	oral	LD_{100}	355	234	1.70	2.77
	Rat	oral	LD_{50}	150	99	0.72	3.14
	Rat	sc	LD_{100}	67	44.2	0.32	3.49
	Rat	iv	MLD	20	13.2	0.10	4.02
	Guinea pig	sc	LD_{100}	55	36.3	0.26	3.58
	Rabbit	sc	LD_{100}	60	38	0.28	3.56
	Rabbit	iv	LD_{100}	15	9.9	0.072	4.14
	Cat	sc	LD_{100}	50	33	0.24	3.62
	Dog	oral	LD_{100}	90	59.4	0.43	3.36
	Dog	sc	LD_{100}	15	10	0.074	4.13
	Human	oral	MLD	80	52.8	0.38	3.42
Barium carbonate $BaCO_3$	Mouse	oral	LD_{100}	200	112.3	0.82	3.09
	Rat	oral	LD_{100}	800	449	3.27	2.49
	Rat	oral	LD_{50}	630	337	2.45	2.61
	Rat	iv	MLD	20	14	0.10	3.99
	Guinea pig	oral	LD_{100}	121	85	0.62	3.21
	Rabbit	oral	LD_{100}	487	340	2.48	2.61
	Human	oral	MLD	100	56.2	0.41	3.39
Barium titanate $BaTiO_3$	Rat	ip	LD_{50}	3000	1770	12.9	1.89
Barium zirconate $BaZrO_3$	Rat	oral	LD_{30}	1980	983	7.16	2.14
	Rat	ip	LD_{50}	420	209	1.52	2.82
Barium fluoroborate BaB_4F_8	Rat	oral	MLD	250			
Barium silicofluoride $BaSiF_6$	Rat	oral	LD_{50}	175	86.0	0.63	3.20
Barium sulfite $BaSO_3$	Rat	oral	LD_{50}	3000	1900	13.8	1.86
Barium acetate $Ba(C_2H_3O_2)_2$	Rabbit	oral	LD_{100}	236	127	0.92	3.03
	Rabbit	oral	LD_{100}	815	438	3.19	2.50
	Rabbit	sc	LD_{100}	96	51.6	0.38	3.43
	Rabbit	iv	LD_{100}	12	6.4	0.047	4.33

1965) and humans (Diengott, 1964). Inhaled fine dusts of $BaSO_4$ form harmless nodular granules in the lungs, an affliction called baritosis. Baritosis produces no symptoms of bronchitis or emphysema, and lung functioning is not affected, although some patients complain of dyspnea upon exertion (Stokinger, 1963); the nodulation disappears if exposure to barium salt is stopped. The exhaust solids emitted from diesel engines which use barium-based smoke-depressant additives contain carbon, $BaSO_4$, and small amounts of $BaCO_3$; these are not toxic to rats at ten times the industrial concentration of airborne exhausts (Miller, 1967).

There is no specific mechanism for Ba toxicity. The abnormal muscle stimulation is believed to be due to the Ba-stimulated catecholamine secretion from the adrenal medulla without prior Ca deprivation (Douglas and Rubin, 1964). Barium is detoxicated only by increased excretion.

Radium (Ra)

Radium is a disintegration product of uranium and occurs in uranium ores such as pitchblende and uranite. The earth's crust contains about $7 \times 10^{-12}\%$ Ra. The human adult with no known occupational exposure contains about $1.7 \times 10^{-8}g$ radium in his body, 99% of it in the skeleton (Hursh and Gates, 1950; Palmer and Queen, 1958).

Radium salts are used in making luminous paints mixed with zinc sulfide, and $RaBr_2$ was formerly used as a radiation source to cure malignant tumors. Radium salts are radiation hazards, and they are not essential for mammals.

Chemistry

Radium, a member of subgroup IIA metals, exists in a number of isotopic forms and is highly radioactive. Radium loses 1% of its radioactivity in about 25 years, emitting α, β, and γ rays to finally form lead. Radium metal is more volatile than barium; it exhibits a valence of $+2$ and forms strong electrovalent salts, most of which are water-soluble. Compared with Th salts, Ra salts are relatively soluble. Radium readily substitutes for Ca in bone.

Metabolism

The gastrointestinal absorption of radium in experimental animals is fairly rapid, and is about 10–35% of a moderate intake. Absorption from sites of injection also is moderate. Intravenously administered Ra^{2+} is quickly cleared from the blood. Part of the absorbed Ra^{2+} is retained in the skeleton; the soft tissues do not retain any Ra^{2+}. The excretion of absorbed Ra^{2+} is both fecal and urinary; in man the excretion is predominantly urinary, while in dogs it is mainly fecal, and in rats it is in equal amounts in urine and feces. Radium lodged in the bone mineral matrix is retained permanently.

There are no reports of Ra involvement in any biochemical systems or normal physiologic functions. The daily intake of ^{226}Ra from the environment ranges from 1.4 pCi to 2.8 pCi/kg Ca intake. Because it is highly radioactive, assessment of the chemical toxicity of Ra is very difficult. Radium causes lung cancer, osteogenic sarcoma, ostitis, skin injury, and blood dyscrasias. On the basis of its atomic structure and relationship to subgroup IIA metals, it is conceivable that Ra is actually more toxic than Ba, but that this chemical toxicity is masked by radiation damage.

SUBGROUP IIB

Zinc, cadmium, and mercury are posttransition metals; thus, as the atomic number increases, the inner electronic orbitals, and not outer orbitals, are gradually filled. There are no d sublevel electrons to act as valence electrons, and thus, there is no ligand-field-stabilizing effect from these metal ions. The large number of inner electrons enable these ions to form stronger bonds with less electronegative elements; bonds in which sharing of electrons occurs. This tendency to form so-called covalent bonds decreases in the order: $Zn > Cd > Hg$. These metals form coordination and chelation complexes; Zn and Cd can form complexes with a coordination number of 4 and a tetrahedral disposition of ligands around the metal; Zn^{2+} and Cd^{2+} are similar to Cu^+ in this respect, and Zn^{2+} is isoelectronic with Cu^+. Both Zn and Hg, and to a certain extent Cd, form water-soluble salts. All three exhibit high affinity for thiol groups, in the order: $Hg > Cd > Zn$.

Cadmium binds to thiol sulfur more strongly than do all essential trace metals except copper but the affinity of Pb^{2+} and Hg^{2+} toward thiol groups is higher than that of Cd. The electropositivity and toxicity of subgroup IIB metals increases with increasing atomic weight.

Mercury toxicity is higher than that of Cd owing to its greater electropositivity and solubility, and its greater penetration and fixation in the tissues, while zinc is the least toxic of the subgroup IIB metals. Among the three metals, Cd is most poorly absorbed from the digestive tract.

Zinc (Zn)

Zinc occurs in nature as the sulfide, blende or sphalerite (ZnS), as the silicate, calamine, willemite, or zinc spar ($ZnSiO_4$), and as the oxide, zincite (ZnO). Zinc minerals are contaminated with cadmium in a Zn:Cd ratio of 200:1. Zinc-containing minerals are also associated with dolomite and other ores. Zinc is considered to be the 17th most abundant element in the earth's crust, with a concentration of 65 ppm; seawater contains 9–21 ppb Zn. Zinc is present in all plant and animal tissues; the human adult body contains about 2300 mg Zn, 65% of it in muscle, 20% in bone, 6% in plasma, 2.8% in the erythrocytes, and about 3% in the liver. Zinc is one of the most abundant of the essential trace metals in the human body, and is oncogenic.

Zinc is used extensively in industry to make alloys such as bronze, brass, German silver, and galvanized sheet iron, and to make protective coatings for other metals; galvanized iron is used for building purposes, dry cells, jar caps, and water pipes. Salts of Zn are also widely used: ZnO in lithopone, rubber goods, linoleum, paints, ceramics, cosmetics, and textiles; $ZnCl_2$ in wood preservation and in dry cell batteries; $ZnBr_2$ in photographic emulsions; and $ZnSO_4$ in rayons, glue, fertilizers, and textiles. Zinc phosphide is a rodenticide, and zinc orthosilicate is used in television screens, zinc chromate in paint pigment, and zinc caprylate in fungicides. The following Zn salts are used therapeutically: zinc stearate in ointments, salves, and pills, $ZnCl_2$ as a topical antiseptic, $Zn_3(PO_4)_2$ in dental cement, and $Zn(OH)_2$ in the preparation of zinc insulin crystals. Zinc itself is not considered to be an industrial health hazard, but the presence of metals such as arsenic, cadmium, manganese, and lead in the atmosphere in factories using zinc, along with dusts of zinc compounds, cause metal fume fever among workers in the zinc industries. Zinc poisoning is mostly accidental from the intake of pesticides, inadvertant therapeutic use of heavy doses of Zn salts, or drinking of acidic juices or brews made in galvanized iron utensils (Brown *et al.,* 1964; Arena, 1974).

Chemistry

Zinc forms stable and water-soluble salts exhibiting a valence of $+2$, but in solution at elevated pH it forms insoluble $Zn(OH)_2$ which is amphoteric in nature and ionizes as follows depending upon the pH of the medium:

$$2H^+ + ZnO_2^- \rightleftharpoons Zn(OH)_2 \rightleftharpoons Zn^{2+} + 2[OH]^-$$

This is due to the compactness of electrons in the completely filled d

suborbital. Zinc ions are small (smaller than the corresponding subgroup IIA metals), and the force of attraction between zinc and oxygen in $Zn(OH)_2$ is as strong as that between oxygen and hydrogen. Zinc forms tetrahedral coordination complexes with a coordination number of 4; Zn complexes with coordination number 6 are rare. Since there are no d-sublevel electrons which could act as valence electrons, there is no ligand-field-stabilizing effect in the Zn ion. Zinc readily complexes to amino acids, peptides, and proteins in biologic media. In both Zn metalloenzymes and Zn–enzyme complexes, Zn undergoes no oxidation–reduction, and thus is not involved in electron transfer. Zinc resembles Cd more than it does Hg. Zinc complexes with nucleotides and is found in ribonucleic acids from a variety of sources; it has an affinity for thiol and hydroxyl groups and for ligands containing nitrogen as donor.

Metabolism

Zinc is an essential trace metal; after K, Ca, and Mg, it is the metal with the highest intracellular concentration. Manifestations of Zn deficiency are noted in several species, including man; these are skin disorders, infantilism, skeletal defects, stunted growth, and failure of skeletal development. Geophagia leads to Zn deficiency in children.

In man the average daily intake of Zn is between 8 and 15 mg/day. Zinc is present in all foods; man's recommended dietary intake is about 0.3 mg Zn/kg per day. The biochemistry and physiology of Zn has been well studied and extensively reviewed (Vallee, 1962; Prasad, 1966; Forbes, 1967; Schroeder *et al.*, 1967*b*; Underwood, 1971; O'Dell and Campbell, 1971; Cuthbertson, 1973). Zinc forms a part of metalloenzymes: peptidases, esterases, carbonic anhydrase, alkaline phosphatase, and dehydrogenases. Zinc ions activate enzymes such as arginase, histidine diaminase, and enolase, and the involvement of Zn in DNA and protein synthesis and insulin storage has been established. Zinc and other divalent metal ions are present in the toxins of many snake venoms (Tu, 1977). The hemorrhagic toxin of the rattlesnake contains one Zn atom per mole of toxin, and the metal is essential for activity (Tu and Bjarnason, 1976).

The gastrointestinal absorption of soluble Zn salts in mammals is highly variable; it averages about 50% of the dietary intake and is dependent upon the Zn level in the diet. Becker and Hoekstra (1971) reviewed the complexity of Zn absorption. When small amounts of Zn are fed to experimental animals and ruminants, the absorption of Zn may increase to 80%; high dietary Ca, P, and Cu reduce Zn absorption, and naturally occurring chelates such as phytates decrease Zn absorption in several species. Calcium is believed to retard Zn absorption only in the presence of phytate, since Ca forms a highly insoluble complex with zinc and phytate (Oberleas *et al.,* 1966; O'Dell, 1969). Phytate inhibition of Zn absorption does not occur in ruminants (Miller *et al.,* 1970); the presence of active phytase in such foods increases Zn absorption. Zinc is less-well-absorbed in soya protein than in animal proteins. Synthetic chelating agents such as EDTA, certain amino acids, and peptides enhance the absorption of dietary Zn. The duodenum has the highest rate of Zn absorption, followed by the jejunum and ileum, a pattern very similar to that of Fe absorption. The absorption of parenterally injected Zn is gradual but complete; the rate of removal depends upon the dose and the solubility of the Zn salts. Insoluble salts remain at the site of injection, causing local necrosis before eventual but slow absorption.

Upon intravenous injection Zn is distributed to the erythrocytes, plasma, and leukocytes; about 80% of the Zn present in the blood is found in the erythrocytes, 12–20% in the plasma, and about 3% in the leukocytes. Most of the Zn in the erythrocytes is present as carbonic anhydrase, and leukocytes also have a Zn-binding protein. About 34% of plasma Zn is firmly bound to globulins, especially α globulin, while the remainder is loosely bound to serum albumin. There is a dynamic exchange of Zn between plasma and erythrocytes. Inhalation of dusts of Zn salts results in a transient accumulation in the lungs before its absorption into the blood. There is little absorption of Zn or its salts through the skin.

Zinc is distributed to all organs and tissues of mammals, and it is found in high concentrations in bone, skin, prostate glands, choroid of the eye, and semen. Within the cell, Zn is distributed in cytosol, the nucleus, and in organelles. Although muscle contains considerable Zn, there is no specific storage site of Zn in mammalian systems.

In mammals Zn is excreted primarily in the feces; fecal Zn consists of unabsorbed dietary Zn and endogenous Zn from bile, pancreatic juice, and other secretions. About 80% of parenterally administered Zn is excreted in the feces, suggesting an enterohepatic circulation for Zn. In humans, about 10% of the absorbed Zn is excreted in the urine, and in tropical climates about 2–3 mg Zn per day may be lost in sweat. However, alcoholism, nutritional imbalance, and malnutrition cause high retention of Zn; decreased protein intake also causes Zn retention. During aging the Zn level progressively increases in the prostate glands, and decreases in the uterus, but there is no overall increase in the total body level of Zn. Quick establishment of equilibrium between plasma Zn and tissue Zn following absorption, and rapid turnover of Zn in tissues such as liver, kidney, and pancreas are due to an efficient Zn homeostatic mechanism in mammals; absorption, tissue uptake, and excretion of Zn are controlled by homeostasis. Miller (1969) suggested the existence of two homeostatic mechanisms acting at intestinal sites, one controlling absorption, and the other controlling excretion. In Zn-deficient animals the control mechanism acts to increase dietary Zn absorption and to decrease the endogenous flow of Zn into the intestines. Soft tissues develop an increased affinity for Zn, and thus the Zn content of brain and muscle tissues does not decrease, even after prolonged deficiency of Zn in the diet. Bone appears to accumulate the Zn left over after the Zn requirements of the soft tissues have been met. The homeostatic mechanism evidently fails at higher levels of Zn intake, leading to Zn intoxication.

Toxicity

Zinc salts are relatively nontoxic owing to an efficient Zn homeostatic mechanism; there is a wide margin between the toxic level of Zn and the normal uptake of Zn by mammals. The natural occurrence of Zn in food is only about 1% of the several thousand ppm Zn required to produce toxicity.

The symptoms of Zn toxicity are lassitude, slower tendon reflexes, bloody enteritis, diarrhea, lowered leukocyte count and depression of the central nervous system leading to tremors, and paralysis of the extremities.

The toxicity of Zn is summarized in Table 2-6. Zinc toxicity by oral administration is low, and water-insoluble Zn salts are practically nontoxic. The toxicity of zinc phosphide is due more to the liberation of phosphine than to the cationic Zn. Salts such as $ZnCl_2$ in large doses are corrosive to the skin and irritate the gastrointestinal tract; they cause serious damage to buccal and gastroenteric mucous membranes and act as emetics.

In rats acute Zn intoxication causes anemia of the hypochromic, microcytic type, decreased erythrocyte production, formation of immature erythrocytes, and increased leukocyte production (Sutton and Nelson, 1937). Following large doses of dietary Zn, fibrosis of the pancreas was observed in cats, while rabbits suffer from glycosuria and albuminuria. Most animals have a high tolerance for Zn, but ruminants are more susceptible to Zn intoxication than monogastric animals. Beef cattle and lambs tolerate 500 μg of Zn/g feed, 0.05% of Zn in the feed, but at the 0.1% level, weight gains and feed conversions are reduced (Ott *et al.*, 1966*a,b*). Pigs and rats tolerate 0.10% Zn without harmful effects, but when fed 0.5% Zn in the diet, the rats become anemic, grow poorly, and have high mortality (Sutton and Nelson, 1937; Lewis *et al.*, 1957).

Dietary intake of 0.25% Zn as metal, chloride, or carbonate has no discernible effects on rats, but 0.5% Zn in the diet in the form of $ZnCl_2$ suppresses growth with heavy mortality. Zinc oxide and carbonate do not retard growth at this level, but zinc oxide retards growth when the diet

TABLE 2-6. Zinc Toxicity

Compound	Animal	Route	Toxicity	Compound mg	Metal mg	Metal mM	pT
Zinc oxide ZnO	Rat	inhal	MLD	400[a]	321[a]	4.91	2.31
Zinc chloride $ZnCl_2$	Mouse	oral	LD_{50}	350	168	2.57	2.59
	Rat	oral	LD_{50}	350	168	2.57	2.59
	Rat	iv	MLD	30	14.4	0.22	3.66
	Rat	iv	LD_{100}	75	36	0.55	3.26
	Guinea pig	oral	LD_{50}	250	120	1.83	2.74
Zinc fluoride ZnF_2	Guinea pig	oral	MLD	200	126	1.93	2.72
	Guinea pig	sc	MLD	100	63	0.96	3.02
Zinc perchlorate $Zn(ClO_4)_2 \cdot 6H_2O$	Mouse	ip	MLD	76	13.4	0.205	3.69
Zinc phosphide Zn_2P_3	Rat	oral	LD_{50}	45.7	34.5	.528	3.28
Zinc sulfate $ZnSO_4 \cdot 7H_2O$	Mouse	ip	LD_{50}	53	12	0.18	3.74
	Rat	oral	LD_{100}	2200	500	7.64	2.12
	Rat	sc	LD_{100}	385	87	1.33	2.88
	Rat	ip	LD_{50}	50	11.3	0.173	
	Rat	iv	LD_{100}	55	12.5	0.19	3.72
	Rabbit	oral	LD_{100}	2100	477	7.29	2.14
	Rabbit	sc	LD_{100}	370	84	1.28	2.89
	Rabbit	iv	LD_{100}	44	10	0.153	2.82
	Dog	sc	LD_{100}	78	17.7	0.27	3.57
	Dog	iv	LD_{100}	66	15	0.23	3.64
Zinc acetate $Zn(C_2H_3O_2)_2 \cdot 2H_2O$	Rabbit	oral	MLD	1400	416	6.36	2.21
	Rat	oral	LD_{50}	2460	732	11.2	1.95
Zinc cyanide $Zn(CN)_2$	Rat	ip	MLD	100	55.7	0.852	3.07
Zinc hexafluorosilicate $ZnSiF_6$	Rat	oral	MLD	100	31.5	0.48	3.32
	Guinea pig	oral	MLD	100	31.5	0.48	3.32
	Guinea pig	sc	MLD	200	63	0.96	

[a]Exposure of 4 hr; number of mg/m^3.

contains 1% Zn (Sadasivan, 1951*a,b*). Widespread metabolic effects were observed in rats fed 1% Zn salt in their diet: urinary nitrogen, uric acid, and creatinine, and fecal nitrogen, phosphate, and sulfate increased while urinary phosphorus levels decreased (Sadasivan, 1952). In pregnant rats, dietary ZnO at 4000 ppm Zn causes resorption and death of fetuses. Excess $ZnCO_3$ retards growth in weanling rats, which produce stillborn pups when they mature (Schlicker and Cox, 1968; Trentini *et al.*, 1969). Fahim *et al.* (1975) noted that Zn injections or feeding (15–50 mg/rat per day) resulted in decreased testes size and infertility. Dietary $ZnCO_3$ reduced liver cytochrome oxidase and catalase activities when the Zn content was 0.4%

(Duncan, 1953). Subcutaneous injections of finely powdered Zn into rabbits cause intercapillary glomerulonephritis, and glycosuria and albuminuria were observed in rabbits following intravenous administration of a toxic dose of $ZnCl_2$. Insoluble zinc silicate caused severe lesions when deposited in the lungs of guinea pigs; the respiratory epithelia of bronchi and bronchioles were largely destroyed, and histiocytes infiltrated the alveolar walls during phagocytosis of insoluble zinc silicate particles.

Zinc is reported to be involved in teratogenesis (Ferm and Carpenter, 1968). When 2 mg of $ZnSO_4$/kg was intravenously injected into pregnant hamsters, teratogenic effects such as fused ribs and encephaly are reported in 4% of the offspring. With 10–25 mg $ZnSO_4$/kg, fetal resorption was observed in 12% of the pregnant animals; 30 mg $ZnSO_4$/kg was fatal to the maternal hamsters.

Human subjects tolerated 660 mg dietary $ZnSO_4$/day (which is equivalent to 165 mg Zn/day or 2.3 mg Zn/kg bodyweight per day) for 26 days without any evidence of hematologic, hepatic, or renal toxicity; adrenocortical function and fertility were not studied (Husain, 1969; Greaves and Skillen, 1970). Vomiting, cramps, renal damage, and hemorrhagic pancreatitis preceded the death of a woman who had taken 6 g Zn in the form of $ZnSO_4$ (Cowan, 1947). A sixteen-year-old boy suffered ill effects after swallowing 12 g metallic zinc: high values for serum amylase and lipase levels indicated pancreatic derangement, and lightheadedness and a staggering gait suggested mild derangement in the cerebellar function; the boy recovered fully after chelation therapy (Murphy, 1970). Development of dermatitis from exposure to zinc chromate and sulfate has been reported, but such instances are rare.

Inhalation of air containing ZnO at 1–34 mg/m^3 causes metal fume fever and pneumonitis in humans; metal fume fever is characterized by fever, malaise and depression, nausea, dryness of throat and body, and headache, symptoms similar to malaria. Temperature and leukocyte count increase, followed by chills. The etiology of the fever is not known, but Lehman (1910) postulated that inhaled Zn particles combine with some protein to modify its structure, and this modified protein invokes the characteristic immune response associated with the injection of a foreign body or protein.

Inhalation of $ZnCl_2$ causes pneumonitis in humans (Johnson and Stonehill, 1961). Dogs and cats were able to survive inhalation of ZnO at 175–1000 mg/m^3, but glycosuria occurred in dogs, and fibrous degeneration of pancreas occurred in cats.

The toxic effects of Zn result from its complex interaction with other essential nutrients such as Cu, Fe, Mg, and Ca. Antagonistic effects of Zn on metabolism in the rat have been demonstrated at 0.5% dietary Zn by various workers. These include reduction of Fe, Cu, ferritin, and hemosiderin levels in liver and other tissues, decreased cytochrome and catalase activities, and reduction in levels of phospholipid P in brain and of Cu and ceruloplasmin in serum (Grant-Frost and Underwood, 1958; Magee and Matrone, 1960; Cox and Harris, 1960, 1962; Lee and Matrone, 1969; Chu and Cox, 1972; Murthy *et al.*, 1974). The severe anemia in rats produced by Zn toxicity is overcome or prevented by the addition of Cu or liver extract to their diet (Smith and Larsen, 1946). Most symptoms of Zn intoxication can be reversed by supplements of soluble Cu salts to the diet; Van Campen and Scaife (1967) postulated that excess dietary Zn reduced the intestinal absorption of Cu. Zinc inhibits ATPase which is Mg-dependent.

The occurrence of a hypochromic, microcytic anemia in rats following the ingestion of excessive Zn and the reversal of this anemia by Fe supplementation demonstrate the interaction between these two metals (Cox and Harris, 1960; Magee and Matrone, 1960). Zinc intoxication affects iron metabolism by increasing the iron turnover, decreasing the life span of erythrocytes and decreasing the hepatic accumulation of iron as ferritin (Settlemire and Matrone, 1967*a,b*). Before the onset of anemia, the ceruloplasmin activity decreases very rapidly in Zn-intoxicated rats (Lee and Matrone, 1969); the ceruloplasmin has ferrioxidase properties and is involved in the mobilization of Fe from the duodenal mucosa and the liver (Lee *et al.*, 1968).

Evidence for Zn involvement in carcinogenesis is increasing. Patients with malignant tumors have subnormal plasma Zn levels (Vikbladh, 1950; Addink and Frank; 1959; Addink, 1960). Leukocyte Zn content also decreases in patients with a variety of neoplastic diseases; Zn concentration in peripheral leukocytes is greatly reduced in patients with chronic leukemia (Szmigielski and Litwin, 1964). Zinc deficiency depresses the growth of leukemia in weanling mice (Barr and Harris, 1973); growth of a hepatoma induced in rats by 3-methyl-4-dimethylaminoazobenzene was significantly reduced by a Zn-deficient diet (Duncan *et al.*, 1974). Zinc is necessary for the growth of both normal tissue and tumors. Mammary carcinomas have been reported in rats that ingested $ZnCl_2$ in their drinking water (Halme, 1961). Neoplasms caused by parenteral implantation is attributed to nonspecific surface oncogenesis. Intratesticular injections of Zn salts in rats produce malignant tumors *in situ* (Riviere *et al.*, 1960). In earlier reports Zn was found at high concentrations in mammary tumors induced by other carcinogens (Tupper *et al.*, 1955), but in hepatic carcinoma in rats, the malignant tumor contained a lower level of Zn than the uninvolved liver tissue (Olsen *et al.*, 1954). The cocarcinogenesis of Zn is suggested by the enhanced growth of experimental sarcomas induced by polynuclear aromatic hydrocarbons (Chahovitch, 1955).

The mechanism of Zn toxicity is not definitely known, but Zn could affect the metabolism of Fe, Cu, and P. Cadmium present as a contaminant in zinc salts could be a contributing factor. Within physiologic limits, Zn toxicity can be alleviated by Cu and Fe.

Cadmium (Cd)

Cadmium does not occur free in nature, and there are no specific ores from which it is mined. Cadmium sulfide, greenockite (78%Cd), occurs as a coating on sphalerite, a ZnS ore, and other ores of Zn, Fe, and Cu. Cadmium is obtained as an important by-product from the refining of zinc and copper. The earth's crust contains about 0.05 ppm Cd, and seawater contains 0.3 ppb. The ratio of Cd to Zn in the soil varies from 1:500 to 1:5000, in seawater it is 1:168. Cadmium is present in trace amounts in plants and marine animals. The middle-aged human adult contains about 50 mg Cd in his body, one-third is in the kidneys, and the liver, lungs, and pancreas are the other sites of Cd storage. Oysters contain about 3.75 μg Cd/g(wet weight); wheat and rice proteins also contain appreciable Cd. Cadmium may be stimulatory in mammals, but it is not considered to be essential. It is oncogenic.

Cadmium is becoming an ever more widely used metal in industry. Cadmium metal is used in protective coatings for iron, copper, and steel; Cd-electroplated parts are used in radios and television sets. Telephone wires are made of Cu–Cd alloys, and Ni–Cd rechargeable batteries are extensively used in electronic equipment. Metallic Cd and CdF_2 act as good neutron absorbers in nuclear reactors. Cadmium oxide is used in storage batteries, Cd–Ag alloys, semiconductors, phosphors, and ceramic glazes, cadmium chloride and bromide in photography, lithography, calico printing, and dyeing, and cadmium tungstate in X-ray screens, scintillation counters, and phosphors. Cadmium salts were used in some parts of the world as antihelminthics, ascaricides, nematocides, and antiseptics in veterinary medicine, but these uses are obsolete. Cadmium sulfide is used in treating seborrheic dermatitis.

Cadmium intoxication is caused mainly by environmental contamination; accumulation of Cd-containing scrap such as batteries, alloys, and paints contaminate water supplies and the air; commercial phosphate fertilizers and cattle manure also contain Cd. Cadmium levels in forages are inversely related to their distance from the highway; these arise from tire erosion. Cigarette-smoke inhalation is another source of exposure. The total Cd in cigarette smoke varies from 15 to 18 μg per 20 cigarettes; this represents 70% of the Cd content of cigarette tobacco (Vandi, 1969). A dietary source of Cd intoxication is seafood, especially oysters. Extensive food processing and refining raise Cd levels in wheat and rice flour. Cadmium is an industrial health hazard, because cadmium dust, fumes, and mists pollute the atmosphere in zinc, copper, lead, and cadmium refinery processes. The industrial waste from an upstream cadmium mine was responsible for the contamination of food and drinking water which caused "itai-itai" disease in Japan.

Chemistry

Cadmium, like mercury, is easily vaporizable, and Cd^{2+} is similar to Ca^{2+} in ionic size and charge. Since Cd is a posttransition metal, the electronic buildup is in the inner orbital positions, while the number of outer valence electrons remains constant. Normally Cd exhibits a maximum valence of $+2$, and forms stable cationic salts. However, with its large number of inner electrons, Cd has a tendency to form outer orbital complexes with less electronegative elements, involving covalent bonds. Cadmium forms tetrahedral complexes with a coordination number of 4. It forms halogen complexes such as $[Cd\ X_3]^-$ and $[Cd\ X_6]^{4-}$; Cd is also amphoteric and forms stable salts such as sodium cadmate, $Na_2[Cd(OH)_4]$, in which the hydroxyl groups can be replaced by water and monovalent groups. Cadmium resembles zinc in its electronic configuration and affinity toward organic ligands, but it has greater affinity than zinc to thiol groups and will replace zinc in some metal enzyme complexes. However, Zn is bound more tightly than Cd to O- and N-containing ligands. The binding of Cd with S is stronger than that of any essential metals except Cu; among the toxic metals only Hg and Pb bind more strongly with S ligands than Cd. Cadmium has a high affinity toward hemoglobin and metallothionein, a protein present in mammalian kidney and liver. Cadmium binds to nucleotides *in vitro* and can presumably bind with nucleic acids. It forms stable chelates with some of the common chelating agents *in vitro* but not *in vivo*. Cadmium could not be removed from tissues with chelating agents which can mobilize zinc from tissues; apparently it is bound very firmly in the tissues.

Metabolism

Cadmium is not essential in human and animal nutrition. The biochemistry, metabolism, and toxicology of Cd have been studied extensively, and excellent reviews are available (Schroeder and Balassa, 1961; Stokinger, 1963; Schroeder *et al.*, 1967*b*; Athanassiadis, 1969*a*; Friberg *et al.*, 1971;

Miller, 1971; Anspaugh and Robison, 1971; Flick *et al.*, 1971). The daily dietary intake of Cd by human adults is about 215 μg; Cd intake by inhalation is about 2 μg by nonsmokers and about 20 μg by smokers. Cadmium is not required for enzymes, although it nonspecifically activates some enzymes. In general, it is inhibitory and acts as an antimetabolite for Zn.

The mammalian gastrointestinal absorption of Cd is low: 2% in rats (Decker *et al.*, 1958) and goats (Miller *et al.*, 1969), 5% in swine (Cousins *et al.*, 1973) and lambs (Doyle *et al.*, 1974), and 16% in cows (Miller *et al.*, 1967). In humans, Cd absorption was computed to be 3–8% (Friberg *et al.*, 1971). It is influenced by the dietary levels of Zn and the solubility of the Cd salts. The mechanism of absorption is not known. The absorption of Cd ions from the sites of parenteral injection into the blood is rapid and complete (Johnson and Sigman, 1971). In the blood Cd^{2+} penetrates erythrocytes and is also distributed in blood plasma; Cd binds to serum proteins, especially α globulins, and is readily distributed to other tissues (Shaikh and Lucis, 1969). The absorption of Cd^{2+} through the skin is very limited. The absorption from the lungs following inhalation of Cd mists or aerosols is rapid and complete, and only insoluble salts such as CdS remain in the lung unabsorbed; these cause local inflammation and ulceration. Most cases of acute Cd poisoning among industrial workers are owing to Cd absorption from the lungs. Retention of Cd in small quantities has been demonstrated in all animal tissues; greater amounts accumulate in the kidneys, liver, reproductive system, and lungs, but the pancreas, aorta, esophagus, and omentum also accumulate Cd under certain circumstances.

Irrespective of the mode of administration, Cd crosses the placental barrier in pregnant rats and hamsters, as shown by detectable amounts in the liver, brain, and digestive tract of the newborn. More than 2.5 times the amount was found in the livers of the young born of test animals dosed with Cd while pregnant than in control animals (Tsevetkova, 1970; Holmberg and Ferm, 1969). However, there is an efficient placental barrier against Cd in goats (Anke *et al.*, 1970).

Excretion of Cd in mammals is slow and is predominantly fecal; however, excretion in the urine can be appreciable (Bonnell, 1955). Using radioactive Cd in tracer quantities in goats, it was found that only 6% of an injected dose was excreted in the first two weeks, most of it in the first three days, and the rest was excreted over a long period of time (Miller *et al.*, 1969). The unabsorbed Cd (80%) of an ingested dose is excreted within five days in the feces. The retention of inhaled doses of Cd salts varies with the species, 10–20% in mice and about 40% in dogs (Potts *et al.*, 1950). The slow excretion and prolonged retention of Cd in the tissues suggest that no homeostatic mechanism exists for Cd in animals (Decker *et al.*, 1958;

Cotzias *et al.*, 1961; Miller *et al.*, 1969). In humans Cd accumulates with age, and the Cd levels in the kidneys, liver, and lungs increase; in newborn babies no cadmium is present, but it is detected at 10 months of age.

Toxicity

Cadmium is toxic to all systems studied in man and animals, exhibiting a variety of toxic effects which are dose-related. Symptoms of acute toxicity following oral ingestion of Cd salts include excessive salivation, persistent vomiting, abdominal pains, diarrhea, vertigo, and loss of consciousness. Severe ulcerative gastroenteritis and congestion along with pulmonary infarcts and subdural hemorrhages also may be noticed. Symptoms of chronic Cd toxicity are growth retardation, impaired kidney function, impaired reproductive function, hypertension, tumor formation, and teratogenic effects. The main symptoms following inhalation are persistent choking, coughing, leading to pulmonary emphysema, and bronchitis; damage to renal tissues and olfactory nerves subsequently occurs.

"Itai-itai byo" disease, encountered in Japan mostly in women aged 45–70 years, was caused by Cd intoxication. The disease is characterized by severe pain in the bones associated with osteomalacia, a waddling gait, amino aciduria, and glycosuria (Roe, 1971); while most of the symptoms are caused by Cd intoxication, other factors such as high dietary Cu, Pb, and Zn and the nutritional and hormonal status of the patients contribute to the disease.

Cadmium has a high toxicologic potential; this is enhanced by cadmium accumulation in mammalian tissues owing to a very poor or nonexistent homeostatic mechanism. The toxicity of Cd is summarized in Table 2-7. The degree of toxicity varies widely and is influenced by external factors; ingested salts which are very poorly absorbed from the digestive tract are, of course, less toxic than are the soluble salts. Following inhalation of Cd salts in experimental animals, the mortality rate is influenced by particle size and the duration of exposure. Cadmium intoxication causes derangment in carbohydrate and mineral metabolism, in renal, hepatic, testicular, and prostate functions, and disturbs the integrity of the central nervous system; it also causes poor lactation and lowered hematocrit values.

Chronic feeding of Cd at low levels to rats, rabbits, lambs, pigs, calves, and poultry causes diminished growth and feed consumption. Rats tolerate 10 ppm Cd as $CdCl_2$ in their drinking water, and show no visible symptoms of toxicity, but their longevity is reduced (Schroeder *et al.*, 1963*b*). Loss of weight was observed in rabbits which were fed 40–44 mg Cd/kg as $CdCl_2$ (Nomiyama *et al.*, 1973). Growth retardation was observed in lambs fed 60 ppm Cd (Doyle *et al.*, 1974), in pigs fed 450 ppm Cd (Cousins *et al.*, 1973),

TABLE 2-7. Cadmium Toxicity

| Compound | Animal | Route | Toxicity | Dosage/kg body weight | | | |
| | | | | Compound | Metal | | |
				mg	mg	mM	pT
Cadmium oxide	Human	inhal	Txc	50^a	43.5^a	0.39	3.41
CdO	Monkey	inhal	LD_{100}	1100^a	962^a	8.56	2.07
Cadmium chloride	Mouse	sc	LD_{100}	20	12.3	0.11	3.96
$CdCl_2$	Mouse	ip	MLD	13.5	8.3	0.074	4.13
	Rat	oral	LD_{50}	88	54	0.48	3.32
	Rat	im	LD_{50}	25	15.3	0.136	3.87
	Rabbit	oral	LD_{100}	70	43	0.38	3.42
	Rabbit	sc	LD_{100}	25	15.3	0.136	3.87
	Rabbit	iv	LD_{100}	2	1.26	0.011	4.95
	Cat	sc	LD_{100}	25	15.3	0.136	3.87
	Dog	iv	LD_{100}	5	3.15	0.028	4.55
Cadmium fluoride	Guinea pig	oral	LD_{50}	150	112	0.996	3.00
CdF_2	Guinea pig	sc	MLD	200	150	1.33	2.87
Cadmium sulfate	Mouse	ip	LD_{50}	69	37.2	0.331	3.48
$CdSO_4$	Dog	oral	LD_{100}	105	56.6	0.503	3.30
	Dog	sc	LD_{100}	27	14.6	0.13	3.89
Cadmium phosphate $Cd_3(PO_4)_2$	Mouse	inhal	LD_{100}	650^a	415^a	3.69	2.43
Cadmium fluoborate CdB_2F_8	Rat	oral	MLD	250	98	0.872	3.06
Cadmium lactate $Cd(C_3H_5O_3)_2$	Mouse	sc	LD_{50}	13.9	5.4	0.048	4.32
Cadmium succinate	Mouse	ip	LD_{50}	270	131.6	1.17	2.93
$CdC_4H_4O_4$	Rat	oral	LD_{50}	660	323	2.87	2.54

[a]Exposure for 4 hr; number of mg/m^3.

and in calves fed 640 ppm Cd (Powell *et al.*, 1964); dietary Cd at 1350 ppm caused complete cessation of growth in pigs, and 2560 ppm caused heavy mortality in calves.

Renal damage is the classic syndrome of chronic Cd intoxication in humans. High incidence of proteinuria, the presence of low-molecular-weight serum proteins in the urine, and amyloid deposits in the kidney are clinical symptoms. Renal tubular damage results in inefficient protein absorption in the glomerular filtrate (Friberg, 1957; Piscator, 1966); similar observations in animals are recorded. The presence of light chains of immunoglobulins in the urine during Cd intoxication is attributed to Cd inhibition of peptidases present in the kidneys and responsible for the normal breakdown of these immunoglobulins (Vigiliani, 1969). Experimental rabbits suffering from renal damage owing to low chronic doses of Cd recovered fully following the cessation of Cd exposure (Piscator and Axelsson, 1970), recovery being attributed to the removal of Cd from the kidney and its detoxication by binding with metallothionein.

Electronmicroscopic studies of renal changes in Cd intoxication show an increased size and number of lysosomes with mitochondrial enlargement from swelling, proportional to the Cd concentration. This suggests a stimulation of detoxication processes and a derangement of energy metabolism (Nishizumi, 1972). Cadmium proteinuria results primarily from impaired tubular reabsorption of proteins caused by increased tubular permeability, deranged mitochondrial metabolism, and the inhibition of peptidase (Jones *et al.*, 1971).

The interactions of Cd with Cu, Zn, and Fe in mammalian tissues can cause symptoms associated with Cu, Zn, and Fe deficiencies and can attentuate Cd toxicity, depending upon which metal ion is in excess; the antimetabolite effects of Cd against Fe, Cu, Zn, and Ca are very pronounced.

Dietary Cd affects the intracellular distribution of Zn and Cu in rats (Banis *et al.*, 1969). Cadmium inhibits enzymes associated with Zn by competing with Zn and displacing it from metalloenzymes such as alkaline phosphatase; alkaline phosphatases in the kidney and prostate of guinea pigs are inhibited by Cd (Ribas-Ozonas *et al.*, 1971). Dietary Cd inhibits Zn uptake by the prostate, testes, and other tissues (Schroeder and Nason, 1974). There are resemblances among the symptoms of Cd intoxication and Zn deficiency in several species; dietary Zn supplements decrease the severity of Cd intoxication and, in some instances, prevent Cd toxicity (Powell *et al.*, 1964; Pond *et al.*, 1966; Bunn and Matrone, 1966). In rats 50 ppm Cd in the diet causes depletion of Fe in the liver (Whanger, 1973); cadmium induced anemia in chicks, due to defective intestinal absorption of iron (Freeland and Cousins, 1973). Cadmium interferes with the metabolism of copper by reducing its absorption (Anke *et al.*, 1970), and by increasing its urinary excretion. Reduced ceruloplasmin activity was observed in rats fed 1.5 μg Cd/g diet with normal Cu content, 3 μg Cu/g diet; increased Cd intake, 6 and 18 μg Cd/g diet, considerably reduced plasma and liver Cu concentrations and serum ceruloplasmin activity (Campbell and Mills, 1974). Bone malformations from simple Cu deficiency were also noted. Similar effects of Cd intoxication on Cu were observed in pregnant ewes and lambs (Mills and Dalgarno, 1972).

Cadmium intoxication in rats fed 100 μg Cd/g diet inhibited Fe absorption and hemoglobin formation leading to anemia, growth failure, and poor bone mineralization (Banis *et al.*, 1969), and similar Cd and Fe interactions occur in chicks and Japanese quail. In birds dietary supplementation with Fe^{2+} or ascorbic acid reverse the effects of Cd intoxication. The inverse relationship between the Cd and Fe contents of duodenal mucosa, following Cd administration, suggests that the interaction involves competition between Fe and Cd for mucosal binding sites (Jacobs *et al.*, 1974). The antimetabolite activity of Cd^{2+} causes destruction of microvilli of the duodenal mucosa and produces extensive cytoplasmic vacuolation in the gastrointestinal tract. Bone disease is a recognized symptom of industrial exposure to Cd (Nicaud *et al.*, 1942). Serum Ca decreases following intravenous injection of $CdCl_2$; Cd concentrations in mice fed dietary Cd are greater when the Ca intake is low (Kobayasha *et al.*, 1971). Tissue Cd concentrations were higher in rats fed Ca-deficient and high-Cd diets than in rats fed a normal-Ca and high-Cd diet (Larson and Piscator, 1971). Chronic Cd toxicity causes poor bone mineralization and increased liver Ca concentrations in quail (Fox *et al.*, 1971). Cadmium is also reported to block the renal synthesis of 1,25-dihydroxycholecalciferol, the metabolically active form of vitamin D (Feldman and Cousins, 1973); this could be indirectly responsible for the skeletal abnormalities found in Cd embryotoxicity or the lack of bone mineralization and osteomalacia in "itai-itai" disease.

In lactating cows fed 3 g Cd as $CdCl_2$ daily, Cd accumulated in the mammary gland, and milk production was decreased drastically, but Cd was not incorporated into milk proteins and could be found only in trace amounts, <0.1 ppm, in the milk (Miller *et al.*, 1967). Generally, ruminants absorb very little ingested Cd, and only trace amounts of absorbed Cd are found in the milk, but in rats ingested Cd accumulates in the mammary tissue, and part of it is expressed in the milk.

In life-term studies, rats with drinking water containing 5 μg Cd acetate/ml became hypertensive and showed thickening of renal arterioles and arteries and changes in the glomeruli which resembled those found in chronic benign hypertension (Kanisawa and Schroeder, 1969), which has been linked with renal injury following chronic Cd exposure (Perry and Schroeder, 1955; Schroeder, 1965, 1966). Carrol (1966) found a significant correlation between the Cd concentration in urban air and diseases of the heart in humans. Cadmium caused enhanced renal retention of Na^+ which contributes to hypertension (Perry *et al.*, 1971). Cadmium given intramuscularly at 1 mg Cd/kg body weight lowered urinary Na excretion by 42 μeq without any change in urinary volume, protein, K, Zn, or even Cd. Renal retention of Na increased with successive increases of Cd administrations.

Excess of Cd causes gonadal tissue damage in experimental animals, and sterility in both sexes. Testicular atrophy, cessation of spermatogenesis, and vascular changes in the ovary lead to sterility, irrespective of the mode of administration. Intratesticular injection of Cd damages the male gonads in all species and permanently stops spermatogenesis (Chiquoine and Suntzeff, 1965); this chemical castration was studied for possible use in practical animal husbandry. A subcutaneous dose of 0.02 mM $CdCl_2$/kg causes testicular atrophy in rats (Malcolm, 1972), but in calves subcutaneous injection of Cd at this level was harmless and produced a local reaction, whereas intravenously administered Cd caused complete cessation of spermatogenesis. Testicular damage can be prevented by prior administration of Se (Mason and Young, 1967; Gunn *et al.*, 1968*b*), Zn (Parizek, 1957), or Co (Gunn *et al.*, 1968*a*). Adult female rats dosed with Cd do not become sterile, while prepubertal females suffer profound vascular changes in the ovary leading to infertility. Pregnant rats are highly susceptible to Cd intoxication. Avian testes are more resistent to parenterally administered Cd than are mammalian testes; egg production is reduced with 50 ppm of Cd in the diet (Anke *et al.*, 1970). Acute Cd intoxication leads to the following changes in the testes which result in atrophy or degeneration: hemorrhagic necrosis, vascular injury, increased permeability, elevated intratesticular pressure, decreased regional blood flow, changes in energy metabolism, and pinocytosis of the testicular endothelium. Cadmium-induced damage is accompanied by high testicular concentrations of Cd; the testicular damage is caused primarily by a local ischemia resulting from an increase in the permeability of the testicular blood vessels (Setchel and Waites, 1970), which is followed by changes in energy metabolism (Harkinen and Kormano, 1970). Zinc uptake by the testes is decreased; glucose concentration and glycolytic activity increase with a decrease in ATP level. The seminiferous tubules and interstitial tissues become hypoxic, which leads eventually to necrosis. Spermatogenesis is known to be extremely sensitive to changes in the temperature of testes; parenteral administration of $CdCl_2$ to rats causes a dose-related increase in the testicular temperature (Johnson *et al.*, 1970), owing to vasodilation. These changes occur at Cd levels which are far above the Cd levels in the environment or the present level of daily intake of Cd by humans.

Cadmium embryotoxicity occurs in experimental rats, mice, and hamsters; 2 mg $CdSO_4$/kg given intravenously to pregnant hamsters on day 8 of gestation increased fetal resorption and caused gross fetal malformations, especially obliteration of the facial architecture, with clefts of the lip and palate in more than 50% of the recovered embryos (Ferm and Carpenter, 1967). Sodium selenite or Zn^{2+} prevented this fetal damage. Lead salts reduce the severity of cadmium teratogenicity, while cadmium accentuates lead teratogenicity. Cadmium was teratogenic to mice at 10 ppm in the drinking water; many abnormal offspring were found in the litters, and the breeding population died out in two generations (Schroeder *et al.*, 1964*b*; Schroeder and Mitchner, 1971*b*). In rats, oral doses of 40 mg Cd/kg body weight per day as $CdCl_2$ given on days 6–19 of gestation increased fetal resorption and caused skeletal, kidney, and heart abnormalities in fetuses, stillborn, and offspring. The incidence and intensity of skeletal defects increased with Cd dosage, but the renal and cardiac abnormalities were not dose-related (Scharpf *et al.*, 1972). Pregnant ewes fed 12–15 ppm dietary Cd and $CdSO_4$ from week 13 to 14 of pregnancy did not produce abnormal offspring (Mills and Dalgarno, 1972), but pregnant goats fed 75 ppm of Cd as $CdCl_2$ in a semisynthetic diet had 50% fetal resorption and produced abnormal offspring which did not survive to maturity (Anke *et al.*, 1970). The dosage and route of administration in these experiments and the poor dietary absorption suggest that environmental Cd should not be a significant factor in human teratogenesis.

Primary carcinogenicity of Cd has been well established in experimental animals but not in humans. However, epidemiologic evidence such as the presence of Cd in the cancer tissues of human victims of fatal Cd intoxication (Butt, 1960; Butt *et al.*, 1961) and cancer fatalities among workers in the Cd battery industry (Potts, 1965) implicates Cd in human cancer. Ingested Cd is not carcinogenic (Sunderman, 1972); chronic exposure of mice to 5 ppm Cd as $CdCl_2$ in their drinking water did not cause carcinogenesis (Schroeder *et al.*, 1964*b*). The subcutaneous implantation of Cd metal or the subcutaneous injection of insoluble Cd salts in rats produced tumors at the sites of injection and implantation (Heath and Daniel, 1964*a*; Kazantzis and Hanbury, 1966). Cadmium oxide suspended in saline solution produces tumors at the site of subcutaneous injection (Malcolm, 1972). When injected parenterally into experimental animals, soluble Cd salts, such as $CdCl_2$ and $CdSO_4$, produce tumors at the sites of injection, especially in rodents (Roe, 1964; Haddow *et al.*, 1964). Subcutaneous, intravenous, or intratesticular injection of Cd salts produces sarcomas in the testes of rats and fowl (Haddow *et al.*, 1964; Guthrie, 1964). Cadmium is reported to accumulate in malignant tumors induced by organic carcinogens (Gorodiskii *et al.*, 1956).

Pulmonary emphysema, the most typical symptom of chronic Cd intoxication caused by the inhalation of dusts of Cd compounds, is characterized by shortness of breath and bronchitis (Bonnel, 1955; Dunphy, 1967); the alveoli of the lungs become dilated, with distention of their walls. These symptoms are observed among industrial workers in the Cd industry and in experimental animals. There are three phases of Cd toxicity from chronic Cd inhalation, depending upon the dose and the duration of exposure: edematous, proliferative, and fibrogenic. In the first phase, inflammation of the irritated pulmonary membrane leads to interstitial edema and the separation of epithelial sheets from their underlying stroma, accumulation of fluid following the progressive distention of interstitial cells. In the second phase, metaplasia of the alveolar epithelium and proliferation of histiocytes and polypoidal lesions cause destruction of alveoli and produce proliferative interstitial pneumonitis. Experimental animals which survive acute Cd exposure show the effects of the third phase, pulmonary fibrosis localized around the bronchi and vascular tubes; Cd evidently binds to respiratory chain components in the pulmonary alveolar cells causing these changes. Some Cd from the lungs is also absorbed into the blood and affects the renal, cardiovascular, and reproductive systems.

Absorbed Cd accumulates in the mammalian kidneys and liver; it is strongly bound to macromolecules in the intracellular compartments. Metallothionein isolated from rat liver contains 4.2% Cd, 2.6% Zn, 0.5% Hg, and 0.3% Sn (Pulido *et al.*, 1966). Similar Cd-binding proteins (Cd-Bp) were isolated from mammary glands and from the cytoplasm of liver and kidney, these Cd-Bp differing from metallothionein in molecular weight (Shaikh and Lucis, 1971, 1972). Cd-Bp synthesis is induced by increases in the body burden of Cd; other metals such as Zn, Cu, Co, Ni, Hg, or Pb do not induce Cd-Bp synthesis (Lucis *et al.*, 1972). Metallothionein synthesis is increased as body levels of Cd decrease. Disappearance of toxic symptoms of Cd some days after Cd ingestion, restoration of normal renal functioning after Cd intoxication, formation of Cd-Bp, and increases in liver metallothionein suggest that animals detoxify Cd by sequestering it in a relatively harmless and unavailable form in the liver and kidney. The poor and slow excretion of Cd supports this suggestion.

Zinc, copper, and iron, when present in more-than-adequate levels in the diet, protect experimental animals from the severe toxic stress of potentially toxic doses of cadmium. Ascorbic acid (Fox *et al.*, 1971; Fox and Fry, 1970), vitamin D (Worker and Migicovesky, 1961), and cysteine, glutathione, and selenium (Gunn *et al.*, 1968*b*) have ameliorative effects on cadmium toxicity.

The mechanism of Cd toxicity in mammals is complex and is yet to be understood. Changes in membrane permeability leading to abnormal transport of metabolites and minerals, antimetabolite effects, derangement of cellular energy metabolism, and binding to cellular respiratory components are considered to be some of the mechanisms of Cd toxicity in mammals. (Schlaepfer (1971) confirms that the primary action of Cd is on tissue endothelium.

Cadmium inhibits ATPase of myosin (Lipkan, 1970), pulmonary alveolar macrophages, and cell membranes (Cross *et al.*, 1970). In mice dietary Cd increased the liver and heart tissue levels of malic dehydrogenase and glucose-6-phosphate dehydrogenase (Weber and Reid, 1969*a*). Cadmium sulfate at 0.5 mg/ml of blood causes *in vitro* multiple cytophagy or phagocytosis of the erythrocytes and thrombocytes by the leukocytes, and alterations in cellular consistency, especially karyoplasmatic alterations of absorbing elements (Granta *et al.*, 1970). In massive Cd doses, multiple cytophagy could also occur *in vivo*. Cadmium chloride significantly increased cAMP-synthesizing enzyme in the hepatic tissues of rats (Merali *et al.*, 1975). Changes in the cAMP levels can affect carbohydrate metabolism in the various tissues that are targets of Cd intoxication.

Mercury (Hg)

Mercury occurs as the native metal, mixed with its ores. Cinnabar (α HgS) is the most common ore; other less important ores are metacinnabarite (β HgS) and livingstonite ($HgS \cdot 2Sb_2S_3$). The earth's crust contains 0.5 ppm, and seawater contains about 0.03 ppb. Mercury is not abundant nor widely distributed, but it exists in highly concentrated ores from which it is readily obtainable. Mercury is found in all tissues of animals at a very low concentration. Some plants appear to concentrate Hg; microdroplets of elemental mercury are sometimes found in plant tissues (Shacklette, 1970). The human adult body contains about 13 mg Hg, about 70% of which is present in fat and muscle tissues; trace amounts of Hg are present in the nails and hair. Mercury is ubiquitous and is found in trace amounts in most foods and water. It is stimulatory under certain conditions, but it is not essential for mammals.

Primary mercury, refined from its ores, and secondary mercury, recovered from industrial wastes, find extensive use industrially, especially in electrical apparatus, electrolytic manufacture of chlorine and caustic soda, and in industrial control instruments. Mercury-arc lamps, neon and fluorescent lamps, arc rectifiers, batteries, switches, and related equipment use elemental mercury. Mercury steam-boilers, rectifiers, and ion thrusters are used in power generation. Mercury is also used in scientific laboratory equipment such as barometers, thermometers, manometers, and hydrometers. Amalgam metallurgy and precision casting used in jewelry and molding processes are some of the newer industries which require Hg. Mercury salts are used as catalysts in the manufacture of organic compounds; mercury compounds are used as fungicides, herbicides, insecticides, bactericides, disinfectants, and antiseptics. Mercuric chloride, corrosive sublimate, is used for preserving wood and anatomic specimens in embalming, leather tanning, intensifiers in photography, mordants for furs, and fungicides in treating seed potatoes. Mercuric nitrate is used in felt manufacture and in mercury fulminates; HgO is used in paints and HgS is a pigment for rubber. Some Hg salts used formerly in therapeutics are: calomel (Hg_2Cl_2) as a cathartic, diuretic, and antisyphilitic; $HgCl_2$ as a topical antiseptic and disinfectant; and mercuric benzoate in the treatment of syphilis and gonorrhea.

Mercury and its salts are considered industrial health hazards, and reports of Hg poisoning due to industrial exposure are numerous. The reports include Hg poisoning from mining and extraction of Hg and from almost all industries that use Hg. A hazard from mercury exists in scientific and medical laboratories, hospitals, schools, and dental offices; spilled Hg vaporizes and causes intoxication in confined laboratories with poor ventilation. Exposure to agricultural Hg-containing insecticides and fungicides causes Hg intoxication, and inadvertant and ignorant misuse of Hg-treated seed cereals also has caused Hg poisoning. Consumption of (organic Hg compound) fungicide-treated seed grain in Iran and Iraq was responsible for disastrous large-scale Hg intoxication among the human population. Consumption of meat from animals fed such grain can also be disastrous. Mercury toxicity is a world wide problem. Improper disposal of the miniature batteries used in hearing aids and similar appliances and of damaged fluorescent and neon lamps also constitute Hg hazards. Although Hg levels in the food supply of the United States are not considered hazardous, exclusive eating of seafood, especially tuna fish, may lead to Hg intoxication. Methyl mercury poisoning from contaminated fish was responsible for the neurotoxic Minamata disease in Japan.

Man's activities have increased the level of Hg in the environment, especially in the water; inadequate and improper disposal of industrial mercury wastes increase the Hg levels in the water and atmosphere. Microorganisms are able to catalyze the interconversions of metallic Hg, Hg^+, and Hg^{2+} ions; some aerobic microorganisms solubilize Hg^{2+} from insoluble HgS and finally reduce it to elemental Hg. This is considered a detoxication mechanism for bacteria, since elemental Hg enters the vapor phase and escapes into the atmosphere. Other microorganisms convert the elemental mercury into methyl mercury salt (CH_3HgCl) and dimethyl mercury, which also escape into the atmosphere. Most of these reactions take place in the sediments of river and ocean beds. In the atmosphere, ultraviolet light and some bacteria free toxic elemental Hg from the more toxic dimethyl (alkyl) mercury. Some reactions in the biologic cycle of mercury are depicted in Figure 2-1. The major source of Hg contamination is the disposal of industrial Hg wastes into the water where the wastes settle as sediment, only to be recycled back into the water and air. Fish trap these methyl mercury salts and dialkyl mercury in their tissues and fat depots; organic mercury compounds at these concentrations are relatively nontoxic to fish which have less-well-developed nervous systems than man and live at a lower environmental temperature, but the contaminated fish are toxic to man; the occurrence of the neurotoxic Minamata disease among people living near Minamata Bay in Japan was caused by eating mercury-contaminated fish. The Hg contamination was traced to the industrial wastes of vinyl chloride factories using elemental Hg as a catalyst; bacterial methyla-

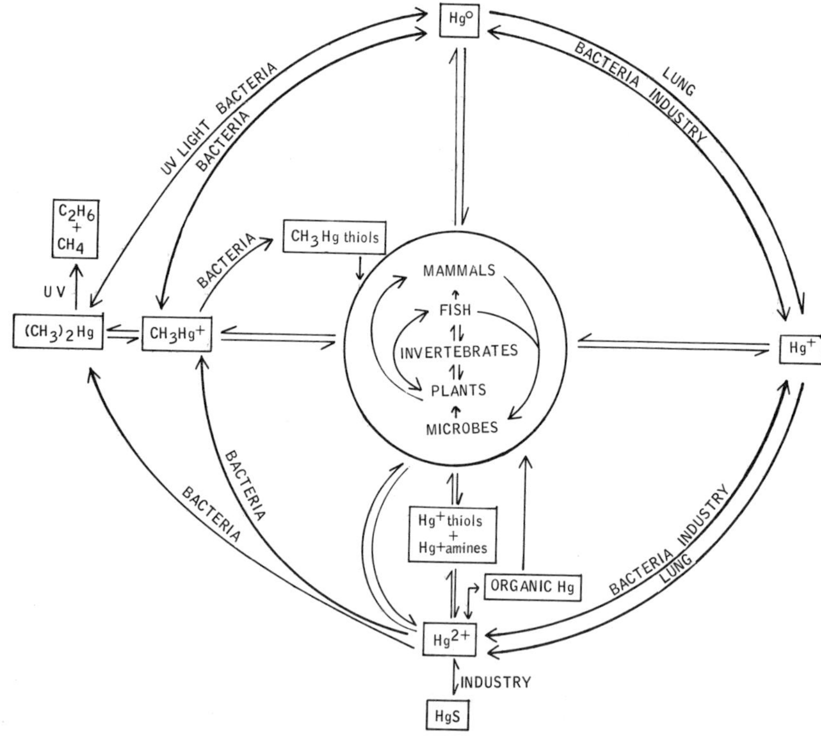

FIG. 2-1. *Ecologic mercury cycle.*

tion is believed to have taken place in the effluent from the factories, before
its discharge into the sea. Except in such isolated problem areas, industrial-
ization may not present an unusual hazard; Cockburn *et al.* (1975) report
that bones from an Egyptian mummy (2000 B.C.) contain about the same
Hg concentration as that in modern man. Activation analysis of hair
samples suggests that King Charles II of England died of chronic Hg
poisoning from his activities as an amateur alchemist (Underwood, 1971).

Chemistry

Mercury, the heaviest among the posttransition elements, exhibits
valences of $+1$ and $+2$ and forms stable cationic salts; the divalent salts are
more soluble in water than are the monovalent (mercurous) salts. The
solubility of Hg in water is about 0.02 mg/liter and in body fat 0.6–2.7 mg/
liter; Hg is volatile at room temperature, and some Hg^{2+} salts sublime.
Mercury cannot form stable ring complexes because the bond angle of Hg

is 180°. Electrons in the inner orbitals of the Hg atoms are involved in coordinate covalent bonds; the more common coordination number is 4 and the geometry of the Hg coordination complexes is tetrahedral. Mercury complexes with primary and secondary amines and forms halogen complexes such as $[Hg\ X_3]^-$ and $[Hg\ X_4]^{2-}$. The affinity of Hg toward reactive groups is: $SH > CONH_2 > NH_2 > COOH > PO_4$. Among the metals, Hg has the greatest affinity towards thiol groups; in biologic fluids and tissues, the thiol groups of proteins and other compounds bind the available Hg ions by forming reversible complexes. This binding causes protein agglutination, inhibition of thiol-group-containing enzymes, and clumping of erythrocytes. Tissues and erythrocytes oxidize Hg and Hg^+ to Hg^{2+} ions.

Methyl mercury salt is reported to be formed from Hg^{2+} salts in rat intestine and in carbonated waters, and is present in the environment as a microbial breakdown product from Hg^{2+} salts and from organic mercurials used as bactericides, fungicides, and insecticides. The pink bread mold, *Neurospora crassa,* methylates Hg as a means of detoxication, and the methylated Hg is then complexed with homocysteine in that organism. Organic Hg salts are more highly soluble in lipids than in water, and CH_3HG^+ is 100 times more soluble in lipids than in water. In CH_3HgBr, the covalent-binding H_3CHg^+ is more firm and stable than the ionic Hg—Br bond; CH_3Hg^+ can exist free only in minute concentrations.

The affinity of CH_3Hg^+ for some ligands is high, especially toward the thiol group; some affinity constants $(\log_{10}K)$ are:

$$RS^- = 17, OH^- = 9.7, I^- = 8.0, Br^- = 7.0, \text{ and } Cl^- = 5.7$$

Methyl mercury can permeate cellular membranes; blood serum proteins form a loosely bound serum-protein–S–Hg–CH$_3$ complex and as a CH_3Hg^+ ion leaves the blood by dissolving in a lipid membrane more dissociates from the protein complex. However, organic mercurials behave differently from alkyl Hg^+ salts.

The metabolism and physiologic behavior of elemental Hg vapor, inorganic Hg salts such as $HgCl_2$ and $Hg(NO_3)_2$, organic alkyl Hg salts such as CH_3Hg^+Cl, and organic mercurials such as diphenyl mercury and mercurochrome differ greatly according to their chemical structure and ionization.

Metabolism

Mercury is not essential to man or animals and is not required for any specific biochemical or physiologic function in living tissues, although mercury is stimulatory at very low levels of concentration. The highly effective bacteriostatic $HgCl_2$ stimulates microbial fermentation at 10^{-6} M levels (Schulz, 1888; Branham, 1929). Schroeder (1973) reported that methyl mercury at 1 ppm Hg in the drinking water of mice stimulated a slight but significant growth increase, while 5 ppm retarded growth. The average daily intake of mercury by human adults is about 0.02 mg, most of it from dietary sources. Fruits, vegetables, and cereal grains contain 0.005–0.035 ppm Hg. Fish are reported to contain 0.02–0.18 ppm Hg.

The gastrointestinal absorption of elemental Hg and Hg^{2+} salts is moderate in mammals; considerable amounts of ingested Hg salts are retained in the alimentary mucosa (Goodman and Gilman, 1965). Insoluble Hg^+ salts are poorly absorbed. Mercurous compounds can be oxidized to Hg^{2+} and absorbed if Hg^{2+} salts remain in the digestive tract long enough or if the subject is exposed to Hg^{2+} salts for a long time, e.g., from the chronic use of calomel as a cathartic agent. The apparent absorption of Hg^{2+} is about 30% in goats (Howe *et al.*, 1972), whereas the absorption of CH_3Hg^+ in lactating cows is about 60% (Neathery *et al.*, 1973); this suggests that there is greater absorption of organic Hg compounds.

The absorption and distribution of Hg^{2+} salts and CH_3Hg^+ is rapid from sites of parenteral administration. Skin readily absorbs Hg salts and even elemental Hg if it is dispersed in a suitable medium. The fecal recovery of 30% of a dose of elemental Hg applied to the skin suggests the high rate of Hg absorption through the skin. Similarly, Hg compounds present in vaginal jellies are rapidly absorbed. Intravenously injected Hg salts are sequestered in equal amounts in erythrocytes and serum albumin; α globulins also bind Hg.

Inhaled Hg vapor is usually oxidized rapidly ensuring a high percentage of absorption. About 80% of the inhaled Hg vapor diffuses rapidly into alveolar membranes and other tissues, including the brains; there is total absorption of Hg^{2+} salts and CH_3Hg^+. Brain tissue retains inhaled Hg vapor and CH_3Hg^+ better than Hg^{2+} given intravenously (Hayes and Rothstein, 1962; Magos, 1968).

The retention of absorbed Hg in the soft tissues is fairly high, and follows the distribution: kidneys > liver > intestinal and colon walls > spleen > brain > heart > lungs > respiratory mucosa > muscle > skin. The kidneys retain the highest concentration of Hg; about 80% of absorbed Hg^{2+} salts accumulates in the proximal renal tubules. Inorganic Hg has greater affinity toward thiol groups of soft tissue proteins, whereas CH_3Hg^+ accumulates in the central nervous system and erythrocytes. In the nervous system of the guinea pig, accumulation of CH_3Hg^+ is as follows: cerebrum > cerebellum > spinal cord (Iverson *et al.*, 1973). In experiments with humans given a single dose of radioactive methyl mercury ($^{203+}HgCH_3$) intravenously, 90% of the CH_3Hg^+ was found in the erythrocytes and 10% in serum proteins; subsequently 10% was found in the brain and about 50% in the liver. The biologic half-life of inorganic Hg in blood (presumably in the erythrocytes) is 70 days, and that of CH_3Hg^+ is about 120 days. The increased half-life of Ch_3Hg^+ is attributed to the reabsorption from the intestines of any CH_3Hg^+ which is excreted into the digestive tract with bile. Hair and nails accumulate Hg compounds; the Hg level in hair ranges from 200 to 2500 ppm, depending upon the intensity and duration of exposure and the nature of the mercury salts. The Hg content of hair in humans exposed to CH_3Hg^+ is 300 times the CH_3Hg^+ level in the blood (Ohta, 1971).

Tissue retention is greater for CH_3Hg^+ than for Hg^{2+} salts; rats fed equal amounts of Hg as CH_3HgCl and $HgCl_2$ retained far higher concentrations of CH_3Hg^+ than $HgCl_2$ in their soft tissues (Potter and Matrone, 1973), and similar observations were reported for chickens. In man the enterohepatic absorption mechanism accounts for the preferential retention of CH_3Hg^+ in the soft tissues, and for its relatively longer biologic half-life.

Inorganic mercuric ions do not cross the placental barrier (Garret *et al.*, 1972), but CH_3Hg^+ can cross these membranes and accumulate in the brain of the fetus and the newborn. In pregnant women CH_3Hg^+ can reach 30% higher concentration in fetal than in maternal erythrocytes due to differences between maternal and fetal hemoglobin in their binding with CH_3Hg^+; the fetal plasma CH_3Hg^+ level is lower than is the maternal plasma CH_3Hg^+ level. Similarly, CH_3Hg^+ can be secreted into the milk of lactating animals (Neathery *et al.*, 1973), and can accumulate in the brain or the suckling young (Yang *et al.*, 1973).

Excretion of Hg compounds is both fecal and urinary; excretion of ingested Hg^{2+} salts is predominantly fecal, while that of CH_3Hg^+ compounds is predominantly urinary. Some of the absorbed Hg^{2+} is excreted in the urine, bound to homocysteine and other SH-containing organic compounds. Since the inorganic Hg^{2+} ion can be methylated in the rat's intestine (Abdulla *et al.*, 1974), it is difficult to speculate on the ratio of organic to inorganic forms of Hg after gastrointestinal absorption of Hg compounds. When CH_3Hg^+ is administered to rats, much of the biliary mercury is present as a methyl mercury–cysteine complex (Norseth and Clarkson, 1970). Practically all of this Hg is subsequently reabsorbed from the intestinal tract, whereas other forms of biliary Hg are not comparably reabsorbed. In rats oral ingestion of polythiol resins inhibits the gastrointestinal absorption of Hg compounds and increases their fecal excretion (Clarkson *et al.*, 1971). About 70% of the average daily dietary intake of Hg is excreted by man; the amount depends on the ratio of organic to inorganic forms (since only about 30% of CH_3Hg^+ is excreted). The excretion of an experimental dose of $Hg(NO_3)_2$ given parenterally (0.2 mg Hg/kg) to rats occurred in three phases: 35% of the dose was excreted in about two weeks, the next 50% in about four weeks, and the last 15% in about three months. Because of this tissue retention Hg is classified as a cumulative poison. There are no indications for the presence of a homeostatic mechanism for Hg.

Toxicity

The toxicity of Hg salts has been known since 1500 B.C. Mercury in all forms is a protoplasmic poison, and in the higher concentrations it is lethal to all species; forms of mercury include elemental Hg and its vapor, inorganic Hg salts such as $HgCl_2$, Hg_2Cl_2, and $Hg(NO_3)_2$, organic Hg salts such as monoalkyl halides (especially CH_3Hg^+), dialkyl mercury compounds, and organic mercurials such as diphenyl mercury and mercurochrome. Organometallic compounds are excluded from this book, but CH_3Hg^+ salts behave like ionic compounds and are formed from inorganic Hg salts by microbial action, and thus toxicity of simple dialkyl Hg compounds and CH_3Hg^+ salts is included. The recent concern about environmental pollution from mercury has led to extensive metabolic and other studies on this metal. The biochemistry and physiology of Hg toxicity is extensively reviewed (Hughes, 1957; Battigelli, 1960; Passow *et al.*, 1961; Stokinger, 1963*b*; Clarkson, 1965, 1972; Brown and Kulkarni, 1967; Anon., Report of an International Committee, 1969; Browning, 1969; Friberg and Vostal, 1971, 1972; Oehme, 1972; Krehl, 1972), and the toxicity of Hg is given in Table 2-8. A more comprehensive and complete list of LD_{50} values for both organic and inorganic compounds of Hg is furnished by Friberg and Vostal (1971). Alkyl mercury salts are more toxic than are elemental Hg (liquid or vapor) and inorganic Hg^{2+} salts. Inhalation of mercury vapor (1.2–8 mg Hg/m^3) by man causes acute Hg poisoning, resulting in permanent damage to the nervous system, and possibly death. Chronic mercury poisoning following inhalation is called mercurialism; it affects the nervous system in an insidious way, so that the toxic effects may not be observed for months following exposure.

Acute inorganic Hg intoxication causes nausea, headache, abdominal pain and diarrhea, a metallic taste in the mouth, and albuminuria. Later, stomatitis and gingivitis develop, with the swelling of salivary glands and ulceration in the buccal cavity; a dark line of HgS forms on the inflamed gums. Hemorrhagic colitis, hemolysis, digital tremors, delirium, and hallucinations follow; death results from extreme exhaustion. Intoxication by Hg vapor inhalation initially causes acute tightness and pain in the chest with breathing difficulties; more serious symptoms occur later.

Symptoms of chronic Hg intoxication by both ingestion and inhalation of inorganic Hg salts are slow in appearing or developing. They are complex and, although less severe, similar to those of acute intoxication. Some of the symptoms, such as nervous anxiety and insomnia, are psycho-

TABLE 2-8. Mercury Toxicity

| | | | | Dosage/kg body weight | | | |
| | | | | Compound | Metal | | |
Compound	Animal	Route	Toxicity	mg	mg	mM	pT
Mercury	Rat	ip	LD_{50}		400	2.0	2.70
	Dog	inhal	LD_{100}		15[a]		
	Dog	inhal	MLD		5[a]		
	Human	inhal	Txc		1[a]		
Mercuric oxide	Mouse	oral	LD_{50}	22	20.4	0.102	3.99
HgO	Rat	oral	LD_{50}	18	16.7	0.083	4.08
	Rat	im	MLD	2.5	2.32	0.011	4.94
Mercuric chloride	Mouse	oral	LD_{50}	10	7.3	0.036	4.44
$HgCl_2$	Mouse	ip	MLD	5	3.7	0.018	4.73
	Mouse	ip	LD_{50}	14	10.3	0.051	4.29
	Mouse	sc	LD_{50}	23	17	0.084	4.07
	Mouse	iv	LD_{50}	7.6	5.6	0.028	4.55
	Rat	oral	LD_{50}	37	27.3	0.136	3.87
	Rabbit	sc	LD_{100}	10	7.3	0.036	4.44
	Rabbit	iv	MLD	2	1.47	0.0072	5.14
	Cat	iv	LD_{100}	5	3.7	0.018	4.73
	Dog	oral	LD_{100}	15	11.1	0.055	4.26
	Dog	sc	MLD	10	7.3	0.036	4.44
Mercuric iodide	Mouse	oral	LD_{50}	80	35.3	0.176	3.75
HgI_2	Mouse	ip	LD_{50}	60	26.5	0.132	3.88
	Rat	oral	LD_{50}	40	17.6	0.087	4.06
	Human	oral	MLD	357	157.6	0.785	3.10
Mercuric nitrate	Mouse	ip	MLD	8	4.9	0.024	4.61
$Hg(NO_3)_2$	Mouse	ip	LD_{50}	4	2.5	0.013	4.90
Mercuric sulfate	Mouse	oral	LD_{50}	40	27	0.134	3.87
$HgSO_4$	Rat	oral	LD_{50}	57	38.5	0.192	3.72
Mercuric cyanide	Mouse	oral	LD_{50}	33	26	0.129	3.89
$Hg(CN)_2$	Rat	oral	MLD	25	20	0.099	4.00
	Rat	ip	MLD	7.5	2	0.009	5.00
Mercuric acetate	Mouse	oral	LD_{50}	62	38.9	0.194	3.71
$Hg(C_2H_3O_2)_2$	Rat	oral	LD_{50}	76	47.7	0.238	3.62
Mercuric lactate $Hg(C_3H_5O_3)_2$	Rat	oral	LD_{50}	200	19	0.095	4.02
Mercurous chloride HgCl	Rat	oral	LD_{50}	210	178	0.887	3.05
Mercurous iodide	Mouse	oral	LD_{50}	110	67	0.334	3.48
HgI	Mouse	ip	LD_{50}	50	33.5	0.167	3.78
	Rat	oral	LD_{50}	110	67	0.334	3.48

(Cont'd)

TABLE 2-8. (Cont'd)

Compound	Animal	Route	Toxicity	Dosage/kg body weight			
				Compound	Metal		
				mg	mg	mM	pT
Mercurous nitrate	Mouse	oral	LD_{50}	388	296	1.47	2.83
$HgNO_3$	Mouse	ip	LD_{50}	5	3.82	0.019	4.72
	Rat	oral	LD_{50}	297	227	1.13	2.95
Methyl mercuric chloride	Mouse		LD_{50}	16	12.8	0.064	4.20
CH_3HgCl	Rat	ip	LD_{50}	11	8.8	0.044	4.36
	Guinea pig	oral	LD_{50}	21	16.5	0.082	4.08
	Guinea pig	ip	LD_{50}	7	5.5	0.027	4.56
	Rabbit	iv	MLD	15	12	0.060	4.22
Ethyl mercuric chloride	Mouse	ip	LD_{50}	16	12.1	0.060	4.22
C_2H_5HgCl	Rat	oral	LD_{50}	30	22.7	0.113	3.95
Phenyl mercuric chloride	Rat	oral	LD_{50}	60	38.4	0.192	3.72
C_6H_5HgCl	Rat	ip	MLD	50	32.0	0.16	3.80
Butyl mercuric chloride C_4H_9HgCl	Rat	sc	LD_{50}	50	34.2	0.17	3.77
Phenyl mercuric acetate	Mouse	oral	LD_{50}	26	15.5	0.077	4.11
$C_6H_5Hg \cdot$	Mouse	ip	LD_{50}	8	4.77	0.024	4.62
$C_2H_3O_2$	Mouse	sc	MLD	37	22	0.11	3.96
	Rat	oral	MLD	40	23.8	0.119	3.92
	Rat	oral	LD_{50}	50	29.8	0.148	3.83
	Rat	oral	LD_{100}	70	41.7	0.208	3.68
Chloromercuric benzoate $C_7H_5O_2HgCl$	Mouse	ip	LD_{50}	25	14.5	0.072	4.14

[a]Exposure for 4 hr; number of mg/m^3.

pathologic, being exhibited also by persons who have had no known exposure to Hg. The more common symptoms of chronic Hg toxicity are ataxia, dysarthria, dysphagia, uncoordinated movements of arms and legs, impaired hearing, recession of gums, loss of teeth, paresthesia, impairment of taste and smell, and ocular lesions (Kazantzis, 1966).

Dialkyl Hg compounds and monoalkyl Hg salts affect the central nervous system, causing gross ataxia, sensory loss in the limbs, impaired hearing and vision, loss of intellectual capacity, and personality changes. Methyl mercury salts also cause teratogenic effects; acute and chronic toxicity symptoms are similar. Organic mercury compounds are metabolized to inorganic Hg compounds, and thus, the toxicity symptoms of organic compounds are similar to those of inorganic Hg intoxication.

Chronic Hg toxicity symptoms in humans are erethism (exaggerated emotional response), gingivitis, and muscular tremors; the symptoms are psychopathological in nature. The toxic effects of Hg in humans are influenced by (1) the physicochemical properties of Hg compounds, (2) the amount and rate of administration or absorption, and (3) individual susceptibility. The route of administration is not considered, because after absorption Hg salts become widely distributed in the body.

In humans the neurotoxic symptoms of methyl mercury salts, the Hunter–Russel syndrome, involve focal cerebral and cerebellar atrophy (Hunter and Russel, 1954). The granular cell layer of the neocerebellum is affected followed by cortical atrophy of the area striata, which leads to blindness. Minamata disease is caused by the toxic effects of CH_3Hg^+ on multiple organ systems including the kidneys, reticuloendothelial system, gastrointestinal tract, and testes. The symptoms associated with this disease had been produced in experimental rats with CH_3HgOH (Klein *et al.*, 1972). Animals appear to exhibit Hg toxicity symptoms similar to those of man, but are more susceptible to Hg intoxication than is man.

Mercuric salts cause extreme local irritation and inflammation of the mucous membranes of the respiratory and gastrointestinal tracts, and at the sites of intramuscular injections. Lung tissue is also affected in this manner. Toxic action is attributed to the crowding of Hg^{2+} ions around the immediately available thiol groups of proteins, and the delay in distribution of these ions among the rest of the thiol groups throughout the body. In experimental animals Hg retards growth in both the Hg^{2+} and CH_3Hg^+ forms. Significant growth reduction in mice occurred when 6 ppm Hg (as $HgCl_2$ or CH_3HgCl) was added to the drinking water (Schroeder, 1973). Daily subcutaneous administration to rats of 10 mg CH_3HgOH/kg body weight for seven days caused loss in weight beginning with day 5 (Klein *et al.*, 1972); growth inhibition was also reported in birds. Mercuric salts induce hypercoagulability in blood when 3 mg/kg is injected subcutaneously in dogs daily for three days; increase in fibrinogen content and shortening of prothrombin time in the blood were observed (Worowski, 1968). Mercuric salts cause agglutination and hemolysis of erythrocytes; Hg^{2+} ions enter into either coordination or chelation complexes with erythrocytes, causing clumping of cells. At low concentrations Hg^{2+} ions initially block glucose entry by complexing with phosphate ligands and by increasing passive alkali-ion permeability, and then enter the cell and accumulate. *In vitro* studies on erythrocytes show the presence of Hg^{2+} binding sites on erythrocyte membranes (Clarkson and Megos, 1966); cellular membrane permeability is affected by the binding of Hg ions to thiol and phosphate ligands.

At low concentrations Hg^{2+} ions accumulate in liver lysosomes, and at high concentrations Hg^{2+} ions rupture the lysosomes and release destructive acid hydrolases (Verity and Reith, 1967). Studies on Hg^{2+}-induced kidney necrosis revealed decreased lysosomal enzyme activity and mitochondrial cytochrome-c activity (Norseth, 1968). The nephrotoxic effect of Hg^{2+} in the female is more prominent than in the male. Mercury intoxication studies with isolated cells indicate changes in cell membrane permeability, decreased electric potential across the cell membrane, loss of cellular potassium, reduced cellular uptake of glucose, and a strong inhibition of cellular respiratory enzymes. Like Cd, Hg induces the formation of metallothionein. Bryan *et al.* (1974) found that Hg accumulated in the euchromatin of liver cells in mice given sublethal doses of $HgCl_2$.

Extensive renal damage is usually the cause of death from mercury intoxication; this damage is due to the binding and retention of Hg^{2+} in renal tissue. Mercury ions induce diuresis by increasing the permeability of renal proximal tubules to Na^+ ions. Renal tissue damage is proportional to the accumulation of Hg; sometimes as much as 80% of the absorbed Hg^{2+} accumulates in the proximal tubules, and severe lesions of kidney necrosis occur. Proteinuria has been observed in acute and subacute Hg intoxication.

Mercuric salts cause reproductive dysfunction in Japanese quail when they are fed $HgCl_2$ (125 ppm Hg) in drinking water; fertility and hatchability of their eggs were reduced (Cogburn *et al.*, 1973). Thinning of egg shells of migratory birds and heavy mortality in their young are attributed to the hens eating Hg-treated seeds; experimental birds fed 1 to 8 ppm $HgCl_2$ also exhibit these symptoms of Hg toxicity. Intravaginal application of mercury antiseptics to pregnant laboratory animals caused increased fetal resorptions. When male rats were treated with CH_3Hg^+ and used for breeding, litter size was appreciably reduced; this was attributed to preimplantation loss (Khera, 1973). Inhibition of the early phase of DNA synthesis is suggested as the cause of the toxic effect of CH_3Hg^+ on spermatogenesis (Sukai, 1972). Reproductive capacity is not adversely affected in humans of either sex following Hg intoxication, as was observed in the Minamata patients (Kojima and Fujita, 1973).

Teratogenic effects of CH_3Hg^+ salts in experimental animals are far-reaching irrespective of the modes of administration, including vaginal application. The neurologic and teratogenic effects of mercury intoxication are manifested more in poisoning from CH_3Hg^+ than from organic Hg, because CH_3Hg^+ easily crosses the blood-brain barrier and placental membrane, and is retained in the tissues for a longer period. When administered to pregnant rodents in the early stages of gestation, CH_3Hg^+ produced growth and developmental retardation, congenital malformation, and death of fetuses (Harris *et al.,* 1972; Rizzo and Furst, 1972; Spyker and Smithberg, 1972). Mercuric methyl chloride, phosphate, sulfide, and phenyl acetate can provoke teratogenesis; some of the observed effects are fetal resorption and death, lowered body weight of offspring, incidence of cleft palate, abnormal tails, and changes in the nervous tissue. Malformation of the spinal cord, retarded growth of the cerebellum and disappearance of cerebellar granular cells, incomplete development of the cerebral granular matter, and diminution and atrophy of cells in the cerebral white matter are some of the changes in Hg-poisoned nervous tissue (Krehl, 1972).

Some of the above teratogenic effects were partially present in fetuses of women suffering from Minamata disease. A cerebral-palsey-like disease was reported in the children born of these women; mild spasticity, ataxia, seizure, and severe mental retardation were some of the clinical symptoms observed. In some cases, the mothers showed no symptoms of Hg intoxication, and the children were never exposed directly to mercury-contaminated diets, but were nursed by the mothers. This indicates the transfer of organic Hg salts through maternal milk, with subsequent deposition in the nervous tissues; a similar observation was made in the case of a boy born to a mother who ingested contaminated pork during her pregnancy (Ordonez *et al.,* 1966). The developing nervous and hematopoietic systems in the fetus are especially susceptible to CH_3Hg^+ intoxication (Krehl, 1972).

The outward neurotoxic symptoms of acute CH_3Hg^+ intoxication in humans depend upon the severity of the exposure. The symptoms extend from simple paresthesia or concentric constriction of visual fields with abnormal blind spots to grotesque uncoordinated movements, tremor paralysis, coma, and death. Neurotoxicity has been observed in humans from indirect CH_3Hg^+ intoxication by eating pork raised on Hg-contaminated feed; in this instance, some of the affected pigs developed blindness, lack of coordination, and posterior paralysis within a month, and died after six weeks. The brain must accumulate a certain level of CH_3Hg^+ before any clinical symptoms of CH_3Hg^+ neurotoxicity are observed. In laboratory animals, this level has been established as 30 μg Hg (as CH_3Hg^+) per gram of wet brain tissue. Evidence from human victims of CH_3Hg^+ intoxication, especially the fatal cases of Minamata disease, indicate that at concentrations greater than 8 μg Hg/g wet brain tissue overt neurotoxic symptoms appear in adult humans; this is equivalent to 60 mg Hg in the human body. A fatal ingested dose in adult humans is 1 mg Hg per day (as CH_3Hg^+) over a period of several weeks.

Experimental CH_3Hg^+ toxicity in rats, from subcutaneous injections of seven daily doses of 10 mg CH_3HgOH/kg, produced extensive damage to the central nervous system (Klein *et al.,* 1972). The sciatic nerve sheaths were swollen and nerve fibers were damaged, followed by extensive loss of myelin. Bundles of nerve cells were lost from the lower portions of the spinal cord, the cerebellum, and some of the brain stem nuclei, but the white matter of the cerebral cortex and the midbrain nuclei showed no degenerative changes. The high and specific neurotoxicity of CH_3Hg^+ is attributed to its binding to cell membranes and reducing the RNA content of nerve cells in the dorsal root ganglia and the cerebellum.

At the molecular level, the effect of Hg intoxication is a strong inhibition of sulfhydryl enzymes. Histochemical study reveals significant decreases in the activity of glucose-6-phosphatase, alkaline phosphatase, ATPase, and succinic acid dehydrogenase, and moderate increases of acid phosphatase activity in the liver, kidneys, and brain of rats four weeks after Hg^{2+} and CH_3Hg^+ intoxication (Chang *et al.*, 1973); the extent of the changes was proportional to the Hg content in each organ. Methyl mercury also inhibits δ-aminolevulinic acid dehydratase and cholinesterase activities (Aberg *et al.*, 1969); the malonic dialdehyde level in the kidneys increases in one hour following parenteral injection of Hg^{2+} or CH_3Hg^+ salts to rats, suggesting the role of mercury in the inhibition of enzymes involved in lipid peroxidations (Taylor *et al.*, 1973).

Dietary Se provides some protection against Hg intoxication, as shown by enhanced growth rate and decreased mortality in experimental rats and quails (Ganther *et al.*, 1972; Potter and Matrone, 1973). Rats fed 25% fish-protein concentrate containing 0.22 ppm Hg, mostly as CH_3Hg^+, showed no symptoms of Hg intoxication, but grew and reproduced normally for five generations, although rats fed this diet accumulated more Hg in their tissues than did control rats fed a casein diet (Newberne *et al.*, 1972); Japanese quail fed a 17% tuna diet were able to withstand 20 ppm Hg in the form of CH_3Hg^+ (Ganther *et al.*, 1972). Both the fish concentrate and tuna contained Se. Simultaneous subcutaneous injections to rats of sodium selenite at 0.5 mg/kg body weight prevented the weight loss, manifestations of specific toxicity symptoms, and mortality which would ordinarily be caused by subacute oral administration of 10 mg CH_3HgI/kg body weight. Methyl mercury accumulated in the brain, and, to some extent, in liver, kidneys, and blood (Iwata *et al.*, 1973). Potter and Matrone (1974) believe that the addition of selenite to either Hg^{2+}- or CH_3Hg^+-containing diets causes a shift in the tissue Hg distribution with a concomitant reduction of Hg in the kidneys and increase of Hg in the liver, thereby reducing the toxic effects. Vitamin E also is reported to reduce the effects of Hg intoxication (Welsh and Soares, 1973).

There are no indications of the existence of a homeostatic mechanism for Hg. A possible but poor detoxication mechanism for Hg is the enhanced formation of liver metallothionein following Hg intoxication, as metallothionein can bind Hg ions.

There are many theories for the mechanism of Hg toxicity. The ultimate effect of the high affinity of Hg for thiol groups is the binding of Hg with thiol ligands of functional proteins and consequent inhibition or inactivation of these proteins. The binding of Hg to phosphate ligands produces changes in cellular membrane permeability, and conformational changes in some biological macromolecules. Mercury disturbs the protein–thio-disulfide cycle; mitotic effects of Hg are attributed to this disturbance. Mercury inhibits I uptake by the thyroid gland, similar to Cd intoxication, and it also acts as an antimetabolite to Zn. Mercury decreases the activity of the liver microsomal detoxication systems and indirectly potentiates the harmful effects of other toxicants.

TOXICITY OF GROUP III METALS

Boron (B), aluminum (Al), gallium (Ga), indium (In), and thallium (Tl) constitute subgroup A and scandium (Sc), yttrium (Y), and the lanthanides and actinides constitute subgroup B of Group III in the periodic chart. The lanthanides, the rare-earth metals, constitute a group of 15 inner-transitional elements with successive atomic numbers and very similar chemical properties: lanthanum (La), cerium (Ce), praseodymium (Pr), neodymium (Nd), promethium (Pm), samarium (Sm), europium (Eu), gadolinium (Gd), terbium (Tb), dysprosium (Dy), holmium (Ho), erbium (Er), thulium (Tm), ytterbium (Yb), and lutetium (Lu). The actinides form a group of heavier inner-transitional metals which exist mostly in highly radioactive isotopic forms: actinium (Ac), thorium (Th), protactinium (Pa), uranium (U), neptunium (Np), plutonium (Pu), americium (Am), curium (Cm), berkelium (Bk), californium (Cf), einsteinium (Es), fermium (Fm), mendelevium (Md), nobelium (No), and lawrencium (Lr). Most actinides are manmade or occur in radioactive fallout from nuclear fission or fusion reactions.

All elements in Group III are metals with the exception of boron. Group III metals exhibit a normal valence of three; some exhibit variable valence. Thallium is monovalent, while some of the lanthanides exhibit valence from 2 to 4; the actinides exhibit valence ranging from 2 to 5. The Group III metals are electropositive, and their basicity increases directly as their atomic weights increase. The simple salts of Group III metals are soluble in water, but readily undergo hydrolysis at and above neutral pH to form insoluble hydroxides; Tl salts are the only exceptions. Depending upon their concentration, the metal hydroxides exist in particulate, colloidal, or radiocolloidal forms in aqueous solutions. Hydroxides of metals of subgroup IIIA undergo further olation and oxalation to form hydrated

Toxicity of Group III Metals

Atomic number Name (Symbol) Atomic weight	(Core) Active electrons Usual valences
Subgroup A	Subgroup B
5 (He) 2, 1 Boron (B) 10.81 +3	
13 (Ne) 2, 1 Aluminum (Al) 26.98 +3	
31 (Ar) 2, 1 Gallium(Ga) 69.72 +3	21 (Ar) 1, 2 Scandium (Sc) 44.96 +3
49 (Kr) 2, 1 Indium (In) 114.82 +3	39 (Kr) 1, 2 Yttrium (Y) 88.91 +3
81 (Xe) 2, 1 Thallium (Tl) 204.37 +1, +3	57–71 *Lanthanides* 89–103 *Actinides*
Lanthanides	*Actinides*
57 (Xe) 0,8,1,02 Lanthanum (La) 138.9 +3	89 (Rn) 0,8,1,0,2 Actinium (Ac) 227 +3
58 (Xe) 2,8,0,0,2 Cerium (Ce) 140.12 +3,+4	90 (Rn) 0,8,2,02 Thorium (Th) 232.038 +2,+3,+4
59 (Xe) 3,8,0,0,2 Praseodymium (Pr) 140.907 +3	91 (Rn) 2,8,1,02 Protactinium (Pa) (231) +4 or +5
60 (Xe) 4,8,0,0,2 Neodymium (Nd) 144.24 +3	92 (Rn) 3,8,1,0,2 Uranium (U) 238.03 +2,+6
61 (Xe) 5,8,0,0,2 Promethium (Pm) (147) +3	93 (Rn) 4,8,1,0,2 Neptunium (Np) (237) +3,+4,+5,+6

(*Cont'd*)

Toxicity of Group III Metals (Cont'd)

Atomic number Name (Symbol) Atomic weight		(Core) Active electrons Usual valences	
Lanthanides		*Actinides*	
62 Samarium (Sm) 150.35	(Xe) 6,8,0,0,2 +2,+3	94 Plutonium (Pu) (242)	(Rn) 6,8,0,0,2 +2,+3,+4
63 Europium (Eu) 151.96	(Xe) 7,8,0,0,2 +2,+3	95 Americium (Am) (243)	(Rn) 7,8,0,0,2 +2,+6
64 Gadolinium (Gd) 157.25	(Xe) 7,8,1,02 +3	96 Curium (Cm) (245)	(Rn) 7,8,1,0,2 +3,+4
65 Terbium (Tb) 158.924	(Xe) 9,8,0,0,2 (249)	97 Berkelium (Bk) 	(Rn) 9,8,0,0,2 +3,+4
66 Dysprosium (Dy) 162.5	(Xe) 10,8,0,0,2 +3	98 Californium (Cf) (251)	(Rn) 10,8,0,0,2 3+
67 Holmium (Ho) 164.93	(Xe) 11,8,0,0,2 +3	99 Einsteinium (Es) (254)	(Rn) 11,8,0,0,2 +3
68 Erbium (Er) 167.26	(Xe) 12,8,0,0,2 +3	100 Fermium (Fm) (255)	(Rn) 12,8,0,0,2 +3
69 Thulium (Tm) 168.934	(Xe) 13,8,0.0,2 +3	101 Medelevium (Md) (256)	(Rn) 13,8,0.0,2 +2,+3
70 Ytterbium (Yb) 173.04	(Xe) 13,8,0,0,2 +2,+3	102 Nobelium (No) (254)	(Rn) 14,8,0,0,2 +2,+3
71 Lutetium (Lu) 174.97	(Xe) 14,8,1,0,2 +3	103 Lawrencium (Lr) (257)	(Rn) 14,8,1,0,2 +3(?)

polymers of hydrous oxides. Group III metals have great affinity toward free and bound phosphate groups and ligands and have the capacity to combine with metabolites such as nucleotide phosphates and nucleic acids. Metabolically these metals are considered to be "bone seekers," as they tend to accumulate in the bone.

No Group III metals are known to be essential in animal or human nutrition, but some are found to be stimulatory. With the exception of Ga, In, Tl, and the actinides, the Group III metals are considered to be practically nontoxic when administered orally, but all Group III metals are moderately toxic following parenteral administration of their soluble salts.

The gastrointestinal absorption of the compounds of these metals is poor with the exception of Ga, In, and Tl; the latter behaves like potassium in its metabolism. The retention of Group III metals in mammals occurs mostly in the skeleton and less in soft tissues, and excretion is both fecal and urinary.

There are no reports of the existence of any homeostatic mechanism which controls the levels of these metals in humans or animals.

SUBGROUP IIIA METALS

Subgroup IIIA metals exhibit degrees of amphoteric character which decrease as the atomic weight increases; however, Tl does not exhibit this character. These metals form coordination complexes and compound salts. Salts of Al, Ga, and In undergo hydrolysis in aqueous solutions, and the metal hydroxides undergo olation and oxalation to form polymers of hydrous oxides. The nearly neutral pH of biologic tissues and fluids favors this hydrolysis and effectively prevents the absorption of these salts from the intestine and from sites of intramuscular and subcutaneous injections. With the exception of Al, subgroup IIIA metals are more toxic than the metals of subgroup IIIB; the monovalent Tl mimics K in its metabolism, and accompanies K into cells, causing toxic effects. The toxicity of subgroup IIIA metals becomes greater with increase in electropositivity and atomic weight, to the limits of solubility of the compounds formed.

Aluminum (Al)

Aluminum occurs as bauxite ($Al_2O_3 \cdot 2H_2O$), cryolite ($3NaF \cdot AlF_3$), spinel ($MgO \cdot Al_2O_3$), orthoclase ($H_2O \cdot Al_2O_3 \cdot 6SiO_2$), muscovite

($H_2O \cdot 3Al_2O_3 \cdot 6SIO_2 \cdot 2H_2O$), and as other complex aluminum silicates or clays (bentonites). The earth's crust contains about 8% Al in undecomposed rock fragments and in secondary aluminosilicate clays, and the Al content of seawater ranges from 3 to 2400 ppb. An adult man's body contains about 61 mg Al, 35% of which is in the skeleton and about 20% in the lungs. Aluminum is not essential for mammals.

The use of metallic Al and its compounds is very extensive. Aluminum metal is an important structural material in the building, canning, automobile, and aviation industries, and Al powder is used in paints. The natural aluminum mineral, bentonite, is used extensively in water purification, sugar refining, and in the brewing and paper industries. Food processing and baking industries use Na_3AlO_3 and $NaAl(SO_4)_2$. Aluminum trioxide is used as an absorbant, abrasive, and refractory material; $AlCl_3$ is used in the cracking of petroleum, in the manufacture of synthetic rubber and lubricants, and in wood preservation. Aluminum isopropoxide is used in the soap and paint industries and in waterproofing of textiles; aluminum silicate is used in dental cement, and aluminum borate in the glass and ceramics industries; and aluminum borosilicate is a constituent of jet and rocket fuel. Aluminum is also used as a tanning agent in the leather industry, in water purification, and in the manufacture of paper and vegetable glue; $Al(NO_3)_3$ is an ingredient of antiperspirants. Aluminum salts find use in human and veterinary medicine: alum and $Al(CH_3COO)_2$ as antiseptics and astringents; aluminum magnesium silicate, and $Al(OH)_3$ as antiacids; and $Al(OH)_3$ as an adjuvant in veterinary vaccines. Alumina gels are used to bind intestinal phosphate. $Al(OH)_3$ is also used to lower the plasma phosphorus level of patients suffering from acute renal failure. Solutions of $AlCl_3$ are used to disinfect stables and slaughterhouses.

Aluminum and its salts are not generally considered to be industrial health hazards except under special conditions; most Al toxicity symptoms are observed under experimental conditions with excessive dose levels, but chronic inhalation and excessive intake of Al compounds for therapeutic purposes can cause Al poisoning. Aluminum oxide particles from corroded air conditioners can raise room air pollution to unacceptably high levels.

Chemistry

Aluminum exhibits trivalence and amphoteric character; its simple cationic and anionic salts are water-soluble. Aluminum forms binary salts of the type $MAl(SO_4)_2 \cdot 12H_2O$ or $M_2SO_4 \cdot Al_2(SO_4)_3 \cdot 24H_2O$ (where M is any mono- or divalent metal); high atomic weight of the M ion confers great stability to Al salts. Halogen anion complexes such as $[AlF_6]^{3-}$ are stable and more soluble than CaF_2. Aluminum can exist as the central atom in salts such as $[K_9Al(MoO_4)_6] \cdot xH_2O$ and in hetero-6-polyacids $H_9[MO_6]$, where M is any trivalent metal such as Al. In dilute aqueous solutions, and especially near neutral pH, simple cationic Al salts are hydrolyzed and olated. The olated compounds can be oxalated, and the oxalated forms aggregate into colloidal, water-insoluble, and chemically inactive hydrous oxides:

$$Al(NO_3)_3 \xrightarrow{[OH]} [Al(OH)(NO_3)_2 \cdot H_2O]_n \longrightarrow [AlONO_3]_n$$

oxolated aggregate

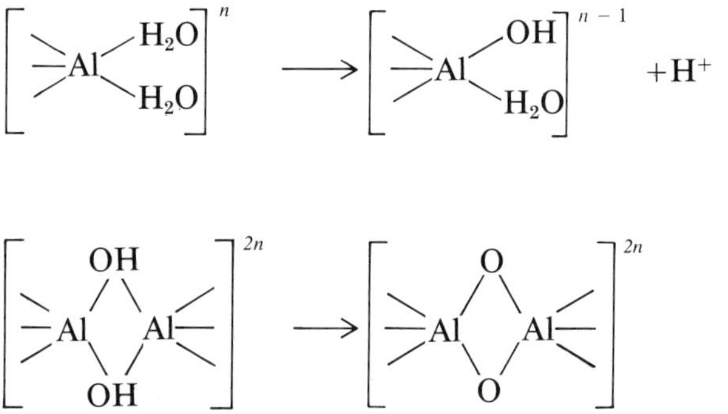

The hydrous oxides can be peptized by formic, acetic, glycolic, and tartaric acids. Continued olation results in aggregate complexes with zero charge and decreased chemical activity. The solubility and chemical reactivity of Al compounds change in the gastrointestinal tract, where the pH ranges from 2 to 7. Enzymes, cofactors, and phosphates present in the digestive tract and in food also affect the solubility of Al compounds. Aluminum has a high affinity to both free and bound phosphate groups in biologic macromolecules, and is capable of forming complex compounds with proteins.

Metabolism

Aluminum is present in foods in varying amounts; the Al content of foods of plant origin depends upon the soil Al content and ranges from 10 to 30 mg/kg fresh weight, whereas the Al content of foods of animal origin is 50 or 100 times less than that of foods of plant origin. The available data on the aluminum content of animal tissues are conflicting, but the aluminum levels in plant and animal tissues are disproportionate to its relative abundance in the earth's crust. There are no reports of attempts to formulate Al-free diets for experimental animals. Negative results in balance studies on Al in man suggest its nonessential nature in humans. There is no conclusive evidence that Al is involved in any essential biologic function in animals, but it appears to possess stimulatory effects on the growth of several pasture plant species (Hackett, 1962); evidence of similar stimulatory effects in animals is lacking. The biochemistry and physiology of Al was reviewed by Underwood (1971), Browning (1969), Stokinger (1963), Campbell *et al.* (1957), Burn (1932), and Underhill *et al.* (1929; Underhill and Peterman, 1929). The daily dietary intake of Al by adults in the United States is about 45 mg.

In mammals the gastrointestinal absorption of ingested Al is poor due to the transformation of Al salts into insoluble $AlPO_4$ in the digestive tract, brought about by pH changes and the presence of phosphate in the diet. At higher doses, such as 200 mg Al/kg as $Al_2(SO_4)_3$, the intestinal absorption is about 10% in rats (Kontus, 1962); the absorption and removal of injected Al compounds such as $Al(OH)_3$ from intramuscular sites and from the peritoneal cavity is slow, but it is complete in a few days, provided the doses are within physiologic limits and the Al salts are not mixed with any irritants such as anions of chloride and nitrate. Absorption is mainly through phagocytosis by macrophages. Subcutaneously injected Al compounds behave similarly. Intravenously injected Al compounds undergo hydrolysis and form colloidal Al hydrous oxides or colloidal $AlPO_4$ and are removed from the blood by phagocytosis; however, a small portion of ionic Al is bound to albumin and is transported out of the blood into soft tissues including the brain. Following inhalation, most of the insoluble Al salts are retained in the lung for a long time; soluble compounds are slowly absorbed into the blood. The Al in the lung tissues of industrial workers in Al refineries increases with duration of exposure. Aluminum is not absorbed readily through the skin.

Following absorption, Al is distributed mainly in the skeleton, liver, testes, kidneys, and brain, and in much smaller amounts in other soft tissues. The retention of Al in bone is prolonged, while it is transient in the soft tissues. Renal failure increases Al deposition in the bone (Thurston *et al.*, 1972). In rats the retention of Al in bone, liver, and testes progressively increases with large ingested doses such as 200 mg Al/kg in the form of $Al_2(SO_4)_3$ (Ondreicka *et al.*, 1966b); the Al level in the brains of these experimental rats increases at a rate tenfold that of the control. In mammals Al excretion is mainly fecal, representing both unabsorbed Al and the Al excreted into the digestive tract from the liver and bile. An enterohepatic circulation for Al is suggested, since most parenterally administered Al is excreted in the feces. When $Al(OH)_3$ is injected intraperitoneally in rats, the highest Al concentration is found in the liver, and muscle, brain, bone, and heart also have high Al levels (Berlyne *et al.*, 1972). Small amounts of Al are excreted in the urine and sweat; urinary Al level increases with increased dietary intake. Dietary fluoride (1 mg F^- per animal) increases the gastrointestinal absorption of Al in rats; Al retention in tissues decreases in the presence of increased F^-, while urinary and fecal excretion of Al increases. Even the permanently deposited Al in the bone can be removed by a prolonged high dietary level of F^-. The increased excretion of Al following high F^- intake is attributed to the formation of the readily soluble halogen complex $(AlF_3)^{3-}$. Phosphates and organic phosphorous compounds in the diet decrease Al absorption from the digestive tract and increase the fecal Al excretion.

The lack of accumulation of Al in animals with age, or of any increase in tissue levels of Al following fairly high dietary Al, intake and efficient excretion are suggestive of a homeostatic mechanism for Al in mammals, but there is no direct experimental evidence for Al homeostasis.

Toxicity

Aluminum compounds are not generally harmful and are considered to be inert toxicologically except in cases of high experimental doses or prolonged inhalation. This is partly owing to poor absorption from the digestive and respiratory tracts. The toxicity of Al compounds is summarized in Table 3-1; Al salts are more toxic when parenterally injected than when orally ingested. The type of anion present and the degree of solubility of Al salts also affect the level of Al toxicity, fluoroaluminate is toxic due to its great solubility and stability. Digestive or gastrointestinal disturbance, skin lesions, and nervous afflictions are common symptoms of acute Al toxicity. The toxic effects of Al are attributed to phosphate depletion in the tissues, lowered phosphate absorption from the digestive tract, negative phosphorous balance, and the subsequent adverse alterations in phosphorylation reactions in the tissues. Growth retardation, perihepatic Al granulomas, and fibrous peritonitis are some of the less pronounced effects of Al toxicity. These toxic effects are more pronounced in nephrectomized animals that in control animals (Berlyne *et al.*, 1972), and, in man, most absorbed Al from large oral doses is readily excreted if the kidney is intact (Berlyne *et al.*, 1970). Uremic patients absorb 0.1 to 0.6 g/day (Clarkson *et al.*, 1972). A 1% solution of an Al salt injected intravenously into the blood precipitates in the blood causing emboli. Intraperitoneally injected $Al(NO_3)_3$ causes marked congestion of liver and lung in mice, and an intense chemical peritonitis with ascites formation and marked congestion of lung in the rat (Hart *et al.*, 1971).

Rats fed 6–10 mg Al [as $Al(OH)_3$] daily showed significant impairment of growth at 3 to 4 weeks (Thurston *et al.*, 1972). Rachitic changes occurred in the bones; addition of phosphate in the diet prevented these changes. Prolonged administration of $AlCl_3$ to mice retarded growth in the second

TABLE 3-1. Aluminum Toxicity

| | | | | Dosage/kg body weight | | | |
| | | | | Compound | Metal | | |
Compound	Animal	Route	Toxicity	mg	mg	mM	pT
Aluminum hydroxide $Al(OH)_3$	Rat	ip	LD_{100}	435	150	5.00	2.30
Aluminum trifluoride	Guinea pig	oral	MLD	600	193	6.44	2.19
AlF_3	Guinea pig	sc	MLD	3000	964	32.2	1.49
Aluminum chloride	Mouse	oral	LD_{50}	3810	770	25.7	1.59
$AlCl_3$	Rat	oral	LD_{50}	3730	755	25.2	1.60
Aluminum nitrate	Mouse	ip	LD_{50}	320	23	0.77	3.12
$Al(NO_3)_3 \cdot 9H_2O$	Rat	oral	MLD	2400	173	5.77	2.24
	Rat	oral	LD_{50}	4280	324	10.81	1.97
	Rat	ip	LD_{50}	327	23.1	0.77	3.11
Aluminum sulfate (anhydrous)	Mouse	oral	LD_{50}	4210	664	22.2	1.65
$Al_2 \cdot (SO_4)_3$	Mouse	ip	LD_{50}	270	42	1.40	2.85
	Rat	oral	LD_{50}	1930	300	10.0	2.00
	Rat	oral	LD_{100}	2410	375	12.5	1.90
	Rabbit	sc	LD_{100}	640	100	3.34	2.48
Aluminum phosphide AlP	Rat	inhal	LD_{100}	1[a]			
Sodium fluoroaluminate	Rat	oral	LD_{50}	200	25.7	0.86	3.07
Na_3AlF_6	Rat	ip	LD_{50}	59	7.58	0.25	3.60
	Rabbit	oral	MLD	9	1.16	0.039	4.41
Aluminum hexafluorosilicate	Guinea pig	oral	MLD	5000	562	18.8	1.73
$Al_2(SiF_6)_3$	Guinea pig	sc	MLD	4000	450	15.0	1.82

[a]Number of ppm in air resulting in death in 30 minutes.

and third generations, without affecting the size of the litters. In humans a dietary intake of 16 g $Al(OH)_3$ daily for six months did not produce any adverse effects. Uremic patients may suffer Al intoxication following prolonged use of alumina gels (Berlyne *et al.*, 1970; Clarkson *et al.*, 1972; Alfrey *et al.*, 1976).

Nephrectomized rats have been shown to be more susceptible than control animals to experimentally induced Al toxicity by oral and subcutaneous or intraperitoneal injections of $Al(OH)_3$ or $AlCl_3$ (Berlyne *et al.*, 1972); the oral doses were equivalent to 90 mg Al/kg per day. Intraperitoneal injection of $Al(OH)_3$ caused heavy mortality with periorbital bleeding in both nephrectomized and control rats within 5 days. Subcutaneous injection of $Al(OH)_3$ caused no mortality but induced periorbital bleeding in nephrectomized rats, and patchy pneumonia with desquamation of the bronchial epithelium was observed in dying rats with periorbital bleeding. Perihepatic granuloma and fibrous peritonitis are the other toxic effects of intraperitoneally injected $Al(OH)_3$. High serum Al levels and lowered oxygen consumption in liver slices were also observed. Mortality was high in nephrectomized animals.

Toxic doses of dietary Al compounds disturb the gastrointestinal absorption and metabolism of phorphorus and carbohydrates. Although Al salts are only partially ionized in aqueous solution and relatively few Al^{3+} ions are present in the digestive tract, these form insoluble and unabsorbable $AlPO_4$, increasing the fecal excretion of P and, thus, depriving the organism of P; Al can also be bound to organic P compounds in the digestive tract (Schmidt and Hoagland, 1912). In rats 5% $Al_2(SO_4)_3$ in the diet decreased inorganic P in the serum and P in femoral bone (Jones, 1938). Aluminum chloride intoxication in chicks decreased the inorganic serum P level and produced muscular dystrophy (Pragay, 1962); depletion of P stores of the body has also led to acute rickets, which is resistant to vitamin D (Deobald and Elvehjem, 1958). Aluminum intoxication affects P metabolism in tissues in addition to P absorption (Ondreicka *et al.*, 1966). Bessarabova (1955) suggested that impaired phosphorylation was the most significant toxic effect of Al compounds. Decreased incorporation of ^{32}P into phospholipids, RNA, and DNA of the rat liver, kidneys, and spleen following chronic and acute oral intoxication from $AlCl_3$ confirmed this view (Ondreicka *et al.*, 1971); ATP levels decreased with a concomitant increase in ADP and AMP levels in mammalian tissues. The decreased ATP/ADP ratio suggested that the equilibrium in the reaction

$$ATP \rightleftharpoons ADP + P_i \rightleftharpoons AMP + P_i$$

is shifted to the right following Al intoxication; the adverse effects of decreased ATP formation are disastrous.

Following chronic Al intoxication in rats, the blood glucose and liver glycogen levels decreased while liver lactate and pyruvate, and muscle pyruvate levels increased (Kortus, 1967; Ondreicka and Ginter, 1966); serum aldolase levels increased followed by hyperglycemia. These toxic effects were observed in rats receiving single intragastric doses ranging from 120 to 200 mg Al/kg. Impaired P metabolism depressed glucose absorption from the gut (Gisselbrecht *et al.*, 1957), and lowered the ATP and coenzyme A levels in the tissues. Aluminum intoxication can also lead to reduced food intake owing to gastrointestinal irritation.

Decrease in plasma parathyroid hormone levels following $Al(OH)_3$ therapy in patients with chronic renal failure appears to be an indirect effect of Al intoxication (Bailey *et al.*, 1971); the binding of Al with plasma inorganic P increased the plasma Ca level with a proportionate fall in parathyroid hormone level. *In vitro*, Al^{3+} can cause sensitization of erythrocytes and their agglutination at 0.1–1.0 μmol Al, when 1 ml of washed cells was suspended in normal saline.

Comparing the carcinogenic activity of subcutaneously implanted iron and alumium in rodents, O'Gara and Brown (1968) reported that, while iron was only weakly carcinogenic, aluminum foil induced sarcomas in 8 out of 18 rats; the particulate smooth surface appears to contribute to the carcinogenic activity.

Following inhalation, dusts of Al, Al_2O_3, and other Al compounds are retained in the lungs. In rats, the initial response to this deposit was the proliferation of macrophages within the alveolar spaces, resulting in lipoid pneumonia (Christie *et al.*, 1963); prolonged exposure caused focal deposits of hyaline material in alveolar walls. Chronic inhalation of extremely fine dusts of Al compounds by humans caused a toxic pulmonary reaction; an interstitial fibrosis of a nonnodular type called Shaver's disease developed in the lungs (Shaver and Riddel, 1947); this is common with workers in industries using Al abrasion. Inhalation of fine particles of Al metal dust in factories caused both encephalopathy and pulmonary fibrosis in humans (McLaughlin *et al.*, 1962). Some fatalities were due to terminal bronchopneumonia; the lung and brain contained 20 times the normal level of Al, and 120 times the normal Al level was found in the liver. The progressive encephalopathy was followed by dementia and convulsions; the brain showed a unique cellular change, neurofibrillary degeneration. Similar changes were observed when hydrated oxides of Al were applied to the surface of the brain or were injected into the cerebral cortex or cisterna magna in apes and monkeys, which subsequently suffered from severe recurrent epilepsy. Cultured neuroblastoma cells are sensitive to Al toxicity (Miller and Levine, 1974).

The changes observed during Al intoxication are similar in part to the symptoms of Alzheimer's disease in humans over 40, progressive senile dementia with neurofibrillary degenerative changes in the brain (Crapper *et al.*, 1973). In the early stages of $AlCl_3$-induced encephalopathy in rats, selective impairment of short-term memory and associative learning occurred, and later the brain tissue suffered neurofibrillary degeneration; experiments with cats injected intracranially with 150–225 μg Al (as $AlCl_3$) revealed the same toxicity symptoms. The high tissue concentration of 9–11 μg Al/g tissue in some regions of the human brain in Alzheimer's disease was similar to the levels of 14 μg Al/g tissue in cat brain found in experimental Al intoxication with neurofibrillary degeneration. Crapper *et al.* (1973) suggest a possible role for Al as a neurotoxic factor in Alzheimer's disease in humans. The massive neurofibrillary degeneration of the neuronal perikarya and marked gliosis produced by injecting sterile $Al(OH)_3$ into the lower lumbar region of spinal cords of cats support the above suggestion (de Estable-ping *et al.*, 1971). Some glial cells exhibited conspicuous intranuclear fibrillary aggregates and intranuclear bodies, attributed to changes in the intracellular fibrous protein metabolism following Al intoxication. A case for the possible neurotoxicity of Al is being slowly established.

Many metals act as electric-charge carriers and as controllers of current flow in axon excitability processes in the squid and lobster; *in vitro,* Al was found to be a very effective charge carrier. The significance of this observation, and the binding of Al to mammalian nerve tissues is not clear.

In mammals, the mechanism of Al intoxication appears mainly to be the impairment of P metabolism and its resultant adverse effects. The neurotoxicity of Al is due to its binding with nervous tissues, with massive neurofibrillary degeneration.

There are no specific mechanisms in mammals for the detoxication of Al, but efficient renal excretion of Al and poor absorption of Al from the digestive tract appear to form part of a poorly defined homeostatic mechanism which can control appreciable doses of dietary Al.

Gallium (Ga)

Gallium does not occur free in nature. There are no specific ores of Ga, but it occurs mixed with Al, Zn, and Sn. Bauxite (Al ore) contains 0.002–0.008% Ga; some Sn ores contain 0.01–0.05% Ga, and sphalerite (ZnS) also contains some Ga. Gallium is extracted mainly from leached zinc ores or from coal ash, which contains about 490 g Ga/ton. The earth's crust contains 5 ppm Ga; no reliable data is available on the Ga content of seawater or the Ga content of the human body, but traces (less than 1 ppm) are reported to be present in the adult skeleton. Gallium is not essential for mammals.

Gallium is used in high-temperature thermometers, in liquid sealant for glass joints and valves, and in the plating of optical mirrors, and it has replaced mercury in rectifiers and mercury–quartz lamps. Gallium is used in making small rugged electronic parts for missiles, satellites, and space probes; GaP solar-cell power plants are likely to be used in space stations. Gallium is also used as a substitute for amalgam for dental purposes. Formerly Ga salts, especially gallates, were used for curative and prophylactic treatment of experimental syphilis and trypanomiasis in animals, and cationic Ga salts are presently used as antitumor agents in medicine. Accidental Ga poisonings are rare, and it is not considered to be an industrial health hazard.

Chemistry

Gallium in the liquid phase has a long range of stability (mp 29.8° and bp 1700°C). Gallium exhibits trivalence, and can exhibit valence of +2 and

+1 under special reducing conditions. It forms two oxides, Ga_2O_3 and Ga_2O, and is amphoteric in nature. Cationic Ga salts undergo hydrolysis to hydroxides in dilute aqueous solutions; anionic gallates are stable and water-soluble. Gallium also forms stable halogen complexes such as $[GaF_6]^{3-}$ and $[GaF_5 \cdot H_2O]^{2-}$; gallium halides are bimolecular in the gaseous state. Gallium complexes are tetrahedral in geometry. Its ions appear to have a greater affinity than does aluminum toward phosphate groups; gallium binds to the microbial protein, transferrin, with the same stoichiometry as iron. Gallium lactate and citrate were used in gallium metabolism studies because they possess great stability; simple cationic Ga salts, such as chlorides, are not stable at physiologic pH; they are hydrolyzed to colloidal $Ga(OH)_3$. The degree of olation is less pronounced for Ga salts than for Al salts.

Metabolism

Gallium is neither essential nor stimulatory to either man or animals, but it may be essential for some molds, such as *Aspergillus niger*. There are few reports about the distribution and presence of Ga in animal and plant tissues; it is reported to be present in minute quantities in bone (Dudley and Levine, 1949). The daily dietary intake of Ga by man is not known. The metabolism of Ga was studied using ^{72}Ga, and the physiologic and pharmacologic properties of Ga were reviewed by Brucer *et al.* (1953).

The gastrointestinal absorption of cationic Ga salts in mammals is less than 1% for physiologic doses, owing to the hydrolysis of the Ga^{3+} salts to insoluble and unabsorbable $Ga(OH)_3$. Gallium lactate and citrate are absorbed slightly better than the simple inorganic gallium salts. Metabolic studies on Ga were made using either lactate or citrate salts, and in rabbits, subcutaneously injected gallium citrate induced edema at the site of injection, and was slowly absorbed into the blood; gallium citrate was more rapidly cleared than was the lactate from blood. Gallium was deposited chiefly in the bones and kidneys, less in spleen and bone marrow, and negligibly in other soft tissues. The skeleton retained 30–48% of the injected Ga citrate, 1–2% was excreted into the digestive tract for fecal excretion, and about 30–55% was excreted in the urine (Dudley *et al.,* 1950*b*). Gallium lactate, when injected intravenously into rabbits, is transported by the plasma to soft tissues and bone within a day; the kidneys and liver absorb appreciable amounts of gallium lactate, which is retained for one month. The deposition in the bones takes place 4 hours following injection, and the retention of Ga in the bones lasts for at least 3 months, the Ga finally being released by osteolysis, or bone demineralization. Similar patterns of Ga distribution were observed in the rat, mouse, and dog. The amount of Ga deposited in bone from parenteral injection of gallium citrate is independent of the route of administration and depends upon the dose within the range of 1–15 mg Ga/kg body weight (Dudley and Marrer, 1952). The ultimate deposition of Ga in bone occurs in any area of osteoblastic activity. Following inhalation of $GaCl_3$ aerosols containing 0.125–0.25 mg Ga/liter, Ga is retained in the alveoli and not absorbed (Stokinger, 1963). There are no reports of absorption of Ga through the skin.

Gallium citrate shows great affinity for neoplasms by its localization in a variety of solid tumors following parenteral injection. Numerous reports of localization of ^{72}Ga in human solid tumors have resulted in the use of ^{72}Ga in locating malignant tumors (Edwards *et al.*, 1970; Edwards and Hayes, 1970; Hayes *et al.*, 1970; Vaidya *et al.*, 1970). Gallium nitrate is active in inhibiting the growth of solid tumors in rodents (Hart *et al.*, 1971). The mechanism for Ga localization and retention in solid tumors and its inhibition of tumor growth is not known. Excretion of dietary cationic inorganic Ga salts is mainly fecal, but gallium lactate and citrate are excreted in both feces and urine. Parenterally injected Ga compounds are excreted mainly in the urine; only about 7.8% of a dose of $GaCl_3$ and 3.3% of a dose of Ga citrate are excreted in the feces. Parenterally administered gallium citrate (100 mg Ga/kg) was retained by rabbits for 10 days; 85% was excreted in the urine over a period of time. There are no reports of any evidence for the presence of a homeostatic mechanism for Ga in mammals.

Toxicity

Orally administered gallium chloride or citrate are not harmful, owing to its hydrolysis in the digestive tract and to the poor absorption of the hydrolyzed product, $Ga(OH)_3$. Rats fed $GaCl_3$ at dietary levels of 500 or 1000 ppm Ga for 26 weeks exhibited no observable toxic symptoms. Autopsy revealed no Ga in soft tissues and only minute traces in bone. However, parenterally injected Ga salts showed variation in the toxicity levels (LD_{50} values) in different species, depending upon the mode of administration. The toxicity of Ga compounds is summarized in Table 3-2. Intravenous administration of $GaCl_3$ or $Ga(NO_3)_3$ was highly toxic, and sometimes immediately lethal, because of colloidal $Ga(OH)_3$ formation in the blood. The toxicology of Ga has been reviewed by Browning (1969) and Stokinger (1963). Mice seem to be resistant to Ga toxicity, while dogs are susceptible. Dogs and rabbits intravenously injected repeatedly with small doses of Ga compounds showed a high mortality rate, although the total of these small doses was much less than that of a single lethal intravenous dose. Parenterally administered $GaCl_3$ and $Ga(NO_3)_3$ were more toxic than gallium citrate or lactate. Acute Ga toxicity symptoms in dogs are vomiting,

TABLE 3-2. Gallium Toxicity

Compound	Animal	Route	Toxicity	Dosage/kg body weight			
				Compound	Metal		
				mg	mg	mM	pT
Gallium metal	Rat	sc	MLD		110	1.58	2.80
Gallium chloride	Mouse	ip	LD$_{50}$	37	24.2	0.35	3.46
GaCl$_3$	Rat	oral	LD$_{50}$	4700	1850	26.53	1.58
	Rat	iv	LD$_{50}$	47	18.6	0.27	3.57
	Rat	sc	LD$_{50}$	306	121	3.20	2.70
	Rat	inhal	MLD	191[a]			
	Rabbit	sc	LD$_{50}$	245	97	1.39	2.86
	Rabbit	iv	LD$_{50}$	43	17.0	0.24	3.61
	Dog	iv	LD$_{50}$	41	16.2	0.23	3.63
Gallium nitrate	Mouse	ip	LD$_{50}$	80	13.4	0.19	3.72
Ga(NO$_3$)$_3$·9H$_2$O	Rat	ip	LD$_{50}$	67.5	11.3	0.16	3.79
	Rhesus monkey	iv	LD$_{100}$	100	27.3	0.39	3.41
Gallium nitrite	Guinea pig	oral	LD$_{50}$	3280	1100	15.8	1.80
Ga(NO$_2$)$_3$							
Gallium ammonium sulfate	Mouse	sc	LD$_{100}$	100	39	0.56	3.25
GaNH$_4$(SO$_4$)$_2$	Rat	sc	LD$_{100}$	200	75.9	1.09	2.96
Gallium citrate	Rat	sc	LD$_{50}$	220	59.3	0.85	3.07
Ga(C$_6$H$_5$O$_7$)	Mouse	sc	LD$_{50}$	2250	600	8.61	2.07
	Dog and goat	sc	LD$_{50}$	27.5–56.3	12.5	0.18	3.75
	Goat	iv	LD$_{100}$	56.3	15	0.22	3.67
Gallium lactate	Rat	sc	LD$_{50}$	540	112	1.61	2.79
Ga(C$_3$H$_5$O$_3$)$_3$	Rat	iv	LD$_{50}$	205	42.6	0.61	3.21
	Rabbit	sc	LD$_{50}$	438	91	1.3	2.88
	Rabbit	iv	LD$_{50}$	192	40	0.57	3.24
	Dog and goat	sc	LD$_{100}$	103	20	0.29	3.54
Gallium arsenide	Rat	oral	LD$_{50}$	4700	2265	32.5	1.49
GaAs							

[a]Exposure of 3 hr; number of mg/m^3.

diarrhea, bloody stools, and anorexia and weight loss; the urea-nitrogen level in the blood was increased to 300 mg/100ml, the hemoglobin level was reduced by about 40%, and the urine contained erythrocytes and much albumin. The autopsies showed swollen lymph nodes, marked nuclear fragmentation with necrosis of the lymphoid tissues, and marked renal damage, similar to that caused by Hg intoxication; bone marrow showed aplastic changes. Rats suffered from photophobia and blindness in addition to the above toxic symptoms (Dudley and Levine, 1949; Dudley *et al.*, 1950*a*). Chronic Ga toxicity symptoms in mice fed 5 ppm Ga as gallium citrate in drinking water from weaning until death were a slight depression of growth and decreased life span in the females (Schroeder and Mitchner, 1971*a*). Clinical trials with intravenously injected ^{72}Ga mixed with stable Ga as gallium citrate produced anorexia and nausea in man; itching, subcutaneous edema, dermatitis, and lymphopenia were followed by lowered total leukocyte counts; the gastrointestinal symptoms, dermatitis, and edema were attributed to the toxicity of the stable Ga (Stokinger, 1963).

Recent studies on Ga toxicity follwoing intraperitoneal administration of $Ga(NO_3)_3$ showed the following symptoms: pneumonitis, chronic inflammatory infiltration of the alveolar septa and capillaries, and renal tubular damage in mice; and chemical peritonitis, interstitial pneumonitis, and moderate damage and lesions of the liver and kidneys in rats (Hart *et al.*, 1971). Subcutaneous implantation of Ga metal or alloy in guinea pigs produced necrosis *in situ*. The mammalian kidney appears to be the main target of parenterally injected Ga salts. Renal tubular damage leads to uncompensated acidosis, decreased blood pH, and decreased serum Na and K levels; the levels of the blood sugar and urea nitrogen increased, followed by albuminuria and glycosuria. Gallium can also be a neuromuscular poison, since it causes hyperexcitability, blindness, and terminal paralysis.

The development of malignant tumors in mice which were fed 5 ppm Ga as citrate in life-term experiments, reported by Schroeder and Mitchner (1971*a*), is paradoxical when considered with numerous reports that parenterally injected $Ga(NO_3)_3$ inhibits growth of solid tumors in man and animals (Hart *et al.*, 1971).

There are no reports of mechanisms by which Ga can be detoxified in mammals. The lack of a homeostatic mechanism and the retention of Ga in the skeleton suggest that the Ga level in animals and man will increase with age, especially if they are subjected to chronic exposure to Ga compounds.

Indium (In)

Indium is widely distributed in nature in very minute quantities. There is no exclusive ore for In, but it occurs with Zn in sphalerite (ZnS), and in marmatite and christophite (FeS:ZnS), and with Sn in siderite. Flue dusts of Zn smelters contain about 0.5% In from which In may be extracted. The earth's crust contains about 0.5 ppm In, and seawater contains traces of In. Indium is reported to be present in traces in animal and human bones, but reports about specific amounts are inconsistent. Indium is not essential for mammals, and it is one of the most toxic metals.

Indium, mixed with highly refined Sb, As, and Ge, is used in transistors and infrared detectors. The surfaces of metals such as Ag and Cu, and of types of bearing alloys using Cd and Pb, are protected against corrosion by electrodeposition of In; In is used to harden certain alloys of Be and Cu, and also dental and bearing alloys. Indium, which reflects equally all spectral wavelengths of color, is used in motion picture screens, mirrors, and cathode oscillographs. In_2O_3 is mixed with sulfur to make yellow colored glass. Indium complexes such as ^{111}In–bleomycin are used in modern medicine to locate and identify cancers of the brain, neck, colon, rectum, ovaries, breast, and lungs, utilizing the unique property of indium to accumulate in a variety of malignant tumors (Thakur *et al.*, 1973).

Indium is a potential industrial health hazard; exposure to In during its refining, while electroplating, and while soldering transistor parts containing In, Sb, and As is considered hazardous. The toxic effects of In, which can be leached from In-plated silver vessels by liquid foods with acidic pH, have not been studied in detail.

Chemistry

Indium resembles tin in physical properties. It exhibits mono-, di-, and trivalence, but the predominant and stable form is the trivalent state. Indium forms stable cationic salts; it is more electropositive and less amphoteric than Ga. The cationic In salts are relatively stable but are hydrolyzed in aqueous solutions to hydrated In_2O_3 or $In(OH)_3$ above pH 4.0, although not to the same extent as are Ga salts. Indium forms halogen complexes such as $[InF_6]^{3-}$; the covalent complexes are geometrically tetrahedral. Indium ions bind to free and bound phosphate groups in biologic fluids and tissues; In binds to transferrin to a lesser degree than does Ga.

Metabolism

Indium is neither essential nor stimulatory in man or animals; it occurs widespread in minute traces in foods of plant origin, but there are no reports of the dietary intake of In by man. Radioactive [114]In was used in animal metabolism studies (Smith *et al.*, 1947, Smith and Scott, 1957), and the biochemistry and physiology of In was reviewed by Stokinger (1963) and Browning (1969).

In mammals the gastrointestinal absorption of In salts is poor; the absorption of In_2O_3 from physiologic doses is about 0.2–0.4% in dogs and rats. The absorption rate is a little higher for $InCl_3$ and $In_2(SO_4)_3$ (Vignoli *et al.*, 1946). The citrate complexes are absorbed better than are In^{3+} salts; subcutaneous injection of In citrate into rats results in 80–90% absorption into the blood within two days, whereas $InCl_3$ is slowly absorbed (Smith and Scott, 1957). Similar absorption phenomena are observed for In salts from the peritoneal cavity. Intravenous injection of In salts results in quick removal from the blood due to the binding of In^{3+} with transferrin and its distribution to the soft tissues and bone; ionic forms of In also bind to α globulin and albumin. Colloidal forms are not bound to plasma proteins but are engulfed by leukocytes and removed to the liver and spleen. In rats, following inhalation or intratracheal intubation, soluble In salts remain in the lung tissues and are slowly absorbed; about 50% of an administered dose is absorbed from the lungs in two weeks, and the rest being retained in the tracheobronchial lymph nodes for as long as two months.

The distribution of absorbed In in the soft tissues and bone depends on the mode of administration and on the nature of the In salt; a large percentage of the absorbed In accumulates in the bones. Following subcutaneous injection of indium citrate in rats, In is distributed evenly to the skin, muscle, and bones; after some days In is deposited in the adrenals and submaxillary glands with transient distribution in the spleen, kidneys, and liver. Intraperitoneal injection of In salts results in an initial high accumulation in the mediastinum and liver, and subsequent redistribution in spleen, kidneys, and bone. In rats given $InCl_3$ intraperitoneally, the content of In in the liver is twice that obtained from a similar intravenous dose. The kidneys absorb the highest amount of intramuscularly injected doses of In salts. Intravenously injected In accumulates initially in the kidneys to a large extent and in the liver to a smaller extent; later this renal and hepatic In^{3+} is redistributed, with ultimate deposition in the skin, bones, and muscle.

In mammals, the excretion of In^{3+} is both fecal and urinary, and it is slow; orally ingested and subcutaneously injected In salts which are transformed into colloidal forms are excreted mostly in the feces and only about 8% in the urine. In rats, the fecal excretion of intratracheally deposited In is about 50% of the dose within two months, and 8% in the urine during the same period, while the rest is retained. Following intravenous or intraperitoneal administration of In^{3+} to rats, 10% of the dose is excreted in 2–3 days, and 23% in 15 days, in the urine, and 6% in 2–3 days, and 16% in 15 days, in the feces (the rest being retained in skin and bones). If the In salts become hydrated and colloidal in the blood, they are quickly removed and deposited in the liver and spleen, and about 60% is excreted in the feces. In contrast to Ga^{3+}, the retention of In^{3+} in mammals is in the skin, bones, and muscles.

Toxicity

Soluble In salts are toxic to man and animals, and the toxicity of In^{3+} is higher by parenteral administration than by oral ingestion; the toxicity of In salts is summarized in Table 3-3. An ionic form of In prepared from $InCl_3$, suitably buffered to prevent the formation of $In(OH)_3$, is less toxic than an aqueous solution of $InCl_3$, which hydrolyzes to hydrated $In(OH)_3$ in bio-

TABLE 3-3. Indium Toxicity

| | | | | Dosage/kg body weight | | | |
| | | | | Compound | Metal | | |
Compound	Animal	Route	Toxicity	mg	mg	mM	pT
Indium	Mouse	sc	MLD		10	0.09	4.06
	Rat	oral	LD_{50}		4200	36.6	1.44
Indium oxide	Rat	ip	MLD	546	451	3.93	2.41
In_2O_3	Rat	ip	LD_{100}	955	790	6.88	2.16
	Rat	iv	MLD	100	82.7	0.72	3.14
Indium oxide, hyrated and stabilized	Mouse	iv	LD_{50}	0.4	0.323	0.0028	5.55
Indium chloride	Mouse	iv	LD_{100}	24.0	12.5	0.11	3.96
$InCl_3$	Mouse	sc	MLD	60	31	0.27	3.57
	Mouse	ip	LD_{50}	3	1.5	0.013	4.88
	Rat	oral	LD_{50}	4200	2110	18.4	1.74
	Rat	sc	MLD	10	5.3	0.046	4.34
	Rat	ip	LD_{100}	50	26	0.23	3.65
	Rabbit	sc	MLD	2.35	1.2	0.017	4.98
	Rabbit	ip	LD_{100}	8.85	4.6	0.040	4.40
	Rabbit	iv	MLD	0.64	0.33	0.0029	5.54
Indium nitrate	Mouse	oral	LD_{50}	3350	975	849	2.07
$In(NO_3)_3$	Mouse	ip	LD_{50}	7.5	2.2	0.019	4.72
	Rat	ip	LD_{50}	5.5	1.6	0.014	4.85
Indium sulfate	Rat	sc	LD_{50}	22.5	10	0.09	4.06
$In_2(SO_4)_3$	Rat	ip	LD_{100}	40.5	18	0.16	3.80
	Rat	sc	LD_{100}	28	12.5	0.11	3.96
	Rat	iv	LD_{50}	5.63	2.5	0.022	4.66
	Rabbit	oral	MLD	1800	800	6.97	2.16
	Rabbit	ip	LD_{100}	9.6	4.3	0.04	4.43
	Rabbit	sc	LD_{100}	2.6	1.1	0.096	5.02
		iv	MLD	0.67	0.3	0.0027	5.58
Indium citrate $In(C_6H_5O_7)$	Mouse	sc	LD_{50}	600	311	2.71	2.57
Indium antimonide InSb	Mouse	ip	LD_{50}	3760	1820	15.9	1.80

logic fluids (Castronovo and Wagner, 1971). Colloidal In_2O_3, prepared specially and stabilized with gelatin, NaCl, and NaOH at pH 7.4, is about 40 times more toxic than is regular In_2O_3. In mice the LO_{50} dose at 4 days was 0.323 mg In/kg. Symptoms of acute In toxicity are reduced food and water intake, localized convulsive motions, leg paralysis, nosebleed, pulmonary edema, necrotizing pneumonia, renal and hepatic damage, and lowered lymphocyte count. Symptoms of chronic In toxicity are more mild than are those of the acute toxicity, and include weight loss and degenerative changes in the urinogenital tract. Autopsy following death from chronic In poisoning reveals damage and inflammation in the spleen, adrenals, heart, and brain, necrotic changes in the convoluted tubules surrounding the glomeruli in the kidneys, and liver necrosis. The physiologic and toxicologic properties of In were studied by McCord *et al.* (1942), Harrold *et al.* (1943), and Downs *et al.* (1959*b;* 1960).

Ingested In_2O_3 is harmless to rats; animals fed 8% In_2O_3 in the diet for 3 months showed no harmful effects, but 2.4% $InCl_3$ in the diet caused a slight growth depression, and 4% caused marked depression. Indium chloride fed to mice at 5 ppm in drinking water throughout the life term affected their growth, but not their survival or mortality (Schroeder and Mitchner, 1971*a*). Indium sulfate fed to rats at 20–30 mg/day per rat for 30 days caused no toxic symptoms, but chronic feeding of the same amount over 80 days caused weight loss and roughened skin owing to indium deposition in the skin.

Inhalation of the dusts of insoluble In_2O_3 for 4 hours daily for 3 months by rats caused only a persistant inflammatory reaction with no fibrosis. Indium-coated silver disks implanted in the peritoneal cavity, under the skin, or intramuscularly in rats induced only foreign-body reactions and no specific symptoms associated with indium toxicity (Harrold *et al.,* 1943). Metallic In strapped to the bare skin in rats showed no irritant properties (McCord *et al.,* 1942).

Parenteral injections of In salts caused passive congestion of all organs, hepatic degeneration with foci of necrosis, severe renal necrosis, occasional spleen and pancreatic necrosis, hemorrhages of the lungs, and degenerative myocarditis. In mice $In(NO_3)_3$ toxicity caused myolipia in the liver, moderate spermatogenic arrest, bone-marrow hyperplasia, and congestion in the peritoneum and pleural cavity. In rats $In(NO_3)_3$ caused interstitial pneumonitis and inflammation of the hepatic serosa and parenchyma (Hart *et al.,* 1971).

Degenerative changes in the kidneys and liver following intravenous In intoxication depend upon the nature of the In compound; ionic In from $InCl_3$ at pH 3.0 is highly nephrotoxic to mice, with an LD_{50} dose of 12.5 mg In/kg at 4 days (Castronovo and Wagner, 1971). Damage to the kidney occurred in stages; swelling in the epithelium of the kidney tubules, occasional pyknotic nuclei, and eosinophilia were accompanied by extensive necrosis of the tubular lining, affecting mainly the proximal tubules. Liver damage was restricted to multiple small necrotic areas of parenchymal cells, and relatively little change occurred in other organs. Clinical symptoms of In toxicity are spasmodic breathing, weight loss, skin abnormalities, total inactivity in three days, and death in four days. Renal damage was minimal, but extensive damage to the liver, spleen, bone marrow, thymus, and lymph nodes occurred, and the reticuloendothelial system was greatly affected. Extensive hemorrhage, marked thrombocytopenia, and fibrin thrombi occurred in the liver. Loss of weight, inactivity, hind-leg paralysis by day 2, muscle tremors, clonic convulsions, and bleeding from the nose, mouth, and ear prior to death on day 3 were the outward symptoms. Loss of glycogen and fatty infiltration of the liver and lymphocyte depletion in the spleen were observed at autopsy. The entire toxic dose of the hydrated oxide was retained in the body at death, while only 60% of the toxic dose of ionized In was retained at death. However, when the reticuloendothelial system (RES) in mice was blocked by administering relatively less toxic $Ga(OH)_3$, the toxicity of In was less, owing to less accumulation of In in the RES (Evodkimoff and Wagner, 1972); thus the LD_{50} of $In(OH)_3$ was found to increase from 12 to 40 μg In/mouse. The greater toxicity of hydrated and colloidal In_2O_3 was tentatively attributed to an unusual activation or biotransformation of the insoluble oxide in macrophages or hepatic lysosomes following its phagocytosis. In addition, hepatic phagocytes accumulated and concentrated the In in the reticuloendothelial cells, from which In leaked out and damaged the hepatic parenchymal cells. This activation is unusual, since hepatic cells are generally involved in detoxicating or inactivating a potential toxicant.

Indium chloride is a powerful "direct calcifier," causing topical calcification wherever introduced into connective tissue (Selye, 1962), but histamine liberators such as polymixin can prevent the calcifying toxic effect of indium (Seramivof and Tuchweber, 1965).

Embryopathic and teratogenic effects of In intoxication were observed in pregnant golden hamsters (Ferm and Carpenter, 1970). With intraperitoneal injection of 2.0 mg In/kg, there was a high percentage of resorption of fetuses; at 0.5–1.0 mg In/kg, teratogenic effects included skeletal defects in the limb extremities, such as polydactaly. Indium appears to be a potent site-specific teratogen which affects enzyme-dependent limb-bud differentiation. Intraperitoneal administration of ferric dextran at 50 mg/animal prevented the embryopathic and teratogenic effects of In intoxication (Ferm, 1970); ferric dextran, which accumulates in the reticuloendothelial system, reduces indium toxicity, especially the calcifying effect and fatal hepatic necrosis (Selye *et al.*, 1962; Gabbiani *et al.*, 1962).

There are no specific detoxication mechanisms for In; similarly there is no evidence for any homeostatic control for In. Although In accumulates in neoplasms, it does not inhibit the growth of solid tumors in man and animals (Hart *et al.*, 1971).

Thallium (Tl)

Thallium is widely distributed in nature, mainly in the minerals crookesite ($TlCuAg_2Se$), lorandite ($TlAsS_2$), and orbaite ($TlAs_2SbS_5$). Commercially, Tl is obtained from Cd, Pb, and Zn smelters and refineries. The earth's crust contains about 0.3 ppm Tl, while the Tl content of seawater is not known; Tl does not occur naturally in plants or animals. The amount of Tl present in humans has not been accurately determined, but Tl is known to accumulate in the human body with age. Thallium is not essential to mammals, but is a highly potent toxicant.

Thallium is used to make stainless steel alloys with Ag, corrosion-resistant alloys with Pb, and Hg–Tl alloy for switches and closures to operate at subzero temperatures and in low-temperature thermometers. Thallium salts are used as catalysts in organic reactions and firecrackers. Thallium oxide, with its high coefficient of refraction, is used in optical lenses, artificial gems, and flint glass; mixed thallium bromide and iodide crystals are used in infrared optical instruments and in making prisms for spectrometers; thallium oxysulfide is used in photoelectric cells because of its great sensitivity; Tl_2SO_4 is a good rodenticide, and Tl_2CO_3 is an effective fungicide; some Tl salts are used as depilatory agents. Thallium and its salts are considered to be industrial health hazards in coal, from its smoke as well as in the mining and refining of coal; Pb, Cd, and As contaminants potentiate Tl toxicity. Most of the toxicology data about Tl in animals and humans were obtained from studies of cases of inadvertant or criminal misuse of Tl rodenticides in poisonings; predator poisoning leads to the recurrent finding of Tl in animal tissues, as does the grazing of herds near Tl outcroppings.

Chemistry

Thallium is a soft metal resembling Sn in its physical properties; it is reactive chemically and exhibits mono- and trivalence. Thallous (Tl^+) salts are more basic and stable than thallic (Tl^{3+}) salts, but Tl^{3+} forms stable halogen complexes such as $[TlF_3]^{3+}$. Complex Tl^+ salts are planar, while Tl^{3+} salts are tetrahedral in configuration. Thallic salts are powerful oxidizing agents and can be reduced to the Tl^+ state by Sn^{2+} and Fe^{2+} salts; Tl does not exhibit amphoteric character. Thallium is the most basic among subgroup IIIA metals and, unlike the other Group III metals, soluble Tl^+ salts are not easily hydrolyzed to the soluble TlOH in biologic fluids at or above pH 4.0. Situated between Hg and Pb in the periodic chart, Tl has an undefined affinity toward thiol groups and S-containing ligands; however,

Tl does not block SH groups in some thiol-containing enzymes and proteins. The affinity between Tl^+ and COOH, NH_2, NH, and PO_4 groups is minimal. Thallium resembles and behaves like potassium in its metabolism, accumulation in erythrocytes, active transport, and excretion (Prick *et al.*, 1955). Monovalent Tl resembles K^+ in its ionic size and charge.

Metabolism

Thallium is neither essential nor stimulatory in either man or animals. It is the most highly toxic cumulative cation. Plants can absorb Tl from Tl-treated soils; livestock, especially pigs, were reported to have been poisoned when reared on Tl-rodenticide-treated soils. Thallium rodenticides at one time were extensively used, but when the data on the high toxicity of Tl became known they were discontinued. There are no reports about the Tl content of different foods. It is presumed that the dietary intake of Tl by man and animals is negligible, but ingestion of Tl from contaminated food, and from unwashed hands of workers in the Tl industry, is a potential risk. The biochemistry and metabolism of radioactive Tl in chicks, rats, and man were studied using radioactive Tl (Barclay *et al.*, 1953; Thyresson, 1951).

In mammals the gastrointestinal absorption of soluble Tl salts is rapid and complete; insoluble salts such as Tl_2S or Tl_2O_3 are not absorbed. The absorption of soluble Tl salts from the peritoneal cavity and from the sites of subcutaneous and intramuscular injections is equally rapid and complete. Following inhalation of thallium oxides and salts, Tl is rapidly absorbed from the mucous membranes of the respiratory tract, mouth, and lungs. Thallium salts are also absorbed through the skin (Thyresson, 1951). After absorption into the blood, and following intravenous injection, Tl^+ is rapidly distributed to the tissues; part of the Tl^+ is absorbed into the erythrocytes. The transport of Tl in the blood appears to be in the ionic form; it has been shown that Tl^+ does not combine *in vitro* with the serum albumin or other proteins. Thallium easily passes through the blood-brain and placental barriers. When fed Tl salts, lactating animals secrete Tl^+ in their milk. The quick absorption and transport of Tl^+ is demonstrated by peak blood levels, and its presence in the urine, 2 and 4 hr, respectively, after oral ingestion. Skin absorption is also a hazard.

Studies of ^{204}Tl showed the widespread distribution of administered Tl in animal tissues (Lie *et al.*, 1960). In rats, the kidneys contained the largest amounts of Tl, followed by the bones, liver, lungs, spleen, and brain (Lund, 1956; Downs *et al.*, 1960). The gastric and intestinal mucosa, pancreas, salivary glands, hair, and eye lens also contained appreciable amounts of Tl. Initially the central nervous system contained trace amounts of Tl, but the levels increased a few days after Tl^+ administration and was higher than the Tl^+ level in the liver. Similar observations were made about Tl^+ distribution following Tl intoxication in human tissues (Reines and Leonard, 1932). Retention of Tl^+ in organ tissues is prolonged; 45% of an ingested dose of Tl_2CO_3 was still present in the human body after 25 days, and Tl^+ could be detected in the cerebrospinal fluid (Prick *et al.*, 1955). Similar Tl^+ retention was recorded in dogs and geese. Retention of Tl in the hair follicles during hair growth was demonstrated by autoradiograph (Thyresson, 1951).

In mammals Tl^+ excretion is slow and prolonged, although Tl^+ appears in the urine a few hours after absorption; Tl^+ excretion is mainly urinary and partly fecal, with lesser amounts in the sweat, milk, and tears. Fecal Tl^+ excretion four days after ingestion, and its complete absorption from the digestive tract, suggest an enterohepatic circulation for Tl^+. The amount of Tl^+ excreted in the urine is about three times that excreted fecally. Excretion of parenterally administered Tl^+ by humans is prolonged; it continues for 3 months in urine, and, following ingestion, for 35 days in feces. Progressively smaller quantities of Tl^+ are excreted with time; the daily rate of urinary excretion was found to be 3.2% of the amount remaining in the body (Barclay *et al.*, 1953). Retention of Tl in bones, and to a smaller extent in the central nervous system and other tissues, increases with age, indicating the absence of effective homeostatic mechanisms for Tl.

Toxicity

The toxicology of Tl compounds has been studied in detail in both man and animals. Symptoms vary with dose, age, and acuteness of Tl intoxication; polyneuritis, gastrointestinal disorders, retrobulbar neuritis, alopecia, and encephalopathy are the general symptoms. Symptoms of acute poisoning depend upon the route of administration (Arena, 1974); large oral doses cause gastrointestinal hemorrhage, gastroenteritis, metallic taste, salivation, nausea, and vomiting. Neurologic symptoms, delirium, hallucinations, convulsions, a tingling pain in the extremities, muscular weakness, and coma appear in 2 or 3 days irrespective of the mode of administration. Death from respiratory failure occurs in 5–7 days. The extreme sensitivity of the lower extremities to touch is a characteristic feature of Tl intoxication in humans; the victim cannot bear even bedclothes on his feet; vasomotor disturbances include puffiness of the cheeks, eyelids, and lips; ataxia, blindness, and failure of individual senses are also common in humans. Symptoms of chronic Tl intoxication are milder, including alopecia, cardiac and renal disorders, endocrine disorders, psychoses, encephalitis, and paralysis of the extremities. Alopecia starts 10 days after ingestion of Tl salts and is complete in 30 days; axillary and facial hair are spared. Cardiac disorders include hypertension, irregular pulse, and angina-like pain. Albuminuria and hematuria are the symptoms of renal dysfunction (Reed *et al.*, 1962).

The physiology and toxicology of Tl have been reviewed (Oehme, 1972; Browning, 1969; Stokinger, 1963; Truhaut, 1959; Prick *et al.*, 1955; Heyroth, 1947; Gettler and Weiss, 1943). The toxicity of Tl is summarized in Table 3-4; guinea pigs and rabbits are the most susceptible laboratory animals, and male rats appear to be more susceptible than female rats. Thallium salts are more toxic when administered parenterally than when given orally, and thallium chloride appears to be the most toxic of the Tl salts studied. Trivalent Tl compounds are somewhat less toxic than are Tl^+ compounds; the parenteral LD_{50} for Tl_2O_3 ranges from 9 to 92 mg Tl/kg for different animals. However, guinea pigs are equally susceptible to both Tl^+ and Tl^{3+} compounds. In humans the lethal dose was computed to be 8–10 mg Tl/kg, on the basis of the death of a 10-year-old boy who was given a therapeutic dose of 166 mg thallium acetate (Browning, 1969).

Thallium acts as a mitotic agent and a general cellular poison. The cellular accumulation of Tl in muscle and other tissues causes derangement of normal cellular metabolism, with resultant toxicity. High Tl^+ concentration in the blood agglutinates erythrocytes and lyses them following Tl^+ accumulation within the erythrocytes. Thallium intoxication can also result in severe damage to gastrointestinal mucosa causing hypoacidity or anacidity. Although the nature of the transient binding between Tl and the nervous tissues is not definitely known, Tl causes swelling and fatty degeneration of the nervous tissue; the damage is severe in the neurons. Neuritis of the vagus causes intestinal paralysis; peripheral neuritis of cranial nerves causes paralysis of ocular muscles, ptosis, and amblyopia (Moeschlin, 1965). Kidney tissue suffers both functional and structural damage. Histological studies show tubular damage involving cytoplasmic protrusions into the lumen (Hermann and Bensch, 1964). The deposition of Tl in hair follicles causes the total atrophy of the follicles and their replacement by scar-tissue collagen or fat; medium doses of Tl^+ inhibit keratinization and damage the sebaceous and sweat glands (Moeschlin, 1965). The hair of human Tl-poisoning victims undergoes changes, with a higher deposition of melanin pigments in the hair (Curry *et al.*, 1969). Moeschlin (1965) reported the teratogenic effect of Tl^+ in humans; chronic Tl intoxication in the first three months of pregnancy caused deformities in the newborn babies. If the intoxication takes place after the third month of pregnancy, the central nervous system of the baby is damaged, showing lesions which cause polyneuritis, anaurosis, and pronounced cachexia in the newborn.

TABLE 3-4. Thallium Toxicity

| Compound | Animal | Route | Toxicity | Dosage/kg body weight | | | |
| | | | | Compound | Metal | | |
				mg	mg	mM	pT
Thallium oxide	Rat	oral	MLD	20.8	20	0.10	4.01
Tl_2O	Rat	oral	LD_{50}	40.6	39	0.19	3.72
	Rat	ip	LD_{50}	75	72	0.35	3.45
	Rat	ip	MLD				
	Rabbit	oral	MLD	31.2	30	0.15	3.83
	Rabbit	ip	MLD	62.4	60	0.29	3.53
	Rabbit	iv	MLD	40.6	39	0.19	3.72
	Dog	oral	MLD	31.2	30	0.15	3.83
	Dog	ip	MLD	31.2	30	0.15	3.83
Thallic oxide	Guinea pig	oral	MLD	5.2	5.0	0.024	4.61
Tl_2O_3	Guinea pig	ip	MLD	34	31	0.15	3.82
Thallium chloride	Mouse	oral	LD_{50}	24	20.5	0.10	4.00
TlCl	Mouse	ip	MLD	5.66	3.7	0.025	4.74
	Mouse	ip	LD_{50}	24	20.5	0.10	4.00
	Rat	ip	MLD	6.9	4.5	0.022	4.66
Thallium bromide	Mouse	oral	MLD	29	20.9	0.102	3.99
TlBr							
Thallium iodide	Mouse	oral	MLD	28	17.3	0.086	4.07
TlI							
Thallium nitrate	Mouse	oral	LD_{50}	32.5	25	0.12	3.91
$TlNO_3$	Rat	sc	LD_{50}	26	20	0.095	4.01
	Rat	sc	LD_{100}	20	15.3	0.074	4.13
	Rabbit	iv	LD_{100}	14	10.7	0.052	4.28
	Dog	oral	LD_{100}	45	34.5	0.17	3.77
	Dog	sc, iv	LD_{100}	15–20	13.5	0.066	4.18
Thallous sulfate	Mouse	oral	LD_{50}	29	23.2	0.11	3.94
Tl_2SO_4	Rat	oral	LD_{100}	25	20	0.098	4.01
	Rat	oral	LD_{50}	15.8	12.6	0.062	4.21
	Man	oral	MLD	3	2.4	0.012	2.93
Thallous carbonate	Rat	sc	MLD	18	7.9	0.039	4.41
Tl_2CO_3							
Thallium selenite	Rat	oral	MLD	50			
Tl_2SeO_3							
Thallium acetate	Mouse	oral	LD_{50}	35.2	27.3	0.13	3.87
$Tl(C_2H_3O_2)$	Rat	oral	MLD	37.4	29	0.14	3.85
	Rat	oral	LD_{50}	41.3	32	0.16	3.81
	Rat	ip	MLD	25.8	20.0	0.098	4.01
	Rat	ip	LD_{50}	29.6	23.0	0.11	3.95
	Guinea pig	oral	MLD	15.5	12.0	0.059	4.23
	Guinea pig	ip	MLD	9.03	7.0	0.034	4.47
	Rabbit	oral	MLD	24.5	19.0	0.094	4.03
	Rabbit	ip	MLD	16.8	13.0	0.063	4.20
	Rabbit	iv	MLD	25.8	20.0	0.098	4.01
	Dog	oral	MLD	25.8	20.0	0.098	4.01

There is no specific and concrete mechanism to explain Tl toxicity. Thallium permeates cells of all tissues since it is similar to potassium in ionic size and charge, and cell membranes may not distinguish between the two ions. Available evidence suggests a vague explanation: general cellular poisoning by combined Na–K antagonism owing to the rapid diffusion of Tl ions through membranes to destroy any specificity of Na and K as inter- and intracellular ions, respectively. Interference with sulfur metabolism or with sulfur-containing compounds or enzymes is also possible. Thallium inhibits *in vitro* the aerobic respiration of isolated cells of rat brain, kidney, and skin; at relatively high concentrations succinic dehydrogenase activity is inhibited. Dietary supplements of cystine, methionine, and betaine afford some protection for rats against experimental Tl intoxication. Sulfur-containing synthetic compounds such as diphenylthiocarbazone also give protection (Stavinoha *et al.*, 1959). These compounds are believed to protect the active sulfhydryl groups of proteins by forming reversible disulfide bonds at sites where thallium is active. Selenium and thallium interact in the tissues, reducing the toxicity of each other; soluble selenates prevent or greatly reduce mortality in rats fed toxic doses of Tl compounds (Rusiecki and Brzezinski, 1966). Thallium inhibits the pulmonary excretion of volatile Se compounds in the rats and increases Se retention in the liver and kidneys in a bound form (Levander and Argrett, 1969). The relationship between this interaction and Tl interference with the metabolism of S-containing compounds is not clear.

There are no reports of specific detoxication mechanisms for Tl; Tl^+ forms a soluble biologic complex with diphenylthiocarbazone when it is administered to Tl^+-intoxicated rats, and this complex is eliminated in the urine very rapidly (Doak *et al.*, 1965). It is not known whether such Tl^+ complexing can occur with S-containing compounds such as cystine or methionine, which are normally present in the tissues. A review of the pathology, diagnosis, and treatment of Tl toxicity is provided by Arena (1974).

SUBGROUP IIIB METALS

The metals of subgroup IIIB exhibit variable valence. Yttrium resembles the heavier lanthanides and is often classified with them. The lanthanides form a special group of inner-transitional elements with highly similar chemical properties which render the chemical separation of these metals very difficult. The actinides form a characteristic group owing to their high radioactivity and short half-lives; otherwise, they resemble subgroup IIIB metals in their chemical properties. Although the soluble salts of the metals of this subgroup readily hydrolyze in aqueous solutions to their respective hydroxides, the separated hydroxides (with the exception of those of Sc and Y) do not undergo olation or oxalation, this in distinction to the salts of subgroup IIIA metals. However, these separated hydroxides undergo polymerization to give rise to radiocolloidal forms. Scandium, yttrium, and the lanthanides are considered to be nontoxic or harmless due to their poor absorption. The chemical toxicity of actinides has not been fully assessed, but it should be equivalent to the rest of the subgroup. Their permanent retention in the skeleton potentiates their high radiation toxicity. Although a few of subgroup IIIB metals may be stimulatory (see Appendix), none are essential nutrients for mammals.

Scandium (Sc)

Scandium occurs widespread in very minute quantities in over 800 mineral ores. The rare mineral thortveitite contains scandium as the principal metal; scandium also occurs with rare-earth metals in yttervite, cerite, and orthite, and with tungsten in wolframite and davidite. Scandium is obtained mostly as a by-product from the residues of tungsten and uranium ores. The earth's crust contains about 5 ppm Sc; the Sc content of seawater is not known. Scandium has been reported to be present in minute traces in the skeleton of man and animals.

Scandium has been little used industrially, since it is an expensive metal and was obtained in 99% purity only in 1960. It is used as a tracer to study the movement of oil and sand. Scandium is of interest to aircraft and spacecraft designers, being a very light metal with a melting point higher than that of aluminum. Scandium is not considered an industrial health hazard; scandium poisonings are rare, because scandium and its salts are not used extensively, and because it has low toxicity.

Chemistry

Scandium exhibits trivalence and forms cationic salts which are hygroscopic and soluble in water. Scandium does not exhibit amphoteric character, unlike the metals of subgroup IIIA, but forms complex halogen compounds such as $(ScF_4)^-$, $(ScF_5)^{2-}$, and $(ScF_6)^{3-}$. Scandium salts undergo hydrolysis and olation in aqueous solution. With increasing pH the olated products undergo polymerization to give charged and hydrated polymers; $Sc(OH)_3$ on exposure to air forms a hard horny mass:

$$H_2O + ScCl_3 \longrightarrow Sc(OH)_3 + 3HCl$$

$$\downarrow \text{olation}$$

$$[Sc(H_2(H_2O)_5 \cdot OH_2]^{3+} \rightleftharpoons [Sc(H_2O)_5OH]^{2+} + H^+$$

$$\downarrow$$

$$[Sc(H_2O)_4OH]_2^{4+}$$

Scandium forms stable chelates with hydroxy acids, such as citric acid, EDTA, and DTPA. It can bind to serum proteins, especially γ globulins, nucleoproteins, amino acids, and nucleic acids, and it has great affinity to phosphate-containing ligands.

Metabolism

Scandium is neither essential nor stimulatory in man or animals. It is present in few plant tissues because it is poorly absorbed by the root system; plants thus serve as an effective barrier against the transfer of Sc from the soil to man and animals. However, neutron activation analysis showed the presence of Sc in trace amounts in commercial feeds for laboratory animals (Hutcheson et al., 1975a,b). The daily intake of Sc by humans is very minute or negligible. The few studies done on Sc metabolism in man and animals involved the use of radioactive Sc. There are no reports about Sc involvement in biochemical or enzyme systems, and there are very few reports about the distribution of Sc and the retention of stable Sc in mammals. The chemistry, biology, and technology of Sc was reviewed by Horovitz (1975).

The absorption of dietary Sc salts is so negligible that Sc_2O_3 is used as a nutritional marker in man and animals (Venugopal et al., 1974; Hutcheson et al., 1975a; Luckey et al., 1977). In mammals the gastrointestinal absorption of soluble and simple Sc salts is very poor, less than 0.05% of a physiologic dose, owing to the formation of colloidal or radiocolloidal forms of olated $Sc(OH)_3$ complexes. However, scandium citrate and other simple complexes of Sc are absorbed in appreciable amounts. The absorption of Sc salts from sites of intramuscular and subcutaneous injections and from the peritoneal cavity is also very poor; Sc salts remain in the lungs without appreciable absorption. Intravenously injected Sc salts undergo hydrolysis and olation to form insoluble colloidal particles of hydrated Sc oxides or hydroxides; the ionized Sc salts bind to γ globulins and are transported from the blood into the kidneys and liver. The colloidal form of $Sc(OH)_3$ or scandium phosphates are phagocytized and taken into the liver and the reticuloendothelial system before final deposition in the skeleton. About 33% of an intravenous dose of $^{46}ScCl_3$ in young cattle remained in the blood after 10 hours (Miller and Byrne, 1970), whereas no ^{46}Sc was detected in the urine or blood following ingestion of ^{46}ScCl$_3$. Ogg *et al.* (1968) used ^{47}Sc to measure gastrointestinal absorption of Ca in humans; ^{47}Ca decays into daughter ^{47}Sc, and the ratio of ^{47}Ca to ^{47}Sc in both feces and the diet gives a measure of ^{47}Ca absorption.

Excretion of Sc salts in mammals is predominently fecal. In rats, orally ingested ^{47}Sc salts are fully recovered in the feces (Pearson, 1966). Intravenous injection of ^{47}Sc resulted in no significant urinary excretion. Part of it was retained permanently in the skeleton, and the rest was excreted in the feces after a transient retention in the liver and the reticuloendothelial system, suggesting an enterohepatic circulation for Sc.

Toxicity

The pharmacology and toxicology of ScCl$_3$ was studied by Haley *et al.* (1962*a*). The symptoms of Sc intoxication in mice following oral or parenteral administration of lethal doses were immediate defecation, abdominal distention, depressed respiration, and sedation. Depression of growth rate was recorded in mice when ScCl$_3$ was added to drinking water at 5 ppm Sc (Schroeder and Mitchner, 1971*a*). However, ScCl$_3$ was scarcely toxic to rats when fed as 1% of their diet for two weeks, and mice did not show any toxic symptoms when they were fed 120 ppm of Sc as Sc$_2$O$_3$ in their diet for three generations. Necroscopy showed no tumor formation, and growth rates, reproduction, lactation, and hematology were normal (Hutcheson *et al.*, 1975*b*). The lethal dose of ScCl$_3$ by oral and peritoneal administration is given in Table 3-5 (Haley *et al.*, 1962*a*); intravenous administration of ScCl$_3$ at 0.2–0.5 mg Sc/kg body weight to cats caused depressant action on all systems, and resulted in death from respiratory paralysis and cardiovascular collapse. Transient ocular irritation was observed when ScCl$_3$ solution was applied to the corneas of rabbits. Reports of tumor formation in

TABLE 3-5. Scandium Toxicity

Compound	Animal	Route	Toxicity	Dosage/kg body weight			
				Compound	Metal		
				mg	mg	mM	pT
Scandium chloride	Mouse	oral	LD$_{50}$	3980	1190	26.5	1.58
ScCl$_3$	Mouse	ip	LD$_{50}$	750	223	4.96	2.30
	Mouse	iv	LD$_{50}$	93	28	0.62	3.24

mice following the ingestion of 5 ppm $ScCl_3$ in drinking water (Schroeder and Mitchner, 1971*a*) were not confirmed by Hutcheson *et al.* (1975*a*); no toxic effect of Sc on other organs of mammals has been reported.

There are no reports about homeostatic or detoxication mechanisms for scandium, but its poor absorption owing to the formation of insoluble and hydrated scandium oxides and hydroxides in the digestive tract can be considered a poorly defined homeostatic control mechanism. Their poor absorption classifies Sc compounds as nontoxic; oral toxicity in mice (LD_{50} value in 7 days, 4.0 g/kg) is perhaps due to mass effect, i.e., the rupture or blocking of the gastrointestinal tract due to the accumulation of inert salts which are either poorly absorbed or poorly metabolized.

Yttrium (Y)

Yttrium occurs with the rare-earth metals in minerals such as xenotime, yttrialite, yttrocrasite, samarskite, gadolinite, bastnasite, and monazite sand. Yttrium-90 occurs in radioactive fallout and in nuclear fission reactions. Lunar rock samples contain a relatively high Y content. Monazite sand contains 3% Y and about 0.2% bastnasite. The earth's crust contains 31 ppm Y; the Y content of seawater is not known but should be very low. The amount of Y present in the adult human body is not known definitely, but traces of Y have been detected in the bones of man and animals.

Yttrium is currently in use industrially as a deoxidizer for vanadium and other nonferrous metals; Y in small amounts (0.1–0.2%) reduces the grain size in Cr, Mo, Zr, and Ti, increases the strength of Al and Mg alloys, increases the ductility of Fe, and functions as a nodulizer in nodular cast iron. Yttrium is used in laser instruments and is a catalyst in ethylene polymerization. Yttrium oxide imparts shock resistance and low expansion characteristics to glass and ceramics. Y_2O_3–Eu phosphors give red color in color television tubes. Yttrium–iron garnet ($Y_3Fe_5O_{12}$) is used in microwave filters and in transmitters and transducers for acoustic energy instruments, and yttrium–aluminum garnet ($Y_3Al_5O_{12}$) is valued as a gem stone. There are no specific therapeutic uses for Y salts. Yttrium and its salts are not considered to be industrial health hazards, and cases of Y poisoning are rare except for experimental toxicology studies. The potential for increased environmental exposure to Y salts is growing, owing to the increased use of Y alloys and compounds. The chances of Y poisoning in man through dietary sources is negligible, although very minute traces of [91]Y occur in barley and wheat as a result of radioactive fallout. The discrimination against Y absorption by root systems prevents accumulation of stable Y in plant tissues (Klechkovsky, 1957). Plants thus provide an efficient barrier against the translocation of Y from the soil to man and animals.

Chemistry

Yttrium exhibits trivalence and forms stable cationic salts; Y resembles the heavier lanthanides (from samarium to lutetium) in their chemical properties; $Y(OH)_3$ is less basic than the hydroxides of the heavy lanthanides, but more basic than most other trivalent metal hydroxides, and it is not amphoteric. In aqueous solution at physiologic pH, YCl_3 undergoes hydrolysis to insoluble $Y(OH)_3$; the solubility of $Y(OH)_3$ is about 10^{-6} M; $Y(OH)_3$ can undergo olation and polymerization which is similar to, but less than, that of Sc salts. $Y(OH)_3$ becomes radiocolloidal at concentrations much below the solubility limit. The radiocolloidal behavior of Y at 10^{-11} M concentration is similar to that of the lanthanides; at this concentration, $Y(OH)_3$, formed from the hydrolysis of soluble Y salts, separates out as a filtrable and centrifugable disperse phase instead of remaining in solution. This tendency of Y^{3+} to polymerize at the pH of blood and other tissues contributes significantly to the chemical and biologic behavior of Y in tissues. *In vitro* studies on the effect of concentration and pH on the hydrolysis and radiocolloidal behavior of Y salts showed that as the concentration of Y^{3+} decreases from 10^{-2} M to 10^{-8} M, the pH at which half the Y is removed by centrifugation decreases from 7.0 to 4.0. At much lower concentrations the radiocolloidal formation is complete below the pH at which higher concentrations of Y^{3+} just begin to precipitate as $Y(OH)_3$ (Schweitzer and Scot, 1955). The chelation complexes of Y with citrate, EDTA, diethylenetriaminepentaacetic acid (DTPA), and nitrilotriacetic acid (NTA) are quite stable and are not hydrolyzed. Ionic Y^{3+} has been shown *in vitro* to bind with nucleoproteins, γ globulin, phospholipids, amino acids, and some purified enzymes such as ATPase. Yttrium has great affinity for phosphate-containing ligands.

Metabolism

Yttrium and its salts are neither essential nor stimulatory in man and animals. However, if tumor formation were considered to be a form of stimulation, Y could be considered a stimulatory agent in mice. Algae and some microorganisms are reported to accumulate Y. Yttrium is present in animal bones (0.49 μg Y/g), but not in other tissues (Brooksbank and Leddicote, 1953); human bones contain similar amounts of Y (Leddicote and Tipton, 1958). The daily dietary intake of Y by man is not definitely known. The metabolism of ^{91}Y has been studied extensively in animals due to the presence of ^{91}Y in nuclear fallout, and several reports are available (Kyker and Anderson, 1956). As the most readily available and most easily studied, Y is the exemplar of the lanthanides; it is remarkably like the heavier lanthanides.

In mammals, the gastrointestinal absorption of soluble Y salts is very poor, about 0.05% of a physiologic dose (J. G. Hamilton, 1947). The poor absorption is owing mainly to the hydrolysis of Y salts into precipitable, colloidal, or radiocolloidal forms of $Y(OH)_3$ or YPO_4 in the digestive tract. Soluble citrate Y chelates and other chelates are more easily absorbed. The absorption of Y from the sites of intramuscular and intraperitoneal injections is very poor for YPO_4 (Norris *et al.*, 1956) and YF_3 (Mayer and Morton, 1956); the insoluble Y salts are completely retained at the sites. The mobilization of soluble Y salts such as YCl_3 is slow and depends upon the concentration: the mobilization in rats is about 55% in three days for concentrations of 10^{-8}–10^{-11} M Y; there is no mobilization at levels of 10^{-4} M (Kyker, 1956). Following inhalation, insoluble Y salts are not fully absorbed, part being fixed permanently in the lungs, while the remainder enters the lymphatic system. Soluble Y compounds behave like intravenously injected Y salts. Yttrium salts are bound to γ globulins or albumin in radiocolloidal forms after absorption into the blood and following intravenous injection; Y^{3+} is not present in the blood. In rats Y does not enter the erythrocytes, although some Y compounds are attached to the surface of erythrocytes. The transport of Y out of the blood is rapid and follows a four-term exponential equation. The colloidal or radiocolloidal Y-protein complexes are removed by phagocytoses in the reticuloendothelial system and liver. The uptake of Y is light in other soft tissues, but Y deposition in the bone marrow and bone is heavy, and Y retention in the bone is prolonged. The distribution of Y in tissues is as follows: bone > liver > kidneys > spleen > lymph nodes > lungs > pituitary > adrenals > thyroid > heart. The mobilization and loss of Y from bone is slow. Yttrium metabolic patterns in the rhesus monkey are similar to those of other species following intravenous administration (Daigneault, 1963); the deposition of ^{91}Y in the bone is mostly in the areas of bone resorption and not in the osteoid matrix (Jowsey *et al.*, 1958). However, when large doses of stable YCl_3 (60 mg/kg body weight) were injected intraperitoneally into rats, there was no increase in Y deposition in the bone (Macdonald *et al.*, 1952), possibly owing to poor absorption. Prolonged retention of Y in the bone is due to strong electrostatic binding with carbamyl and sulfate groups, as shown by the *in vitro* binding of Y^{3+} with an acidic glycoprotein and with chondroitin-sulfate–protein complex, isolated from bovine cortical bone (Williams and Peacocke, 1967). Yttrium retention in the tissues can be decreased by administering chelating compounds. Absorbed Y compounds do not cross the placental membrane in pregnant rats and cattle but are secreted into the milk; colostrum contains relatively large amounts of Y when administerd to the mother (Hood and Comar, 1956).

Excretion of Y is both fecal and urinary, and depends upon the form of dosage. Tracer studies in rats with carrier-free ^{88}Y, parenterally injected or inhaled in microgram levels, reveal urinary excretion; when mixed with inactive Y carrier, the excretion is predominantly fecal (Wenzel *et al.*, 1969). Excretion of ingested Y is mainly fecal due to poor absorption. The excretion rate for injected Y in rats is slow; only 40% of an administered dose is excreted in a week, one-third in the urine and the rest in feces.

Toxicity

The physiology and toxicology of Y was reviewed with rare-earth metals (Haley, 1965; Kyker, 1962; Stokinger, 1963). Symptoms of acute Y toxicity in experimental animals are anorexia, asthenia, and a progressive depression of general activity; death is due to cardiac and respiratory failure. Small but chronic doses of YCl_3 are not toxic to rats; intraperitoneal doses of 60 mg YCl_3/kg given every alternate day for 5 months were tolerated by rats (MacDonald *et al.*, 1952). The toxicity of Y salts is summarized in Table 3-6; Y salts appear to be moderately toxic to experimental animals. The higher toxicity of Y_2O_3 in HCl solution was due to the acidity of the solution, because acidity causes a high but transient depression in the respiratory and cardiac functions. Chloride citrate complexes of Y are more toxic than YCl_3; there was no significant inhibition of growth in rats which were fed YCl_3 at 1% dietary level, but Schroeder and Mitchner (1971*a*) reported growth depression in mice which were fed 5 ppm YCl_3 in their drinking water during life-term experiments.

Subcutaneous implantation of metallic Y in mice induced growth of granulomatous tissue (Talbot *et al.*, 1965). Blood coagulation time was significantly longer than in control mice six months after implantation. Decreases in total leukocyte numbers and incidence of neoplasms were observed. Schroeder and Mitchner (1971*a*) also reported malignant tumor formation in mice, but no tumors were found by Hutcheson *et al.* (1975*b*) after three generations of feeding Y_2O_3.

The liver is the main organ affected by subtoxic, parenterally administered Y salts in rats. Swelling of the liver with focal necrosis in the caudal or midzone lobes was reported (Magnusson, 1962). Due to hepatic dysfunction, liver glycogen and blood glucose levels were decreased, and blood ornithine carbamyl phosphatase activity was increased. Yttrium, like the heavy lanthanides, lowers blood pressure when injected intravenously in experimental animals.

TABLE 3-6. Yttrium Toxicity

| | | | | Dosage/kg body weight | | | |
| | | | | Compound | Metal | | |
Compound	Animal	Route	Toxicity	mg	mg	mM	pT
Yttrium oxide	Rat	ip	LD_{50}	500	395	4.44	2.35
Y_2O_3							
Yttrium oxide, dissolved	Mouse	ip	LD_{100}	112	88	0.99	3.00
in dilute HCl	Rat	ip	LD_{50}	57.2	45	0.51	3.30
	Rat	iv	MLD	5	4	0.045	4.35
	Rat	iv	LD_{100}	12.7	10	0.11	3.95
Yttrium chloride	Mouse	ip	LD_{50}	88	40.1	0.45	3.35
$YCl_3 \cdot 6H_2O$	Rat	ip	LD_{50}	45	13.2	0.148	3.83
	Guinea pig	ip	LD_{50}	85	38.7	0.44	3.36
Yttrium citrate	Mouse	ip	LD_{50}	254	79	0.89	3.05
$Y(C_6H_5O_7)$	Mouse	ip	LD_{50}	79	25.1	0.28	3.55
	Guinea pig	ip	LD_{50}	44	14.0	0.16	3.80
Yttrium nitrate	Mouse	sc	MLD	1660	403	4.53	2.34
$Y(NO_3)_3$	Mouse	ip	LD_{50}	1710	415	4.67	2.33
	Rat	ip	LD_{50}	350	117	1.32	2.88
	Rat	iv	MLD	75	25	0.28	3.55
	Rabbit	ip	LD_{50}	515	125	1.41	2.85

Blood lactic acid and leukocyte counts increased in dogs following Y_2O_3 inhalation. Hypertrophy, hyperplasia, desquamation of the alveolar epithelial cells, and leukocyte infiltration were seen in the lungs (Reece *et al.*, 967).

Yttrium salts inhibit *in vitro* ATPase and succinic dehydrogenase enzymes. There are no reported mechanisms for Y detoxication in mammals.

LANTHANIDES, THE RARE-EARTH METALS

Lanthanides, or rare-earth metals, constitute a series of fifteen metals starting with lanthanum (atomic number, 57) and ending with lutetium (atomic number, 71); they are subgrouped as follows: (1) The light lanthanides, or Ce group, comprise lanthanum (La), cerium (Ce), praseodymium (Pr), neodymium (Ne), promethium (Pr), and samarium (Sa); (2) the medium lanthanides, or Tb group, comprise europium (Eu), gadolinium (Ga), terbium (Tb), dysprosium (Dy), and holmium (Ho); and (3) the heavy lanthanides, or Er group, comprise erbium (Er), thulium (Tm), ytterbium

(Yb), and lutetium (Lu). Yttrium (atomic number, 39) typifies the Er group owing to its similarity in physicochemical properties. The electronic configurations of the lanthanides suggest they should all be put into the La space in subgroup IIIB of the periodic chart. The name *rare earth* is misleading and inaccurate because these are metals and not earths, and are far from rare; some lanthanides, such as Ce, occur in the earth's crust in larger amounts than do Sn, Co, Ag, and Au. These lanthanides occur together as oxides in the following minerals: xenotime, fergusonite, gadolinite, cerite, lanthanite, euxenite, polycrase, samarskite, and monazite; monazite sand serving as the major source. Promethium does not occur in nature, being a man-made element.

Radioactive isotopes of these metals, including Pm and Y, are found in radioactive fallout as products of nuclear reaction. The abundance of these metals in the earth's crust and in igneous rock is summarized in Table 3-7. Cerium and lanthanum occur predominantly in rare-earth minerals, in concentrations of about 40% and 25% respectively. Promethium metal is now available in a pure form; some of its salts are known, and Pm chemistry has been reviewed (Boyd, 1959). The lanthanide content of seawater is not yet accurately determined. Normally these metals do not occur in plant or animal tissues, but traces of these metals (0.5 μg/g) were detected in the bones of man and animals exposed to them.

In the past, the use of lanthanides was restricted to five main areas; glass polishing, carbon arcs, cracking catalysts, flints for lighters, and nodularizing agents for iron, but recently, their uses have increased. Europium-activated YVO_4 (1 Eu atom to 19 Y atoms) is used extensively in color television tubes as a red phosphor, because europium is the only element which emits the right part of the spectrum for red phosphor. Lanthanide polishes are used on ophthalmic and camera lenses, television tubes, mirrors, face plates, and other glass products. Lanthanum oxide is used in making camera lenses, binoculars, and gunsights, because of its high refractive index and low dispersion; lanthanide glasses are used in ophthalmic lenses and sun glasses for their selective absorption of certain wavelengths of light. Carbon electrodes for arc lights contain lanthanide cores; the light from these arcs has a continuous spectrum almost identical to sunlight, which makes them useful in the movie industry. Recently these arcs have become energy sources for solar simulation in space-environmental chambers. Mischmetal, a mixture of lanthanide chlorides, is used in petroleum cracking and in the gas flint industry. Isotopes of Sm, Eu, Dy, and especially Gd have very high thermal neutron cross sections, and are used as control rods in nuclear reactors and power plants; Dy is considered a burnable poison in atomic-fuel elements. Yttrium- and gadolinium-con-

TABLE 3-7. Lanthanides in the Earth's Crust and in Igneous Rock[a]

Metal	Earth's crust	Igneous rock
Cerium	22.0	46.0
Neodymium	12.0	24.0
Samarium	6.5	6.5
Gadolinium	6.3	6.4
Lanthanum	5.0	18.0
Dysprosium	5.0	4.5
Erbium	4.0	2.5
Praseodymium	3.5	5.5
Ytterbium	2.6	2.7
Terbium	1.0	0.9
Holmium	0.7	1.2
Lutetium	0.7	0.8
Thulium	0.5	0.2
Europium	0.14	1.1
Promethium	—	—

[a]Values given in ppm.

taining garnets are used in lasers. Organic Ce compounds [cerium (2,2,6,6-tetramethyl-3,5-heptane-dionate)$_4$] are good antiknock agents which act similarly to lead tetraethyl. Some lanthanide salts are used as drying agents for paints, in glazing ceramics, and in waterproofing textiles. Lanthanide cobalt oxides are used to make cheap auto exhaust oxidation catalysts (Pederson and Libby, 1972). In the past, some cerium compounds were used to treat tuberculosis, and as antinausea agents during early pregnancy. Neodymium compounds act as anticoagulant agents for the prevention of thrombosis. Lanthanides are not considered industrial health hazards except in lanthanide refining and in the lithographic industry which used cored arc lights. Therapeutic use of lanthanides has been discontinued in some countries. Accidental lanthanide poisonings are rare, but the potential for exposure to these metals leading to intoxication is growing.

The low solubility of their oxides and oxysalts makes lanthanides useful as nutritional markers; radioactive lanthanides are frequently used. Our laboratory prefers the natural isotopes for experimental work, followed by neutron activation analysis of appropriate samples for quantification. Their nuclear properties make most lanthanides very good for activation analysis—a large thermal neutron cross section, and a high abundance of radioactive isotope with distinctive γ radiation of convenient half-life.

Chemistry

The lanthanides are inner-transitional elements. The atomic weights of lanthanides increase in small units; their chemical properties, metabolism, and involvement in physiologic and biochemical systems are quite similar. The electrons, introduced one by one as the nuclear charge increases up the series, go into inner electron shells, leaving the outer shell of three electrons intact in each of the lanthanides. The addition of electrons into lanthanum $4f$ shells increases the binding energy of the valence electrons; Ce and Gd have electrons in $5d$ orbitals. The increase in binding energy and nuclear charge causes a progressive decrease in the atomic and ionic sizes and in the basicity of these lanthanides, as their atomic mass increases. The lanthanide atomic contraction accounts for the gradation in various properties of the lanthanides: (1) decreasing basicity, thermal stability, and salt solubility, and (2) increasing acidity, stability of complex ions, and covalent characteristics. The salts of heavy lanthanides are less soluble than the salts of light and transitional lanthanides. The molar solubilities of lanthanide hydroxides range from 8.8×10^{-5} to 5×10^{-7} M from $La(OH)_3$ to $Lu(OH)_3$. Poor solubility confers adsorptive capacity, since adsorption is an inverse solubility function. The adsorptive property of the lanthanides is exhibited at concentrations much lower than the molar solubility levels.

The lanthanides exhibit trivalence; La, Eu, Sm, and Yb exhibit divalence in addition, and Ce, Pr, and Tb also exhibit tetravalence. These metals do not exhibit amphoteric character; they do form cationic salts. Lanthanide hydroxides are more basic than those of other trivalent metals. Within the series, $La(OH)_3$ is more basic than $Lu(OH)_3$. The lanthanide cations are hydrolyzed in aqueous solutions, a tendency highly favored at physiologic pH. The low solubility of these hydroxides favor their precipitation unless other compounds such as citrates or lactates, present in living tissues, chelate with the metal ions to keep them in solution. Unlike Sc salts, the lanthanides do not undergo further olation following hydrolysis in aqueous solutions, owing to their greater basicity. In biologic fluids, especially blood plasma, the injected lanthanide cations separate out as insoluble particles if the concentration is greater than the molar solubility of the compounds (above 8.8×10^{-5} M).

If the concentration of the compounds approximates or slightly exceeds the molar solubility of the lanthanide hydroxides (from 8.8×10^{-6} to 5×10^{-7} M), the hydroxides separate out as colloidal particles; aggregates of metal hydroxides become surrounded by a protein coat or similar stabilizing agent; this applies to chemical doses and to radioisotopic preparations with large amounts of stable carriers. If the concentration of the injected lanthanide compound is far below the molar solubility of the corresponding hydroxide (10^{-7}–10^{-11} M), the lanthanides separate out not as ions but as radiocolloidal particles, owing to the polymerization of the lanthanide hydroxides at the pH of the blood; this polymerization increases with decreased basicity of the hydroxides. The interaction between these aggregates and the plasma proteins stabilizes them (Kyker, 1962; Schweitzer, 1956; Schweitzer and Jackson, 1952a,b). The adsorptive capacity of the lanthanides is quite effective at this radiocolloidal concentration.

Lanthanide cations show a strong tendency to complex with synthetic and naturally occurring chelating agents such as citric and lactic acids, nitrilotriacetic acid, EDTA, DTPA, etc. Lanthanide–EDTA complexes are stable and are not hydrolyzed in biologic fluids; the stability constants of lanthanide–EDTA increase gradually from La to Lu.

Lanthanides react *in vitro* with various tissue components, which include nucleoproteins, plasma proteins, amino acids, phospholipids, enzymes, intermediary metabolites, and inorganic phosphate. The affinity of lanthanides toward phosphate is high; the capacity of the lanthanides to form insoluble complexes with phosphates enables them to function at low concentrations as nonspecific "phosphatases," cleaving off phosphate groups from ATP, glycerophosphate, nucleotides, and nucleic acids (Trapmann, 1959). Lanthanides can bind to serum proteins, especially globulins and amino acids, when the lanthanides are in the ionic form and not in the colloidal form. Transferrin has been reported to bind with Tb, Eu, Er, Ho, Nd, and Pr (Kaluk, 1971). Heavy lanthanides complex with proteins more readily than do light lanthanides (Graca *et al.*, 1962), but with amino acids the opposite is true; the medium lanthanides behave more like the heavy lanthanides in binding to proteins, and like lighter lanthanides toward amino acids. Lanthanides can precipitate *in vitro* DNA from aqueous solutions by combining with the phosphate groups of adjacent chains.

Metabolism

The lanthanides are not essential for plants and animals, although cerium has been found to be stimulatory to microorganisms and animals. Although radioactive isotopes of lanthanides from fallout have been found in some barley and wheat, plants in general do not absorb the lanthanides from the soil, owing to discrimination against their absorption by the roots. Yeast takes up La selectively; marine algae and other organisms absorb some lanthanides preferentially over Sr. The negligible accumulation of lanthanides by plants effectively blocks the dietary transfer of lanthanides from the soil to man and animals. Although lanthanides were found at 0.5 μg/g in bones and Gd, Ho, Er, and Tm were found at 10.3 μg/g in the kidneys of men exposed to these metals, the dietary intake of these lanthanides by man is not known. Owing to the presence of radioactive lanthanides in fallout, the biochemistry and physiology of these metals were studied using mostly radioactive tracers; Kyker and Anderson (1956), Kyker (1962), Stokinger (1963), and Marchetti *et al.* (1967) have reviewed the chemistry and physiology of the lanthanides.

In mammals, the gastrointestinal absorption of soluble simple lanthanide salts is poor, less than 0.05%. In the intestines the lanthanides, because of their basicity, separate out as insoluble hydroxides in precipitable, colloidal, or radiocolloidal forms, depending on the dose and on intestinal pH. The heavier lanthanides separate out more than do light lanthanides and are less well absorbed. The lanthanides also combine to form insoluble phosphates or adsorb to particulate food ingesta. Poor absorption of lanthanides is recorded in all species studied: rats and mice (Cochran *et al.*, 1950; Haley, 1965; Hutcheson *et al.*, 1975*b*; Luckey *et al.*, 1975), goats (Ekman and Aberg, 1961), cows (Garner *et al.*, 1960), calves (Miller *et al.*, 1971), monkeys (Hutcheson *et al.*, 1975*b*), and humans (Luckey *et al.*, 1977). This is the reason the lanthanides are good markers for nutrition studies. However, dietary absorption of water-soluble and stable lanthanide–citrate and lanthanide–EDTA complexes is rapid and complete in experimental animals. Following intramuscular and subcutaneous injections into experimental animals, insoluble lanthanide fluorides and phosphates and colloidal preparation of oxides and hydroxides are retained almost comletely at the sites of injection. Soluble salts injected in minute doses are absorbed slowly; about 55% is absorbed when levels of 10^{-11}–10^{-8} mol/kg are administered; at 10^{-4} mol/kg retention at the intrapleural and peritoneal cavities is almost complete. Following inhalation, about 50% of lanthanide aerosols are absorbed slowly from the lungs, the rest being retained for a long time; the aerosols contain mostly soluble salts (Semenov *et al.*, 1967; Berke, 1969; Morgan *et al.*, 1970).

The transport and removal of lanthanides from the blood to the reticuloendothelial system, soft tissues, and bones, following parenteral injections, depend on the concentration and nature of the lanthanide compounds. The light lanthanides (soluble salts) at low concentrations exist in the ionic form and bind to transferrin and other γ globulins, to erythrocyte cell walls, and to free inorganic phosphates or phosphate-containing metabolites; these ions do not penetrate erythrocytes. These light lanthanides react with other body components and are distributed rapidly into the liver and kidneys, with a gradual uptake and retention in the bones. At intermediate concentrations, most of the absorbed light lanthanides separate out as colloidal hydroxides or phosphates, and are taken into macrophages and the reticuloendothelial system by phagocytosis, removed to the lymph nodes, bone marrow, and liver, and excreted into the digestive tract through the enterohepatic circulation. Medium and heavy lanthanides do not exist as ions in the blood, but as coloidal or radiocolloidal forms of either hydroxides or phosphates. The removal of colloidal particles of lanthanide compounds from the blood is rather slow. The colloidal aggregates of lanthanide complexes are not deposited to a great extent in bone; the absorption and distribution of stable chelate complexes of lanthanides depend upon the metabolism of the chelate part of the complex. Lanthanide–EDTA complexes are stable, are quickly and completely absorbed into the blood, and are excreted both in the feces and urine with little permanent retention in the tissues. Lanthanide–citrate complexes are transported from the blood to the bones and are retained permanently in the nonmineral portion of the bones; all the lanthanides exhibit this property. However, osteolysis can mobilize these lanthanides from the bone. Young animals absorb and retain the injected lanthanides better than old animals; females accumulate more lanthanides in the liver, males more in the skeleton.

The excretion of lanthanides in mammals is both fecal and urinary, depending upon the anion, dose, and mode of uptake. Intermediate concentrations of the light lanthanide salts and all doses of the salts of medium and heavy lanthanides given orally are excreted in the feces. Parenterally administered heavy lanthanides are excreted in the feces, while the light lanthanides are excreted mostly in the urine. Ionic forms, administered at *very* low levels, are excreted in the urine, and colloidal forms in the feces (Takada *et al.*, 1970). Following parenteral administration the excretion of lanthanides in the feces and urine is slow, and about 45% of the dose is retained permanently in the skeleton. Fecal secretion of parenteral doses occurs partly via the bile and partly by direct secretion through the intestinal wall (Magnusson, 1962).

Skeletal deposition and retention of lanthanides depend upon the dose, the route of administration, and the adomic number of the lanthanide, when uniformly low levels were administered parenterally, the heavier lanthanides accumulated rapidly and in larger amounts (50–60%) of the dose than did the lighter lanthanides (Kyker, 1962); the elimination half-time was calculated to be twenty-five years. Deposition of the lighter lanthanides (from La to Sm) was 50% in the liver and 25% in the skeleton, the skeleton retaining ⅔ of the initial deposit after eight months. The medium lanthanides were deposited equally in the liver and skeleton.

Toxicity

The pharmacologic and toxicologic effects of the lanthanides in mammals were studied exhaustively by Haley, Kyker, Graca, and DuBois and their groups. Haley (1965) reviewed the pharmacology and toxicology of the lanthanides and considered these metals to be slightly toxic; the toxicity data of the lanthanides are summarized in Table 3-8. These data do not contraindicate rational use of the lanthanide oxides in digestion studies (Hutcheson *et al.*, 1975*a*). Simple compounds, especially the oxides, are practically nontoxic to experimental animals when fed orally; the mortality following oral ingestion is mainly owing to mass effect. Intravenous injection appears to be the most toxic among parenteral administration routes, and intraperitoneal administration is more toxic than either intramuscular or subcutaneous injection. Simple cationic salts appear to be less toxic than are chloride–citrate complexes. The light lanthanides are slightly more toxic than the heavy lanthanides on the basis of LD_{50} values, and intermediate lanthanides are the least toxic. Mice appear to be less susceptible to lanthanide intoxication than other animals, and males are less susceptible than females. Acute lanthanide toxicity causes a delayed mortality; a biphasic mortality response was observed in mice and guinea pigs following intraperitoneal administration of lanthanide nitrates (Graca *et al.*, 1962).

Symptoms of acute lanthanide intoxication in rats are immediate defecation, writhing, ataxia, sedation, labored respiration, and reduced activity. Death is mainly due to respiratory and cardiac failure. Autopsy reveals dyspnea, pulmonary edema, hyperemia, liver edema and necrosis, portal congestion, pleural effusion, and granulomatous peritonitis with serous and hemorrhagic ascites. Chronic lanthanide intoxication causes derangement of the renal and hepatic functions with markedly increased prothrombin and coagulation times. Liver damage is more prominent in males than in females. Symptoms following chronic inhalation of lanthanide oxides and fluorides by experimental animals are slow in appearing; clinical hyperemia, cellular eosinophilia, and vascular granulomata occur after a few months, and acute chemical pneumonitis, subacute bronchitis, and focal hypertropic emphysema also occur following inhalation of lanthanide fluorides. Intratracheal intubation of lanthanide oxides in rats caused granulomas after a few months. The assessment of precise lanthanide toxicity effects is rendered difficult by the protein-precipitating capacity of the lanthanides, the influence of the anion part of the compound, and the differing concentrations and rates of injection of the compounds.

Subcutaneous implantation of Y, Ce, Pr, Gd, Dy, or Yb metal pellets produce granulomatous tissue in mice (Talbot *et al.*, 1965); increased coagulation time and decreased leukocyte count were observed in the lanthanide-implanted mice six months after implantation. Lesions were found in human skin following injection of lanthanides in patients with sarcoidosis and anthracosilicosis (Shelly *et al.*, 1958).

Lanthanide salts (1 mg per animal) cause soft tissue calcification at the sites of subcutaneous injection; these are termed calcergens. Systemic calcification in the spleen in rats is induced by intravenous injection of soluble lanthanide salts; the dose ranged from 8 to 15 mg lanthanide/kg body weight; La and Sc did not cause splenic calcification. The doses for the rest of the lanthanides, expressed as mg of the metal given as chlorides, were: Tb, 8; Y and Tm, 10; Lu, Er, Yb, Sm, and Dy, 12; and Ce, Eu, Gd, Pr, and Nd, 15. Calcification starts in the macrophages of the marginal zone and invades the red pulp forming a ring around the follicles (Gabbiani *et al.*, 1966; Tuchweber and Savoie, 1968).

Lanthanide chloride solutions (0.1 M) cause blisters when applied to corneas of rabbits. A diffuse gray opacity appeared after several hours, leading to dense yellowish plaques, and finally to spontaneous perforations (Grant and Kern, 1956). The extent of injury was correlated with the extent to which the metal bound to the cornea.

Growth inhibition was observed in mice and rats fed lanthanide salts at 1% of their diet. Feeding oxides of several lanthanides (La, Dy, Eu, Yb, Tb, and Sm) at a total dietary level of about 0.5% to mice through three generations revealed no signs of toxicity (Hutcheson *et al.*, 1975*a*).

The liver is the target organ in lanthanide intoxication; fatty liver degeneration, increase in neutral fat esters, and mitochondrial damage occur following parenteral injection of lanthanide compounds. Liver glycogen and blood glucose levels decrease while free fatty acids increase in plasma following intravenous injection of lanthanides (Snyder *et al.*, 1959). Subcellular changes in the endoplasmic reticulum include dilations of cisternae and dissociation of ribosomes. The drug-metabolizing activity of the liver is decreased. The light lanthanides cause greater hepatic damage than do the medium or heavy lanthanides; this metabolic disorder varied in degree among different strains of rats and was more intense in the females. Extrahepatic lipids were the source of increased neutral lipids in the liver, suggesting enhanced mobilization from lipid depots.

The activity of the reticuloendothelial system in rats was depressed by intravenous administration of lanthanide chlorides at 0.2 mg metal/100 g body weight, and the activity was completely stopped by administration of 0.5 mg/100 g (Lazar, 1973); the depression is caused by saturation of the reticuloendothelial system by lanthanide hydroxide colloids.

At low concentrations, the light lanthanides exerted *in vitro* toxic effects on the electric potentials of nerve and neuromuscular function, and on the strength of contraction of skeletal, cardiac, and smooth muscles of

TABLE 3-8. Toxicity of Lanthanides

Compound	Animal	Route	Toxicity	Compound mg	Metal mg	Metal mM	pT
Lanthanum metal La	Rat	iv	LD_{50}		3.5	0.025	4.60
Lanthanum oxide La_2O_3	Rat	oral	LD_{100}	10,000	8500	61.2	1.21
Lanthanum chloride (anhydrous)	Mouse	ip	LD_{50}	121	68.5	0.49	3.31
$LaCl_3$	Mouse	sc	MLD	3500	1980	14.2	1.85
	Rat	oral	LD_{50}	4200	2380	17.1	1.77
	Rat	ip	LD_{50}	106	60.04	0.43	3.36
	Guinea pig	ip	LD_{50}	129	73.1	0.53	3.28
	Rabbit	iv	MLD	200	113	0.813	3.09
Lanthanum chloride (hydrated)	Mouse	ip	LD_{50}	211	78.9	0.568	3.25
$LaCl_3 \cdot 7H_2O$	Rat	ip	LD_{50}	197	73.7	0.53	3.28
	Rat	iv	MLD	91	34	0.24	3.61
	Rabbit	iv	MLD	84	31.4	0.226	3.65
Lanthanum nitrate (anhydrous)	Mouse	ip	LD_{50}	410	175	1.26	2.90
$La(NO_3)_3$	Rat	oral	LD_{50}	4500	1920	13.9	1.86
	Rat	ip	LD_{50}	450	192	1.38	2.86
Lanthanum sulfate	Rat	oral	LD_{50}	5000	2460	17.6	1.75
$La_2(SO_4)_3$	Rat	ip	LD_{50}	275	135	0.972	3.01
Lanthanum ammonium nitrate	Rat	oral	LD_{50}	3400	1170	8.4	2.07
$LaNH_4(NO_3)_4$	Rat	ip	LD_{50}	625	214	1.54	2.81
Lanthanum acetate	Rat	oral	LD_{50}	10,000	440	3.16	2.50
$La(C_2H_3O_2)_3$	Rat	ip	LD_{50}	475	209	1.50	2.82
Lanthanum citrate	Mouse	ip	LD_{50}	78	44.9	0.323	3.49
$La(C_6H_5O_7)$	Guinea pig	ip	LD_{50}	71	40.9	0.294	3.53
Lanthanum chloride citrate	Guinea pig	ip	LD_{50}	61	34.6	0.25	3.60
Cerium dioxide CeO_2	Rat	oral	MLD	1000	814	5.8	2.24
Cerium chloride	Mouse	sc	MLD	5000	2840	20.3	1.69
$CeCl_3$	Mouse	ip	LD_{50}	98	55.7	0.397	3.40
	Rat	sc	MLD	2000	1140	8.11	2.09
	Rat	iv	LD_{50}	50	28.4	0.20	3.69
	Guinea pig	ip	LD_{50}	56	31.8	0.23	3.64
	Guinea pig	sc	LD_{50}	2130	1210	8.62	2.06
	Rabbit	iv	LD_{50}	35	19.9	0.14	3.85
Cerium fluoride	Guinea pig	oral	MLD	5000	4070	29.0	1.54
CeF_3	Guinea pig	sc	MLD	5000	4070	29.0	1.54
Cerium nitrate	Mouse	ip	LD_{50}	470	152	1.08	2.96
$Ce(NO_3)_3 \cdot 6H_2O$	Mouse	ip	LD_{50}	149	63.4	0.45	3.34
	Rat	oral	LD_{50}	4200	1360	9.71	2.01

(Cont'd)

TABLE 3-8. (Cont'd)

Compound	Animal	Route	Toxicity	Compound mg	Metal mg	Metal mM	pT
Cerium nitrate	Rat	ip	LD_{50}	290	93.6	0.67	3.18
$Ce(NO_3)_3 \cdot 6H_2O$	Rat,♂	iv	LD_{50}	50	16.1	0.115	3.99
Cerium citrate	Rat,♀	iv	LD_{50}	4.5	1.45	0.01	4.96
$Ce(C_6H_5O_7)$	Rat	ip	LD_{50}	147	62.6	0.45	3.35
	Guinea pig	ip	LD_{50}	104	44.3	0.316	3.50
	Guinea pig	ip	MLD	83	35.3	0.25	3.60
Praseodymium chloride	Rat	ip	LD_{50}	2000	1140	8.1	2.09
$PrCl_3$	Rat	iv	MLD	9.5	5.41	0.038	4.42
	Mouse	oral	LD_{50}	4500	2560	18.2	1.74
	Mouse	ip	LD_{50}	359	205	1.46	2.84
	Mouse	sc	MLD	900	513	3.65	2.44
	Guinea pig	ip	LD_{50}	125	71.2	0.51	3.30
	Rabbit	ip	MLD	200	114	0.81	3.09
Praseodymium chloride (hydrated)	Mouse	ip	MLD	550	207	1.47	2.83
$PrCl_3 \cdot 7H_2O$	Mouse	ip	LD_{50}	600	226	1.61	2.79
	Mouse	sc	LD_{50}	2500	944	6.71	2.17
	Guinea pig	ip	LD_{50}	190	71.7	0.51	3.29
Praseodymium nitrate	Mouse	ip	LD_{50}	290	98.0	0.696	3.16
$Pr(NO_3)_3 \cdot 5H_2O$	Rat	oral	LD_{50}	3500	1180	8.41	2.08
	Rat	ip	LD_{50}	245	83	0.59	3.23
	Rat	iv	MLD	10.8	3.65	0.026	4.59
	Rat,♂	iv	LD_{50}	6.4	2.16	0.015	4.81
	Rat,♀	iv	LD_{50}	77	26.0	0.185	3.73
	Guinea pig	ip	LD_{50}	97	32.8	0.233	3.63
Praseodymium citrate	Mouse	iv	LD_{50}	3500	1370	9.73	2.01
$Pr(C_6H_5O_7)$	Mouse	ip	LD_{50}	141	60.0	0.43	3.37
Praseodymium propionate $Pr(C_3H_5O_2)$	Guinea pig	ip	LD_{50}	53	22.6	0.16	3.79
Neodymium metal Nd	Guinea pig	iv	MLD		70	0.48	3.31
Neodymium oxide Nd_2O_3	Rat	ip	LD_{50}	800	270	1.87	2.73
Neodymium chloride (anhydrous)	Mouse	ip	LD_{50}	347	200	1.39	2.86
$NdCl_3$	Mouse	sc	MLD	4000	2300	16.0	1.80
	Rat	ip	MLD	150	86.3	0.598	3.22
	Guinea pig	ip	MLD	140	81.0	0.56	3.25
	Guinea pig	iv	MLD	70	40.3	0.28	3.55
	Rabbit	iv	MLD	200	115	0.80	3.10
Neodymium chloride (hydrated)	Mouse	oral	LD_{50}	5250	2110	14.6	1.84
$NdCl_3 \cdot 6H_2O$	Mouse	ip	LD_{100}	600	241	1.67	2.77
	Rat	ip	LD_{50}	375	151	1.05	2.98

(Cont'd)

TABLE 3-8. (Cont'd)

Compound	Animal	Route	Toxicity	Compound mg	Metal mg	Metal mM	pT
Neodymium chloride	Rat	iv	MLD	8.7	3.50	0.024	4.62
(hydrated)	Guinea pig	ip	MLD	202	81	0.56	3.25
$NdCl_3 \cdot 6H_2O$	Guinea pig	iv	MLD	100	40.2	0.28	3.55
Neodymium nitrate	Mouse	ip	LD_{50}	270	89	0.62	3.21
(hydrated)							
$Nd(NO_3)_3 \cdot 6H_2O$	Rat	oral	LD_{50}	2750	905	6.27	2.20
	Rat	ip	LD_{50}	270	89	0.62	3.21
	Rat, ♀	iv	LD_{50}	6.4	2.1	0.015	4.82
	Rat, ♂	iv	LD_{50}	66.8	22.0	0.153	3.83
Neodymium nitrate	Mouse	ip	LD_{50}	138	59.7	0.414	3.38
(anhydrous)							
$Nd(NO_3)_3$	Guinea pig	ip	LD_{50}	41	17.8	0.123	3.91
Neodymium citrate	Rabbit	iv	MLD	50	21.7	0.15	3.82
$Nd(C_6H_5O_7)$							
Samarium chloride	Mouse	oral	LD_{50}	2000	840	5.59	2.25
$SmCl_3$	Mouse	ip	LD_{50}	365	153	1.02	2.99
	Rat	sc	MLD	2000	840	5.59	2.25
	Guinea pig	sc	MLD	500	210	1.40	2.85
	Guinea pig	sc	LD_{50}	1000	420	2.79	2.55
Samarium nitrate	Mouse	ip	LD_{50}	315	106	0.705	3.15
$Sm(NO_3)_3 \cdot 6H_2O$	Rat	oral	LD_{50}	2900	980	6.52	2.19
	Rat	ip	LD_{50}	285	96.3	0.64	3.19
	Rat, ♀	iv	LD_{50}	9.0	3.04	0.020	4.69
	Rat, ♂	iv	LD_{50}	59.0	19.9	0.132	3.88
	Guinea pig	sc	MLD	500	169	1.12	2.95
Samarium citrate	Mouse	ip	LD_{50}	164	72.7	0.48	3.32
$Sm(C_6H_5O_7)$	Guinea pig	ip	LD_{50}	75	33.2	0.22	3.66
Samarium acetate	Mouse	iv	LD_{50}	1000	460	3.06	2.51
$Sm(C_2H_3O_2)_3$							
Europium chloride	Mouse	oral	LD_{50}	5000	2940	19.3	1.71
(anhydrous)							
$EuCl_3$	Mouse	ip	MLD	387	228	1.5	2.82
	Mouse	ip	LD_{50}	550	324	2.13	2.67
	Guinea pig	ip	LD_{50}	156	91.8	0.60	3.22
Europium nitrate	Mouse	ip	LD_{50}	320	109	0.717	3.14
(hydrated)							
$Eu(NO_3)_3 \cdot 6H_2O$	Rat	oral	LD_{50}	5000	1700	11.2	1.95
	Rat	ip	LD_{50}	210	72.0	0.47	3.32
Europium citrate	Mouse	ip	LD_{50}	187	83.3	0.55	3.26
$Eu(C_6H_5O_7)$	Guinea pig	ip	LD_{50}	72	32.1	0.21	3.68
Gadolinium oxide	Rat	oral	LD_{50}	7500	6500	41.3	1.38
Gd_2O_3	Rat	ip	LD_{50}	1000	868	5.52	2.26

(Cont'd)

TABLE 3-8. *(Cont'd)*

Compound	Animal	Route	Toxicity	Compound mg	Metal mg	mM	pT
Gadolinium chloride (anhydrous) $GdCl_3$	Mouse	ip	LD_{50}	378	226	1.44	2.84
Gadolinium chloride (hydrated)	Mouse	oral	LD_{50}	2000	846	5.38	2.27
$GdCl_3 \cdot 6H_2O$	Mouse	ip	LD_{50}	550	233	1.48	2.83
Gadolinium nitrate (hydrated)	Mouse	ip	LD_{50}	300	105	0.67	3.18
$Gd(NO_3)_3 \cdot 6H_2O$	Rat	oral	LD_{50}	5000	1740	11.1	1.96
	Rat	ip	LD_{50}	230	80	0.51	3.29
Gadolinium citrate	Mouse	ip	LD_{50}	153	69.5	0.442	3.35
$Gd(C_6H_5O_7)$	Guinea pig	ip	LD_{50}	60	27.4	0.174	3.76
Terbium oxide Tb_2O_3	Rat	oral	MLD	1000	868	5.46	2.26
Terbium chloride (hydrated)	Mouse	oral	LD_{50}	5100	2170	13.6	1.86
$TbCl_3 \cdot 6H_2O$	Mouse	ip	LD_{50}	550	234	1.47	2.83
Terbium chloride (anhydrous) $TbCl_3$	Mouse	ip	MLD	332	199	1.25	2.90
Terbium citrate	Mouse	ip	LD_{50}	121	55.2	0.34	3.46
$Tb(C_6H_5O_7)$	Guinea pig	ip	LD_{50}	74	33.8	0.21	3.67
Dysprosium chloride (anhydrous)	Mouse	ip	LD_{50}	343	207	1.27	2.89
$DyCl_3$	Guinea pig	ip	LD_{50}	196	118	0.73	3.14
Dysprosium chloride (hydrated)	Mouse	oral	LD_{50}	7650	3300	20.3	1.69
$DyCl_3 \cdot 6H_2O$	Mouse	ip	LD_{50}	585	252	1.55	2.81
Dysprosium nitrate (hydrated)	Mouse	ip	LD_{50}	310	110	0.68	3.17
$Dy(NO_3)_3 \cdot 6H_2O$	Rat	oral	LD_{50}	3100	1110	6.80	2.17
	Rat	ip	LD_{50}	295	105	0.65	3.19
Dysprosium citrate	Mouse	ip	LD_{50}	113	52.2	0.32	3.49
$Dy(C_6H_5O_7)$	Guinea pig	ip	LD_{50}	54	24.9	0.15	3.81
Holmium chloride (hydrated)	Mouse	oral	LD_{50}	7200	3130	19.0	1.72
$HoCl_3 \cdot 6H_2O$	Mouse	ip	LD_{50}	560	243	1.47	2.83
Holmium chloride (anhydrous) $HoCl_3$	Mouse	ip	MLD	312	190	1.15	2.94
Holmium nitrate (hydrated)	Mouse	ip	LD_{50}	320	115	0.697	3.16
$Ho(NO_3)_3 \cdot 6H_2O$	Rat	oral	LD_{50}	3000	1080	6.54	2.18
	Rat	ip	LD_{50}	270	97	0.59	3.23

(Cont'd)

TABLE 3-8. (Cont'd)

Compound	Animal	Route	Toxicity	Compound mg	Metal mg	Metal mM	pT
Holmium citrate	Mouse	ip	LD$_{50}$	117	54.5	0.33	3.48
Ho(C$_6$H$_5$O$_7$)	Guinea pig	ip	LD$_{50}$	63	29.3	0.18	3.75
Erbium oxide	Rat	ip	MLD	600	525	3.14	2.50
Er$_2$O$_3$							
Erbium chloride (anhydrous)	Mouse	ip	LD$_{50}$	226	138	0.83	3.08
ErCl$_3$	Guinea pig	ip	LD$_{50}$	128	78.4	0.47	3.33
Erbium chloride (hydrated)	Mouse	oral	LD$_{50}$	6200	2720	16.2	1.79
ErCl$_3 \cdot$6H$_2$O	Mouse	ip	LD$_{50}$	535	234	1.40	2.85
Erbium nitrate (hydrated)	Mouse	ip	LD$_{50}$	225	81.6	0.49	3.30
Er(NO$_3$)$_3 \cdot$6H$_2$O	Mouse	iv	LD$_{100}$	83	30.1	0.180	3.71
	Rat	ip	LD$_{50}$	230	83.0	0.50	3.30
	Rat	iv	LD$_{100}$	90.0	32.6	0.195	3.74
	Rat	iv	LD$_{50}$	30.0	10.9	0.065	4.19
	Rat,♀	iv	LD$_{50}$	35.8	13.0	0.077	4.11
	Rat,♂	iv	LD$_{50}$	52.4	19.0	0.114	3.94
Erbium citrate	Mouse	ip	LD$_{50}$	122	57.3	0.34	3.47
Er(C$_6$H$_5$O$_7$)	Guinea pig	ip	LD$_{50}$	63.0	29.6	0.177	3.75
Thulium chloride (hydrated)	Mouse	oral	LD$_{50}$	6250	2630	15.6	1.81
TmCl$_3 \cdot$7H$_2$O	Mouse	ip	LD$_{50}$	485	204	1.21	2.92
Thulium chloride (anhydrous)	Mouse	ip	LD$_{50}$	335	205	1.21	2.92
TmCl$_3$	Guinea pig	ip	LD$_{50}$	144	88.3	0.523	3.28
Thulium nitrate (hydrated)	Mouse	ip	LD$_{50}$	255	94	0.56	3.25
Tm(NO$_3$)$_3 \cdot$6H$_2$O	Rat	ip	LD$_{50}$	285	104	0.62	3.21
Thulium citrate	Mouse	ip	LD$_{50}$	80.0	37.7	0.223	3.65
Tm(C$_6$H$_5$O$_7$)	Guinea pig	ip	LD$_{50}$	55.0	25.9	0.153	3.81
Ytterbium chloride (hydrated)	Mouse	oral	LD$_{50}$	6700	2990	17.3	1.76
YbCl$_3 \cdot$6H$_2$O	Mouse	ip	LD$_{50}$	395	176	1.02	2.99
Ytterbium chloride (anhydrous)	Mouse	ip	LD$_{50}$	285	176	1.02	2.99
YbCl$_3$	Guinea pig	ip	LD$_{50}$	132	81.7	0.47	3.33
Ytterbium nitrate (hydrated)	Mouse	ip	LD$_{50}$	250	93	0.54	3.27
Yb(NO$_3$)$_3 \cdot$6H$_2$O	Rat	oral	LD$_{50}$	3100	1200	6.90	2.16
	Rat	ip	LD$_{50}$	255	94	0.54	3.27
Ytterbium citrate	Mouse	ip	LD$_{50}$	143	68.3	0.39	3.40
Yb(C$_6$H$_5$O$_7$)	Guinea pig	ip	LD$_{50}$	69	32.9	0.19	3.72

(Cont'd)

TABLE 3-8. (Cont'd)

| Compound | Animal | Route | Toxicity | Dosage/kg body weight | | | |
| | | | | Compound | Metal | | |
				mg	mg	mM	pT
Lutetium chloride (hydrated) $LuCl_3 \cdot 6H_2O$	Mouse	oral	LD_{50}	7100	3190	18.2	1.74
Lutetium chloride (anhydrous)	Mouse	ip	LD_{50}	315	196	1.12	2.95
$LuCl_3$	Guinea pig	ip	LD_{50}	161	100	0.57	3.24
Lutetium nitrate (anhydrous)	Mouse, ♀	ip	LD_{50}	350	170	0.97	3.01
$Lu(NO_3)_3$	Mouse, ♂	ip	LD_{50}	290	140	0.80	3.10
	Rat	ip	LD_{50}	335	162	0.926	3.03
Lutetium citrate	Mouse	ip	LD_{50}	135	64.9	0.37	3.43
$Lu(C_6H_5O_7)$	Guinea pig	ip	LD_{50}	81	38.9	0.22	3.65

dogs (Krasnow, 1972). Lanthanide salts inhibit Ca binding, Ca uptake, and Ca^{2+}-activated Mg^{2+}-ATPase. Lanthanide salts inhibit *in vitro* erythrocyte glucose-6-phosphate dehydrogenase, tissue succinic dehydrogenase, and ATPases. Mitochondria from lanthanide-intoxicated livers of rats showed evidence of uncoupling of oxidative phosphorylation.

The light lanthanides are very effective anticoagulants. An intravenous injection of 50 mg metal (as chloride)/kg in rabbits is very effective for many hours (Beaser *et al.*, 1942); 250 mg/100 g *in vitro* completely prevents clotting. Lanthanides are believed to act as antiprothrombic agents.

There are no specific mechanisms for the detoxication of lanthanides; administration of synthetic chelating agents such as EDTA or DTPA increase the urinary secretion of the absorbed lanthanide, but EDTA also causes Ca depletion (Koval and Kondrashev, 1967). There is no direct evidence for any homeostatic mechanism for lanthanides.

Toxicity of Light Lanthanides

The toxicity of La and Ce will be discussed, as they typify the light lanthanides. Lanthanum typifies the general metabolic properties of the light lanthanides, and exhibits all the mild effects of toxicity associated with them. La-EDTA chelates are rapidly absorbed from the digestive tract and from sites of parenteral administration; they are excreted rapidly through the feces and urine without causing any apparent signs of La intoxication (Hart *et al.*, 1955). Lanthanum differs from the other light lanthanides by its greater retention quantity and time. As the most electropositive among the lanthanides, it is also the most toxic. Its high charge density and tendency to form complexes through electrostatic bonding contribute to this toxicity.

In vitro studies show the binding of La^{3+} to phospholipids of rat liver mitochondrial membrane and the resultant inhibition of Ca^{2+} and Mn^{2+} uptake by the mitochondria; La^{3+} inhibits the reactions of Ca^{2+} with mitochondrial membranes, especially its activation of respiration and of changes in the steady states of the respiratory carriers (Mela, 1969a,b). Lanthanum inhibits respiration-dependent accumulation of Ca^{2+} by rat liver mitochondria by binding and inhibiting the Ca carrier; La^{3+} is not transported by the Ca carrier, but is very rapidly bound to the external sites on liver mitochondria in a respiration-independent process which releases H^+ ions into the medium (Lehninger and Carafoli, 1971). Lanthanum competes with Ca^{2+} for binding sites on the surface of smooth intestinal muscle in rats, and inhibits muscle contraction (Weiss and Goodman, 1969). Similarly La^{3+} inhibits Ca binding to the sarcoplasmic reticulum of skeletal muscles of rabbits (Chevallier and Butow, 1971), and Ca-dependent Mg^{2+} ATPase activity is inhibited. In dogs, La^{3+} in low concentration was found to depress the electric potentials of nerve and neuromuscular function; the power or force generation, the contractility of skeletal cardiac and smooth muscles, was depressed (Krasnow, 1972). In addition to the general mild toxicity associated with light lanthanides, La inhibits the normal activities of Ca^{2+} ions in muscle tissues.

Lanthanum, cerium, and samarium have been used as multiple markers in nutrition, with neutron activation analysis used to provide appropriate γ radiation isotopes for quantification (Luckey *et al.*, 1975, 1977). Their safety was indicated by feeding one-thousandfold excess (100–1000 ppm) to mice through three generations without toxic effect (Hutcheson *et al.*, 1975a,b).

Cerium has been extensively studied, using ^{144}Ce, for its metabolism and toxicity in a number of species. It is also a typical member of the light lanthanides and shows the same general metabolic properties.

The gastrointestinal absorption of Ce is poor in all mammalian species studied; at levels corresponding to radiocolloidal formation, Ce compounds adsorb to particulate food ingesta more than other lanthanides. The poor absorption of Ce from the digestive tract of cattle (Garner *et al.*, 1960; Miller *et al.*, 1967, 1971; Ellis and Huston, 1968), sheep (Buldakova and Burov, 1967; Huston and Ellis, 1968), and rats (Luckey *et al.*, 1975) led to the use of radioactive Ce as a nutritional marker. Cerium–EDTA complexes are absorbed to a greater extent than Ce^{3+} simple salts (Miller and Byrne, 1970). Inhalation of Ce^{3+} aerosols by mice and other experimental animals results in extensive absorption from the pulmonary tissues; about

40% remains in the tissue for a long time before it is slowly cleared (Semenov *et al.* 1967; Morgan *et al.,* 1970). Dietary deficiency of Ca, P, or vitamin A increases Ce absorption.

Intravenously injected Ce salts are rapidly distributed from the blood to the liver, skeleton, spleen, adrenals, and excreta. The distribution to liver and skeleton depends upon the concentration; at low concentrations Ce exists as ions which are deposited mostly in the skeleton; at higher concentrations colloidal forms of Ce are deposited in the liver, lymph nodes, and reticuloendothelial system. Accumulation of Ce in the liver is rapid and removal is slow, whereas uptake by bone is gradual, and retention prolonged. The deposition of Ce in young rats is not affected by the sex of the animals, but in adults, males accumulate Ce^{3+} in the skeleton and females in the liver (Kulikova, 1966). Skeletal retention of Ce is high in young rats (Shysh *et al.,* 1969), and in young lambs (Buldakova and Burov, 1967).

Thus, the retention and subcellular distribution of Ce in rat liver is influenced by age, sex, and dosage. Intravenously administered doses of Ce^{3+}, 2 mg Ce/kg, accumulate in mitochondrial and lysosomal fractions in males and in the nuclear fraction in the females; however, accumulation of Ce takes place in the nuclear fraction irrespective of sex at higher doses, 4 mg Ce/kg, (Mahlum, 1967*a*). Vitamin-A deficiency (Shysh *et al.,* 1969) and administration of hepatic carcinogens such as 4-dimethylaminobenzene (Mahlum, 1967*b*) increase the retention of Ce in rat liver.

Excretion of Ce is both fecal and urinary, and the concentration of administered Ce influences the excretion. Addition of stable Ce carrier to ^{144}Ce suppresses both urinary and fecal excretion of ^{144}Ce (Takada *et al.,* 1970). Low concentrations of injected ^{144}Ce, <1 mg/kg, are excreted mainly in the urine; higher concentrations of Ce given orally, 10 mg Ce/kg, and intravenously, 5 mg Ce/kg, are excreted mainly in the feces. When intermediate concentrations, 3 mg Ce/kg, are given intravenously, Ce excretion is both fecal and urinary, but a slower rate is noted. When 50 μCi ^{144}CeCl$_3$ at pH 4.0 was intravenously injected into pregnant rats, substantial accumulation of Ce took place in the placenta and associated membranes but not in the fetus, suggesting an effective placental barrier for Ce in rats (Mahlum and Sikov, 1968).

Cerium and other light lanthanides are comparatively nontoxic when compared with metals in other periodic groups, but within the lanthanide group, lanthanum, cerium, and other light lanthanides are more toxic than are medium and heavy lanthanides. Insoluble Ce compounds, such as the oxides, are nontoxic when ingested orally; CeCl$_3$ toxicity in rats appears to be nonspecific. A biphasic mortality response was found for intraperitoneally injected Ce–EDTA complex in mice and guinea pigs (Graca *et al.,*

1962). Vascular lesions, noticed in the digestive tract of rats fed [144]Ce, were attributed more to radiation injury than to chemical toxicity (Lebedeva, 1967).

Fatty liver degeneration, with increased liver levels of neutral fats from extrahepatic sources, liver mitochondrial damage, and decreased liver glycogen levels are the symptoms of Ce toxicity in the liver following an intravenous injection of $CeCl_3$, 3 mg Ce/kg, to rats (Magnusson, 1962). Serum glucose levels decreased and serum ornithine carbamyl transferase activity increased, suggesting impaired carbohydrate and lipid metabolism. Following an intravenous dose of 2 mg Ce/kg in rats, subcellular changes such as dilations of cisternae and dissociation of ribosomes were observed in the liver endoplasmic reticulum (Arvela and Karki, 1970). The drug-metabolizing and detoxifying capacity of the liver was reduced, hexabarbitol-metabolizing activity by 30%, and nitrogen-methylaniline-dealkylating activity by 57%. At the same time liver glycogen and blood glucose levels decreased, while the serum free fatty acid level increased. In sheep, degenerative changes in hepatocytes and death of bone marrow cells were noticed following an intravenous injection of 10 mCi [144]Ce-[144]Pr; the bone marrow was regenerated two days postinjection (Sullivan *et al.*, 1969).

Cardiopolygraphic studies in rats following parenteral injection of [144]Ce as $CeCl_3$ revealed a slight cardiotoxic effect of Ce (Sel'tser, 1967). Slight detrimental changes in cardiac contraction, an increase in the inhibitory effects of ether anesthesia on cardiac rhythm, and reduced sensitivity of the rats to epinephrine were some of the effects noted.

Cerium salts, such as the nitrate and chloride, appear to have marked adreno- and sympathomimitic action; this is especially evident when epinephrine or norepinephrine is first administered to the experimental animals (Bagirov, 1968). This has been attributed to an increase in the catecholamine content of the tissues of the brain, adrenals, and heart caused by the Ce stimulation of the peripheral and central adrenoreactive centers.

Cerium causes direct calcification in soft tissues when injected subcutaneously; however, the systemic calcification in spleen caused by intravenous injection of cerium salts can be reversed by injecting sodium pyrophosphate.

Studies on the pharmacology and toxicology of Pr and Nd (Haley *et al.*, 1964*b*), and of Sm (Haley *et al.*, 1961) revealed their similarity to Ce and La. The metabolism and distribution of Pr, Nd, and Sm following absorption was found to be similar to those of La and Ce. These three lanthanides exhibited a delayed acute toxicity with symptoms similar to those of Le and Ce, but the toxicity levels (LD_{50} values) were lower than were those of La and Ce, as seen in Table 3-8.

The chlorides of Pr, Nd, and Sm are less toxic than the nitrates both intraperitoneally and orally (Bruce *et al.*, 1963, Haley *et al.*, 1964*b*). Feeding the chlorides to rats as 1% of the diet did not result in any inhibition of growth, changes in the hemogram, or in any evidence of histopathologic damage to the internal organs; acute toxic doses of these chlorides by parenteral administration caused depressant action and death by cardiovascular collapse and respiratory failure. Intradermal injection produced nodules containing foreign-body giant cells and crystals, and instillation into the eye resulted in corneal ulceration. In cats the chlorides of Pr, Nd, and Sm caused transient hypotension and decreased peripheral blood flow following intravenous injection. At doses of 10–40 mg, the halides of Pr, Nd, or Sm produced an increasing depression of tonus and contractility of isolated rabbit ileum. Chronic intoxication, by both oral and parenteral administrations of Pr, Nd, or Sm, caused damage to the livers of experimental animals.

Promethium does not exist free or in any ores of lanthanides, but it is present in the products of nuclear reactions. It is an artificial man-made element; the halflife of the most stable radioisotope is about 2 years. Although simple salts and EDTA chelates have been prepared, there are no reports of studies on Pm metabolism in man or animals, but it is assumed that Pm exhibits the characteristic metabolic properties of the light lanthanides.

Toxicity of Medium Lanthanides

Among the medium lanthanides, Eu, Gd, Tb, Dy, and Ho, only Eu and its metabolism have been studied extensively, because of the extensive use of Eu vanadate phosphor in color television tubes. Europium, terbium, and dysprosium oxides have been found to be most useful as multiple nutrient markers in humans (Luckey *et al.*, 1977); their safety is indicated by the health of three generations of mice fed excessive quantities, 20–1000 ppm (Hutcheson *et al.*, 1975*a*). Europium typifies the medium lanthanides, with toxicity lower than that of lighter and heavier lanthanides.

In common with those of the other lanthanides, the distribution and excretion of soluble Eu salts following intravenous administration depends upon their concentration; when administered in very low doses, Eu^{3+} is rapidly cleared from the blood and largely deposited in the skeleton; with higher doses and with colloidal forms, Eu compounds are slowly cleared from the blood and deposited more in the kidneys and liver than in the skeleton (Berke, 1968).

The symptoms of acute parenteral Eu toxicity in rats and mice are ataxia, lachrymation, and depressed respiration and movement, followed by death in 24 hours from cardiovascular collapse and respiratory paralysis;

females are more susceptible than males. Toxicity from ingestion of simple Eu compounds is unusual because of their poor absorption; there was no growth inhibition in rats fed 1% $EuCl_3$ for three months, nor was there any histopathological damage in any internal organ. In addition, there was no inhibition of growth in mice fed a mixture of Eu, Tb, and Dy oxides at 100 ppm for three generations (Hutcheson *et al.*, 1975*a*). Topical application of $EuCl_3$, 0.1 M solution, to rabbit corneas caused transient ocular irritation, while intradermal injection caused local necrosis and tissue calcification (Haley *et al.*, 1965). An intravenous dose of 10–20 mg Eu/kg as $EuCl_3$ caused a transient fall of blood pressure in cats and dogs. A still higher dose, 23–76 mg/kg, in rats caused marked modification in their electrocardiographic patterns, especially when the rats were anesthetized with Na phenobarbitol prior to Eu^{3+} injection (Zanni and Ramos, 1965); sinusal bradycardia, auriculoventricular dissociation, extra systoles, negative T waves, and increase of PR and QT segments were observed, and a slight and transient parasympatholytic effect was noticed in the autonomic nervous system.

Deposition, retention, and clearance of inhaled Eu_2O_3 (^{152,154}Eu) in rats were determined over a period of 30 days (Johnson and Ziemer, 1971), and the Eu_2O_3 was retained in the pulmonary tissues. Inhalation of soluble Eu salts resulted in partial absorption and deposition in the pulmonary lymph nodes and progressively increased the ratio of large to small lymphocytes (Berke *et al.*, 1968). *In vitro* studies on the effect of $EuCl_3$ on isolated guinea pig ileum, rat and rabbit duodenum, and guinea pig and rat uterus showed depression in tonus and contractility (Haley *et al.*, 1965). Europium chloride appears to have teratogenic effects in the chicken; when injected into the yolk sac of 8-day-old chick embryos, $EuCl_3$, 20 mg Eu as $EuCl_3$, produced leg deformities, joint injuries, inhibition of feather formation, edema, and subcutaneous blisters (Zanni, 1965).

Gadolinium, a medium lanthanide, appears to be similar to samarium in its toxicologic properties (Haley *et al.*, 1961). Symptoms of acute parenteral toxicity of Gd include lethargy, abdominal cramps, diarrhea, decreased respiration, muscular spasms, and death due to respiratory collapse. Feeding 1% Gd as $GdCl_3$ to rats resulted in perinuclear vacuolization of parenchymal cells of the liver. Soluble Gd salts compete with Ca^{2+} and disrupt its uptake by binding with the Ca carrier in isolated cardiac and smooth muscles (Krasnow, 1972). Implantation of a 200-mg pellet of metallic Gd in mice induced sarcoma at the site of implantation (Ball *et al.*, 1970).

Terbium toxicology was studied using stable isotopes (Magnusson, 1962; Haley *et al.*, 1963). The absorption, distribution, and excretion of Tb salts from the digestive tract of mammals and from the sites of parenteral administration were found to be similar to those of other lanthanides. Skeletal deposition of Tb was higher than that of the light lanthanides.

Symptoms of acute Tb toxicity in laboratory animals are similar to other lanthanides. Feeding 1% Tb as $TbCl_3$ to rats caused inhibition of growth rate; however, male animals eventually attained the same weight as control animals, although females did not. Aqueous solutions of $TbCl_3$ caused damage to rabbit cornea and iris, with ulceration of the conjunctiva, and $TbCl_3$ is irritating to both intact and abraded skin, intradermal injection causing focal necrosis and topical calcification in the soft tissues.

Following intravenous injection, Tb, 35 mg Tb/kg, produced enlargement of liver and focal necrosis in liver (Magnusson, 1962). The focal necrosis was mostly in the caudal lobe and was owing to poor blood circulation.(The lanthanides are known to lower blood pressure.) The liver mitochondria did not undergo any damage. The dilation of the cisternae and dissociation of ribosomes was partly caused by the binding of Tb to nucleic acids. There was no fatty degeneration of the liver, but liver glycogen levels decreased; blood ornithine carbamyl transferase activity increased, with a fall in blood glucose. *In vitro* studies indicate that Tb salts can depress the tonus and contractility of isolated rabbit ileum, leading to paralysis.

Metabolic and toxicologic studies on Dy (Haley *et al.*, 1966) and Ho (Magnusson, 1962; Haley *et al.*, 1966) revealed the similarity in their general properties of these medium lanthanides with other lanthanides. Radioactive Dy (^{165}Dy) has been tried in the treatment of brain tumors (Brownell *et al.*, 1960).

The metabolism and general symptoms of acute toxicity of Dy and Ho are similar to those of Eu, the first member of the medium lanthanides. These two lanthanides and their salts are not toxic when administered orally to rats, and they cause no inhibition of growth rate. Parenterally administered Dy and Ho salts caused an increase in serum ornithine carbamyl transferase activity, and damage to the liver in rats; focal necrosis without fatty degeneration and without damage to mitochondria was reported.

Following parenteral administration, dysprosium differs from other lanthanides by its greater accumulation in the lungs than in other tissues. Inhalation of Dy vapor by humans causes increased insensitivity to heat, itching, and a sharper sense of taste and smell (Gehrcke *et al.*, 1939). The conjunctivitis caused by topical application of $GdCl_3$ to rabbit corneas is not persistent. Formation of necrosis and soft-tissue calcification is quick following intradermal injections of $GdCl_3$.

Holmium differs from dysprosium in producing a persistent conjunctivitis and in the slow development of necrosis and soft-tissue calcification following administration of Ho. Intense irritation to intact and abraded skin is caused by $HoCl_3$ application.

Toxicity of Heavy Lanthanides

Ytterbium and lutetium salts have been used as multiple heavy markers in rat feeding (Luckey *et al.*, 1975), and ytterbium and thulium oxides have been used as markers in the diets of mice and monkeys (Hutcheson *et al.*, 1975), and humans (Luckey *et al.*, 1977).

Metabolic and toxicologic studies on Er (Haley *et al.*, 1966), Tm and Yb (Magnusson, 1963, Haley *et al.*, 1963), and Lu (Haley *et al.* 1964*a*) reveal the similarity in their properties for absorption, distribution, excretion, deposition in the skeleton, and toxicity. These heavy lanthanides resemble Y, another subgroup IIIB metal, more than the other lanthanides. The toxicity of the heavy lanthanides is considerably less than that of Y, and is significantly less than that of the light lanthanides; however, the heavy lanthanides exhibit most of the symptoms of acute lanthanide toxicity in mammals, but chronic toxicity symptoms are milder than those of the light and medium lanthanides.

Thulium chloride when fed at 1% of the diet inhibits growth in mice and rats; the chlorides of Er, Yb, and Lu do not inhibit growth. Thulium oxide, 80 ppm Tm, had no effects upon growth, reproduction, or survival when fed to mice (Hutcheson *et al.*, 1975*a*); the insoluble oxides are practically nontoxic. Kay (1976) found no [160]Tb absorption when one million times the detection level of [160]Tb_2O_3 (10 mg/kg) was fed to rats. However, Tuchweber and Savoie (1968) report that ingestion of $YbCl_3$ caused gastric hemorrhages in rats, especially in females, and also caused perinuclear vacuolization of the liver. Topical application of chlorides to rabbit cornea caused no damage to the cornea, but transient ulceration in the conjunctiva. Intradermal injection of heavier lanthanide chlorides caused focal necrosis and topical calcification. Intravenous administration of the chlorides, 3 mg metal/kg, in rats induce topical calcification at the site of injection of polymixin, a histamine liberator; cerium salts induce this calcergy without any histamine liberator (Tuchweber and Savoie, 1968).

Intravenous injections of the soluble salts of heavy lanthanides at subtoxic levels cause hepatic damage; enlargement of the liver and focal necrosis develop with damage to mitochondria. The focal necrosis is mostly in the caudal lobe and is caused by poor blood circulation. The dilation of cisternae and dissociation of ribosomes is due to the binding of the lanthanides to the nucleic acids. There is no increased deposition of neutral fats in the liver, but the liver glycogen level is decreased, the serum glucose level is decreased, and the serum ornithine carbamyl transferase activity is increased.

Implantation of approximately 200 mg of metallic Yb pellets in mice induces sarcomas or promotes their formation at the site of implantation by "uncovering latent oncogenic factors" (Ball *et al.*, 1970), but no such data was found for other heavy lanthanides.

Isolated rabbit ileum undergoes depression of tonus and contractility when suspended in media which contain soluble heavy lanthanide salts. Erbium salts are very effective as true antiprothombin and thrombokinase antagonists (Dyckerhoff and Grunewald, 1943). Inhalation of Er vapor by humans causes strong itching, a sharp sense of taste and odor, and increased sensitivity to heat (Gehrcke *et al.*, 1939).

Nodule formation following intradermal injection of Er salts is attributed to the formation of crystalline $ErPO_4$, which is later surrounded by histocytes and foreign-body giant cells; fibroblasts and granulomatous tissue extend into the nodule.

ACTINIDES

The actinides constitute a series of fifteen highly radioactive metals, starting with Ac (atomic number, 89) and ending with Lr (atomic number, 103); other elements to continue the series are being made. These metals are inner-transitional elements, very similar to the lanthanides in their electronic configuration. The elements beyond U are man-made elements and are discussed separately as transuranium elements. Actinium and protactinium are the most highly radioactive of the actinides, about 150 times as active as radium. Actinides occur in the uranium mineral pitchblende; little is known of their chemical toxicity. Thorium and uranium occur in a number of less radioactive isotopes, and their toxicology is discussed separately.

Thorium (Th)

Thorium, the second member of the actinide series, occurs in thorite, thorianite, orangite, and yttrocrassite, and in monazite along with other lanthanides. The earth's crust contains about 15 ppm Th and seawater about 0.001 ppb. There are no authentic reports about Th in humans.

Thorium is increasingly used in nuclear breeder reactors in conjunction with ^{235}U. The lightweight Th–Mg alloy containing 3% Th and 0.7% Y or Zr is used in making frames for missiles, satellites, and supersonic bombers because of its capacity to withstand high temperature. Because of its low work function and high electron emission, Th is coated on tungsten filaments in vacuum tubes and incandescent lamps. Thorium oxide is used in incandescent mantles and as an X-ray contrast medium. Glass with ThO_2 has a high refractive index and a low dispersion which is useful in high-quality camera lenses and scientific equipment; ThO_2 is used as a catalyst in sulfuric acid manufacture, petroleum cracking, and in the manufacture of nitric acid from ammonia. Thorium is an industrial health hazard since its radiodisintegration product, ^{230}Th, is a radiation hazard, but chemical poisonings due to Th or its compounds are rare.

Chemistry

Thorium exists as a number of radioactive isotopes; the one which occurs most abundantly in nature is ^{232}Th, which has a half-life of 1.41×10^{10} years and disintegrates finally into lead. Thorium exhibits tetravalence, and its cationic salts, such as the nitrate and chloride, are water-soluble. With salts of more strongly basic metals Th forms double (complex) salts which are similar in structure to potash alum. Highly charged but weakly basic Th^{4+} has a strong tendency to complex formation with both inorganic anions and organic ligands:

$$\left[\text{=Th} \begin{array}{c} H_2O \\ OH \end{array} \right]^{n} + an^- \longrightarrow \left[\text{=Th} \begin{array}{c} an \\ OH \end{array} \right]^{n-1} + H_2O$$

$$\left[\text{=Th} \begin{array}{c} OH \\ OH \end{array} Th \right]^{n} + an^- \longrightarrow \left[\text{=Th} \begin{array}{c} anHO \\ OH \end{array} Th\text{=} \right]^{n-1}$$

Aqueous solutions of highly acid salts such as $Th(NO_3)_4$ precipitate as hydroxides. However, when tetravalent Th undergoes extensive hydrolysis above pH 3 it behaves more like a colloid or radiocolloid than a true solute; this olation takes place in biologic fluids and tissues. If Th^{4+} is complexed with citrate, olation and interaction of Th^{4+} with other complexing agents is minimized. Tetravalent Th exists in biologic fluids as a monomeric ion only at extremely low concentrations; more commonly it is a particulate or colloidal polymer of olated Th. Phosphate groups in the bone matrix have an affinity for Th.

Metabolism

Thorium is neither essential nor stimulatory in man or animals. Thorium is ubiquitous in nature, and hence minute quantities are presumed to be present in the environment and in food, but there are no specific reports about the dietary intake of Th by man. Studies on Th metabolism are not extensive and are restricted to the toxic effects. Albert (1966) reviews some aspects of Th metabolism while discussing the industrial hygiene aspects of Th; the metabolic behavior of Th is not analogous to any normal body constituent.

The gastrointestinal absorption of Th salts is very poor owing to olation of Th salts in the intestine, and depends upon the solubility and dose of the compound. The absorption in mice of doses of 25 mg ^{232}Th/mouse as Th$(NO_3)_4$ is negligible. Thorium citrate is absorbed a little better than the other Th salts (Salerno and Mattis, 1951). The absorption of Th$(NO_3)_4$ in rats is about 0.06% for a dose of 5–30 mg/kg, and is less than 0.001% for a dose of 500–800 mg/kg, and that of ThCl$_4$, Th$(SO_4)_2$, and ThO$_2$ is almost negligible. The absorption of Th salts from the sites of intramuscular injection is also poor, but soluble Th salts at very low concentration are absorbed in one day. In rats, about 6% of a dose of 2–5 mg/kg of Th$(SO_4)_2$ was absorbed in two weeks; ThO$_2$ remains at the site for a very long time, and only small quantities are gradually removed by lymphocytes to the regional lymph nodes. The absorption of Th salts from the lung into the blood is very slow; Th$(SO_4)_2$ deposited intratracheally in rats remained in the lung, and only 2% of a 4 mg/kg dose of Th$(OH)_4$ was absorbed; Th citrate complex was absorbed a little more rapidly (Boecker *et al.*, 1963). Following inhalation, the absorption of a very low dose, 1.75×10^{-5} mg, of ThCl$_4$ was about 15%; with Th citrate the absorption of a similar dose from the lung increased to 30% (Boecker, 1963).

There is rapid removal of Th salts from the blood following intravenous injection; about 30% remains in the blood after 6 hours. The distribution of Th salts from the blood to different organs depends on the concentration and nature of the salt; 50–70% of an intravenous dose of 2–2.5 mg/kg of either Th$(NO_3)_4$ or Th$(SO_4)_2$ is deposited in the liver, and 10–20% in the bone marrow. About 90% of a very low dose of 2×10^{-9} mg/kg of Th$(NO_3)_4$ or of $1 \times 10^{-3} - 1 \times 10^{-6}$ mg/kg thorium citrate is deposited in the skeleton, and about 7% is deposited in the liver. Still larger doses might be deposited differently. Thorium citrate is retained in the liver and skeleton for a long time. Thorium dioxide is taken up by the cells of the reticuloendothelial system, spleen, liver, and bone marrow, and none of the ThO$_2$ from the reticuloendothelial system is transferred to the skeleton. The slow excretion of absorbed Th salt is both fecal and urinary; the unabsorbed and polymeric colloidal Th from the liver, spleen, reticuloendothelial system, and bile is excreted in the feces, while absorbed Th in the monomeric form is excreted in the urine. The fecal/urinary ratio of orally ingested Th salt is about 45, while the ratio is 1.6 for intravenously injected, carrier-free Th. The distribution in the tissues and the excretion of Th salts depend on the physical and chemical state in which Th is introduced into the body.

Toxicity

Thorium and its isotopes are toxic radiologically, but the chemical toxicity of Th is low and depends on the nature of the Th salt. Chronic subcutaneous administration, prolonged feeding, or inhalation of Th compounds such as ThO_2 at low levels have no significant adverse effects in animals, but the demonstrable lethality of thorium nitrate and chloride has been extablished by intraperitoneal, intravenous, and intratracheal administrations. When $ThCl_4$ or $Th(NO_3)_4$ is given intraperitoneally, animals die of chemical peritonitis, abdominal distension, and rigidity, possibly owing to the toxicity of anions (Mattis, 1950). Toxicity data of Th is summarized in Table 3-9. Mice appear to be more susceptible than rats to Th toxicity from gavage, and Th salts are more toxic when given intravenously than by other modes of administration. Most of the toxicology data on Th are from U.S.A.E.C. reports (Acherman, 1948; Mattis, 1950; Downs *et al.*, 1959*a*).

The weight loss in animals during chronic feeding of $Th(NO_3)_4$ is mostly owing to reduced food intake. Rats tolerate about 3% $Th(NO_3)_4$ in their diet; above 5%, moderate to severe growth depression is observed. Dogs fed 1 g $Th(NO_3)_4$/kg body weight daily for 45 days showed a 15% decrease in growth (Downs *et al.*, 1959*a*). Dogs, rabbits, guinea pigs, and mice exposed to inhalation of the oxide, fluoride, oxalate, and nitrate of Th at 10—80 mg of the metal/m^3 for 60–270 days, showed no mortality, but leukocyte levels increased, and abnormal bone marrow and lung lesions of uncertain etiology were noticed (Hall *et al.*, 1951). Thorium nitrate aerosol at 113–330 mg Th/m^3 was not toxic to mice exposed for 40 minutes per day for 40 days (Patrick, 1948). Abnormal leukocytes were seen in dogs following the inhalation of ThO_2 particles 1 μm in diameter at a concentration of 11–76 mg Th/m^3 (Hodge *et al.*, 1960). Nonetheless, workers in a Th refinery did not show any symptoms of acute Th toxicity following inhalation exposure to various Th compounds for more than six months (Albert *et al.*, 1955). Reports about the chemical toxicity of Th compounds and their involvement in any physiological and biochemical functions in living tissues are scarce and fragmentary. Poor gastrointestinal absorption and the formation of particulate or colloidal forms of Th following absorption, and their subsequent removal through phagocytosis by the reticuloendothelial system, may partially account for the comparatively low chemical

TABLE 3-9. Thorium Toxicity

Compound	Animal	Route[a]	Toxicity	Compound mg	Metal mg	Metal mM	Metal pT	Death in
Thorium oxide ThO$_2$	Rat	it	LD$_{50}$	> 1140	1000	4.31	2.37	24 hr
Thorium chloride ThCl$_4$	Mouse	iv	LD$_{50}$	51.5	32	0.14	3.86	Immediate
	Mouse	sc	LD$_{50}$	4000	2500	10.8	1.97	
	Rat	ip	LD$_{50}$	1900	1180	5.09	2.29	24 hr
	Rat	g	LD$_{50}$	5480	3400	14.7	1.83	24 hr
	Rat	iv	LD$_{100}$	33.8	21	0.094	4.03	Immediate
	Rat	iv	LD$_{50}$	28	17.5	0.076	4.12	Immediate
	Rabbit	iv	LD$_{50}$	50	31	0.13	3.87	Immediate
Thorium nitrate Th(NO$_3$)$_4$·4H$_2$O	Mouse	iv	LD$_{50}$	45	19	0.081	4.09	Immediate
	Mouse	g	LD$_{50}$	1480	620	2.67	2.57	24 hr
	Mouse	g	LD$_{50}$	4000	1680	7.24	2.14	14 days
	Rat	ip	LD$_{50}$	68	28.5	0.12	3.91	14 days
	Rat	ip	LD$_{50}$	1760	650	2.80	2.55	24 hr
	Rat	g	LD$_{50}$	3800	1600	6.90	2.16	24 hr
	Rat	it	LD$_{50}$	88	37	0.16	3.80	24 hr
	Rat	iv	LD$_{50}$	47.6	20	0.088	4.06	Immediate
	Rabbit	iv	LD$_{100}$	57	24	0.10	3.99	Immediate
Thorium citrate Th$_3$(C$_6$H$_5$O$_7$)$_4$	Rat	iv	LD$_{50}$	> 60	> 29	0.13	3.89	Immediate

[a] it = intratracheal; g = gavage.

toxicity of Th salts. There are no reports about any homeostatic mechanism or any detoxication mechanism for Th.

Uranium (U)

Uranium occurs as oxides in pitchblende (UO$_2$, UO$_3$, PbO, ThO$_2$, etc.), carnotite (K$_2$O, 2U$_2$O$_3$, V$_2$O$_5$), and tobernite (CuUO$_2$·P$_2$O$_8$). The earth's crust contains about 2 ppm U, which is higher than the combined total of Sb, Bi, Au, Ag, and Hg; seawater contains about 3.3 ppb U. The bodies of human adults contain about 0.2 mg U, of which 65% is present in the bones.

Uranium is used as nuclear fuel and in nuclear weapons since U can be converted into fissionable Pu. Uranium, as a nuclear explosive, is used in mining, in power plants, and in isotope production. Uranyl acetate is a laboratory reagent and activator in bacterial processes, and is used in dry-copying ink. Uranyl nitrate is a good intensifier in photography and is also used in the ceramics and glass industries; particularly, it may be found in orange pottery. Unexpectedly, we found U in feces during preliminary studies on heavy metals in mice fed a commercial diet. The source was traced to U in the porcelain food container. It is also present in porcelain teeth. Uranium is an industrial health hazard both chemically and radiologically; the chief hazard in mining is from radiation, the emission of α particles from U, radon gas, and its particulate daughters, RaA, and RaC. The chemical toxicity of U is mainly owing to contaminants such as Pb, Th, and V. The potential for environmental pollution is high unless controlled by stringent safety precautions.

Chemistry

All the isotopes of U are radioactive. The naturally occurring and most abundant (99.25% of the total) is ^{238}U with a half-life of 4.51×10^9 years; it is not as highly radioactive as ^{235}U (0.7%) and ^{233}U (0.004%). Uranium exhibits variable valence from $+3$ to $+6$ in the crystalline state; in aqueous media, U^{4+} and U^{6+} are the stable forms, U^{6+} existing as the uranyl (UO_2^{2+}) ion. Uranium forms numerous intermetallic compounds as well as cationic and anionic salts. Uranium oxides and fluorides, and uranyl nitrate and acetate are uranium compounds commonly used in industry and studied in toxicology. Uranyl ions readily form stable complexes with carbonate and phosphate ligands in biologic fluids and tissues; carboxyl and hydroxyl groups are involved to a lesser degree. The water-soluble uranium bicarbonate complex $[(UO_2HCO_3)^+]$ is very stable and will not dissociate at the pH of the soft tissues, and hence is not deposited in these tissues. Lower pH will dissociate the complex, which happens in bone, leading to U accumulation or surface deposition in the bones. The phosphoryl group of DNA can bind uranyl ions, and albumin also readily complexes with uranyl ions; serum albumin binds 18 $(UO_2)^{2+}$ ions, while egg albumin can bind 8 $(UO_2)^{2+}$ ions per mole.

Metabolism

Uranium is ubiquitous in the environment, but it is neither essential nor stimulatory in man or animals. Uranium is an adventitious metal, and it can enter body fluids through processes of anionic and cationic substitution. The average daily dietary intake of U by man is 1–1.5 μg; the U content of some essential foods (expressed in ng U/g) is: fat and oil, 2.0; starchy tubers, 1.0; vegetables, 0.8; and meats and eggs, 0.4 (Hamilton, 1972). The metabolism and toxicology of uranium compounds have been reviewed (Tannebaum, 1951; Voegtlin and Hodge, 1953; Hodge, 1956; Stokinger, 1963). Reports published by the U.S. Atomic Energy Commission and by similar agencies in other countries provide extensive information on U metabolism.

The gastrointestinal absorption of small doses of soluble uranyl salts in mammals is about 10%; insoluble salts are poorly absorbed. The absorption of U salts from the sites of intramuscular injection and from the peritoneal cavity is poor; the U salts remain at the sites, causing an intense local reaction before their eventual, slow absorption. Soluble uranyl salts are also absorbed through the skin. Following inhalation the absorption of U salts from the lung tissues into the blood depends upon their solubility and particle size. Dusts of oxides and fluorides remain in the lung for a longer period of time than do soluble uranyl salts, which are absorbed very quickly. Upon intravenous injection uranyl ions form two fractions in the blood, a nondiffusible uranyl–albumin complex and a diffusible ionic uranyl bicarbonate, which remain in equilibrium with each other. Tetravalent U salts either attach themselves to erythrocytes or form polymeric particulates which are engulfed by phagocytosis. From the circulating blood, part of the uranyl compound is removed immediately, and 20% is deposited in the kidneys and 10–30% in the bones. While 60% of the uranyl compound deposited temporarily in the kidney is excreted in the urine within 24 hours, the bound uranyl compound in bone is mobilized very slowly. The excretion of intravenously injected uranyl compounds in the feces is insignificant. The distribution of tetravalent uranium salts is different; about 50% of the U^{4+} dose is deposited in the liver and is mobilized later to be excreted in the feces. Very little U^{4+} is deposited in the kidney, with much being deposited in the bones. Urinary excretion is slow for U^{4+}, whereas fecal excretion is heavy at 2–4 days. The deposition and retention in the bones is higher for uranyl compounds than for U^{4+} salts. Lungs retain particulate and insoluble U salts, but the retention of U in other soft tissues is insignificant. The U level in the lungs and bones increases with age.

Toxicity

Uranium salts cause harmful effects chemically and radiologically; the acute toxicity data of uranium is summarized in Table 3-10. Uranyl salts are more toxic than U^{4+}; insoluble salts are the least toxic chemically, although they may be highly toxic radiologically when lodged in lung tissues. Rabbits and guinea pits are most susceptible to U toxicity. Orally ingested U salts are of low toxicity due to their poor absorption. Acute toxicity develops when soluble salts are given intravenously. Intraperitoneal toxicity of soluble U salts is considerably less than is their intravenous toxicity, and they are also toxic and lethal when applied to the skin. Dusts and mists of UF_6, UO_2F_2, UCl_4, and uranyl nitrates at 20 mg compound/m^3 (particle size ≤ 1 μm) were generally fatal when laboratory animals were exposed daily for one month. The oral toxicity of these compounds in rats is far less than

TABLE 3-10. Uranium Toxicity

| | | | | Dosage/kg body weight | | | |
| | | | | Compound | Metal | | |
Compound	Animal	Route	Toxicity	mg	mg	mM	pT
Uranium oxide UO_2	Mouse	ip	LD_{50}	400	332	1.39	2.86
Uranyl fluoride	Rat	iv	LD_{50}	40	31	0.13	3.89
UO_2F_2	Rat	ip	MLD	40	31	0.13	3.89
Uranium tetrachloride	Rat	ip	LD_{50}	335	210	0.88	3.05
UCl_4	Rat	iv	LD_{50}	6	3.75	0.108	4.80
Uranyl nitrate	Mouse	ip	LD_{50}	50	23.7	0.10	4.00
$UO_2(NO_3)_2 \cdot 6H_2O$	Mouse	iv	LD_{100}	30	14.2	0.060	4.22
	Rat, ♂	ip	LD_{50}	305	144	0.60	3.22
	Rat, ♀	ip	LD_{50}	135	64	0.27	3.57
	Rat	iv	LD_{100}	2.1	1.0	0.0042	5.38
	Cat	oral	LD_{100}	238	100	0.42	3.38
	Dog	oral	LD_{100}	1400	660	2.77	2.56
	Dog	sc	LD_{50}	14	6.6	0.03	4.56
	Dog	iv	LD_{100}	6.75	3.2	0.01	4.87
	Rabbit	iv	LD_{50}	0.21	0.1	0.00042	6.38
	Guinea pig	iv	LD_{100}	0.63	0.3	0.0013	5.80
Uranyl acetate $UO_2(C_2H_3O_2)_2 \cdot 2H_2O$	Mouse	ip	LD_{50}	400	245	0.030	2.99

the toxicity by inhalation, 30-fold less for UO_4 and 3300-fold less for UF_4 and U_3O_8. The solubility of U salts is the dominant influence on their toxicity, and growth retardation, renal dysfunction, and consequent abnormal hepatic function are the symtoms of chronic chemical toxicity of U compounds. Acute U toxicity is caused mainly by destruction of renal tissues, leading to death. The comparative acute toxicity of some U salts is: $UF_2 > UO_2F_2 > UCl_4 > UO_2(NO_3)_2$; UO_2 and U_3O_8 are the least toxic U salts. Symptoms of chronic toxicity and the results of chronic-toxicity studies become confused with the radiation toxicity of uranium and other actinides.

Rats tolerate diets which contain either 0.5% soluble U salts or 20% insoluble U salts for nearly 30 days with no toxic symptoms, but soluble salts at 1–4% of the diet caused 50% mortality within 30 days. Growth depression was observed when rats were fed 0.5% uranyl nitrate for a period of 6 months or more, and uranyl nitrate also caused adverse effects on the growth of dogs fed 0.2 mg of compound/kg per day for a year. Abnormal increases in urinary protein and sugar values were noticed, with histological changes in the renal cortex.

Renal injury is the major symptom of both acute and chronic U toxicity. Hepatic damage is due to the acidosis, azotemia, and cachexia induced by renal dysfunction; severe renal insufficiency and renal tubular necrosis have been observed in dogs, rats, and rabbits following U intoxication (Bencosme *et al.*, 1960; Nomiyama and Foulkes, 1968). High urinary levels of protein and the appearance of catalase and phosphatase in the urine are the early symptoms of renal injury. Decreased inulin clearance and strong inhibition of creatine clearance suggest depressed glomerular filtration, due to the uranyl cytotoxic effect on the renal tubules. Ryan *et al.* (1973) confirmed the above observations in rats; extensive renal tubular necrosis did not cause azotemia, while renal tubular epithelial dysfunction tended to parallel the changes in renal tubular pathology. The uranium–bicarbonate complex dissociates in the tubules following glomerular filtration, causing the deposition of the uranyl salt on the renal tubular cell walls, which progessively leads from the inhibition of vital enzyme function to renal necrosis. Acute renal failure is due to back-diffusion of glomerular filtrate through the damaged tubular epithelium.

Uranium-235 and -238 are known radiocarcinogens; ^{238}U produces sarcomas in experimental rats following parenteral injections, especially in the femoral marrow and in the chest wall, suggesting that ^{238}U is a chemical carcinogen (Hueper *et al.*, 1952).

Experimental animals develop a tolerance to acute U intoxication following exposure to very small chronic doses of U. Despite this evidence of adaptation, there are no specific mechanisms for U detoxification, and there is no evidence for a homeostatic mechanism for U.

The mechanism of chemical toxicity of U is attributed to changes in cellular membrane permeability due to the binding of uranyl ions to phosphate ligands and to the inhibition of cellular carbohydrate metabolism. The principal effect is inactivation of phosphate-containing cofactors such as ATP.

TRANSURANIUM ELEMENTS

In the actinide series, metals with atomic numbers higher than uranium are called transuranium metals. They are neptunium (Np), plutonium (Pu), americium (Am), curium (Cm), berkelium (Bk), californium (Cf), einsteinium (Es), fermium (Fm) mendelevium (Md), nobelium (Nb), and lawrencium (Lr). Their electronic configuration is similar to the other group of inner-transitional elements, the rare-earth metals. The electronic buildup takes place in either the $5f$ or $6d$ subshells; the two valence electrons are in the $7s$ subshell. These metals are made artifically and do not occur in nature except following nuclear explosions. All are highly radioactive; and some have very short half-lives.

Transuranium metals and their salts are not used for any general purpose; they are extremely dangerous radiation hazards since even 1 μg of some isotopes such as ^{239}Pu causes cancer in laboratory animals. Although the present levels of these metals in food or other components of the environment are far below their toxic levels, future use of Pu in reactors is a matter of concern. At present, their levels of occurrence are insignificant compared with those of other toxic and hazardous materials; exposure to these metals and their radiation is possible in some chemistry and physics laboratories, nuclear reactors, in nuclear power plants, and from fallout after nuclear explosions.

Chemistry

The chemistry of some of the transuranium metals such as Np, Pu, Am, Cm, and Bk has been partially investigated. Transuranium elements generally exhibit a group valence of +3; Np, Pu, and Am exhibit valences ranging from +3 to +6, and Cm and Bk exhibit valences of +3 and +4. The ionic radius and electronegativity of these metals decrease progressively with the increase in their atomic number. The soluble cationic salts will hydrolyze in near-neutral solutions forming insoluble polymeric hydrated oxides of variable composition.

$$M^{4+} \rightarrow M(OH)_3^+ \rightarrow M(OH)_2^{2+} \rightarrow [M(OH)_2^{2+}]^n$$

This tendency to undergo hydrolysis and to associate with water or other coordinating groups to form strong anionic complexes is marked among the smaller tetravalent transuranic ions. These ions readily form stable and soluble chelates with synthetic chelating agents such as diethylenetriamine pentaacetic acid (DTPA), and citrate and lactate complexes of these ions can exist in biologic fluid. Transuranic ions have a great affinity toward phosphate ligands, especially in the bone matrix.

Metabolism

The metabolism of Np, Pu, and Am has been studied; most of the information contained in the following paragraphs, and the biological implications of the transuranium elements were discussed at the Hanford Symposium in 1971 and reported by Lindenbaum and Rosenthal (1972). Of the transuranium metals, plutonium has been most studied; the biochemistry and radiobiology of plutonium has been reviewed (Durbin, 1972; Bair and Thompson, 1974). Transuranium metals are not essential to man or animals, and there are no reports about any stimulatory action.

The gastrointestinal absorption of these metal salts is very poor; since these metal ions cannot exist in free ionic form at physiologic pH, they form polymeric and hydrated colloidal or particulate species. The soluble salts are removed very slowly from intramuscular injection sites, and subcutaneously implanted metals are absorbed in negligible amounts; only about 1–2% is absorbed during the life span of the experimental animals. In miniature swine, intradermally injected salts are removed and translocated to regional lymph nodes and to the liver in about a week. Following inhalation, the particles of metal salts that are aerodynamically capable of reaching the alveoli will be retained in the lung for a long period; later a portion of the particles begins to accumulate in the lymph nodes before final translocation to the liver, bones, and other tissues. Following intravenous injection, the distribution of transuranium metal salts from the blood depends upon the dose and the nature of the complex salt; polymeric colloidal forms of the metal salts are phagocytized and rapidly removed from the blood to the liver, while monomeric forms of these salts or their citrate complexes are removed slowly from the blood for deposition in the bone. Complexes of these metal ions with lactates and serum glycoproteins occur in the blood; serum transferrin binds Pu especially, and this Pu–transferrin complex is preferentially deposited on the endosteal surface of bone.

The polymeric forms of these metal complexes or compounds bind with proteins, especially ferritin in the liver, and accumulate in the lysosomes of the liver and the reticuloendothelial system (Taylor, 1972). The level of deposition in the liver depends upon the ionic radius of the metal ions; metals with decreased ionic radii are deposited more in the bone than in the liver. Skeletal retention is prolonged, whereas clearance from liver and other tissues vary widely.

The excretion of transuranium metal salts is both fecal and urinary; unabsorbed oral doses are excreted in the feces. In humans parenterally administered doses are excreted very slowly in the urine; only about 5% of the dose is excreted in the first 20 days, the next 10% in about two years after injection; it has been indirectly estimated that the next 35% of the dose will be excreted over a period of forty years. Thus, transuranium metals are permanent guests in mammalian tissues, but can be partially removed by therapeutic administration of DTPA as a transuranium-metal–DTPA complex.

Transuranium metal complexes do not easily penetrate physiologic membranes, and passage across cell membranes is mostly by pinocytosis.

In pregnant rats these metal compounds are retained in placentas during early pregnancy, and transplacental transfer to fetuses takes place later in pregnancy (Ovcharenko, 1972); transfer of polymeric forms to the fetuses is 15–20 times lower than that of citrate complexes.

Toxicity

Transuranium metals are extremely toxic radiologically, but their chemical toxicity has not been studied well because of their high radiotoxicity; the toxic effects are both somatic and genetic, the most devastating somatic effects being leukemia and other forms of cancer, cataract, and shortened life span. The genetic effects caused by mutation in gametes are observed in the offspring of exposed mammals. The radiotoxicity of some transuranium metals is summarized in Table 3-11; the LD_{50} radiobiologic dose of ^{237}Np in rats is 1.56 mg Np/kg body weight (Moskalev, 1972). Neptunium appears to be the most radiotoxic of the transuranium metals studied, but there are no data on chemical toxicity.

Transuranium metals are avid bone-surface seekers, accumulating especially in the endosteal surface, from where they irradiate the nearby cells, the presumed sites of cancer induction; bone necrosis and osteogenic sarcoma result from this irradiation. $Metal^{4+}$–transferrin complex remains on the bone surface until removed by resorption. The resorbing surface has an acid pH which dissociates the $metal^{4+}$–transferrin complex, and the resultant hydrated polymeric metal oxide complexes are removed by macrophages and transferred back to the circulating blood. The calcifying surfaces with neutral or alkaline pH do not favor the dissociation of $metal^{4+}$ from the transferrin complex; instead, the $metal^{4+}$–transferrin complex is buried by deposition of new bone.

Following inhalation, the transuranium metal complexes are engulfed by the alveolar epithelium and sequestered in the foci of fibrosis, accounting for their long retention in the lungs. The accumulation of these transuranium metals in the subpleural fibrotic areas of the lung provides the "hot spots" of irradiation which cause epithelial metaplasia and alveolar or bronchiolar carcinoma (Sanders, 1972).

The lodgment of transuranium elements both in monomeric and polymeric forms in the lysosomes of the liver and the reticuloendothelial system is very harmful. In addition to being sources of ionizing radiation, these metal compounds lyse the lysosomes releasing the hydrolases which cause much cellular damage, including chromosomal damage. The transuranium metals also affect the hematopoietic system, causing leukemia and blood abnormalities.

TABLE 3-11. Radiologic Toxicity of Some Transuranium Metals in Rats

Compound	Route	Toxicity[a]	Mortality in days
[237]Neptunium	iv	0.00205	30
[239]Plutonium	iv	0.0014	360
		0.059	30
		0.0245	120
		0.011	360
[241]Americium	iv	0.11	30
		0.043	120
		0.01	360
[244]Curium	iv	0.11	30
		0.038	120
		0.0044	360

[a]LD_{50} dose expressed as $\mu Ci/g$ body weight.

The transient accumulation of the transuranium metal complexes in the ovary causes follicular or lutean cysts, thecomas, and adenomas. Their initial retention in the placenta causes disorders of the developing placenta and intrauterine deaths of embryos, and their final transfer to the offspring causes postnatal disorders such as delayed physical development, variations in weight, abnormalities in blood formation, and decreased viability.

Plutonium (Pu)

Plutonium occurs in trace quantities in uranium ores. Since it was produced in 1940, about 0.5×10^6 curies have been released into the atmosphere from weapon testing, and the adult human probably has about 2 pCi Pu in his total body (Bair and Thompson, 1974).

Plutonium has become an increasingly important and abundant industrial metal among the transuranium elements; the prospects for use of plutonium as commercial nuclear fuel in many countries are increasing. Thus, chances of exposure to Pu radiologic toxicity through inhalation and through contaminated skin wounds among the industrial workers in the Pu nuclear fuel industry are also increasing. Marine vertebrates are able to remove [239]Pu from aquatic sediments and retain it in their bones and liver (Noshkin, 1972), and it is possible for this Pu to contaminate the food chain. However, the levels of Pu in food and in the environment are not significant as a radiation hazard, despite its being more toxic than radium as a potent carcinogen; an internal dose of 1 μg Pu will cause cancer in laboratory animals.

The United States Transuranium Registry maintained at Hanford Environmental Health Foundation, Hanford, Washington, collects and provides information on the epidemiology of toxicity due to transuranium metals. As nations move toward the commercialization of Pu, the current occupational tolerance standards for Pu, set up in 1949, may have to be lowered because its great potential value is matched by its enormous radiobiological hazard.

Chemistry

Plutonium exists as a number of isotopes. The common isotope ^{239}Pu is an α emitter (5 meV) with a half-life of 24,390 years; the much used ^{238}Pu is an α emitter with a half-life of 86.4 years; ^{240}Pu and ^{242}Pu are also α emitters and will become abundant in breeder reactors, while ^{241}Pu is a short-lived β-particle emitter.

Plutonium exhibits valences +3– +6 in solutions, but in most common biological and physiological conditions it occurs as Pu^{4+}. Pu^{4+} cannot exist as a free ion in organisms, but is always complexed with citrates, lactates, or proteins in biologic fluids and tissues. Weakly complexed Pu^{4+} will hydrolyze in near-neutral solutions to form polymeric hydrated oxides of variable composition; Pu^{4+} complexes with the iron-binding proteins, ferritin and transferrin, and these Pu^{4+}–ferritin or Pu^{4+}–transferrin complexes are stable at neutral pH.

Metabolism

The metabolism of Pu in experimental animals has been studied and reviewed (Katz *et al.*, 1955; Durbin, 1972; Bair and Thompson, 1974). Although the practice of extrapolating animal data to humans is questionable, information regarding tissue distribution and retention of Pu in experimental animals and in humans who were accidentally exposed to lethal doses of Pu are similar, suggesting that the general metabolism of Pu in mammals could be the same.

The absorption of Pu^{4+} as the nitrate from the gastrointestinal tract of mammals is extremely poor, about 0.002–0.05% of the ingested dose (Romney *et al.*, 1970). Absorption is about 0.1% with plutonium citrate, and increases to about 2% with hexavalent plutonium complexes and chelates. The poor absorption is due to the hydrolysis of Pu^{4+} at neutral pH and the consequent formation of insoluble hydrated polymeric Pu complexes. The absorption of soluble Pu^{4+} salts from sites of intramuscular and subcutaneous injections is also poor and slow, and absorption from subcutaneously implanted Pu metal is extremely slow, while the absorption of Pu^{4+} from intradermal sites is greater than from intramuscular sites. The absorption of Pu from the alveolar regions of the lung into the blood following inhalation of Pu compounds depends upon the nature of the compound; the uncommon hexavalent complexes and chelates of Pu are quickly absorbed, while the more commonly occurring Pu^{4+} salts are retained in the lungs for a long time before being slowly cleared. About 25% of inhaled particles are exhaled, and about 25% and 50%, respectively, are deposited in the lower and upper respiratory passages.

The distribution of Pu in blood and its subsequent clearance depends upon the Pu compound and the rapidity with which it is hydrolyzed in the blood to form polymeric complexes, shown in the equation

$$Pu^{4+} \xrightarrow{[HOH]} Pu(OH)_3^+ \rightarrow Pu(OH)_2^{2+} \rightarrow [Pu(OH)_2^{2+}]^n$$

Monomeric forms of Pu^{4+} complex with citrate, lactate, and glycoproteins of serum, and Pu^{4+} readily binds with transferrin, the maximum binding occuring at neutral pH. However, the Pu^{4+}–transferrin complex is less stable than the Fe^{3+}–transferrin complex. The particulate polymeric forms of Pu are engulfed by macrophages or leukocytes and are rapidly taken from the blood into the liver and cells of the reticuloendothelial system; ferritin is involved in the binding and retention of polymeric Pu in the liver. The monomeric Pu^{4+}, plutonium citrate, or Pu complexed to transferrin is slowly removed from the blood for eventual deposition in the bone. The distribution of Pu in dogs and humans is: bone, 65%, and liver, 23%. The reticuloendothelial cells of the bone marrow take up the polymeric forms of Pu. The deposition of ^{239}Pu on the bone surface depends upon the route of administration, with deposition, from largest to smallest, as follows: intravenous > intraperitoneal > subcutaneous > intramuscular > intratracheal > inhalation > oral > cutaneous (Lee, 1972). The acid pH at the resorbing surface of the bone (osteoclastic activity) releases Pu^{4+} from Pu^{4+}–transferrin complex and binds it to glycoproteins in and around the cells; the neutral or alkaline pH of the calcifying surface of the bone does not favor the dissociation of the Pu^{4+}–transferrin complex, but buries the complex under successive layers of new bone. Plutonium deposited in the skeleton is retained more or less permanently, while Pu deposited in the liver can be excreted very slowly.

The excretion of Pu is both fecal and urinary (Ballou and Hess, 1972). The unabsorbed Pu^{4+} from ingested doses is excreted rapidly in the feces, but the absorbed Pu^{4+}, from whatever route of administration, is excreted very slowly; about 5% of an intravenously injected dose is excreted in the first three weeks following the injection, and the next 10% may be excreted over a period of two years, depending upon the animal, while the rest is retained. Citrated and chelated complexes of Pu^{4+} are excreted in the urine, and other forms in the feces. Therapeutic injection of DTPA results in greater urinary excretion of absorbed Pu in the form of a soluble Pu–DTPA complex.

Toxicity

There are no reports about the chemical toxicity of Pu; the radiologic toxicity of Pu involves bone necrosis, bone and lung cancer, and detrimental effects on the reproductive system; also, toxic effects are observed in offspring of pregnant animals exposed to Pu. Among the Pu isotopes, the short-lived ^{241}Pu is highly toxic owing to its β-ray emissions, and the long-lived ^{239}Pu is equally toxic owing to the high energy associated with its α-particle emission. Generally, the harmful radiobiologic effects mentioned under the radiotoxicity of transuranium metals are exhibited by Pu (Thompson and Bair, 1972).

Neptunium (Np)

Neptunium, the second most highly radiotoxic of the transuranium metals tested experimentally, occurs in a number of isotopic forms. The most stable isotope, ^{237}Np, exhibits variable valence ranging from +3 to +6 and emits both α and γ radiation; it has all the characteristic physicochemical properties of the transuranium metals.

The gastrointestinal absorption of Np compounds varies from 0.001% to 2% when mg doses of the compounds are used. The absorption decreases in this order: $Np^{6+} > Np^{5+} > Np^{4+}$; the absorption from other modes of administration, such as intratracheal deposition and parenteral injection, follows the same order. The distribution of Np compounds from the blood following intravenous injection is similar to that of Pu; Np is similarly retained in the bones and liver (Ballou *et al.*, 1962). About 30% of an injected dose of Np–citrate complex is excreted in about three weeks.

The higher radiotoxicity of Np is attributed to its greater absorption and to its being retained longer in the skeleton than is found for other transuranium metals.

Americium (Am)

Americium exists in a number of isotopic forms, the most stable of which is ^{241}Am with a half-life of 258 years; it emits α particles. Americium exhibits variable valence from +3 to +6; the stable forms are penta- and hexavalent Am. It exhibits physicochemical properties characteristic of the transuranium metals, and it is a product of ^{241}Pu.

Americium differs somewhat from plutonium and neptunium in its metabolism and radiotoxicity. Among the three metals, Np, Pu, and Am, Am has the lowest radiotoxicity. The absorption of Am from the digestive tract and from sites of intramuscular and subcutaneous injections is similar to that of Pu. In adult baboons, Am is lost rapidly from the blood following intravenous injection, less than 1% of the dose remaining in the blood after 48 hours (Resen *et al.*, 1972). The distribution of Am in the tissues differs among different experimental animals and depends upon the route of administration. In baboons, 60% of an intravenous dose of Am becomes lodged in the skeleton and about 30% is distributed in the lungs, liver, and kidneys. In Chinese hamsters, the liver retains 50% of an intraperitoneally administered dose of Am, while the skeleton retains only about 20%. The excretion of Am is somewhat more rapid than that of either Np or Pu; about 10% of the injected dose is excreted in the first week and about 15% in the next three weeks; also, the removal of Am from the liver is more rapid than that of Np or Pu.

Americium exhibits all the harmful effects which are attributed to the radiotoxicity of transuranium compounds. However, ^{241}Am is more damaging than ^{239}Pu in its harmful effects on the reproductive system and development of the placenta, and in causing the intrauterine death of embryos (Ovcharenko, 1972), and the postnatal detrimental effects caused by ^{241}Am in the newborn are also more damaging than those caused by ^{239}Pu.

TOXICITY OF GROUP IV METALS

The elements of Group IV of the periodic chart are characterized by tetravalence and their ability to form both electrovalent and covalent compounds. The nonmetals carbon (C) and silicon (Si) and the metals germanium (Ge), tin (Sn), and lead (Pb) form subgroup IVA, and the metals titanium (Ti), zirconium (Zr), hafnium (Hf), and rutherfordium (Rf) form subgroup IVB.

Carbon is a major essential element in biological systems, and silicon occurs in mammalian tissues and may be considered essential by several criteria, but neither element is considered further herein. Germanium, tin, lead, titanium, and zirconium are stimulatory in low doses, and an essential role for Sn in the growth of rats is being advanced (Schwarz, 1971). The other Group IV metals are not essential in biological systems. The salts of Sn and Pb are more toxic than are the salts of Ge, Ti, Zr, and Hf, and organometallic compounds of Ge, Sn, and Pb are more toxic than are the inorganic salts.

The position of the metals in the periodic chart indicates their electroneutral and amphoteric character. Within this group of metals the comparative electropositivity, stability, and solubility of the cationic salts increase as atomic weights increase. These metals exhibit valences of +2 and +4, tetravalence being common with the anions of the lighter metals. The cationic salts of the metals Ti, Zr, and Hf undergo olation; Zr and Hf readily form basic oxy salts; however, Ti salts are readily hydrolyzed to oxy acids such as titanic acid, which then forms stable salts. Germanium and tin form unstable hydrides; Ge, Ti, and Sn form stable coordination compounds.

With the exception of Pb, the stable cationic salts of metals of this group are not readily absorbed from the digestve tract of mammals, and hence are not generally considered toxic. Salts of Pb and Sn are distributed mostly in soft tissues, and Pb is deposited also in bones. All Group IV metals are retained to some extent in the soft tissues.

Organic derivatives of the metals of Group IV are more toxic than are the inorganic salts, because of their greater lipid solubility, stability in biological fluids, and penetration into tissues such as brain and lodgment in the central nervous system. Salts of Sn and Pb are more toxic than are salts of the metals of subgroup IVB—Ti, Zr, and Hf. Of these metals Pb has been the most studied; it diffuses readily across blood-brain, placental, and mammary barriers, and accumulates, causing interference with cellular function.

Group IV Elements

Atomic number		(Core) Active electrons
Name (Symbol)		
Atomic weight		Usual valences
Subgroup A		Subgroup B
6	(He) 2,2	
Carbon (C)		
12.011	+2, +4, −4	
14	(Ne) 2,2	
Silicon (Si)		
28.09	+2, +4, −1	
32	(Ar) 2,2	22 (Ar) 2,2
Germanium (Ge)		Titanium (Ti)
72.59	+2, +4	47.9 +2, +3, +4
50	(Kr) 2,2	40 (Kr) 2,0,2
Tin (Sn)		Zirconium (Zr)
118.69	+2, +4	91.22 +4
82	(Xe) 2,2	70 (Xe) 2, 02
Lead (Pb)		Hafnium (Hf)
207.2	+2, +4	+4
		104 (Rn) ?
		Rutherfordium (Rf)
		?

SUBGROUP IVA

The inherent toxicity of metals of subgroup IVA increases with increased electropositivity; germanium is less toxic than tin and lead. Germanium hydride is the most toxic of the inhaled Ge compounds, and soluble Sn salts are more toxic than Pb salts on the basis of LD_{50} values. Lead salts are cumulative poisons. Lead and tin cations complex biologically active S^- compounds, and lead complexes also with phosphate and carboxyl groups.

Germanium (Ge)

Germanium occurs naturally as argyrodite [$(Ag_2S)_4GeS_2$] and as germanite ($7CuSFeSGeS_2$). It is present in very small amounts in coal; the flue dust from combustion of producer gas contains about 3% Ge. The earth's crust contains 2 ppm Ge and seawater about 0.05 ppb. Data about the total Ge content in humans are not available, but blood contains about 0.5 ppm. Germanium occurs widely in food; seafoods such as canned tuna and dried pan fish contain 3 ppm Ge, and canned tomato juice and baked beans contain about 5 ppm. While Ge is not considered an essential nutrient for mammals, it is stimulatory.

Industrially, Ge is used extensively in electronics, in transistor diodes and rectifiers. Germanium glass has a high refractive index and is used in wide-angle camera lenses, microscope objectives, and infrared lenses. Germanium is also used as a low-temperature catalyst in the hydrogenation of coal. In the past Ge compounds were used therapeutically to stimulate erythropoiesis. Germanium is not considered an industrial health hazard; accidental poisonings in humans are rare, in spite of exposure to appreciable amounts of germanium in foods and flue dust, and in the electronics industry.

Chemistry

Germanium exhibits valence of $+2$ and $+4$, and forms unstable cationic salts with halogens but not with oxy acids. Tetravalent anionic compounds are more stable, giving rise to soluble germanates. One form of GeO_2 is soluble and the other has low solubility. In its tetravalent compounds Ge is in coordinate valence with neighboring atoms; Ge forms octahedral chelates with a coordination number of 6. Germanium forms combustible volatile hydrides such as GeH_4 and Ge_2H_6. It does not form stable complexes with SH, or with OH groups containing biologically active molecules.

Metabolism

In spite of its ubiquitous presence in the earth's crust, in the soil, in plants and in food, Ge does not accumulate in human or animal tissues. The average daily intake of Ge by humans is about 4 mg. Germanium is not considered an essential metal, although the position of Ge in the periodic chart is within the range of biologically active trace metals, and it is stimulatory to animals. At low concentrations Ge stimulates mammalian cellular respiration (Read, 1949). The oxygen uptake of rat tissue slices is inhibited at 0.09 mM Ge, but is stimulated at 0.009 mM; Luckey (1975*b*) reviewed the stimulatory action of Ge on hematopoiesis in mammals and on respiration in rat tissue slices. The growth of chicks is stimulated by Ge (Kovar and Luckey, 1975). Data regarding change in the Ge content of human tissues with age or regarding Ge involvement in biochemical enzyme systems are not available. Schroeder and Balassa (1967*a*) have reviewed the biochemistry of Ge as an abnormal trace metal in man. Plants readily absorb Ge, and *Aspergillus niger* can store appreciable quantities of Ge, indicating that it has no antimicrobial activity.

Most metabolic studies on Ge were made with either GeO_2 or soluble germanates, such as GeO_4^{2-}, the most stable of Ge compounds under physiologic conditions. Gastrointestinal absorption of oxides and cationic Ge salts is poor; however, the absorption from small oral doses of germanates is about 75% in 4 hours and 96% in 8 hours. Parenterally administered germanates are absorbed rapidly and completely into the blood. Plasma proteins and erythrocytes carry the germanates in the blood where it is distributed in soft tissues. Also, inhalation studies reveal quick absorption of germanates from the lungs into blood (Dudley, 1953). The spleen, heart, lungs, and kidneys retain the germanate temporarily, and tracer quantities of Ge are retained temporarily in the skeleton (Durbin, 1960); organic Ge is stored in rat skeleton for a longer time than are inorganic Ge salts (Caujolle *et al.*, 1963); female rats retain more germanate in the spleen than do males (Schroeder and Balassa, 1967*b*). The excretion of germanate by animals from any mode of entry is mostly urinary, the urinary excretion of Ge by normal human adults ranging from 0.56 mg to 3.0 mg per day, according to the Ge content of foods eaten. There are no reports of Ge accumulation, nor any supporting evidence of a homeostatic mechanism for Ge in mammals, as was implied in earlier investigations.

TABLE 4-1. Germanium Toxicity

| Compound | Animal | Route | Toxicity | Dosage/kg body weight | | | |
| | | | | Compound | Metal | | |
				mg	mg	mM	pT
Germanium (powder)	Rabbit	sc	LD_{100}	—	586	8.07	2.09
Germanium dioxide GeO_2	Rat	sc	LD_{100}	>180	>125	1.72	2.76
	Rat	ip	LD_{50}	750	520	7.16	2.14
	Rabbit	sc	LD_{50}	845	586	8.07	2.09
	Guinea pig	ip	MLD	400	275	3.78	2.42
	Guinea pig	ip	LD_{100}	1440	1000	13.8	1.86
Germane GeH_4	Mouse	inhal	LD_{100}	610[a]	578[a]		
Triethyl germanium acetate	Rat	oral	LD_{50}	250	99.3	1.37	2.86
$(C_2H_5)_3 GeC_2H_3O_2$	Rat	iv	LD_{50}	50	19.8	0.272	3.56
Dimethyl germanium sulfide $(CH_3)_2GeS$	Mouse	ip	LD_{100}	250	135	1.86	2.73

[a]Exposure for 4 hr; number of mg/m^3.

Toxicity

Among subgroup IVA metals, Ge is considered the least toxic; soluble Ge compounds are more toxic by oral than by parenteral routes. Pharmacological inertness, diffusibility, and rapid excretion are responsible for the low Ge toxicity. Reviews of the toxicology of Ge include Rosenfeld and Wallace (1953), Rosenfeld (1954), Stokinger (1963), and Browning (1969). The symptoms of acute Ge toxicity in animals are profound hypothermia, listlessness, diarrhea, and cyanosis; other toxic effects include edema and hemorrhage of lungs, petechial hemorrhage in the walls of the small intestines, and a peritoneal effusion which is rich in protein. Edematous changes are also seen in heart muscle and in the parenchymal cells of the liver and kidneys (Bailey *et al.*, 1925). The symptoms of chronic Ge toxicity are mainly inhibition of growth and fatty degeneration of liver. The toxicity data of some germanium compounds are summarized in Table 4-1; by inhalation, gaseous germanium hydride is the most toxic of the Ge compounds; there are no reports of oral toxicity. Organic Ge compounds are more toxic than are inorganic compounds.

Germanium dioxide at 10 ppm Ge in the diet of rats stimulated growth; at 1000 ppm growth was depressed with 50% mortality, whereas at 100 ppm there was no observed effect. Giving soluble germanates in the drinking water for four weeks, 100 ppm Ge, was fatal, but rats acquired a tolerance for the dietary lethal dose if injected with subtoxic doses of germanium for some time (Rosenfeld and Wallace, 1953). Unpublished chick data from our laboratory confirm both the low-level stimulation and the high-level toxicity of Ge. In life-term experiments, mice and rats fed 5 ppm Ge as germanates in the drinking water showed no toxic symptoms other than decreased life span and mild fatty degeneration of the liver. Mice fed Ge had fewer tumors than control mice (Schroeder, 1967; Schroeder *et al.*, 1968*a,b*). Germanium tetrachloride fumes are irritating to the eyes and to mucous membranes of the respiratory tract. Germane is toxic and, like other metal hydrides such as AsH_3, its action is hemolytic. Rosenfeld and Wallace (1953) believe the cathartic action of Ge salts is due to the disturbance of water metabolism, leading to dehydration, decrease in blood pressure, and hypothermia. Organic compounds, such as dimethyl germanium oxide, are only slightly toxic to rats (Rochow and Sindler, 1950), but the triethyl germanium acetates are toxic (Cremer and Aldridge, 1964). The lack of gross tissue damage in acute Ge toxicity indicates that Ge somehow interferes with normal utilization of food.

Tin (Sn)

Tin is not abundant in most parts of the earth's crust, occurring mainly as tinstone or cassiterite (SnO_2) and in complex sulfides such as stannite (Cu_2FeSnS_4); tin is present up to 3 ppm in the earth's crust and up to 3 ppb in seawater. The adult human body contains about 17 mg tin, of which 25% is distributed in the fat and skin, 3.2% in erythrocytes, 0.8% in blood plasma, and the rest in soft tissues; tin has not been found in the bones or brain. Tin is present in fresh vegetables in proportion to the tin content of the soil. Fresh asparagus and dried split peas contain about 9 ppm Sn, and canned food products contain more Sn than do fresh products. Tin is stimulatory, may be essential for animals, and is oncogenic.

Tin is used industrially for tin plate and solder, and in the manufacture of alloys, such as bronze, brass, and pewter, in processes developed by the Sumerians by 2000 B.C. Tin plates are widely used as containers for food and for equipment in the dairy industry, and tin is also used in foil and collapsible tubes; electrodeposited coatings are used in electronic and engineering components. Tin salts are used in calico printing, as mordants in dyeing, and in the weighting of silk. Stannous fluoride is used in tooth pastes and in dental washes. Organic Sn compounds are used as stabilizers in polyvinyl plastics and in chlorinated rubber paints, and as fungicides, insecticides, and antihelminthics. Tin is not considered a serious industrial health hazard, but tin poisoning can occur from accidental ingestion, and formerly from eating exclusively canned food products from nonlacquered cans; Knorr *et al.* (1974) found tin accumulation in tomato puree stored in lacquered cans for more than two months. Studies on organic tin insecticides, extensively used in the past as agricultural chemicals, contributed much information about Sn poisoning. Tin-containing packing materials and fungicides and bactericides are considered major sources of Sn poisoning.

Chemistry

Tin exhibits variable valence from $+2$ to $+4$, but the trivalent forms are unstable. Cationic Sn^{2+} and Sn^{4+} are stable, and anionic stannites and stannates are insoluble and stable. In pure aqueous solutions stannous hexachloride forms octahedral complexes with a coordination number of 6, but their existence in biologic tissues is doubtful. Divalent Sn combines weakly with dithiol groups, whereas Sn^{4+} combines more readily. Cationic Sn is absorbed onto proteins, changing their conformation, or is incorporated into proteins; rat liver metallothionein contains about 0.3% Sn. The solubility of cationic Sn salts in acid pH and anionic salts in alkaline pH indicates that Sn may not function at physiologic pH, at which anionic Sn salts may undergo hydration, producing a series of insoluble complex stannic acids. Tin resembles Fe more than Pb in its chemical properties.

Metabolism

Man and animals have long been exposed to Sn, which accumulates in human tissues in widely varying amounts, according to the geographic region and the kinds of food ingested. Inorganic Sn salts are not generally used in metabolism studies because they produce colloidoclastic shock when injected into animals. Schwarz *et al.* (1970) and Schwarz (1971) claim an essential role for Sn in the growth of rats; both organic and inorganic Sn compounds were studied. Schroeder (1973) reported that low levels of Sn give transitory growth stimulation in rats, but data on the metabolism of Sn^{2+} and Sn^{4+} at physiologic intake levels are fragmentary. The daily average intake of Sn by humans in industrialized countries such as the United States and West Germany is about 3–4 mg, while in other geographic areas it is much less. Schroeder *et al.* (1964*a*) reviewed the biochemistry of Sn.

The gastrointestinal absorption of inorganic Sn compounds by animals is poor because most of the commonly ingested Sn salts are insoluble. In man, about 90% of a dose of soluble Sn compounds, such as sodium stannous tartrate, was found in the feces, indicating poor absorption (Calloway and McMullen, 1966), which is attributed to the absorption of Sn by proteins coagulated in the gastrointestinal tract (Barnes and Stoner, 1959). Tin sulfides, oxides, and soaps are negligibly absorbed.

Parenterally administered Sn salts remain at the site of injection, causing intense local irritation and tissue damage before being slowly absorbed into the blood stream. Intravenous injection of cationic Sn salts, which are not complexed with either citrate or tartrate, produces colloidioclastic shock, the breaking up of the physiologic equilibrium of the colloids of the body owing to the absorption of uncharged colloids into the blood. Very fine dusts of inhaled Sn compounds become lodged in the lungs and are slowly absorbed into the blood stream, and lipid-soluble alkyl derivatives of Sn are rapidly and completely absorbed from the gastrointestinal tract. Excretion is mainly fecal, as less than 5% of the daily intake is excreted in the urine (Van Kien and Thai tuong, 1955). The liver and spleen contain most of the Sn absorbed from ingested doses. Injected Sn salts are widely distributed in many tissues; as the tissue concentrations gradually decrease, the Sn levels in the liver and spleen increase. It is possible that colloidal particles of Sn salts in the tissues are phagocytized and transported to the spleen and liver. Most of the injected and absorbed Sn is ultimately excreted in the urine; the daily excretion of about 2.23 mg in human sweat reported by Schroeder and Nason (1971) seems to be high since the dietary intake is only about 4.0 mg and absorption is poor.

The accumulation of Sn in the tissues of normal American adults is probably due to their greater exposure to its compounds and to canned food products, since tissues from humans in other geographic areas do not appreciably accumulate Sn. In 1940, balance studies of Sn using tracer amounts of radioactive Sn showed the absence of any steady accumulation of Sn in American adults (Kehoe *et al.*, 1940); since then, exposure to it has been considerably reduced in the United States. On the basis of the above evidence, Schroeder *et al.* (1964*a*) believe that intestinal transport and barrier systems are present in man and that an efficient homeostatic mechanism is operative for Sn; however, this homeostatic mechanism evidently does not function at the higher doses. The bodies of newborn children contain no Sn, but it is deposited rapidly during their early life.

Toxicity

Metallic Sn and its inorganic salts are generally considered of low oral toxicity due to their poor alimentary absorption. Reports of tin toxicity are restricted to experiments with laboratory animals and to limited observations of humans. Organic Sn compounds are more highly toxic than are

TABLE 4-2. Tin Toxicity

| Compound | Animal | Route | Toxicity | Dosage/kg body weight | | | |
| | | | | Compound | Metal | | |
				mg	mg	mM	pT
Stannous chloride	Mouse	ip	LD_{50}	41	25.6	0.216	3.67
$SnCl_2$	Mouse	ip	LD_{100}	66	41.3	0.348	3.46
	Mouse	oral	LD_{50}	1200	750	6.32	2.20
	Rat	oral	LD_{50}	700	440	3.71	2.43
	Rat	oral	LD_{50}	700	440	3.71	2.43
	Rat	ip	LD_{50}	52	32.5	0.274	3.56
	Dog	sc	LD_{100}	159	100	0.843	3.07
	Dog	iv	LD_{50}	22	13.7	0.115	3.94
Stannic chloride	Mouse	ip	LD_{50}	46	21	0.177	3.75
$SnCl_4$							
Stannic iodide	Rat	iv	MLD	200	38	0.32	3.49
SnI_4							
Sodium stannous							
tartrate	Rat	iv	LD_{100}	52.4	20	0.168	3.77
$NaSn(C_4H_4O_6)$	Cat	iv	LD_{100}	105	40	0.337	3.47
	Rabbit	sc	LD_{100}	65.1	25	0.21	3.68
Dimethyl tin dichloride	Rat	oral	MLD	160	86.4	0.728	3.14
$(CH_3)_2SnCl_2$	Rat	iv	LD_{50}	40	21.6	0.18	3.74
Triethyl tin chloride	Rat	ip	LD_{50}	5	2.8	0.0236	4.63
$(C_2H_5)_3SnCl$							
Tetraethyl tin	Mouse	ip	MLD	32	16.1	0.136	3.88
$(C_2H_5)_4Sn$	Rat	oral	LD_{50}	16	8.1	0.068	4.17
	Rabbit	oral	LD_{50}	7	3.5	0.0295	4.53
	Guinea pig	oral	LD_{50}	37	18.7	0.158	3.80
Dibutyl tin oxide	Rat	ip	LD_{50}	40	15.5	0.131	3.88
$(C_4H_9)_2SnO$							
Dibutyl tin dichloride	Rat	ip	LD_{100}	7.5	2.5	0.021	4.68
$(C_4H_9)_2SnCl_2$							

inorganic compounds, but only the simplest organic compounds will be considered here; the toxicology of Sn has been extensively reviewed (Barneb and Stoner, 1959; Cheftel, 1967; Browning, 1969; Knorr, 1975). The toxicity of some soluble inorganic Sn compounds is summarized in Table 4-2. Divalent Sn salts are more toxic than Sn^{4+} salts, and divalent Sn can form more stable complexes with SH^- compounds than can Sn^{4+}, accounting for the greater toxicity of Sn^{2+} compounds. The toxicity of organo-tin compounds is believed to be due to their ready lipid solubility and their stability in biologic fluids at tissue pH, allowing extensive penetration into the brain and lodgment in the central nervous system.

Metallic Sn is toxic only when given orally in large doses. The unstable stannic hydride gas is more toxic than arsine and primarily affects the central nervous system. Soluble salts of Sn are gastric irritants; the acidity of Sn salt solutions and their irritant properties complicate interpretations of Sn toxicity. Canned food and drink containing 250–700 ppm Sn have caused nausea, vomiting, and diarrhea in humans (Cheftel, 1967); Benoy *et al.* (1971) report that 1400 ppm Sn in fruit drinks, or 4–5 mg Sn/kg body weight as soluble Sn^{2+} salts, could cause these symptoms. Rabbits fed $SnCl_2$ or stannous acetate or tartrate at the level of 1 g/animal per week died in 2 months after suffering from gastritis, degeneration of liver and kidneys, and paralysis of hind limbs (Eckhardt, 1909). In rats, there are marked differences in the oral toxicity of various Sn compounds; insoluble stannous compounds, such as the oxides, sulfide, and oleate, and stannic oxide are relatively harmless at levels up to 1% of the diet. However, cationic Sn^{2+} compounds of sulfate, chloride, and tartarate are toxic at dietary levels above 0.3% (de Groot *et al.*, 1973, de Groot, 1973). Severe growth retardation, decreased food efficiency, anemia, and a slight degeneration in the liver (homogeneous liver cell cytoplasm and hyperplasia of bile ducts) were observed in rats fed 66–99 mg Sn/kg per day or more as stannous chloride, orthophosphate, sulfate, oxalate, or tartrate. Reduced food intake may be responsible for the lowered growth rate, because the compact cytoplasm of hepatic parenchymal cells, observed in Sn toxicity, is a characteristic feature of partial starvation (Herdson *et al.*, 1964). Anemia symptoms include decreased hematocrit, hemoglobin level, and red cell counts. The activities of alkaline phosphatase, glutamic-oxalacetic transaminase, and glutamic-pyruvic transaminase in the blood decreased. De Groot (1973) suggests that some inorganic Sn salts inhibit hemopoiesis in rats by interfering with the intestinal absorption of Fe, because moderately excessive Fe in the diet prevents the toxic effects of Sn. However, the growth depression caused by Sn is not alleviated by supplementing the diet with Cu or Fe. According to de Groot *et al.* (1973) Sn intoxication affects growth and hemoglobin synthesis by entirely different mechanisms. Lifetime intake of 5ppm Sn^{2+} in their drinking water by mice and rats causes fatty degeneration of liver and increased incidence of vacuolar changes in their renal tubules (Schroeder *et al.*, 1967*b*, 1968*a*,*b*). Chronic Sn poisoning in rats causes moderate testicular degeneration, severe pancreatic atrophy, acute bronchopneumonia, and enteritis. Tin has been found to inhibit glutathione transferase. Tin intoxication and death are reported from unusual drug uses (Barnes and Stoner, 1959), and from consuming canned food (Warburton *et al.*, 1962; Kwantes, 1966; Benoy *et al.*, 1971).

Parenteral administration of Sn salts causes spasms and fatal paralysis, with bleeding in the liver and central nervous system, and intravenous injection of cationic Sn salts causes immediate colloidoclastic shock and death. Haddon (1958) demonstrated the nephrotoxicity of intraperitoneally injected Sn^{2+} salts, the LD_{50} at 72 days being 32.5 mg Sn/kg; one-tenth of this toxic dose given at 3-day intervals for two months produced no toxicity.

Inhalation of dusts or fumes of Sn metal and salts produces mild, benign pneumoconiosis in man and animals (Barnes and Stoner, 1959). Metallic tin causes nonspecific smooth-surface carcinogenesis in experimental animals where small tin disks are implanted subcutaneously or intramuscularly, and the dietary intake of soluble Sn salts, such as sodium chlorostannate, induces hepatic sarcomas in rats (Walters and Roe, 1965; Roe *et al.*, 1965). There are no reported mechanisms by which Sn is detoxicated.

Lead (Pb)

Lead has been known to mankind since 2500 B.C. Lead toxicity, *saturninism*, was recorded by ancient Greek and Arab physicians; Great painters suffered Pb toxicity, and thus, Pb poisoning is not a by-product of modern technology. Lead occurs chiefly as its sulfide, galena, and also as cerussite ($PbCl_2$). Although Pb is present at 15 ppm in the earth's crust, it is not uniformly distributed and is found in very high concentrations wherever it occurs; in seawater 5 ppb Pb is present; it is present in all living organisms. Thus, it is widely distributed in the environment and food. The human adult body contains about 120 mg Pb, 96% in the bones. The concentration of Pb in humans increases with age, and the total body content may increase to 400 mg (Barry and Mossman, 1970). Lead is stimulatory and oncogenic, but it is not essential for mammals.

Metallic Pb and its compounds have been used extensively during the last 4000 years. Metallic lead is used in the manufacture of pipes, cisterns, containers for sulfuric acid production and storage, lead shots, bullets, and linotype metal, and in alloys with antimony, tin, and copper for the manufacture of accumulator plates in storage batteries. In the pottery industry, Pb is used in low-solubility lead glaze. Lead monoxide (litharge), red lead, and lead sulfate, chromate, and titanate are used as pigments in paints, prints, and varnishes. Lead arsenate is used in insecticides and lead borate in plastics. In the Northern Hemisphere over 5 million tons of Pb have been used from 1920 through 1970 in automoboile engines as tetraethyl lead. This widespread contamination of the environment amounts, over the 50-year period, to 120 pounds of Pb per square mile; zoo animals in metropolitan areas have died from Pb in polluted air (Bazell, 1971). The exposure of humans and animals to lead contamination in the environment, and the ingestion of lead from cooking and eating utensils causes Pb poisoning. Acidic food, fruit juices, and wines stored in lead-lined bronze utensils caused high Pb toxicity among the Romans; Cockburn *et al.* (1975)

report that bones of an Egyptian mummy contained about one-tenth the amount of lead fround in the bones of modern man. The eccentric behavior of famous painters of the past is attributed to their accidental ingestion and inhalation of lead pigments in the paints; Francisco Goya revealed his plumbism in his self-portrait (Stoll, 1972). Improperly glazed pots have caused Pb poisoning in recent times. Ingestion of Pb paints as a result of pica, a craving for nonnutritive material, is a cause of Pb poisoning among ghetto children; newsprint in the colored pages of magazines contains 4000 ppm Pb, and children making spitballs out of these magazines are exposed to plumbism; children living within one-quarter mile of a Pb smelter may also be affected (Roberts *et al.*, 1974). Where annual hunting is allowed in the great flyways of waterfowl, the marshes and lakes are lined with lead shot which is poisonous to plants, birds, and mammals. The level of Pb in animals is highest in those living closest to the highways. Workers in industries involving Pb metallurgy and utilization are constantly exposed to the hazard of Pb poisoning. The Pb level in primitive man was calculated to be only 2 mg, compared with 120 mg in today's American adult; in some city dwellers the Pb level is as much as 200 mg. This level may increase by a factor of three or four with the continued use of tetraethyl lead in gasoline. Surprisingly, Wilkinson and Palmer (1975) found less Pb in the teeth of urbanites than of suburbanites.

Chemistry

Lead exhibits valences of $+2$ and $+4$, the more stable form being $+2$; tetravalence is shown in organometallic compounds and in basic lead acetate. The divalence of Pb is owing to ready availability of only two valence electrons from the $6SP$ shells; the charge density of the nucleus holds the other two electrons very firmly. Lead thus resembles the divalent subgroup IIA metals in chemical behavior more than its own subgroup IVA metals; it differs from subgroup IIA metals in the poor solubility of its halides, hydroxides, and phosphates. In physiological media the Pb^{+2} concentration may reach 10^{-4} M; interaction between Pb^{+2} and cellular phosphate groups is obvious, since the solubility product of $Pb_3(PO_4)_2$ is about 10^{-30} M. Microcrystalline precipitates of $Pb_3(PO_4)_2$ are formed in cell cytoplasm.

A divalent Pb ion could be either a hard or soft acid (HSAB theory) depending upon the biologic medium; thus Pb forms stable complexes with free thiol-, carboxylate- and phosphate-carrying ligands of biopolymers and membranes; imidazole and amino groups are of very little importance for Pb^{+2} complex formations. Lead nucleoside, especially the cytidine complex, is very stable. Although lead chelates can be formed *in vitro,* their existence in biological fluids is not confirmed.

Metabolism

Lead is a nonessential element which is relatively abundant in the body of mammals, although Schwarz (1974) reported some evidence for a possible essential function for Pb.

Under certain unusual conditions Pb is stimulatory, causing enhanced protein synthesis, and increased erythropoiesis, respiration, DNA synthesis, cell replication, and reproduction (Luckey, 1975*b*); small doses of intraperitoneally injected Pb stimulate DNA synthesis in rat kidney (Choie and Richter, 1973). The average dietary intake of Pb by modern man is about 0.45 mg. The sources of Pb in the average daily intake of Pb by man are: food, 0.22–0.4 mg; water, 0.01 mg; inhalation, (urban air) 0.03 mg or (rural air) 0.01 mg; and cigarette smoking, 0.04 mg (if one smokes about 30 cigarettes a day). Intake by inhalation is rather difficult to assess, owing to the variation in Pb concentration in different environments. However, Rabinowitz *et al.* (1973) found that one-third of the daily intake of Pb of man came from inhalation and the rest was of dietary origin.

Lead accumulates in the body to varying degrees, depending upon the nature of the Pb compound and the environmental load of Pb. Lead is similar to cadmium in its deposition and in its mobilization from bone; it may remain immobilized for years in the skeleton without causing adverse reactions. Any metabolic disturbance resulting in osteolysis will liberate Pb from its skeletal storage; thus, there could be an indefinite time-lag between Pb intake and manifestation of chronic toxicity. The metabolism and toxicology of Pb have been studied extensively by researchers in different but related scientific disciplines, and excellent reviews are available (Passow *et al.,* 1961; Kehoe, 1962; Chisolm, 1964, 1971; Griggs, 1964; Hammond, 1969; de Bruin, 1971; Goyer, 1971*a,b;* Jernigan, 1971; Ammerman *et al.,* 1973). Industrial Pb poisoning was reviewed by Kehoe (1962) and Browning (1969).

Absorption of Pb salts from the gastrointestinal tract of mammals is poor, about 1.5% over a dosage range of 2–108 mg Pb/day; rate of absorption is not influenced by dosage in cattle, sheep, and rabbits, but in man the absorption increased to 7% with increased intake (Huysh and Suomela, 1968), while in rats, intestinal absorption decreased with increased dosage. Absorption of insoluble lead sulfate, sulfide, or chromate is poor, but that of lead acetate or nitrate is slightly greater. In rats, absorption of Pb from the gastrointestinal tract is influenced by age; the absorption increases until weaning, with most of the administered dose being absorbed; then the absorption drops to the level of the adult (Forbes and Reina, 1972). In the rat the age–absorption curve of Pb is similar to Ca and Sr (Taylor *et al.,* 1962); during maturation the rat intestine evidently loses its capacity to pinocytose ingested particles. Other factors influencing Pb absorption are the levels of dietary Ca and vitamin D, and the motor activity of the digestive tract; high dietary Ca and increased peristalsis of the digestive tract decreases Pb absorption, and high dietary vitamin D increases it. Soluble Pb salts administered intramuscularly, intraperitoneally, and subcutaneously are not fully absorbed, as part is retained at the injection sites (Choie and Richter, 1972*b*). Intravenously injected Pb salts are either sequestered in erythrocytes or deposited on erythrocyte stroma membrane; about 5% remains in the plasma, mostly as colloidal lead phosphates and diffusible lead peptides or Pb–organic-acid complexes. The absorption of inhaled Pb is more complete and rapid, about 90%, than by any other route; the lungs and the respiratory tract, including the nasal passages, absorb Pb. Lead in urban air occurs as $PbCl_2$, $PbBr$, or PbO_2 particles derived from tetraethyl lead; 40% of these particles are about 0.45 μm in diameter. About 40% of the inhaled Pb is trapped in the upper respiratory tract and swallowed; of the Pb retained in the lung, 90% is absorbed if the particles range from 0.01 to 0.1 μm in size. The daily absorption of Pb by man through inhalation varies from 20 to 30 μg. Skin does not absorb inorganic lead, but readily absorbs tetraethyl lead.

Absorbed Pb is deposited mostly in bone, and some in soft tissues, but the initial pattern of distribution depends on the mode of Pb uptake and on the vascularity and intrinsic affinity of the tissues. Intravenously administered Pb is distributed mostly in liver and bone, while ingested Pb is distributed as follows: bone, 60%; liver, 25%; kidney, 4%; reticuloendothelial system, 3%; intestinal wall, 3%; and other tissues, traces. The kidneys contain a high concentration of Pb per gram tissue.

Under conditions approximating the steady state, i.e. with about 400 μg Pb being absorbed daily, 90% of the Pb in humans is present in the skeleton. The mechanism of Pb deposition in bone is not definitely known, except that Pb accumulates in areas of active bone formation, suggesting the deposition of Pb salt crystals and subsequent binding to the organic matrix. This potential pool of exchangeable Pb in the bones becomes less mobile with the passage of time, but it is readily mobilized during bone fracture, chronic infection,and metabolic disturbances. The retention of Pb in mammalian soft tissues is as follows: liver > kidneys > aorta > muscle > brain. Lead permeates the placental barrier, so that the distribution of lead in fetal tissues and organs is similar to that in the adult rat (Horiuchi *et al.*, 1959). The subcellular distribution of Pb within the kidneys and other tissues was determined by injecting [203]Pb intraperitoneally into rats (Barltrop *et al.*, 1971); the lysosomal fraction contained most of the [203]Pb, and renal cell organelles contained Pb in the following order: supernatant > microsomal > mitochondrial = nuclear. The mitochondrial uptake of [203]Pb was reduced by 35% if the rats had been previously subjected to ordinary lead.

Inorganic Pb is excreted in both the feces and urine, fecal Pb representing the unabsorbed portion of ingested Pb and Pb from the bile. About

80% of intravenously injected small doses of Pb is recoverable from the bile through the enterohepatic pathway. Lead is excreted through the kidneys by both glomerular filtration and transtubular flow, and it is also eliminated through sweat and milk, although the total quantity excreted is small. Urinary excretion of lead is increased in acidosis and after vitamin D, parathyroid hormone, or iodide administration. The kidneys can excrete large amounts of Pb without excessive damage to the renal tubular lining by keeping the renal Pb in a nondiffusible, nontoxic inclusion body from within the renal cell and excreting the inclusion body (Goyer, 1971*a*).

Evidence for a homeostatic mechanism for regulating the mammalian body load of Pb is indirect. Balance studies with rats showed that the gastrointestinal absorption of Pb decreases progressively with increased Pb ingestion, a situation similar to that of Ca or Fe. But in man a daily intake of 1 mg or more of Pb is followed by a net retention and progressive increase of Pb stores (Kehoe, 1963). However, intake and excretion of Pb in adult humans remain balanced when the intake is less than 1 mg/day, and the average intake of Pb by modern man is well within the limits of the excretory mechanism.

Tetraethyl lead (TEL) is readily absorbed from the digestive and respiratory tracts and through the skin, owing to the solubility of TEL in lipids and to its diffusibility. Intraperitoneally injected TEL is quite toxic (Table 4-3). Although part of TEL is metabolized and the released inorganic Pb is distributed in other soft tissues, the major portion accumulates in the brain owing to a special affinity between the organic Pb and the lipids of nerve tissues. The chemical nature of the organic Pb found in the brain of Pb-poisoned victims (12–20 μg/g) is not definitely known.

Toxicity

The toxicology of lead is well-known and well-documented. Lead toxicity is more related to the levels of diffusible lead and to the lead content of soft tissues, such as liver, kidneys and brain, than to the total body content. The partitioning of Pb among target organs and the concentration of diffusible Pb in the soft tissues are directly responsible for toxic symptoms. Sometimes Pb toxicity symptoms are not immediately seen after oral intake of a toxic dose, making it difficult to assess the lethal dose. Lead poisoning is cumulative, and the experimentally toxic and lethal doses are reached at low levels. Acute toxicity symptoms are lassitude, vomiting, loss of appetite, uncoordinated body movements, convulsions, and stupor, eventually producing coma and death. Chronic Pb toxicity symptoms include loss of appetite and vomiting, and long-term effects are renal malfunction, hyperactivity, mild anemia, liver cirrhosis, brain damage, and general intellectual and psychological impairment. Children surviving acute encephalopathy suffer permanent damage to the central nervous system.

TABLE 4-3. Lead Toxicity

Compound	Animal	Route	Toxicity	Compound mg	Metal mg	Metal mM	pT
Lead metal (powdered)	Rat	ip	LD_{100}		1000	4.8	2.32
	Guinea pig	ip	MLD		100	0.48	3.32
Lead monoxide (PbO)	Rat	ip	LD_{50}	400	371	1.79	2.75
Lead dioxide PbO_2	Guinea pig	ip	LD_{50}	200	173	0.835	3.08
Lead tetroxide	Rat	ip	LD_{50}	220	19.9	0.96	4.02
Pb_3O_4	Guinea pig	ip	LD_{50}	450	408	1.97	2.71
Lead difluoride PbF_2	Guinea pig	oral	MLD	4000	3380	16.3	1.79
Lead chloride $PbCl_2$	Guinea pig	oral	LD_{50}	2000	1490	7.19	2.14
Lead perchlorate $Pb(ClO_4)_2 \cdot 6H_2O$	Mouse	ip	MLD	275	110	0.53	3.28
Lead nitrate	Rat	ip	MLD	270	169	0.815	3.09
$Pb(NO_3)_2$	Guinea pig	oral	MLD	1330	832	4.01	2.40
Lead sulfate $PbSO_4$	Guinea pig	ip	LD_{50}	300	205	0.99	3.00
Lead sulfide	Guinea pig	ip	LD_{50}	1600	1390	6.68	2.17
PbS	Rat	ip	LD_{50}	1600	1390	6.68	2.17
Lead orthophosphate $Pb_3(PO_4)_2$	Guinea pig	ip	MLD	131	100	0.48	3.32
Lead arsenate	Rat	oral	LD_{50}	100	59.3	0.286	3.54
$Pb_3(AsO_4)_2$	Rabbit	oral	LD_{50}	125	74.2	0.358	3.45
	Human	oral	MLD	1.4	0.84	0.405	3.40
Lead silicate $PbSiO_3$	Guinea pig	ip	LD_{50}	136	99.5	0.48	3.32
Lead chromate $PbCrO_4$	Guinea pig	ip	LD_{50}	400	252	1.21	2.92
Lead titanate $PbTiO_3$	Rat	ip	LD_{50}	2000	1360	6.57	2.18
Lead carbonate	Guinea pig	oral	MLD	1000	801	3.86	2.41
$(PbCO_3)_2 Pb(OH)_2$	Guinea pig	ip	MLD	124	99	0.48	3.32
Lead cyanide $Pb(CN)_2$	Rat	ip	MLD	100	80	0.386	3.41
Lead acetate	Mouse	ip	LD_{50}	140	82	0.396	3.40
$Pb(C_2H_3O_2)_2 \cdot 3H_2O$	Rat	oral	MLD	11000	6000	28.9	
	Rat	ip	LD_{50}	150	82	0.395	3.40
	Rat	iv	LD_{50}	140	76	0.36	3.44
	Rabbit	iv	LD_{100}	58	31.7	0.153	3.82
	Dog	iv	LD_{100}	350	191	0.921	3.04
Lead lactate	Guinea pig	oral	MLD	1000	537	2.59	
$Pb(C_3H_5O_3)_2$	Guinea pig	oral	LD_{50}	3000	1610	7.77	
Tetraethyl lead	Rat	oral	LD_{50}	35	22.3	0.107	3.97
$Pb(C_2H_5)_4$	Rat	ip	MLD	10	6.4	0.0309	4.51
	Rabbit	sc	MLD	320	204	0.984	3.01
	Rabbit	iv	MLD	22	14	0.067	4.17
Lead fluosilicate $PbSiF6 \cdot 2H_2O$	Rat	oral	MLD	250	134	0.649	3.19
Lead fluoborate PbB_2F_8	Rat	oral	MLD	50	27.2	0.131	3.88

Lead toxicity is manifested in different areas of physiologic functions. Chronic subtoxic levels also cause subtle changes in the permeability of cells of the testes, kidneys, liver, and brain, leading to their reduced function or complete breakdown. Toxic effects of Pb are seen as reduced growth and longevity, impaired renal function and reproduction, and reduced functioning of the hematopoietic system and the central and peripheral nervous systems; other effects include premature loss of teeth and reduced immunity.

The toxicologic data of Pb compounds are shown in Table 4-3; soluble Pb salts are more toxic than insoluble salts, and rabbits and guinea pigs are more susceptible to Pb poisoning than rats and mice. Increased levels of dietary protein, vitamin D, ascorbic acid, nicotinic acid, calcium, and phosphates decrease the susceptibility of animals to Pb intoxication, but other heavy metals, or excessive alcohol intake seem to increase susceptibility. Low levels of Fe and Ca tend to increase Pb toxicity by increasing the concentration of diffusible mobile Pb in the soft tissues, and vitamin-E deficiency increases the susceptibility of animals to the *in vivo* hemolytic effect of Pb poisoning (Levander *et al.,* 1975). Exposure to small nontoxic doses of Pb develops resistance to higher doses, since mice developed resistance to lethal doses of Pb following pretreatment with very small quantities of soluble Pb salts (Yoshikawa, 1970; Garber and Wei, 1973).

Rats and mice fed 25 ppm Pb as soluble Pb salt (to match the human tissue concentration) over a long period of time suffered early mortality, weight loss, and shortened life span. About 200–400 mg Pb/kg body weight causes heavy mortality in calves, but in older cattle the toxic doses are higher, 600–800 mg/kg body weight. Horses are more susceptible than cattle to Pb toxicity from chronic ingestion, since an estimated daily intake of 1 mg Pb/kg body weight produces toxicity symptoms in horses. Swine seem to be relatively resistant to the toxic action of orally ingested Pb salts (Ammerman *et al.,* 1973).

Impaired reproductive capacity was observed in rats and mice fed 25 ppm soluble Pb salt (Schroeder and Mitchner, 1971*b*). In mice the strain died very rapidly within two generations; there were 69 runts of 72 live offspring in the F_2 generation, and only four grew to breeding age; fetal males were more vulnerable than fetal females to Pb intoxication. Male rats suffered from impotence and prostatic hyperplasia; after a month of feeding 100 μg/Pb per kg diet, the weight of their prostate glands increased, and secretion decreased (Fahim *et al.,* 1972). The abnormal multiplication of cells in prostate glands of Pb-exposed rats could be due to the stimulation of DNA synthesis (Choie and Richter, 1972*a*). When the blood concentration of Pb increased to 50 μg/100 ml, spermatogenesis was inhibited due to testicular damage, and as the blood-Pb level increased from 14 to 50 μg/100 ml in female rats, irregularity in the estrus cycle with persistent vaginal estrus was followed by the development of ovarian follicular cysts and reduction of the corpora lutea (Hilderbrand *et al.*, 1972).

A possible link between stillbirths and unexplained sudden infant deaths and Pb and Cd pollution was established by an epidemiology study (Coumbis, 1975). Large concentrations of both Pb and Cd were found in several body organs of the victims, and the level of 113 μg Pb/100 g brain tissue was more than the lethal Pb concentration. It is presumed that toxic levels of Pb were passed through the placental barrier of the exposed mother.

Lead encephalopathy is not seen in suckling rats whose mothers are fed a diet containing 4% $PbCO_3$; instead, a hyperactivity is observed starting at four weeks of age. Hyperactivity in Pb intoxication is associated with an altered ratio of dopamine to norepinephrine in the brain, as the catechol amines regulate behavioral functions (Sauerhoff and Michaelson, 1973). An eightfold increase in the concentration of Pb and a 20% decrease in dopamine content of the brain was noticed, and a subsequent study revealed slight increases in the norepinephrine level.

Damage to the kidneys from Pb toxicity is less dramatic than Pb-induced anemia and neurologic effects. Nonetheless, chronic and steady low-level exposure to Pb causes nephritis, and scarring and shrinking of the kidney tissues in rats, whereas short and intense exposure causes functional breakdown. Kidney malfunction, structural damage to mitochondria and the proximal renal tubules, occurs at a blood-Pb concentration of 150 μg/100 ml, resulting in loss of amino acids, glucose, and phosphates in the urine. The mitochondria, whose membranes are damaged, consume more oxygen, suggesting derangement of energy metabolism (Goyer, 1968; Cardona *et al.*, 1971); reduced cytochrome content of renal cells was reported to cause respiratory abnormalities (Rhyne and Goyer, 1971). Intranuclear inclusions, mainly lead and nonhistone proteins, are found in the nuclei of renal cells lining the proximal tubules. These inclusion fibrils may serve to detoxicate the Pb for eventual urinary excretion (Choie and Richter, 1972*b*). The immature rat kidney is more susceptible to Pb toxicity than is the adult kidney.

The toxic effects of Pb on the hematopoietic system were studied extensively and reviewed (de Bruin, 1971; Arena, 1974); anemia is not of serious consequence in Pb toxicity and can be reversed. The toxic effects of Pb on hemopoiesis are: (1) abnormal circulating erythrocytes, (2) impairment of the production of hemoglobin owing to Pb activity on the bone marrow, and (3) stimulation of erythropoiesis in the bone marrow. Lamola and Yamane (1974) have suggested a screening test based upon measurement of blood fluorescence from the zinc protoporphyrin in erythrocytes of a nonanemic subject.

Lead readily binds to Sh-, PO_4-, and COOH-containing ligands of the erythrocyte membrane to increase the mechanical, but decrease the osmotic, fragility of erythrocytes. *In vitro* human erythrocytes exhibit osmotic fragility when exposed to less than 200 μg Pb/100 ml; with concentrations exceeding 900 μg Pb/100 ml the cells develop a mild resistance to lysis (Lessler and Walters, 1973). Lead alters membrane permeability (Bishop and Surgerer, 1964), and binds to active sites involved in K permeability, and inhibits active transport by blocking K–Na transport ATPase. Lead seems to liberate normally inactive carrier molecules which are highly specific with respect to K and Rb and which give rise to countertransport (Passow *et al.*, 1961). Although the general metabolism within the erythrocyte cells is not very disturbed, the cells shrink from loss of K and water, and their survival is shortened.

Lead impairs the biosynthesis of heme in the bone marrow and causes extensive excretion of the precursors of porphyrin and hemoglobin in urine; it interferes with the incorporation of divalent Fe into protoporphyrin, inactivates aminolevulinic acid dehydratase (ALAD) which is involved in the conversion of δ-aminolevulinic acid to porphobilinogen, and also inactivates porphyrinogen decarboxylase (Waldron, 1966). Lead intoxication decreased the activity of blood ALAD by 90% and increased the excretion of δ-aminolevulinic acid in the urine. The δ-aminolevulinic acid synthetase (ALAS) activity in blood increased threefold (Kao and Forbes, 1973).

Lead initiates stimulation of erythropoiesis, with a subsequent delay in maturation of erythrocytes. In Pb toxicity the erythropoietic cells of the bone marrow undergo morphological changes which cause increased production of abnormal red cells, basophilic cells with nuclear abnormalities and inadequate hemoglobin (de Bruin, 1971).

Lead is toxic to the central and peripheral nervous systems, affecting the cerebellum, spinal cord, and motor and sensory nerves; both the structure and function of neural tissue are very sensitive to Pb toxicity. Lead encephalopathy is a disease of neonatal man or animals caused by the transfer of Pb through the maternal milk; the mothers are unaffected. Much damage occurs in both the cerebellar and cerebral cortex; the vascular damage is due to altered permeability of the capillary walls, causing brain edema, serous exudation, and derangement of cellular energy metabolism. The abnormal vascular permeability is especially seen in the cerebellum striatum, occipital lobes, and spinal cord. When maternal rats are fed 10% $PbCO_3$, the growth of young is subnormal; paraplegia of the hind legs is observed 20 days after birth, and death occurs after another week (Lambert *et al.*, 1967). After prolonged exposure to Pb, peripheral neuropathy is encountered in adults, and demyelination and axonal degeneration is observed. It is presumed that Pb acts as a competitive inhibitor to Ca in acetylcholine release. Exposure of mice (Hemphill *et al.*, 1971) and rats (Selye *et al.*, 1966) to low concentrations of Pb leads to inactivation of antibodies, which reduces resistance to bacterial infections. Interference with the phagocytic activity of polymorphonuclear leukocytes reduces immune resistance in Pb toxicity in humans (Rakimova, 1968). In rats and primates nontoxic levels of Pb may induce lethal effects from doses of bacterial endotoxins which are otherwise nonlethal by a factor of 10^5 (Di Luzio, 1972). Intravenous administration of acute doses of lead acetate increases one-thousandfold the susceptibility of rats to intravenous challenge with *Escherichia coli,* which was attributed to the endotoxin level in *E. coli* (Cook *et al.,* 1975). Lead also retards prothrombin synthesis in the liver and decreases the level of the thromboplastic factor in rat blood. Chronic exposure of mice to Pb produces significant decreases in antibody synthesis; IgM-antibody formation is stimulated while IgG-antibody formation decreases (Koller *et al.,* 1976). Other toxic effects of Pb in mammals include reduced thyroid activity in sheep. In Pb poisoning, the nicotinic acid content of mammalian blood is decreased owing to greater utilization of nicotinic acid, and the mucoid and sialic acid content of blood also is decreased (de Bruin, 1971). Basic lead acetate in the diet induces malignant neoplasms in the kidneys of rats (Boyland *et al.,* 1962), and parenteral injection of insoluble lead sulfate in rats also produces renal adenomas and carcinomas (Zollinger, 1953); tetraethyl lead is not a primary carcinogen, but its breakdown products can cause cancer.

Minerals such as Ca, Fe, Zn, Cu, and P interact with Pb and influence Pb metabolism and toxicity in mammals. Premature loss of teeth in rats from experimental Pb poisoning is attributed to the demineralization caused by chemical displacement of jawbone Ca by the colloidal secondary lead phosphate of the circulating blood (Koelsch, 1959). The susceptibility of rats to Pb intoxication is increased several times by a low intake of Ca (Mahaffey *et al.,* 1973), which is mainly due to increased retention of Pb in body tissues, such as the kidneys, femur, and blood. Transfer of Pb to newborn and to weanling pups increases in Ca-deficient rats, and a low intake of P also results in increased Pb retention in the tissues; the effects of Ca and P deficiencies are additive. Phosphorus deficiency increases Pb absorption from the intestine, while Ca deficiency affects Pb retention. Parathyroid hormone increases the urinary excretion of Pb and Ca (King, 1971), P influences Pb and Ca distribution in the body (Potter *et al.,* 1971), and Ca and P occur with Pb in the intranuclear inclusion bodies in the kidneys and osteoclasts (Carroll *et al.,* 1970).

Iron deficiency markedly increases the adverse effects of Pb intoxication in rats (Six and Goyer, 1972). Synergistic effects of Pb intoxication and Fe deficiency in the impairment of hemopoiesis were reported (Waxman and Rabinowitz, 1966; Borsook *et al.,* 1957). Children and pregnant women with marginally inadequate dietary intake of Ca, Fe, and P are more susceptible to Pb intoxication than are others with normal intake. Lead poisoning in children from pica is attributed to Ca deficiency, as this syndrome was produced experimentally in rats with Ca deficiency (Snowdon and Sanderson, 1974).

In horses, increased dietary intake of Pb and Zn alters the tissue distribution of Pb; the level of Pb decreases in the bones, with a concomitant increase in the hepatic and renal Pb concentration (Willoughby *et al.*, 1972). These changes in Pb metabolism could be owing to the effect of Zn on bone mineralization, since increased Zn intake would increase the metallothioneine content of the liver and kidneys and increase Pb binding to this protein (Bremner, 1974*b*). A mutual antagonism may exist between Cu and Pb, because rats subjected to Pb intoxication have decreased plasma Cu and ceruloplasmin levels, and decreased dietary intake of Cu increases erythrocyte Pb concentrations (Klauder *et al.*, 1972).

Lead toxicity is mainly the result of the concentration of diffusible Pb in the soft tissues; the organism detoxifies it by converting it into a nondiffusible form to protect the cellular organelles and metabolites. The large intranuclear inclusion bodies that are formed in the proximal tubular cells of the kidneys have the function of binding lead into a nondiffusible form (Goyer *et al.*, 1970). This protein is evidently synthesized in the cytoplasm and moves into the nucleus; during the course of transtubular flow a portion of the Pb enters the nucleus where it forms a Pb–protein complex. Similar Pb inclusion bodies are also found in the nuclei of osteoclasts in pigs with experimental Pb poisoning (Hsu *et al.*, 1973). Another possible detoxication mechanism for Pb could be increased formation of metallothionein, as Pb shows a high affinity for the sulfhydryl groups of metallothionein.

SUBGROUP IVB

Cationic Ti and soluble titanates are generally considered to be nontoxic, although soluble titanates cause slight impairment of reproduction functions in experimental rats. Cationic Ti salts are poorly absorbed from the mammalian alimentary tract.

Zirconium salts are considered to be nontoxic via dietary sources, owing to their poor absorption. Intravenously injected cationic salts form insoluble colloidal polymers and are phagocytized, and bone-seeking Zr is deposited and immobilized in the skeleton. Parenterally injected Zr salts are more toxic than are ingested Zr salts.

Hafnium salts are also considered to be nontoxic or mildly toxic, but excessive intake causes hepatic damage and respiratory paralysis.

Titanium (Ti)

Titanium occurs extensively as ilmenite ($FeTiO_3$) and rutile (TiO_2); as the eighth most abundant element, it is widely distributed in the earth's crust at about 4400 ppm, and some soils contain 0.5–1.0% Ti; seawater contains 1–9 ppb. The leafy parts of plants contain appreciable amounts of Ti, and the bodies of human adults contain about 9 mg Ti, of which 49% is present in the lungs and lymph nodes, 15% in the blood, and the rest in soft tissue and bone. Titanium in the form of titanate is stimulatory, but Ti is not essential for mammals.

Titanium is a constituent of many heat-resistant alloys; titanium tubes are used in aircraft and missiles and in flexible casings for control wires in atomic reactors. Titanium carbide is mixed with cobalt carbides and used in making cutting tools. Titanium dioxide is used as a pigment in ceramics, roofing, and plastics, and is also used as a nutritional marker. The fibrous form of titanate is extensively used as a thermal insulator, and tetraisopropyl titanate is used for heat-resisting surface coatings in the paint industry. Metallic Ti is used in surgical apppliances because of its lightness and tensile strength. Titanium oxides, peroxides, salicylates, and tannates are used therapeutically for skin disorders and in cosmetic preparations. Titanium is nonessential in man and is relatively nontoxic, but in very high doses Ti salts are slightly toxic, more owing to mass effect than to any specific toxic action.

Chemistry

Titanium exhibits a variable valence of +2, +3, and +4, but tetravalent compounds are the most stable. Its tendency to lose electrons makes Ti a strong reducing agent. Titanium forms both cationic and anionic salts. Halides of Ti are hydrolyzed by water to titanyl $(TiO)^{2+}$ or oxy salts; complex salts, such as sesquisulfates, are stable. Trivalent Ti forms octahedral complexes with a coordination number of 6. Tetravalent Ti, because of its polarizing nature, forms a large number of covalent compounds with coordination numbers varying from 4 to 8 which have different geometric structures. In the tetrahedral structure the stable forms are cyclopentadienyl derivatives such as titanocene $(C_{10}H_{10}Ti)^+$; one of the π electrons of each cyclopentadienyl ion is involved in the covalent bonding, giving it an aromatic character. "Charge transfer chelates" of Ti with flavoquinones are possible. Other stable coordination complexes are: (1) distorted trigonal or bipyramidal structure of a coordination number of 5, e.g., $K_2TiF_2O_5$, and (2) octahedral structures with a coordination number of 6, e.g., K_2TiF_6. Due to the preference of Ti for oxygen donors, at physiologic pH its salts undergo olation and oxolation and form stable and water-insoluble polymeric octahedral structures.

Metabolism

Titanium is the only metal in the first transitional group which is not an essential nutrient. Titanium occurs widely in food: shrimp, 7.38 ppm; lettuce, 2.5 ppm; corn oil and butter, 2.0 ppm; cloves, 6.3 ppm; and ground black pepper, 15.9 ppm. Increases in their Ti content are noticed in corn, wheat, and canned tomato after they are processed. The average daily intake of Ti by human adults is about 0.85 mg; it has not been found in the newborn. Titanium can be absorbed and concentrated from the environment, and its concentration in humans rises after the third decade of life.

There is little evidence for the biologic activity of Ti. In spite of the presence of Ti in the adult human lung and its abundance in earth's crust, the biochemistry of Ti has not been well studied; however, Schroeder *et al.* (1963*c*) reviewed the biological chemistry of Ti as an abnormal trace metal in man, and Luckey (1975*b*) reviewed the stimulatory action of Ti in phagocytosis and growth of young mice. The gastrointestinal absorption of Ti oxides and cationic simple salts is poor in rats and rabbits; in rats, about 95% of an ingested dose of TiO_2 is recovered from the feces (Lloyd *et al.*, 1955), although none could be recovered from the rat tissues; in another report, there was 98–100% recovery from the feces in rats and sheep (Kotb and Luckey, 1972). Soluble titanates are absorbed only to a small extent, and parenterally administered salts remain at the site of injection for a long time before they are absorbed. Intravenously injected soluble salts are distributed to the lungs, heart, liver, spleen, and other tissues; part of the dose is transiently retained in the spleen and heart before being finally excreted in the urine; mice retain part of a soluble Ti salt, given in drinking water, in the heart tissues (Schroeder *et al.*, 1963*c*). Intravenously injected TiO_2 in rats at 250 mg/kg was quickly cleared from the blood; 80% was found in the liver with subsequent accumulation in the lymph nodes with significant retention for one year (Huggins and Froehlich, 1966). About 40% of the daily intake of Ti by human adults is excreted in the urine. There seems to be no homeostatic mechanism for Ti.

Toxicity

Titanium is essentially nontoxic in the amounts and forms which are normally ingested, due to its nonabsorption from the mammalian alimentary tract. Toxicity data on Ti are few, and, because of the lack of specific and definitive data, Ti compounds are considered nontoxic. Animals tolerate Ti both orally and by parenteral injection, but inhalation causes slight lesions in the lungs. The physiological inertness of TiO_2 in animals was reported by early workers (Lehman and Herget, 1927; Vernetti Blina, 1928; Spink, 1961).

Large oral doses of TiO_2 and thiotitanic acid, up to 5 g Ti/kg, were tolerated by rabbits for several days with no toxic symptoms (Sollman, 1963), but impaired reproductive capacity was observed in life-term studies in rats fed 5 ppm Ti as titanate in the drinking water; there was also a reduction in the male:female ratio and in the number of animals surviving to the third generation (Schroeder and Mitchner, 1971b). Inhalation or exposure of dogs to $TiCl_4$ caused severe bronchitis, edema, and death, toxicity attributed more to the liberated free HCl following its hydrolysis than to the $TiOCl_2$ or $TiCl_4$ (Lawson, 1961). Exposure to titanium carbide (TiC_2) aerosol for five months caused lung changes in humans similar to silicosis (Mogilevskaya, 1956). Circulating thrombocytes decreased in rabbits after an intravenous injection of a suspension of TiO_2 (Bloom and Swensson, 1958).

The *in vitro* inhibition of serum alkaline phosphatase, yeast invertase, and amylase by Ti has been reported (Gould, 1936); titanium compounds are known to inhibit tyrosinase *in vitro* by competing with the enzyme for the substrate (Cavanaugh *et al.*, 1955). Changes in sulfur metabolism *in vitro* have also been attributed to titanium salts. Conversion of cysteine into cysteic acid, and oxidation of ethyl mercaptan and thioglycollic acid by rat liver slices were inhibited by Ti (Bernheim and Bernheim, 1939); it was suggested that intraperitoneal injection of Ti salts caused inhibition of the oxidative enzymes. The reported presence of TiO_2 in leukocytes may be the result of TiO_2 accumulation by phagocytosis (Carrol and Tullis, 1968). The therapeutic value of Ti compounds in skin disorders is owing to their capacity to stimulate the phagocytic activity of the endothelial cells of the capillaries; small quantities of TiO_2 in the rat diet enhance immunologic activity (Antonova *et al.*, 1968). Titanium complexes, such as titanocene, given intramuscularly are carcinogenic. There is no reported mechanism for Ti detoxication.

Zirconium (Zr)

Zirconium occurs in nature as zircon or zirconium orthosilicate ($ZrSiO_2$) and baddeleyite (ZrO_2). The concentration of Zr in the earth's crust is 170 ppm, 300 ppm in the soil, and the concentration in seawater is 4 ppb. The adult human body contains about 420 mg Zr, of which 67% is present in fat, 2.5% in blood, and the rest in the skeleton, aorta, lungs, liver, brain, kidneys, gall bladder, and other tissues (Schroeder and Nason, 1971). Zirconium is also widely distributed in plant material. Zirconium is not an essential metal, but it is stimulatory.

Zirconium is used extensively in steel and cast iron manufacture, in alloys to be used in noncorrosive chemical apparatus, and in electric furnaces as refractory linings. In the electronics industry, Zr is used in manufacturing vacuum, radio, and television tubes, and as an igniter in photoflash bulbs. Owing to its low cross section for neutron capture, transparency to thermal neutron cross section and high resistance to corrosion and heat, Zr is used as a shield in nuclear reactors; Zirconium is also used as a reflective surface agent on satellites. Zirconium powder alloyed with lead is used in cigarette lighter flints. Zirconium tetrachloride is used in water repellents for textiles; its oxides and hydroxides are used as dyes and pigments and the silicate is used in the glass and ceramic industries. Insoluble zirconium silicate is an ingredient in cosmetic creams and powders, and $ZrOCl_2$ is used as an antiperspirant. Zirconium carbonate and oxide are used therapeutically as topical agents for dermatitis from poison ivy. Zirconium poisoning may occur due to excessive exposure to Zr salts, but industrially, Zr is not considered a health hazard (Schroeder and Nason, 1971).

Chemistry

Zirconium exhibits a variable valence of +2 to +4; tetravalent Zr is the most stable; its electropositivity and electronegativity are about equal, as reflected by its position in the periodic chart. Zirconium forms unstable simple cationic salts such as halides and sulfates and stable anionic zirconates. Cationic Zr salts are hydrolyzed to insoluble but stable zirconyl (ZrO^{2+}) or zirconium oxy salts, although only a few reports are available about the zirconyl ion. However, the ZrO^{2+} ions readily polymerize, and this hydrolytic polymerization increases with increasing pH, allowing extensive olation and oxolation. Two sets of thiol bridges are bonded to each Zr atom in the tetramer, lying in planes perpendicular to each other. Since each tetramer contains a large number of olation sites, the polymers can grow in many directions, simultaneously forming colloidal polymers with molecular weights up to 1000. Zirconium forms coordination complexes with coordination numbers 6, 7, and 8; it forms stable chelates with bidentate ligands, such as lactic acid, which contain oxygen donors. Zirconium can also form labile inner-orbital complexes. Basic nitrogen from biologically active molecules is less effective as a donor towards tetravalent Zr than is oxygen, and the coordinating tendency of Zr is in the following increasing order: $R_3\!\!\equiv\!\!N < R\!\!-\!\!O\!\!-\!\!R << R\!\!-\!\!COO^- < RO^- < OH^- <<O_2$. Acids such as the citrate and tartrate can penetrate the colloidal hydrolytic polymers of ZrO_2; Zr does not exist in aqueous solution as a monoatomic ion but as a central atom of complex anions and cations. During canning or processing, soluble Zr complexes in foods are olated to form insoluble colloidal polymers.

Metabolism

Zirconium is not essential for humans and animals although it possesses some attributes of essential metals. Zirconium has the requisite electronic structure to compete with Cr and Fe, and could be compatible with some physiological functions, but Zr complexes are stable and are formed by irreversible reactions. The occurrence of Zr is widespread in foods: oats, 6.6 ppm; wheat, 2.8 ppm; rice, 3 ppm; butter, 6 ppm; green peas, 4 ppm; paprika, 9 ppm; ginger, 6 ppm; instant tea, 12 ppm; lamb chops, 4 ppm; and corn oil, 4 ppm. However, the analytical methods may be imprecise, so these values must be used with caution. Although Zr is ubiquitous, meat products and seafood contain only trace amounts. The average daily intake of Zr by humans is about 4.2 mg, according to Schroeder and Nason (1971). Zirconium is present in human brain tissue which usually does not contain adventitious trace metals.

Studies on Zr metabolism center mostly on ^{95}Zr, a product of nuclear fission (Hamilton, 1949; Kawin *et al.*, 1950; Mealey, 1957). Schroeder and Balassa (1966*b*) reviewed the biochemistry of Zr as an abnormal trace metal in man. There are no specific reports of direct zirconium involvement *in vivo* in any biochemical systems except the inhibition of ATPase *in vitro* by zirconyl chloride (Cochran *et al.*, 1950). Zirconium complexes have strong anticaries activity in rats (Bobyleva, 1968; Muhler *et al.*, 1970), and injection of Zr compounds promotes the excretion of Pu in experimental animals (Schubert, 1949).

Gastrointestinal absorption of simple cationic Zr salts in laboratory animals is negligible (Hamilton, 1949), because olation and insoluble phosphate formation prevent absorption. However, citrate and tartrate complexes are readily absorbed. Parenterally administered Zr salts are slowly absorbed from injection sites, and simple cationic salts cause local irritation. Intravenously injected cationic salts form insoluble colloidal polymers and are phagocytized by macrophages. Injected and ingested citrate complexes are retained in the blood for some time, transported to other tissues, and then metabolized (Blackstrom *et al.*, 1967). Retention in rat lungs is high following intratracheal intubation of Zr (Morishige, 1968). Bone-seeking Zr is deposited more in the skeleton than in the soft tissues, where the deposition in decreasing order is: lungs > spleen > liver > kidneys. Young rats absorb more parenterally injected Zr salts than adult or old animals, and young rats retain them longer in their skeleton because of vigorous metabolism in the bone marrow (Ishinishi and Morishige, 1969). Excretion of Zr is mainly through the feces, owing to poor alimentary absorption of orally ingested Zr salts and to the accumulation of soluble Zr salts in the liver with their subsequent return to the alimentary tract via the bile. Less than 1% of the daily intake of Zr of humans is excreted in the urine (Schroeder and Nason, 1971), although soluble citrate complexes retained in the kidneys are evidently excreted in the urine. The absorbed Zr is either sequestered in the skeleton or excreted very rapidly. A mechanism of zirconium homeostasis is apparently present in man.

Toxicity

Zirconium salts are of low toxicity to animals. The toxicology of Zr has been reviewed by McClinton and Schubert (1948), Spink (1961), and Browning (1969). Zirconium is more toxic when given parenterally than *per os*. The low Zr toxicity is evidently due to poor absorption from the alimentary tract; the pH of the small intestine favors olation of Zr salts. However, intraperitoneal injection of zirconyl sulfate results in considerable toxicity; the toxicity of zirconium is summarized in Table 4-4.

There are no well-described characteristic physiological or pathological changes in animals owing to acute Zr toxicity; progressive depression and a decrease in activity precede death by approximately one week. Growth rate, survival, and longevity in rats and mice are not affected by feeding 5 ppm Zr in the drinking water in life-term studies; the spleen accumulates most of the Zr present in the soft tissues (Schroeder *et al.*, 1968*c*, 1970*c*). There are no fatalities in rats fed up to 10 g Zr oxide/kg. Intraperitoneally, zirconium gluconate was more toxic than the citrate in rats (McClinton and Schubert, 1948). Rabbits inhaling zirconium lactate aerosol develop pulmonary granulomata (Prior *et al.*, 1960), and zirconium granulomas in humans have been reported after cutaneous exposure to deodorant sticks which contain approximately 0.5% zirconyl sodium lactate (Rubin *et al.*, 1956; Shelly and Hauley, 1957, 1958); they seem to be of allergic epithelial origin, and cause hypersensitivity with granulomatous lesions resembling those caused by Be. There is no known specific detoxication mechanism for Zr.

TABLE 4-4. Toxicity of Titanium, Zirconium, and Hafnium

Compound	Animal	Route	Toxicity	Dosage/kg body weight			
				Compound	Metal		
				mg	mg	mM	pT
Titanium chloride	Rat	im	MLD	50	12.6	0.263	3.58
$TiCl_4$	Hamster	im	MLD	83	21.0	0.438	3.36
Potassium hexafluorotitanate	Guinea pig	oral	MLD	200	40	0.835	3.08
K_2TiF_6	Guinea pig	sc	MLD	450	90	1.88	2.73
Titanocene dichloride	Mouse	ip	LD_{50}	245	47.2	0.984	3.01
(Titanium dichlorodi-π-	Mouse	im	LD_{50}	250	48.1	1.00	3.00
cyclopentadienyl)	Rat	ip	LD_{50}	30	5.77	0.120	3.92
$TiC_{10}H_{10}Cl_2$	Hamster	im	MLD	83	16.0	0.333	3.48
Zirconium fluoride	Mouse	iv	LD_{50}	98	53.5	0.59	3.23
ZrF_4							
Zirconium chloride	Mouse	oral	LD_{50}	655	256	2.81	2.55
$ZrCl_4$	Rat	oral	LD_{50}	1690	660	7.24	2.14
Zirconyl chloride	Mouse	ip	LD_{50}	171	87.5	0.96	3.02
$ZrOCl_2$	Rat	oral	LD_{50}	3500	1790	19.6	1.71
	Rat	ip	LD_{50}	400	204	2.24	2.65
	Rat	sc	MLD	500	256	2.81	2.55
Zirconium nitrate	Rat	oral	LD_{50}	3170	853	9.35	2.03
$Zr(NO_3)_4$							
Zirconyl nitrate	Rat	oral	LD_{50}	2500	986	10.81	1.97
$ZrO(NO_3)_2$	Rat	ip	LD_{50}	1250	493	5.40	2.27
Zirconium sulfate	Rat	oral	LD_{50}	3500	1130	12.4	1.91
$Zr(SO_4)_2$	Rat	ip	LD_{50}	175	56	0.61	3.21
	Rat	sc	MLD	500	161	1.76	2.75
Zirconyl sulfate	Rat	oral	LD_{50}	3500	1570	17.2	1.76
$ZrOSO_4$	Rat	ip	LD_{50}	175	78.4	0.86	3.07
Zirconyl sodium sulfate	Rat	oral	LD_{50}	10,000	2290	25.1	1.60
$ZrOSO_4\ Na_2SO_4$	Rat	ip	LD_{50}	4100	939	10.29	1.99
Zirconyl acetate	Rat	oral	LD_{50}	4100	1660	18.2	1.74
$ZrO(C_2H_3O_2)_2$	Rat	ip	LD_{50}	300	112	1.23	2.91
Zirconium lactate	Rat	ip	LD_{50}	500	102	1.12	2.95
$Zr(C_3H_5O_3)_4$							
Zirconyl lactate	Rat	ip	LD_{50}	670	214	2.35	2.63
$ZrO(C_3H_5O_3)_3$							
Zirconium citrate	Rat	oral	LD_{50}	1690	449	4.92	2.31
$Zr_3(C_6H_5O_7)_4$							
Zirconyl sodium citrate	Rat	ip	LD_{50}	1710	511	5.60	2.25
$ZrONa(C_6H_5O_7)$							
Zirconium gluconate	Rat	ip	LD_{50}	247	78.8	0.86	3.06
Zirconyl sodium glutamate	Rat	ip	LD_{50}	247	95	1.04	2.98
Zirconocene	Rat	ip	LD_{50}	30	9.36	1.10	3.99
(Zirconium, dichlorodi-π-							
cyclopentadienyl)							
$ZrC_{10}H_{10}Cl_2$							
Hafnyl chloride	Mouse	ip	LD_{50}	112	75.3	0.42	3.37
$HfOCl_2$	Rabbit	ip	LD_{50}	112	75.3	0.42	3.37

Hafnium (Hf)

Hafnium does not occur free in nature and does not form any of its own mineral ores; it occurs as up to 1–3% of the total Zr in Zr minerals, because Hf can penetrate the crystal lattice of Zr minerals. Hafnium is a widely scattered element and resembles Zr very much in its chemical properties. Special analytical procedures are needed to separate the two. There are no reports of the occurrence of Hf in human tissues, but it occurs with Zr in a ratio of about 1:100. It is not essential for animals.

Hafnium is extensively used as shielding material and as control rods in thermal nuclear reactors, because its thermal neutron cross section, 105 barns, is nearly 1000 times that of Zr; Hf can absorb and give up heat twice as fast as Zr and Ti. It is also used as a construction material in space technology and in jet engines, but there are no other reported uses for Hf compounds.

Chemistry

Hafnium exhibits valences from $+2$ to $+4$, but the most stable is tetravalence. Hafnium is more electropositive and basic than Zr, and cationic Hf salts are more stable than similar Zr compounds. Anionic Hf compounds are not common. Olation is less pronounced with Hf than with Zr salts. Hafnyl (HfO^{2+}) compounds are formed by the hydrolysis of salts such as $HfCl_4$; hydrolytic polymerization leads to the formation of insoluble colloidal polymers of $HfO_2 \cdot H_2O$.

Metabolism

Hafnium is not essential to man or animals, and it is not reported to be involved in any biochemical systems; there are very few published reports about Hf metabolism. Gastrointestinal absorption of Hf is poor owing to the low solubility of hafnium oxides and salts, and absorption from intraperitoneal and subcutaneous administration sites is also poor because of hydrolysis and retention of hafnia (HfO_2) at the site (Van Niekerk, 1937). Excretion following oral administration is mainly fecal. Injected radioactive Hf differs from Zr by accumulating more in the liver than in the bones.

Toxicity

Toxicology studies on Hf are also rare; Hf is not toxic when fed orally to rats, since they tolerated up to 1% of $HfCl_4$ in their diet with no mortality for 12 weeks. There was no significant effect on growth rate or hematology, but histopathologic examination showed hepatic damage consisting of perinuclear vacuolization of the parenchymal cells and coarse granularity of the cytoplasm (Haley *et al.*, 1962*c*). There are no characteristic toxic symptoms associated with Hf; when administered intraperitoneally to mice the animals showed general lethargy. Van Niekerk (1937) observed that death could be caused by hypotension and respiratory paralysis. The LD_{50} of $HfCl_4$ injected in mice by intraperitoneal route is 75 mg Hf/kg, but the toxicity could be due to free HCl and $HfOCl_2$, because $HfCl_4$ reacts violently with water. Complete cardiovascular collapse occurred in cats when 10 mg $HfCl_4$/kg (6.72 mg Hf/kg) was injected intravenously; death was due to respiratory paralysis. Topical application of $HfCl_4$ on skin caused nonhealing ulcers. Stable Hf salts were not examined for their toxicity.

Lack of toxicology data on Hf leads to speculation on Hf toxicity. Owing to its similarity with Zr and Ti in chemical properties, one may assume that its metabolism in living tissues is similar but less traumatic because of the increased stability of the oxides formed. The higher electropositivity of Hf may render the chelation complexes of Hf with OH-containing ligands very stable, and thereby confer somewhat higher toxicity to Hf than Ti or Zr.

5

TOXICITY OF GROUP V METALS AND METALLOIDS

The nonmetals nitrogen (N) and phosphorus (P), the metalloid arsenic (As), and the metals antimony (Sb) and bismuth (Bi) constitute subgroup VA of the periodic chart; the metals vanadium (V), niobium (Nb, formerly columbium), and tantalum (Ta) constitute subgroup VB. The metal with atomic number 105 is highly radioactive and is yet to be identified and characterized. All these elements exhibit a characteristic maximum valence of 5.

Nitrogen and phosphorus are essential elements for all life, while arsenic, antimony, and bismuth are not essential and are toxic for animals. Arsenic is the most toxic of the three; although toxicity is expected to increase with increase in atomic weight and electropositivity, the insolubility of antimony and bismuth salts masks their greater inherent toxicity. Trivalent compounds of subgroup VA elements are more toxic than pentavalent compounds. Vanadium, within normal physiologic concentration, may be considered essential for mammals, but at doses slightly greater than the proposed essential level it is highly toxic. It interferes with the metabolism of a number of biologically active compounds, inhibits enzyme systems, and retards the biosynthesis of active molecules. Niobium is neither essential nor toxic within normal physiologic levels; it occurs in human and animal tissues at levels which are equivalent to those of copper. Tantalum is neither essential nor is it considered toxic, atlhough toxicity is noticed at phenomenally high levels. Among subgroup VB metals, V is probably essential and certainly the most toxic, despite being less electropositive than Nb or Ta, because V forms a large number of active water-soluble compounds. Pentavalent salts of subgroup VB metals are more toxic than the trivalent salts.

205

SUBGROUP VA

All elements of subgroup VA except Bi exhibit variable valence from -3 to $+5$; Bi exhibits valences from $+3$ to $+5$. The stability of the gaseous hydrides of subgroup VA elements decreases, but the toxicity of the hydrides increases as atomic weight increases with the subgroup. Antimony and bismuth differ from the other elements in having an empty orbital, and their soluble cationic salts hydrolyze to form water-insoluble SbO^+ and BiO^+ cationic salts. The solubility of their cationic salts decreases with increased atomic weights, owing to the formation of oxy salts. The affinity of As, Sb, and Bi to thiol groups of active molecules is

Group V Elements

Atomic number	Name (Symbol)	(Core) Active electrons
Atomic weight		Usual valences
Subgroup A		Subgroup B
7	(He) 2, 3	
Nitrogen (N)		
14.007	$+1-+5, -1$	
15	(Ne) 2, 3	
Phosphorus (P)		
30.974	$+3, +5, -3$	
33	(Ar) 2, 3	23 (Ar) 3, 2
Arsenic (As)		Vanadium (V)
74.92	$+3, +5, -3$	50.94 $+2-+5$
51	(Kr) 2, 3	41 (Kr) 4, 1
Antimony (Sb)		Niobium (Nb)
121.75	$+3, +5, -3$	92.906 $+3, +5$
83	(Xe) 2, 3	73 (Xe) 3, 2
Bismuth (Bi)		Tantalum (Ta)
209.98	$+3, +5$	180.95 $+5$
		105 (Rn) ?
		Hahnium
		(260) ?

high, and the stability of these bonds increases with increased atomic weights. These metals readily form complex anions in which they are the central atoms; halogen complexes are fairly stable. Arsenates, due to their isomorphic similarity with phosphates, can replace phosphates in energy-capturing reactions and prevent the trapping of energy. The biologic activity and toxicity of these metals decreases: As > Sb > Bi.

Arsenic (As)

Elemental As occurs in nature as a brittle metal, and is also abundant in the form of arsenides and arsenosulfides of heavy metals, e.g., mispickel (FeSAs). Arsenic is present in seawater at 5 ppb and the earth's crust at 2 ppm. An adult human body contains about 18 mg As distributed in the tissue (Schroeder and Nason, 1971). Arsenic may be oncogenic; it is not considered essential, but is stimulatory.

White arsenic and other As compounds are used extensively in rodenticides, insecticides, fungicides, wood preservatives, soil sterilants, defoliants, feed additives, and weed-killers; they are also used in the glass, textile, and tanning industries, and in the manufacture of pigments and antifouling paints. Arsenic has been used therapeutically for 2000 years (Frost, 1967); its compounds are toxic and have been used in intentional poisoning, but its beneficial effects have also been recognized. Arsenic is present in most food products because of its use in agricultural chemicals. It is considered to be a serious health hazard in industry, and exposure to toxic amounts of As is carefully avoided. A wide variety of organic arsenicals has been developed for agricultural, industrial, and therapeutic uses. Generally inorganic As is more toxic than organic As; in this book the toxicology of only inorganic forms of As will be discussed.

Chemistry

Arsenic is a metalloid, exhibiting variable valences -3, $+3$, and $+5$; it is capable of forming both cationic and anionic salts. Cationic As halides are reactive and unstable, but hydrolyze to arsenic oxy-acid salts which are quite stable. Volatile AsH_3 is fairly stable and highly toxic. As^{3+} and Sb^{3+} are unique in that they can be either the central atom in a complex anion species or the donor atom in complexing with other metals, e.g., arsenomolybdates. Arsine complexes with metal salts to form stable double bonds between the metal and donor atoms, e.g., $AgI \cdot AsH_3$ or $CuI \cdot 2AsH_3$. Inorganic As compounds readily form reversible complexes with sulfhydryl and hydroxyl groups of proteins and amino acids.

Metabolism

Arsenic has been extensively investigated, but the nutritional status of this metal is still controversial, because it is very difficult to prepare a diet without traces of it. Nielsen (1975) reported evidence that As may be an essential nutrient for rats; deficiency symptoms are poor growth, fragile blood cells, and abnormal Fe metabolism. Seafoods contain appreciable amounts of As compounds (0.3–3 ppm); tissues of animals contain an average of less than 0.5 ppm As. The average daily intake of As by humans is 0.2–1 mg, and about 0.20 mg is excreted daily in the urine. The blood of an average human adult contains about 2.5 mg As. Arsenic accumulates with age in human and animal tissues such as the aorta, spleen, and hair. The physiology and biochemistry of As have been extensively reviewed (Frost *et al.*, 1955; Vallee *et al.*, 1960; Johnstone, 1963; Peoples, 1964; Schroeder and Balassa, 1966*a;* Frost, 1967; Underwood, 1971).

The gastrointestinal absorption of As in animals depends upon the solubility and polarizability of the compounds; anionic As and soluble compounds are well absorbed. Removal from the sites of parenteral injection is complete within 24 hours (Stewart and Stolman, 1960), but cationic As causes necrosis and ulcerations of the area of contact in the tissues before it is fully absorbed. Inhaled As compounds are fully absorbed from the lung and mucous surfaces of the respiratory tract, and As compounds are also absorbed through the skin; absorption through the skin is enhanced if the As compound is lipid-soluble or distributed in lipid-soluble ointment. Following intravenous injection or absorption into the blood, most of the inorganic As is bound to the globin of hemoglobin in erythrocytes; As is also bound to serum proteins. The distribution of As is rapid from the blood to the liver, kidney, lung, intestinal mucusa, spleen, muscles, and some nervous tissues. With chronic intake, As accumulates in an inactive and tightly bound form in hair, and other tissues such as skin, nails, and bones also accumulate As. Excretion of arsenic is mainly in the urine irrespective of the mode of administration, but some is excreted in the feces owing to both its incomplete absorption from the alimentary tract and its enterohepatic circulation. Sweat glands excrete As in small but significant amounts, and As is also present in exhaled breath. Excretion of As from humans and animals is slow, requiring about 10 days following a single parenteral dose. The efficient urinary excretion seems to serve as a satisfactory homeostatic mechanism in preventing As accumulation.

Toxicity

The chemical toxicity of As compounds has been well studied because of the extensive therapeutic use of As compounds in the past; the toxicity of inorganic As compounds is summarized in Table 5-1. Arsine is the most toxic of these compounds; trivalent arsenic trioxide or arsenite are more toxic than pentavalent As owing to the slower excretion and greater retention of arsenic trioxide in the tissues. The toxicology of As has been reviewed (Buchanon, 1962; Browing, 1969; Arena, 1974), and symptoms of acute As toxicity include nausea, vomiting, diarrhea, acute and severe irritation of the nose, throat, and conjunctiva, severe abdominal pain, skin eruptions, and inflammations. Chronic overexposure to As_2O_3 dust or vapor leads to general weakness, prostration, muscular aching, perforation of the nasal septum, ulceration in the alimentary tract, peripheral neuritis, and tremors. Colloidal As is toxic both by ingestion and parenteral injection, but otherwise elemental As is nonpoisonous when ingested, although a small amount may be converted into the irritant As_2O_3.

Arsine (AsH_3), a hazard from lead storage batteries if arsenic-free lead and sulfuric acid are not used, is a potent toxic agent; a 30-minute exposure at 250 ppm in the air may be lethal to humans and animals; the maximum concentration inhalable for one hour without toxic symptoms is 5–20 ppm. Chronic inhalation of AsH_3 at subtoxic levels by laboratory animals leads to hemolysis of erythrocytes, leading to death from chemical asphyxia. Irritation from arsine also results in pulmonary edema.

Rats tolerate As^{3+} at 50 ppm in their diets, but at 300 ppm significant growth depression, chronic hepatitis, and cirrhosis of the liver were noticed (Schroeder *et al.*, 1968b). Systemically, As^{3+} relaxes the capillaries, increasing the permeability of capillaries and arterioles; dilation introduces circulatory changes, causing inflammation and altered function. Arsenites also accumulate in the leukocytes. The acitivity of thiol-containing enzymes, including DNA polymerase, is decreased by arsenite intoxication.

Pentavalent As, slightly less toxic than the trivalent form, inhibits enzyme systems such as α-glycerophosphate dehydrogenase and cytochrome oxidase. It inhibits ATP synthesis by uncoupling oxidative phosphorylation in the liver mitochondria because of its ability to replace the stable phosphoryl group; the unstable arsenate complex hydrolyzes spontaneously without ATP formation or trapping of energy. Pentavalent As affects Se metabolism by preventing Se from reaching a target site or by increasing Se excretion (Hill and Matrone, 1970). When given to pregnant mice intraperitoneally at 45 mg arsenate/kg, pentavalent arsenic produces teratogenic effects, increased fetal resorption, and encephaly and micro-

TABLE 5-1. Arsenic Toxicity

Compound	Animal	Route	Toxicity	Compound mg	Metal mg	Metal mM	pT
Arsenic	Rat	im	MLD		25	0.33	3.48
Arsine AsH_3	Rat	inhal	LD_{100}	250^a	240^a		
Arsenic trioxide As_2O_3	Mouse	sc	LD_{100}	12	9.0	0.12	3.92
	Mouse	oral	LD_{50}	43	32.6	0.44	3.36
	Rat	sc	LD_{100}	8	6.0	0.080	4.10
	Rat	oral	LD_{50}	143	120	1.58	2.80
	Guinea pig	oral	MLD	30	22.6	0.30	3.52
	Guinea pig	sc	MLD	13	9.75	0.13	3.88
	Guinea pig	ip	MLD	16	12.0	0.16	3.80
	Rabbit	oral	MLD	15	11.3	0.15	3.82
	Rabbit	sc	MLD	8	6.0	0.080	4.10
	Rabbit	iv	MLD	6	4.5	0.060	4.22
	Cat	sc	MLD	4.7	3.5	0.047	4.33
	Dog	oral	LD_{100}	50	37.8	0.47	3.30
	Dog	sc	LD_{100}	6	4.6	0.062	4.21
	Dog	iv	LD_{100}	4	3	0.040	4.40
	Man	oral	MLD	1	0.76	0.010	4.99
	Man	inhal	MLD	0.2^a	0.16^a		
Arsenic pentoxide As_2O_5	Rat	oral	LD_{50}	8	2.6	0.035	4.46
	Rabbit	iv	MLD	6.0	2.0	0.027	4.57
Arsenic trifluoride AsF_3	Guinea pig	sc	LD_{50}	20	11.4	0.15	3.82
Arsenic trichloride $AsCl_3$	Mouse	inhal	MLD	2500^a	1030^a		
	Cat	inhal	MLD	100^a	41.4^a		
Sodium arsenite Na_3AsO_3	Rat	ip	MLD	10	7.6	0.10	3.99
Disodium arsenate Na_2HAsO_3	Rat	ip	MLD	30	12.1	0.16	3.79
Potassium arsenite K_3AsO_3	Rat	oral	LD_{50}	14	7.2	0.10	4.02
	Man	oral	MLD	665	341	4.55	2.34
Sodium arsenate Na_3AsO_4	Mouse	ip	LD_{50}	9	3.7	0.049	4.31
	Rat	ip	MLD	35	14.4	0.19	3.72
	Rat	sc	MLD	2	0.82	0.011	4.96
	Man	inhal	MLD	4^a	1.6^a		
Calcium arsenate $Ca_3(AsO_4)_2$	Mouse	oral	LD_{50}	794	112	2.79	2.53

[a]Exposure of 4 hr; number of mg/m^3.

gnathia in the surviving fetuses. Defects caused by As^{5+} are more numerous than those caused by As^{3+}, but the latter causes fetal death at lower levels, 10–12 mg As^{3+}/kg (Hood and Bishop, 1972; Hood, 1972).

The biologic toxic action of As is due more to impaired functional activity than to any damage to the structural integrity of the tissues. The affinity of As^{3+} for keratin of nail and hair and its *in vitro* inhibition of enzymes, such as *d*-amino acid oxidase, pyruvate oxidase, choline oxidase, and transaminase, suggest that As acts by binding to hydroxy or thiol groups of biologically active molecules. The binding of As to thiol-group-containing enzymes and their resultant inactivation has been confirmed (Westernhagen, 1970). Arsenic reacts with two thiol groups per molecule (of a thiol compound) to form stable ring compounds, in preference to reacting with two thiol groups of two separate molecules. Frost (1972) suggested that As may interfere more with Se-dependent enzymes than with thiol-containing enzymes. Arsenic–selenium interaction in tissues seems to attenuate the toxicity of both metals. When a large dietary intake of either As_2O_3 or As_2O_5 overwhelms the delicate homeostatic mechanism, a high nutritional supplement of Se can prevent As intoxication (Frost, 1972), and, similarly, As counters Se poisoning at 12–15 ppm in the diet of cattle and chickens and at 5 ppm in dogs, pigs, and rats. The presence of relatively nontoxic dimethyl arsenic (25 times less toxic than As_2O_3) and methyl arsenic acids in human urine suggests detoxication of arsenic by methylation (Braman and Foreback, 1973).

Involvement of arsenic in carcinogenesis is not unequivocally established, as the epidemiological and experimental evidence seem to be in complete contradiction. Many reports lead to the conclusion that As is a cocarcinogen and may be a primary carcinogen, but the following experimental evaluations of arsenic for carcinogenicity proved negative: As_2O_3 in mice (Finkell and Brues, 1960), As_2O_3 in rats and mice (Hueper and Payne, 1962), As_2O_3 and Na_3AsO_4 in mice (Baroni *et al.*, 1963), K_3AsO_3 in mice (Boutwell, 1963), and Na_3AsO_3 and Na_3AsO_4 in rats and dogs (Byron *et al.*, 1967). A few epidemiological studies also failed to implicate As as a carcinogenic agent; there was no evidence for systemic cancer in humans following chronic As_2O_3 exposure among smelter employees (Pinto and Bennet, 1963), and handling arsenicals in a metallurgical industry did not result in significant change in cancer mortality (Snegireff and Lombard, 1951). Arsenite was the first chemotherapeutic agent against cancer, especially leukemia, and in other animal experiments, a relatively high intake of organic arsenicals decreased the spontaneous occurrence of cancer (Schrauzer and Ishmael, 1974; Kanisawa and Schroeder, 1967). Arsenicals also minimized the induction of cancer in experimental animals (Boutwell, 1963; Milner, 1963). The influence of SO_2 and unidentified chemicals which are associated with As exposure should be considered in assessing As involvement in cancer; earlier reports and reviews strongly assert the carcinogenicity of As (Currie, 1947).

Among the innumerable epidemiologic reports on As carcinogenesis, a few typical studies will be briefly reviewed. Lee and Fraumeni (1969) reported on the incidence of malignant neoplasms of the respiratory tract in a group of eight thousand employees of an American smelting works; the increase in cancer fatalities were as follows: sevenfold when exposed to high concentrations of As, fivefold for medium exposure, and 2.5-fold for light exposure. However, sulfur dioxide fumes in the smelter could have also contributed to the incidence of cancer. In Dow Chemical's study of its workers, exposure to As_2O_3 vapors increased the incidence of cancer in the respiratory and lymphatic systems. The toxic effects of As in the British beer-poisoning episode correlated with the increased incidence of pulmonary cancer among beer drinkers (Satterlee, 1960). The correlation between the increased use of As sprays in agriculture and the increased prevalence of bronchial cancer among the spray handlers provides further epidemiologic evidence (Roth, 1958). Prevalence of skin cancer, hyperpigmentation, and keratosis has been reported in endemic areas of chronic arsenicism where the drinking water contained 0.8–2.6 ppm As. The use of internal arsenical medicines for many years causes skin cancer; the first and the most common is an epithelioma developing at the site of keratosis which is characterized by a long latent period extending to years between As exposure and skin-cancer induction; the second is Bowen's disease or precancerous dermatitis, characterized by multiple epitheliomatosis of low malignancy. The use of Fowler's solution (1% As_2O_3) for asthma treatment can cause Bowen's disease (Novey and Martel, 1969), and a strong correlation has been established between As therapy and bronchial carcinoma in patients treated with As compounds (Robson and Jelliffe, 1963). Prolonged arsenite therapy can deplete the body stores of selenium, thereby increasing the susceptibility of the subject to cancer (Neubaur, 1946; Osswald and Goerttler, 1971).

Experiments *in vitro* have shown that As^{5+} could act as a cocarcinogic agent by suppressing host resistance factors, to increase the probability of survival of a spontaneously arising tumor (Stone, 1969). Arsenate inhibited DNA-repair mechanisms to a considerable degree in *in vitro* experiments involving skin grafts (Jung and Trachsel, 1970); Westernhagen (1970) showed histochemically that chronic As poisoning in guinea pigs decreased the activity of DNA polymerase and other thiol-containing enzymes. Stable thioarsenic compounds are metabolized in mammalian tissues to active aresenoxides and other products (Voegtlin, 1925; Reines and Leonard, 1932), but the involvement of these compounds and other stable arsenic ring compounds in the toxicity, teratogenesis, and carcinogenesis of arsenic must be studied further to establish mechanisms.

Antimony (Sb)

Antimony occurs as the native metal, and in oxides, sulfides, and complex copper, lead, silver, and murcury sulfides. Stibnite (Sb_3S_3) and valentinite (Sb_2S_3) are important ores. Antimony occurs in the earth's crust at about 0.2 ppm and in seawater at about 0.2 ppb; the human adult body contains about 7–9 mg, 25% in the bones and 25% in blood. Antimony and its compounds are not essential for mammals, but antimony has been found to be stimulatory.

Antimony is used extensively in industry; Pb–Sb alloy is used in bearing metals, storage battery grids, lead electrodes, type metal, sheet and pipe, cable covering, solder, and ammunition; antimony sulfides are used in the rubber industry; oxides of Sb are used in the ceramics and glass industries as opacifiers and decolorizers. Other Sb compounds are used in flameproof textiles, paints, and lacquers, and in abrasives, phosphors, and matches; indium antimonide is used as intermetallic compound in electronics. Tartar emetic or potassium antimonyl tartrate was used extensively as a chemotherapeutic agent against schistosomiasis and leishmaniasis in man. Acute Sb poisoning can occur from ingestion, whether accidental or by deliberate poisoning, from therapeutic use, from exposure in industrial operations, and from contamination of food containers. Stibine, the volatile hydride, is a serious exposure risk when storage batteries are charged in a closed area.

Chemistry

Antimony exhibits variable valence, -2, -1, $+3$, and $+4$; it is more electropositive and more basic than arsenic and forms stable tri- and pentavalent cations, antimonyl ion ($SbO+$), and fairly stable antimonate anion (SbO_4^{3-}). However, the halides, though stable under acidic pH, are readily hydrolyzed at neutral pH to unstable free H_3SbO_3 and H_3SbO_4 acids, which in turn form stable Sb_3O_3 and Sb_2O_5. Trivalent antimony can exist as the central atom in Sb complexes or as a donor atom in complexing with another metal ion, e.g., hexahalo anions of Sb; K_2SbF_5 exists in octahedral form. Antimony forms no basic salt; it olates to a small extent to form insoluble complexes. Trivalent Sb forms thioantimonites with SH groups of cellular constituents.

Metabolism

Antimony and its salts are not essential for plants and animals; they are toxic, but can be stimulatory at low levels. Most metabolic studies of Sb are with organic antimonial compounds and are of doubtful significance with respect to inorganic salts; the metabolism of Sb has been reviewed (Fairhill and Hyslop, 1947, 1957), but there have been few recent reports. Schroeder and Nason (1971) report an average daily dietary intake of < 0.15 mg Sb by humans, and a urinary excretion of < 0.07 mg and a loss of 0.01 mg in sweat. Soluble Sb compounds such as tartrates and antimonites are slowly absorbed from the gastrointestinal tract of man and animals; these salts cause local irritation in the stomach and induce severe vomiting and sloughing of mucosal tissues. Antimony halides are hydrolyzed to oxides which are not absorbed but cause severe irritation and caustic action on the alimentary mucosa. Parenterally injected Sb salts are absorbed slowly from injection sites, depending on the solubility of the salt and the local toxic action. Part of the intravenously administered Sb salts are absorbed by erythrocytes, and the rest is distributed to other tissues, predominantly the liver, adrenals, spleen, and thyroid. In rats trivalent Sb is absorbed by erythrocytes, distributed to other tissues, and retained in the liver for a short time before it is gradually excreted in feces; pentavalent Sb remains in the plasma for a short time, and most of it is excreted in the urine, while a part is retained in the hair. In the livers of mice and humans pentavalent Sb is reduced to the trivalent form (Stolman and Stewart, 1960). Inhalation studies with dogs suggest rapid absorption from the lungs and enhanced urinary excretion (Brieger *et al.*, 1954). Ingested ^{124}Sb-labeled potassium antimonyl tartrate by laboratory animals is excreted more in feces than in urine, indicating poor alimentary absorption (Weitz and Ober, 1965); but 96% of an intramuscular injection of ^{122,124}Sb in the form of $HSbO_3$ was excreted in the urine of rats (Durbin, 1960). Excretion in both studies was slow, and a slow fecal excretion pattern is observed following intramuscular injection of Sb^{3+} into humans (Hassan, 1938). The lack of Sb accumulation in the tissues suggests a poorly defined homeostatic excretory mechanism.

Toxicity

Soluble Sb salts are more toxic than similar Pb or As compounds, and trivalent Sb salts are ten times more toxic than pentavalent salts. Among laboratory animals, dogs and rabbits are most susceptible to Sb toxicity; the toxicity of Sb compounds is summarized in Table 5-2. Stibine (SbH_3), the most toxic of the Sb compounds, causes nausea, vomiting, hemolysis,

TABLE 5-2. Antimony Toxicity

| Compound | Animal | Route | Toxicity | Dosage/kg body weight | | | |
| | | | | Compound | Metal | | |
				mg	mg	mM	pT
Antimony metal	Rat	oral	LD_{50}		100	0.82	3.09
	Rat	ip	LD_{50}		100	0.82	3.09
	Guinea pig	ip	LD_{50}		150	1.23	2.91
Stibine	Mouse	inhal	MLD	0.1^a	0.097^a		
SbH_3	Rabbit	iv	LD_{50}	8	7.81	0.06	4.19
Antimony trioxide	Rat	oral	LD_{50}	>20,000	>16,700	160	0.69
Sb_2O_3	Rat	ip	LD_{50}	3250	2720	22.3	1.65
Antimony pentoxide	Rat	ip	LD_{50}	4000	3010	24.7	1.61
Sb_2O_5							
Antimony trifluoride	Rat	sc	LD_{50}	22.9	13.9	0.11	3.94
SbF_3							
Antimony trichloride	Rat	oral	LD_{50}	675	275	2.26	2.65
$SbCl_3$	Man	inhal	MLD	131^a	70.1^a	0.58	
Antimony trisulfide	Rat	ip	MLD	1000	717	5.88	2.23
Sb_2S_3	Man	inhal	MLD	0.565^a	0.203^a		
Antimony pentasulfide	Rat	ip	LD_{50}	1500	905	7.43	2.13
Sb_2S_5							
Antimony trisulfate	Rat	ip	LD_{50}	1000	458	3.76	2.42
$Sb_2(SO_4)_3$							
Potassium antimonyl tartrate	Mouse	oral	LD_{50}	600	222	1.82	2.74
$K\ SbC_4H_4O_7$	Mouse	ip	LD_{50}	50	18.5	0.15	3.82
	Mouse	iv	LD_{50}	42	15.7	0.13	3.89
	Mouse	sc	LD_{50}	55	20.1	0.17	3.78
	Rat	oral					
	Rat	oral	LD_{50}	115	42.0	0.34	3.46
	Rat	im	LD_{100}	33	12.0	0.098	4.01
	Rat	ip	LD_{50}	11	4.02	0.036	4.48
	Guinea pig	im	LD_{100}	55	20.1	0.17	3.78
	Guinea pig	ip	LD_{50}	15	5.48	0.050	4.35
	Rabbit	im	LD_{100}	58	21.2	0.17	3.76
	Rabbit	oral	LD_{50}	115	42.0	0.34	3.46
	Rabbit	iv	LD_{100}	15	5.5	0.050	4.35
	Man	oral	MLD	2	0.73	0.003	5.22
Sodium antimonyl tartrate	Mouse	iv	LD_{50}	25	10	0.048	4.32
$NaSbC_4H_4O_7$	Mouse	ip	LD_{50}	60	43.4	0.21	3.68
	Mouse	sc	LD_{50}	48	34.7	0.17	3.78

^{a}Exposure of 4 hr; number of mg/m^3.

hematurea, acute abdominal pain, and death. In guinea pigs, SbH_3 is absorbed by erythrocytes, which undergo crenation. While less toxic than SbH_3, Sb metal is more toxic than its salts, both by inhalation and intraperitoneal administation, but symptoms of acute and chronic Sb poisoning occur more readily with oral than with other modes of administration. Intravenously administered doses of small amounts of Sb salts are tolerated better than ingested doses; the latter cause acute local irritation of alimentary mucosa and induce vomiting. The general symptoms of acute oral toxicity of Sb compounds in animals and humans are similar to those of As poisoning; owing to high local irritation of the gastric mucosa, violent vomiting occurs, eliminating sloughed-off mucosal cells and most of the toxic Sb salts. Continuous diarrhea and decreased body temperature occur, followed by lowered respiratory rate and death. Chronic oral toxicity is shown by albuminuria, jaundice, and damage to heart, liver, and kidney tissues. Symptoms of Sb toxicity by parenteral administration include dyspnea, loss of weight, loss of hair, general weakness, evidence of myocardial inefficiency, hyperplasia of spleen, glomerular nephritis, papular eruptions on the skin, an increase in erythrocytes, dysfunction of erythrocytes, and a decrease in white cells. Myocardial edema, hyperemia, and capillary engorgement are causes of death from Sb toxicity. In experimental animals, Sb and its compounds are equally toxic whether administered orally or intraperitoneally (Bradley and Frederick, 1940).

Small oral doses of antimony oxides, up to 100 mg Sb daily in the diet of rats, accelerate their growth, but doses of about 200 mg retard growth, and cats fed 450 mg daily of antimony oxides lose weight. The severe toxicity of $SbCl_3$ in inhalation studies is owing to the formation of both HCl and Sb_2O_3 after absorption of antimony halides. The onset of gastrointestinal disturbance is slightly delayed, but the respiratory irritation is immediate. Chronic inhalation of subtoxic levels of Sb compounds causes extensive interstitial pneumonitis, intraalveolar lipoid deposits, and cardiac and liver damage. Antimony halides cause dermatitis, keratitis, conjunctivitis, nasal septum ulceration, and pneumonitis (Levina and Chekunova, 1965).

The mechanism of Sb intoxication is not known; however, the glutathione and nonprotein nitrogen content of blood, and the epinephrine content of the adrenals increase during chronic Sb toxicity in animals, suggesting increased protein catabolism. *In vitro* inhibition of succinic oxidase and pyruvate oxidase suggests interference by antimony with cellular respiratory metabolism by combination with sulfhydryl enzymes. The direct action on the heart muscle is owing to the combined effects of functional disorder of autonomic systems, caused by the inhibitory effect on the cerebral cortex and by hyperexcitability of the myocardium (Ming-Hsin *et al.*, 1958; de Cotton and Logan, 1966).

There are no reports of detoxication mechanisms for Sb in mammalian tissues; administration of chelating compounds, such as dimercaprol (BAL), to rabbits suffering from Sb intoxication increases the urinary excretion of Sb (Eagle *et al.*, 1947).

Bismuth (Bi)

Bismuth occurs in the free state and also as bismuth sulfide (bismuth glance), bismuth oxide (bismuth ochre), and bismuth carbonate (bismutite).

Bismuth occurs in the earth's crust at 3.4 ppm; there are no reports of the presence of bismuth in human tissue or in plants. In mammals, Bi is neither essential nor stimulatory, and it shows little biologic activity.

Bismuth is used industrially in the manufacture of low-melting aluminum and steel alloys. Purified Bi metal is used in the preparation of a number of pharmaceutical products. In the past, bismuth metal and bismuth potassium tartrate has been used as an antisyphilitic agent; bismuth oxy salts are astringents and protective agents in gastrointestinal ulcers in humans. Arena (1974) warns that $BiCO_3$ should be used in therapy in preference to $Bi(NO_3)_2$, since the nitrate reduction by intestinal microbes will form the nitrite, which may have caused nitrite poisoning to have been erroneously attributed to Bi. In veterinary medicine Bi salts are used in eczema, removal of warts, and protection of the gastrointestinal mucosa. Bismuth compounds are highly toxic, but low solubility limits the extent of toxicity. Accidental Bi poisoning can result from therapeutic use of Bi salts, but there are no recent reports of such Bi poisoning.

Chemistry

Bismuth is more basic and electropositive than other metals in Group V; it exhibits variable valence $+3$, $+4$, and $+5$, and, rarely, -3 when it forms bismuthine (BiH_3). Its stable salts are $3+$ salts. The unstable anion, BiO_3^-, is formed under alkaline conditions. Trivalent Bi salts are soluble only in dilute acids; the addition of water or an increase in pH result in the formation of insoluble bismuth oxy salts ($BiOCl$). Olation converts these oxy salts into polymers of monohydroxy compounds which are insoluble in water and quite stable. Trivalent Bi also forms hexahalo anions and gives rise to salts such as K_2BiF_5 and K_4BiI_7, which are used as precipitants for thiamine hydrocholoride and some antibiotics. Bismuth can form stable bonds with sulfhydryl groups. Bismuthyl compounds (BiO^+) are not soluble in water unless they are coupled to highly polar groups.

Metabolism

The biochemistry and metabolism of Bi has not been thoroughly investigated; Sollman (1948) summarized the pharmacological effects of Bi salts, but the involvement of Bi in biochemical or enzyme systems is not known.

Stable and water-soluble Bi salts, including potassium bismuth tartrate, sodium bismuth citrate, sodium bismuth thioglycollate, and sodium bismuthate, are readily absorbed from the gastrointestinal tract in man and laboratory animals. These compounds bind the plasma proteins and are distributed to all tissues, predominantly the kidneys and liver. Trivalent Bi salts readily undergo hydrolysis and olation in the alimentary tract; they become only sparingly soluble and are poorly absorbed. Parenterally injected Bi salts are cleared very slowly from injection sites, depending upon the solubility of the Bi salt in the biologic fluids. Intravenously injected soluble Bi salts are distributed to various tissues, But Bi^{3+} salts undergo hydrolysis and olation to form uncharged colloids in the blood, breaking up the physical equilibrium of the colloids and producing a colloidoclastic shock. Retention in the kidneys is prolonged, and most of the absorbed Bi salts is excreted in the urine (Sollman *et al.*, 1938); unabsorbed Bi salts from the alimentary tract are excreted as black Bi_2S_3 in the feces. In rat studies, intramuscularly injected radioactive BiOCl was excreted mostly in the urine (Durbin, 1960). Parenterally administered Bi salts were found either as metallic Bi or Bi_2S_3 bound to the tissues as a black "lead line" at the alveolar margin or as black spots on the buccal mucosa and throat. In its absorption and retention in the tissues, Bi resembles Pb^{2+} and uranyl (UO_2^+) ions more than its does any subgroup VB metal. Bismuth, like lead, may be released from tissue storage depots during acidosis (Peters, 1942); there are no reports suggesting a homeostatic mechanism.

Toxicity

Chronic Bi intoxication produces alimentary symptoms; the mouth shows Bi deposition in epithelial cells, turning mucosa blue-gray. Diarrhea, vomiting, anorexia, and digestive pains can be caused by Bi salt embolism of mesenteric capillaries (Arena, 1974). These symptoms may be preceded by severe headache, and may be accompanied by peripheral neuritis and liver damage, and they may be followed by various skin lesions, nephrotoxic action, and blood dyscrasias. Available data on the toxicology of Bi is summarized in Table 5-3; owing to their slower excretion rate, dogs are more susceptible to Bi intoxication than rabbits.

Orally ingested cationic Bi salts are not toxic, and are, in fact, used therapeutically to coat the intestinal mucosa in ulcers. However, bismuth oxynitrate can be reduced to nitrite in the gastrointestinal tract by intestinal microflora, and subsequent absorption of nitrite can cause methemoglobinemia in a child (Wallace, 1947).

A single intravenous injection of a Bi salt has caused fatal Bi intoxication in human infants (Sterne *et al.*, 1955). When cationic Bi salts are administered intravenously to rats, they die quickly from colloidoclastic shock and asphyxial convulsions (Sollman and Seifter, 1942); toxicity is influenced by the rate at which soluble Bi salts are injected intravenously. Fast injection (5 ml within 10 min) results in rapid mortality, whereas slower injection (250 ml in about 8 hr) is tolerated. High doses in slow

TABLE 5-3. Bismuth Toxicity

Compound	Animal	Route	Toxicity	Compound mg	Metal mg	Metal mM	pT
Sodium bismuthate	Rat	oral	LD_{100}	720	552	2.64	2.58
$NaBiO_3$	Rat	im	LD_{100}	250	192	0.92	3.04
	Rat	iv	LD_{100}	25	19.2	0.09	4.04
	Rabbit	oral	LD_{100}	510	391	1.87	2.73
	Rabbit	im	LD_{100}	110	84.3	0.40	3.39
	Rabbit	iv	LD_{100}	9	6.9	0.033	4.48
Potassium bismuth tartrate	Rabbit	iv	LD_{100}	9.8	5.0	0.024	4.62
$KBiO(C_4H_4O_6)$	Rabbit	im		295	150	0.72	3.14
Sodium bismuth citrate	Rabbit	iv	LD_{100}	10.6	5.0	0.024	4.62
$NaBi(C_6H_5O_7)$	Dog	iv	LD_{100}	2.25	1.0	0.0048	5.32
Sodium bismuth thioglycollate	Rabbit	iv	LD_{100}	7.36	2.8	0.011	4.87
Sobisminol (sodium bismuthate complex)	Rabbit	iv	LD_{100}	52.5	10.5	0.050	4.30

injections cause nephritis and eventual mortality. Intravenously injected Bi salts also cause renal injury, albuminuria, diarrhea, and ulcerative stomatitis.

Hepatitis and nephrotoxicity in humans involving Bi salts have been reported (Kulcher and Reynolds, 1942; Czerwinski and Gin, 1964; Randall *et al.*, 1972). Degenerative changes occur in the proximal region of the renal tubules in both rats and humans; nuclear inclusion of Bi associated with a lipoprotein has been reported in rat and human epithelial cells (Pappenheimer and Maechling, 1934; Beaver and Burr, 1963; Burr *et al.*, 1965). Nuclear inclusion and inhibition of thiol-containing enzymes could be responsible for the degenerative changes.

There are no reports of any well-defined detoxication mechanisms for Bi, but nuclear inclusions and the formation of inert black bodies in the tissues may serve as detoxication mechanisms.

SUBGROUP VB

Among the subgroup VB metals, V exhibits variable valence +2–+5, Nb +3 and +5, and Ta only +5. The distribution of valence electrons in these metals differs among themselves and from the elements of subgroup

VA. Vanadium has no analog in the periodic table, and differs from Nb and Ta in its biologic properties, solubility, and the stability of its salts. None of these metals form toxic and volatile hydrides. The electronic structure of these metals explains the difference in their properties. Vanadium, niobium, and tantalum do not readily combine or coordinate with thiol-group-containing compounds, although V interferes with thiol-group-containing compounds and with sulfur metabolism in general.

Vanadium (V)

Vanadium occurs naturally as patronite (V_2S_5), and in vanadinite ($V_2O_5 \cdot PbO \cdot PbCl_2$), uravanite (uranyl potassium vanadate), and petroleum oils. It is widely distributed in the earth's crust, up to 300 ppm, and up to 5 ppb in seawater. As a contaminant of fuel oil it becomes an element of air pollution. Plants contain about 1 ppm V; the average V content of the root nodules of legumes is 4–12 ppm by dry weight. About 18 mg V is present in the adult human body, most in the fatty tissues (Schroeder and Nason, 1971). Vanadium is present in invertebrate tissues averaging up to 1.2 ppm and in vertebrate tissues up to 0.22 ppm on dry weight basis; some echinoderms and coelenterates contain about 57 ppm (Noddack and Noddack, 1940). Vanadium is stimulatory and may be considered essential for mammals, but specific requirements are not known.

Vanadium is used extensively in the steel industry to increase hardness, malleability, and resistance to fatigue. Vanadium oxide is used as a catalyst in sulfuric and nitric acid manufacture; V is also used in photography, and in the manufacture of insecticides, dyes, inks, paints, and varnish dryers. Vanadium compounds have been used therapeutically for the treatment of syphilis, tuberculosis, anemia, neurasthenia, and rheumatism. It is essential for the growth of fungi and algae; it stimulates photosynthesis in higher plants and is an oxygen-carrying metal in some invertebrates. Vanadium poisoning occurs from exposure to industrial operations, accidental overdoses in chemotherapy, and from eating food stuffs of high V content.

Chemistry

Vanadium is the least basic of the subgroup VB metals, and exhibits variable valence from +2 to +5; the pentavalent compounds are the most stable. Vanadium forms both cationic and anionic salts, but the exact nature of many of its ions in aqueous solution is still unknown. The dibasic vanadyl ion (VO^{2+}), formed when the trivalent cationic compounds are hydrolyzed, is more stable than the other cations. Vanadium pentoxides dissolve in water to give a mixture of oxy acids:

$$3\ V_2O_5 + H_2O \rightarrow H_3V_6O_{16} \qquad \text{hexavanadate}$$
$$2\ V_2O_5 + H_2O \rightarrow H_2V_4O_{11} \qquad \text{tetravanadate}$$
$$V_2O_5 + H_2O \rightarrow 2\ HVO_3 \qquad \text{metavanadate}$$
$$V_2O_5 + 2\ H_2O \rightarrow H_4V_2O_7 \qquad \text{pyrovanadate}$$
$$V_2O_3 + 3\ H_2O \rightarrow 2\ H_3VO_4 \qquad \text{orthovanadate}$$

The orthovanadates are isomorphic with phosphates. Soluble vanadites (VO_2^{2-}) are also stable, as are salts of the oxy acids of V. Vanadium also forms labile coordination chelates with ligands such as the phenoxy, hydroxy, imidazole, nitro, and cyanogen groups. The preferred ligands are the oxygen-containing groups; the tartarate ion with its two adjacent hydroxyl groups forms a more stable complex than does the citrate ion. The chemistry of the naturally occurring organic derivative of V is unknown, but low-molecular-weight lipoid materials contain V, suggesting that organic V derivatives must be low-molecular-weight forms. Vanadium readily chelates with the Ca salt of EDTA, and also with disodium catechol disulfonate (Braun and Lusky, 1959). Pentavalent V is a powerful oxidizer, and reducing compounds such as ascorbic acid can reduce the pentavalent V to lower valence forms.

Metabolism

Sufficient evidence has accumulated to justify the inclusion of V as an essential trace metal for some animals and humans, although it has not yet been shown to fulfill all the criteria of an essential nutrient. Most of this evidence was collected in experiments done in restricted environments (Nielsen and Ollerich, 1972; Schwarz and Milne, 1971; Hopkins and Mohr, 1971; Strasia, 1971). Vanadium is toxic to animals when fed in excess, and toxic effects are noted at intake levels only somewhat higher than therapeutic levels. The average daily intake of V by adult humans is about 2 mg, with 15–30 μg V excreted daily in the urine. The blood contains about 1 μg V/100 ml (Schroeder *et al.*, 1963*d*).

The biological effects of vanadium have been observed in many systems, both *in vivo* and *in vitro,* including effects on the metabolism of sulfur-containing amino acids, coenzyme A, thiotic acid, cholesterol, choline, lipid generally, and monamine oxidases, and effects on hemopoiesis, red cell maturation, and inhibition of dental caries. The biochemistry of V has been extensively studied and reviewed (Schroeder *et al.,* 1963*d*; Faulkner-Hudson, 1964; Underwood, 1971).

The gastrointestinal absorption of V compounds by animals depends upon their solubility and chemical nature. Water-soluble anions are absorbed fairly well, about 10% of the ingested dose, but cationic V is poorly absorbed; radioactive tracer studies confirm the poor absorption of dietary V (Comar and Chevallier, 1967). The absorption of parenterally injected V compounds into the blood from injection sites is rapid and complete. After inhalation or intratracheal deposition, V absorption from the lungs into the blood is more rapid than either intestinal absorption or absorption from parenteral sites. After rabbits were exposed to V_2O_5 dust, V was found almost immediately in the urine (Molfino, 1938; Massman, 1956), confirming the quick absorption of V compounds by mammals.

Excretion of V is mostly in the urine, with a smaller amount in feces. However, unabsorbed V is excreted mostly in the feces irrespective of the chemical nature of the V compound; ammonium orthovanadate is excreted mostly in the feces in rats (Ballotta, 1931) and in rabbits (Scott *et al.,* 1951). Some absorbed V could also be excreted in the feces from the enterohepatic circulation, but the absorbed V is excreted mainly in the urine, with a ratio of urinary to fecal V excretion of 5:1. Renal excretion is rapid, and the rate increases with each superimposed injection. Kent and McCance (1941*b*) reported that 81% of V doses given intravenously to humans was excreted in the urine and 9% in the feces. Dietary V as vanadyl sulfate is retained in rat tissues: heart > spleen > thyroid > lung > kidney (Schroeder and Balassa, 1967*b*); About 10% of the injected dose is retained in the tissues in experimental animals. Bone retains about 50% of it, presumably because vanadates are isomorphic with phosphates. Small V deposits are found in both the bones and kidneys of rats which inhaled V_2O_5 dust, but the largest deposits remain in the lungs (Wilson *et al.,* 1953). Pentavalent V resembles hexavalent uranium in the rapidity of its urinary excretion. Schroeder *et al.* (1963*d*) suggest the existence of a V homeostasis in both man and animals; tolerance of dietary V in rats suggests that this homeostasis may function at the absorption level.

Toxicity

Vanadium toxicity has been known since 1876, when Priestly reported that neurotoxicity caused somnolence, convulsions, respiratory failure, and gastrointestinal irritation with diarrhea in experimental animals (Faulkner-Hudson, 1964).

The toxicology of V has been studied extensively and reviewed (Faulkner-Hudson, 1964; Athanassiadis, 1969*b*: Browning, 1969; and Ammerman *et al.,* 1973); acute V poisoning in animals causes nervous disturbance, paralysis of hind legs, breathing difficulties, convulsions, hemorrhagic enteritis, and finally death; there are neurotoxic effects on the vasomotor center and intercardiac nervous mechanism, leading to a general weakening of heart function. Postmortem findings include acute desquamative enteritis of the digestive tract, fatty degeneration of the liver, parenchymatous degeneration in renal convoluted tubules, reduced spleen size with congestion, hemorrhage, and lipid decrease in the adrenal cortex, chromaffinolysis in the adrenal medulla, congestion and focal hemorrhage in the lungs, and congestion in the brain and spinal cord. Sublethal dosage symptoms are similar but less marked. Subtoxic levels of dietary V decrease food intake and cause emaciation. Inhalation of toxic doses causes nasal bleeding and acute bronchitis. Symptoms of V toxicity from inhalation in man include salivation, lacrimation, diarrhea, lowered body temperature, and respiratory and cardiac failures. The toxicology data of V compounds is summarized in Table 5-4; anionic V is more toxic than cationic V, and pentavalent compounds are the most toxic, followed by tetra- and trivalent compounds. There is little margin of safety between the therapeutic and toxic doses of pentavalent V. Dietary V seems to have no cumulative action in rats; if the daily tolerance dose is not exceeded, rats survive several weeks in spite of ingesting, over a period of time, a total dose of V in excess of a single oral lethal dose (Daniel and Lillie, 1938).

Among laboratory animals, rabbits and guinea pigs are the most susceptible to V toxicity, and horses are equally susceptible to intravenously administered tetravanadate (Faulkner-Hudson, 1964). As little as 25 ppm V as dietary metavanadate was toxic to rats in ten days (Franke and Moxon, 1937); rats are more susceptible than mice (Pham-Huu-chanh, 1965). Young ruminants tolerate up to 10 mg V/kg as ammonium metavanadate in their diet, but toxic symptoms are observed after three days with 20 mg V/kg in the diet (Platonow and Abbey, 1968). Vanadium contamination found in commercial samples of tricalcium phosphate depressed chicken growth (Berg, 1963; Berg *et al.,* 1963). Vanadium given intravenously as metavanadate to dogs at subtoxic levels produces intense vasoconstriction in the spleen, kidneys, and intestine and elevates the blood pressure (Jackson, 1912).

Proescher *et al.* (1917) suggested that an intravenous dose of 30 mg V_2O_4 (16.8 mg V) would kill an adult human. Exposure to V, and the resultant poisoning among humans, is highest in industrial operations (Dimond *et al.,* 1963; Somerville and Davis, 1972). Inhalation is the major route of entry, and skin contacts are also harmful. In addition to workers in mining and industrial processing of V and its compounds, those engaged in the cleaning of oil-fired boilers are exposed to V compounds. Crude oils contain up to 0.1% vanadium by weight, probably as an oil-soluble por-

TABLE 5-4. Vanadium Toxicity

Compound	Animal	Route	Toxicity	Compound mg	Metal mg	Metal mM	pT
Vanadium pentoxide	Rat	inhal	LD_{100}	70[a]	40[a]		
V_2O_5	Rabbit	iv	LD_{100}	10	5.8	0.11	3.94
	Rabbit	sc	LD_{100}	20	11.2	0.22	3.66
Vanadium pentoxide (colloidal)	Rabbit	iv	LD_{100}	2	1.12	0.022	4.66
	Guinea pig	sc	LD_{100}	24	13.5	0.27	3.58
	Mouse	sc	LD_{100}	102	57.4	1.13	2.95
Vanadium dichloride VCl_2	Rat	oral	LD_{50}	540	22.6	4.40	2.35
Vanadium trichloride	Rat	oral	LD_{50}	350	114	2.24	2.65
VCl_3	Rabbit	sc	LD_{100}	20	6.5	0.13	3.89
Vanadium tetrachloride VCl_4	Rat	oral	LD_{50}	160	42.5	0.83	3.08
Vanadium oxytrichloride $VOCl_3$	Rat	oral	LD_{50}	140	41	0.80	3.09
Vanadium tribromide VBr_3	Rabbit	sc	LD_{100}	20	3.5	0.070	4.16
Vanadyl sulfate	Mouse	sc	LD_{100}	300	77	1.51	2.82
$VOSO_4 \cdot 2H_2O$	Rat	sc	LD_{100}	380	97.4	1.91	2.72
	Guinea pig	sc	LD_{100}	87.4	22.4	0.44	3.36
	Rabbit	iv	LD_{100}	41.5	10.6	0.21	3.68
Sodium vanadite	Mouse	ip	LD_{100}	165	70	1.37	2.86
Na_3VO_3	Rat	ip	MLD	11	4.7	0.092	4.04
	Rat	ip	LD_{100}	19.6	8.4	0.16	3.78
	Guinea pig	ip	LD_{100}	45.8	19.6	0.38	3.41
Sodium metavanadate	Rabbit	oral	LD_{100}	200	70	1.37	2.86
$NaVO_3$	Rabbit	iv	LD_{100}	17	6.0	0.12	3.93
	Dog	sc	LD_{100}	17	6.0	0.12	3.93
	Dog	iv	LD_{100}	11	3.8	0.074	4.13
Ammonium metavanadate	Mouse	sc	LD_{100}	41	17.8	0.35	3.46
NH_4VO_3	Rat	sc	LD_{100}	37	16	0.31	3.50
	Rat	oral	LD_{50}	160	69.6	1.37	2.86
	Guinea pig	sc	LD_{100}	3.0	1.31	0.027	4.59
	Rabbit	iv	LD_{100}	3.0	1.31	0.027	4.59
	Rabbit	sc	MLD	15	6.53	0.13	3.89
Sodium orthovanadate	Mouse	sc	LD_{50}	300	42	0.82	3.08
$Na_3VO_4 \cdot 10H_2O$	Rat	sc	LD_{50}	220	31	0.61	3.22
	Guinea pig	sc	LD_{50}	8	1.12	0.022	4.66
	Rabbit	oral	LD_{100}	100	14.0	0.27	3.56
	Rabbit	sc	LD_{100}	15	2.1	0.042	4.38
	Rabbit	iv	LD_{50}	12	1.68	0.033	4.48

(Cont'd)

TABLE 5-4. (Cont'd)

Compound	Animal	Route	Toxicity	Dosage/kg body weight			
				Compound	Metal		
				mg	mg	mM	pT
Sodium pyrovanadate	Mouse	sc	LD_{100}	126	42	0.82	3.09
$Na_4V_2O_7$	Rat	sc	LD_{100}	75	25	0.49	3.31
	Guinea pig	sc	LD_{100}	3.36	1.12	0.022	4.66
	Rabbit	iv	LD_{100}	6.75	2.25	0.045	4.35
Sodium tetravanadate	Mouse	sc	LD_{100}	43.5	21.0	0.41	3.38
$Na_2V_4O_{11}$	Rat	sc	LD_{100}	40.5	19.6	0.38	3.41
	Guinea pig	sc	LD_{100}	22	10.6	0.21	3.68
	Rabbit	iv	LD_{100}	9.3	4.5	0.090	4.05
Sodium hexavanadate	Mouse	sc	LD_{100}	145	70	1.37	2.86
$Na_3V_6O_{16}$	Rat	sc	LD_{100}	57.8	28	0.55	3.26
	Guinea pig	sc	LD_{100}	52.0	25.2	0.49	3.31
	Rabbit	iv	LD_{100}	40.5	19.6	0.38	3.41

[a]Exposure of 4 hr; number of mg/m^3.

phyrin complex, and the ash of these oils contains about 65% vanadium as V_2O_5, lower oxides, or sulfides; sulfur compounds are contaminants in the oil which also contribute their share of toxicity. The prominent early symptoms of V toxicity through inhalation are severe conjuctivitis with a purulent eye discharge, rhinitis, soreness of the pharynx, bronchitis, and a green-coated tongue. Chronic exposure leads to pneumonia and other pathologic symptoms associated with V toxicity, and dermatitis of a dry eczematous type on the hands and face.

Vanadium toxicity is intensified by high dietary zinc (Molfino, 1938), and is alleviated by vitamin C. Chromates and Mn^{3+} ions decrease the oral toxicity of vanadium by retarding its gastrointestinal absorption. Wright (1968) studied the V–Cr interrelationship in chicks and suggested that these elements compete for membrane transport sites. High-protein diets decreased the oral toxicity of V in rats; metavanadate at a dietary level of 100 ppm V is either toxic or nontoxic, depending on the dietary protein content. Vanadium at toxic levels depletes the tissues of their ascorbic acid. Mitchel and Floyd (1954) countered V toxicity in rats and dogs at LD_{70} and LD_{95} levels by subcutaneous injection of 1 g ascorbic acid/kg; a low dose of 125 mg ascorbic acid/kg protects mice against V toxicity at the LD_{75} level. Ascorbic acid either forms a chelate, which is unlikely, or reduces the highly toxic pentavalent V to the less toxic tetra- and trivalent forms. The pH of injected V salt solutions also influences the toxicity;

lower pH reduces the toxicity, while administration of sodium bicarbonate prior to V salt injection increases the mortality rate in mice (Mitchel, 1953).

The toxicity of excess V is attributed to its ability to lower serum cholesterol (Mountain et al., 1953; Curran *et al.*, 1959; Sommerville and Davies, 1962), to inhibit the synthesis of phospholipids and other lipids (Dimond *et al.*, 1963), to interfere with enzyme systems such as ATPase (Rifkin, 1965), and tyrosinase (Pham-Huu-chanh, 1964), and to decrease the activities of coenzymes A, NAD (Mascitelli-Coriandoli and Citterio, 1959*a,b*), and Q (Aiyar and Sreenivasan, 1961). Vanadium retards the synthesis of coenzyme A by blocking the decarboxylation of pantothenyl cysteine, which requires pyridoxal phosphate. Inhibition of glucose metabolism (Meekes *et al.*, 1971) and xanthine dehydratase activity (Soremark, 1967) in rats by ammonium metavanadate may be related to vanadium involvement with coenzyme A. Vanadium inhibits sulfhydryl activity, decreases corticosteroid excretion, and disturbs acetyl choline metabolism. The inhibition of cholinesterase by V leads to choline deficiency, which causes growth retardation, fatty degeneration of the liver and myocardium, anemia, kidney atrophy, and muscular dystrophy. The interference of V in the formation and utilization of mevalonic acid inhibits the formation of cholesterol, the precursor of the adrenocortical hormones. Monoamine oxidase activity is enhanced or retarded by V; at 1.0 mM, V^{3+} and V^{4+} but not V^{5+} activate the enzymes, and V has been suggested as a cofactor for this enzyme, while, at higher levels, V in all valence forms inhibits this enzyme. Inhibition of monoamine oxidase, which catalyzes the oxidation of serotonin to 5-hydroxyindolacetic acid, may result in serotonin accumulation in the central nervous system, which could account for some of the clinical neurologic symptoms noted in V toxicity. The etiological mechanism is yet unknown. Vanadium compounds are also reported to decrease immune resistance and allergic reactivity in mammals (Roshchin, 1967); the leukocyte phagocytic activity in guinea pigs is considerably decreased by an oral dose of metavanadate (Kulieva, 1971). Vanadium is indirectly involved in a variety of metabolic disorders, but it is difficult to link the above effects to specific defects.

Stocks (1960) in a statistical survey suggests the involvement of V together with As and Zn in lung cancer. Vanadium ambient-air concentrations showed statistical correlations of high V levels with increased lung cancer in human males. Roshchin (1967) noted that rabbits and rats exposed to VCl_3 had lowered RNA and DNA contents in cells of liver, kidney, myocardium, stomach, and lung. Other scattered observations suggest the possible involvement of V in lung cancer, although the connection is not unequivocally established. Hathcock *et al.* (1964) suggested that EDTA protects against V toxicity.

Niobium (Nb)

Niobium, formerly referred to as columbium, occurs with tantalum as either tantalites or niobites [Fe(NbTaO$_3$)$_2$]; if Nb is present in large amounts the ore is called niobite, and otherwise it is called tantalite. Niobium is present in the earth's crust at 24 ppm, which is more than many common elements; it ranks 20th in abundance among all elements. In seawater Nb occurs at about 0.5 ppb, and the human adult body is estimated to contain about 112 mg Nb (Schroeder and Balassa, 1965). Niobium has little biologic activity and is not an essential nutrient for mammals.

Niobium is used industrially in stainless steels and in high-temperature steel alloys. It is used in jet engines, guided missiles and atomic reactors, and also used in the manufacture of electrolytic rectifiers and condensers. Niobium salts are not used in heavy industry or in therapy; however, niobium will be important in the future in superconduction applications.

Niobium salts are considered to be of low toxicity and are not industrial health hazards; cases of niobium poisoning are rare.

Chemistry

Niobium and tantalum are strikingly similar in their chemical properties; niobium exhibits variable valence from +2 to +5; the common and most stable is +5. Like V, Nb is a strong reducing agent. Niobium forms stable cationic and anionic salts; Nb salts are less soluble but more basic than V salts. Cationic Nb salts do not readily hydrolyze to niobium oxides or oxy acids. Niobium forms both square planar bonds of coordination number 4 and octahedral bonds of coordination number 8 in forming stable complexes. Cationic Nb chelates with oxalates, citrates, and ascorbates to form water-soluble compounds.

Metabolism

Niobium is not essential to man or animals, despite its widespread occurrence in food products of both plant and animal origin, and it has not been found to be stimulatory. The average daily dietary intake of Nb in man is about 600 μg, of which about 360 μg is excreted in the urine and about 3 μg in sweat. Human hair contains about 2.2 μg/g. Reports of *in vivo* Nb involvement in biochemical or enzyme systems are rare.

Studies on the metabolism of Nb compounds are few and are restricted to ^{95}Zr–^{95}Nb since they occur together in fissionable material. However,

Nb is pentavalent and more reactive than Zr, and Nb salts accumulate in the blood, whereas Zr accumulates in the bones (Ramasastry *et al.,* 1964). Niobium metabolism was reviewed by Schroeder and Balassa (1965).

In rats and other laboratory animals the gastrointestinal absorption is higher for anionic niobates than for cationic niobium salts such as halides; the absorption of dietary niobate is about 50%. The absorption of cationic Nb is increased if Nb^{5+} is mixed with citrates or complexed with ascorbic acid. The absorption of small doses of Nb salts from parenteral injection sites is similar; when complexed with citrate, Nb^{5+} salts are absorbed rapidly and maximally. Intravenously injected Nb, both anions and cations, forms stable complexes with serum proteins and remains in circulation for some time before distribution to the bones, liver, spleen, and kidneys. Colloidal Nb salts accumulate in the spleen and bone marrow; retention of ^{95}Nb in tissues is prolonged (Durbin, 1960). Most of the absorbed niobate (60%) is excreted in the urine as niobate; the remaining absorbed niobates and unabsorbed niobium salts are excreted in the feces. Schroeder and Nason (1971) have reported that Nb did not accumulate with age in the rat. There are no specific excretory mechanisms for Nb, but the constant levels of Nb in urine, blood, and tissues of animals, in spite of high oral intake, suggest an efficient homeostasis for Nb in animals (Schroeder and Balassa, 1965).

Toxicity

Niobium is not toxic at physiologic levels, as shown by its presence in the human body at concentrations equivalent to that of Cu. The dietary intake of Nb is higher than Cr, Ti, Pb, or Cd. There are no specific clinical symptoms of Nb intoxication.

The available toxicity data on Nb are summarized in Table 5-5. Of laboratory animals, mice are the most susceptible to Nb toxicity, and cationic salts are more toxic than niobates, perhaps because of their instability, since they hydrolyze to form colloidal Nb compounds. At sublethal levels animals suffer renal and hepatic injury. Rats given 30 mg Nb/kg as niobate consume more water, and the functional ability of their kidneys is lowered. This nephrotoxicity is not observed when Nb is administered with ascorbic acid (Downs *et al.,* 1965). Rats suffer hepatic damage when fed 0.1–1 ppm Nb as the halide in their diet, and perinuclear vacuolization of the parenchymal cells and coarse granulation of the cytoplasm were noted (Haley *et al.* 1962*b*), and at 5 ppm Nb, mice showed hepatic fatty degeneration, and Nb accumulated in the spleen (Schroeder *et al.,*

TABLE 5-5. Niobium Toxicity

| | | | | Dosage/kg body weight | | | |
| | | | | Compound | Metal | | |
Compound	Animal	Route	Toxicity	mg	mg	mM	pT
Niobium chloride	Mouse	ip	LD_{50}	61	21	0.23	3.65
$NbCl_5$	Mouse	oral	LD_{50}	940	325	3.50	2.46
	Rat	ip	LD_{50}	40	14	0.15	3.82
	Cat	iv	LD_{50}	16.0	5.0	0.054	4.27
Potassium niobate	Mouse	ip	MLD	28	11	0.12	3.93
$4K_2O\cdot3Nb_2O_5\cdot16H_2O$	Mouse	ip	LD_{50}	34	13	0.14	3.85
	Rat	ip	MLD	147	56	0.60	3.22
	Rat	ip	LD_{50}	225	86	0.93	3.03
	Rat	oral	MLD	1550	590	6.35	2.20
	Rat	oral	LD_{50}	1900	725	7.80	2.11
	Rat	oral	LD_{100}	3000	1140	12.3	1.91
	Guinea pig	ip	MLD	99.5	38	0.41	3.39
	Cat	ip	MLD	14.4	55	0.59	3.23
	Dog	ip	MLD	227	85	0.91	3.04
Potassium oxopentafluoniobate K_2NbOF_5	Mouse	oral	LD_{50}	130	40.2	0.43	3.36

1970*c*). Cats die of respiratory paralysis after an intravenous injection of 5 mg Nb/kg as $NbCl_5$. Niobate inhibits succinic dehydrogenase *in vitro* (Cochran *et al.*, 1950) and oxidizes 5-hydroxytryptophan.

Tantalum (Ta)

Tantalum occurs with other metals in the form of tantalite, e.g., columbium tantalite $[(Fe,Mn)(NbTa)_2O_5]$. Tantalum is present at 1 ppm in the earth's crust. Reports on the presence of tantalum in normal mammalian tissues are conflicting; tantalum is not essential for mammals.

Tantalum is used industrially to make corrosion-resistant Ta–W–Co and Ta–W–Mo alloys for use in acid-proof equipment in chemical industries, and is also used in filament, grid wires, rectifiers, and electrolytic capacitors in electronic industries, and in surgical and prosthetic appliances. Tantalum oxide is used as a nonradioactive tracer in glass research. This metal is not considered an industrial health hazard. There are few reports of the presence of Ta in food material, and accidental poisoning, if any, is rare.

Chemistry

Tantalum exhibits variable valence ranging from $+2$ to $+5$. The most stable is the pentavalent form. Cationic Ta salts are hydrolyzed to stable Ta_2O_5 or mixture of soluble tantalic acids, which include the meta-(TaO_3^-), ortho-$(Ta_6O_9^{8-})$, and per-tantalic (TaO_8^{3-}) acids. Tantalum also forms complex halides such as Ta_6Br_{14} and fluorocomplexes $(TaF_8)^{3-}$ with coordination numbers of 6 and 8, but these complexes are not likely to exist in biologic tissues and fluids.

Metabolism

Tantalum is not essential to animals. Its salts do not show stimulation and are not considered toxic. There are not reports of Ta involvement in any specific biochemical or enzyme systems, and studies on Ta metabolism are few (Schepers, 1955c; Durbin *et al.*, 1957; Fleshman *et al.*, 1971) and use mostly minute quantities of radioactive Ta.

In rats the gastrointestinal absorption of soluble Ta salts is relatively low; microgram quantities or less are absorbed maximally in guinea pigs. The absorption from sites of parenteral injection is about 15% of the dose administered. The insoluble salts are very poorly absorbed from ingested and parenterally administered doses, but when mixed with soluble citrates, tantalic acids are readily absorbed from sites of parenteral injections. Intravenously administered cationic salts hydrolyze into colloidal and complex tantalic acids at physiologic pH. Intratracheal introductions of particulate Ta_2O_5 (with diameter of less than 3 μm) are readily absorbed from the lungs of guinea pigs. The poor absorption of Ta oxide and other insoluble salts from the alimentary tract results in complete fecal excretion. However, Ta excretion depends upon dose; small quantities such as radioactive tracer amounts, which are maximally absorbed, are excreted in the urine, and intramuscularly injected $^{182}Ta_2O_5$ mixed with soluble citrate is excreted mainly in the urine (Durbin, 1960). Soluble potassium tantalate (8 mg) mixed into NaCl was introduced by stomach tube into rats; ^{182}Ta was eliminated in the feces in three phases: rapid excretion of unabsorbed tantalate, slow elimination of loosely bound tantalum in the tissues, and a 3–4 day elimination of minute quantities of tightly bound tantalum. Retention of Ta in the tissues is more prolonged than the other metals of Group V, in the following order: bones > gastrointestinal tract > pelt > muscle > liver > kidneys > testes > spleen. Tantalum held in the bones accounts for 40% of the body content (Fleshman *et al.*, 1971). There are no reports of the accumulation of Ta in tissues or of the existence of any homeostatic excretory mechanism for Ta.

TABLE 5-6. *Tantalum Toxicity*

| Compound | Animal | Route | Toxicity | Dosage/kg body weight | | | |
| | | | | Compound | Metal | | |
				mg	mg	mM	pT
Tantalum oxide Ta_2O_5	Rat	oral	LD_{50}	>8000	>6560	>36.25	>1.44
Tantalum fluoride TaF_5	Mouse	iv	LD_{50}	110	72.1	0.40	3.40
Tantalum chloride $TaCl_5$	Rat	oral	LD_{50}	1900	958	5.29	2.28
	Rat	ip	LD_{50}	75	38	0.21	3.68
Poatssium tantalum fluoride	Mouse	oral	LD_{50}	110	50.7	0.28	3.55
$KTaF_7$	Mouse	ip	LD_{50}	97	45	0.25	3.60
	Rat	oral	LD_{50}	2500	1150	6.36	2.20
	Rat	ip	LD_{50}	375	173	0.96	3.02

Toxicity

Tantalum oxide is practically nontoxic to rats (Cochran *et al.*, 1950); the available Ta toxicity data are summarized in Table 5-6. Metallic Ta is relatively inert in contact with tissue, and it is used in surgery as bone supports (Koontz and Kimberly, 1953). Powdered Ta stimulates wound epithelialization. Transient bronchitis and interstitial pneumonitis with hyperemia and residual focal hypertrophic emphysema are observed in guinea pig lungs when Ta_2O_5 is introduced (Schepers, 1955). The effect on the lungs varies according to the Ta compound inhaled. Potassium fluoro-tantalate causes a thickening of alveolar septa and blood vessels, and the proliferation of histiocytes (Stokinger, 1963). The toxicity of $TaCl_5$ could be due to the combined action of its hydrolysis products, free HCl and tantalic acid.

6

TOXICITY OF GROUP VI METALS AND METALLOIDS

The elements of Group VI of the periodic chart are characterized by their hexavalence and by the stability of their anions. The nonmetals oxygen (O) and sulfur (S), the metalloid selenium (Se), and the metals tellurium (Te) and radioactive polonium (Po), and atom 106 form subgroup VIA; the metals chromium (Cr), molybdenum (Mo), and tungsten (W) form subgroup VIB. Oxygen and sulfur are major essential elements. The metals of Group VI are active physiologically. Selenium, chromium, and molybdenum are essential trace metals for man and other mammals, but at high doses all these metals are toxic. Tellurium, tungsten, polonium, and element 106 are not essential. Polonium and the metal with atomic number 106 are highly radioactive and are not considered further in this book. Group VI elements exhibit variable valence, ranging from -2 to $+6$; many form stable anions in the $4+$ and $6+$ oxidation states, e.g., selenite and selenate.

Selenium and tellurium of subgroup VIA are stable only in anionic forms; chromium of subgroup VIB exists in stable cationic and anionic forms, while cationic molybdenum and tungsten salts are less stable and are partially hydrolyzed to their respective oxides or oxy acids, e.g.,

$$2\,WCl_5 + 5\,H_2O \rightleftharpoons W_2O_5 + 10\,HCl$$

The anionic salts are readily absorbed from the gastrointestinal tract, and their excretion is both fecal and urinary. All Group VI metals are retained to some extent in the soft tissues. Salts of Group VIA metals, Se and Te, and more toxic than are salts of Cr, Mo, and W, on the basis of LD_{50} values. The metals of Group VI are industrial toxicants, and Cr and Se are carcinogens. The order of toxicity of Group VI metals is: $Se > Te > Mo > Cr^{6+} > W > Cr^{3+}$. Homeostatic excretory mechanisms, which are not fully

233

understood, maintain the tissue level of these metals, but these mechanisms are limited.

SUBGROUP VIA METALS

Selenium is essential for mammals, while tellurium is not, But selenium is the most toxic of Group VI metals on both molar and therapeutic index bases. Subgroup VIA metals act like S, forming stable anions in the 4+ and 6+ oxidation states; the cationic salts, such as halides, are unstable and are readily hydrolyzed to stable oxy acids, e.g.,

$$SeCl_4 + 3H_2O \rightarrow H_2SeO_3 + 4HCl$$

Selenium and tellurium form stable gaseous hydrides which are very toxic. Hexavalent Se and Te are reduced to tetravalent anions in animal tissues. Selenium and tellurium anions readily complex with biologically

Group VI Elements

Atomic number Atomic weight	Name (Symbol)	(Core) Active electrons Usual valences			
Subgroup A			Subgroup B		
8 16.00	Oxygen (O)	(He) 2, 4 −2			
16 32.06	Sulfur (S)	(Ne) 2, 4 +4, +6, −2			
34 78.96	Selenium (Se)	(Ar) 2, 4 +4, +6, −2	24 52.00	Chromium (Cr)	(Ar) 5,1 +2. +3. +6
52 127.60	Tellurium (Te)	(Kr) 2, 4	42 95.94	Molybdenum (Mo)	(Kr) 5, 0, 1 +6
84 (210)	Polonium (Po)	(Xe) 2, 4 +2, +4	74 183.85	Tungsten (W)	(Xe) 4, 0, 2 +6

active thiol compounds to render them partly inactive, and both elements prefer oxygen- or sulfur-containing ligands; selenide complexes are more stable than telluride complexes. Selenium and tellurium form few stable coordination complexes in biological systems, although compounds such as selenite pentamines and selenopersulfides are known, and selenate functions as a bridging group in polynucleate compounds. Selenium can also bind to the myofibrillar proteins of muscle. Selenium and tellurium form stable and soluble anions which are rapidly absorbed from the mammalian digestive tract, and which are retained in the soft tissues and excreted in urine and feces. The inherent toxicity of metals of this subgroup decreases directly with increased atomic weight and electropositivity.

Selenium (Se)

Selenium occurs as metallic selenides in very small quantities. It is deposited unevenly but widely in the earth's crust, at an average of 0.09 ppm. Selenium is present in soils as elemental selenium, basic ferric selenite [$Fe_2(OH)SeO_3$], calcium selenate, and as organic selenium compounds from decayed plant and animal tissues. It is present in seawater at about 4 ppb. The normal human adult body contains about 13 mg Se, about 40% of which is in the muscle. Selenium is considered to be an essential nutrient.

Selenium is used in electronic industries for the manufacture of rectifiers, photocells, and xerography equipment, in the glass industry as a decolorizer and pigment, in the ceramics industry to introduce colored glazes, in the rubber industry as a sulfur supplement and in other industries as a coating for stainless steel and copper to improve the machinability and homogeneity of fine steels, and as a plasticizer. It was formerly used in insecticide sprays.

Highly seleniferous soils and selenium accumulator plants are the source of potential Se intoxication for animals in the northwestern United States. Selenium-accumulator plants (*e.g.,* Astragalus) form compounds such as selenocystathionine. Selenium poisoning occurs in animals which feed on seleniferous legumes; this leads to different forms of selenosis called blind staggers or alkali disease. In humans, Se intoxication can occur from consuming cereals, grains, and vegetables grown on soils containing up to 5 ppm Se and meat of animals reared in seleniferous areas. Wheat in which Se is present in methionine fractions up to 12 ppm is toxic, but extensive processing removes Se from the food. The reported Se content (ppm fresh weight) of some American foods are: 0.99 in seafoods, 0.92 in meats, 0.37 in dairy products, 0.15 in cereals and grains, according to

Schroeder *et al.* (1970*b*), or, according to Morris and Levander (1970), 0.01 in vegetables, 0.006 in fruits, 0.069 in dairy products, 0.532 in seafoods, 0.387 in grains and cereals, 0.224 in meats, and 0.068 in baby foods. The levels of Se (ppm) in some German foods are: 0.16 in vegetables, 1.5 in fish, and 1.01 in egg powder (Oelschläger and Menke, 1969). Animal feeds vary greatly in Se content, depending upon where the ingredients were grown.

Chemistry

Selenium resembles sulfur, but is a metalloid. With a $4s^2$ and $4p^4$ electronic configuration, Se exhibits a variable valence -2 to $+6$; the more common forms are Se^{2-}, Se^{4+}, Se^{6+}. Selenium di- and trioxides are stable and water-soluble, giving rise to the stable anions, selenite and selenate; Se^{2-} forms volatile gases with two electron-pair bonds, such as $(CH_3)_2Se$ and H_2Se. Among the organic Se compounds of biological importance are selenomethionine and selenocysteine, the Se analogues of S amino acids. Selenate can function as a bridging group in polynucleate ions. Selenite pentamine complexed with metals such as cobalt is stable. Selenomercaptides and ethers form complexes with Hg^{2+}; diarylselenoxides coordinate to Hg^{2+} through the selenium atom. Selenopersulfide linkages have been reported in proteins (Ganther, 1971).

Metabolism

Selenium was first recognized for its toxicity; its essential nature in animals was discovered and established later. That Se is essential to human nutrition has yet to be confirmed, but it appears that very small amounts are required by humans, and that an average daily intake of about 70 μg may meet the requirements with an adequate α tocopherol intake. The daily intake of Se by man is estimated to be 60–150 μg from food, and about 1 μg from water and air. Selenium deficiency may occur when the diet contains <0.04 ppm, or when Se-depleting factors are present. Selenium is essential for the growth and reproduction of animals (Underwood, 1971; Schwarz, 1971), and deficiency of it results in stunted growth, liver necrosis, pancreas atrophy, kidney damage, infertility, and white muscle disease. The amount of Se required to prevent deficiency symptoms depends on the intake of antioxidants and S amino acids. With the most biologically active form, Factor 3, 0.02 ppm Se in the diet is adequate for rats.

Selenium is a component of glutathione peroxidase which prevents the accumulation of toxic peroxides within the cell. Antioxidants, such as tocopherols, are synergistic in preventing peroxide accumulation. Apart from lowering vitamin-E requirements, Se has other biochemical and physiologic functions, such as a sparing effect on biologically active sulfur compounds. Selenium metabolism has been extensively studied and reviewed (Scot, 1962, 1967; Muth *et al.,* 1967; Schroeder *et al.,* 1970*b*; Underwood, 1971; O'Dell and Campbell, 1971; Frost, 1972; Frost and Lish, 1975).

On a molar basis Se salts are the most toxic among the essential minerals; the therapeutic to toxic dose ratio is about 1:100. Depending on the intake of As, tocopherols, and protein, 3–10 ppm Se in the diet will cause chronic intoxication in animals. Selenium is a paradox because the essentiality of Se ends at about 0.1 ppm, while toxicity may begin at 0.4 ppm diet. There is only a two- to one-hundredfold safety margin between the nutrient and toxic levels; this margin is very critical when the nutrient level is 0.1 ppm. Estimated dietary threshold levels at which Se can cause physiologic and pathologic symptoms in animals are categorized in Table 6-1.

The gastrointestinal absorption of Se in the form of soluble selenites, selenates, and seleniferous organic compounds by mammals (both ruminant and monogastric) is rapid and efficient, but metal selenides and elemental Se are poorly absorbed. The absorption of Se in monogastric animals occurs in the lower small intestine, with a slight excretion in the duodenal portion. Absorption and excretion are negligible in the stomach, cecum, or colon. The chemical form of the element, the presence of other metallic salts with similar electronic configurations such as As, and the animal species influence the degree of intestinal absorption. Parenterally administered Se compounds are quickly absorbed and taken into the blood. Absorption through the lungs is rapid, but selenium compounds are not well absorbed through the skin. In the blood Se is associated both with erythrocytes and with plasma albumin and globulins. Albumin appears to be the immediate receptor and is involved in the transport of Se to more stable binding sites in the blood and tissues (Stolman and Stewart, 1960). From the blood Se is quickly distributed to all tissues, especially the liver,

TABLE 6-1. Biologic Spectrum of Dietary Selenium

Characterization	ppm
Deficiency	<0.03
Physiological	0.03–0.4
Pathological	0.4–3.0
Clinical	3–20
Lethal	>20

ne, nails, and hair. The intracellular distribution varies with the Se level; in the liver 50% of the total is in the soluble fraction, mitochondria, 11% in microsomes, and 2% in nuclei (Ganther,

Retention of Se in tissues depends on the dietary intake and existing body level of the metal. Rats fed a Se-supplemented diet retain proportionately less than rats fed a Se-deficient diet. In rats organic Se compounds such as selenomethionine (0.5–1 ppm Se in the diet) are retained longer than selenites or selenates. Retention in mammalian tissues is owing partially to selenium replacing sulfur in cystine and methionine; these may exist in structural and functional proteins or as free selenoamino acids (McConnell, 1963), or selenium may be bound to sulfur compounds (Cummins and Martin, 1961). Some Se occurs in enzymes (Thompson *et al.*, 1975).

Selenium is excreted in feces and urine, and exhaled, depending upon its mode of intake, its chemical form, the presence of metal ions such as As, Hg, Cd, and Tl, and the species involved. Under physiologic conditions about 2% of an injected dose of selenium is exhaled as gaseous dimethyl selenide; the percentage increases with increase in dose. Fecal Se is mostly the unabsorbed portion of the ingested dose. In mice most intravenously injected Se is excreted: about 65% in urine, 8% in feces, and 6% in exhaled air. Dimethyl selenide constitutes 60%, and an unidentified Se metabolite constitutes 30% of the urinary Se in rats. In ruminants fecal Se excretion is higher, owing to the reduction of dietary Se to poorly absorbed forms by rumen microorganisms; in monogastric animals urinary excretion of ingested Se is high (Paulson *et al.*, 1968). Selenium is also excreted in sweat and in milk. The elimination of organic Se is slow, while selenite is eliminated rapidly. Sulfate is a competitive antagonist and increases the urinary excretion of selenate but not selenite, while arsenite inhibits $(CH_3)_2$ Se formation but increases the fecal excretion of selenite.

Although there is no direct evidence for Se homeostasis, there are indications that within physiological limits animals are able to maintain nontoxic levels of Se by controlled excretory mechanisms.

Toxicity

Selenium compounds are the most toxic of the Group VI elements (Franke and Moxon, 1936, 1937; Allaway, 1969). Selenium toxicity, or selenosis, is caused by both organic and inorganic forms of Se; the toxicology of Se has been studied extensively and reviewed (Dudley, 1936; Buchan, 1947; Rosenfeld and Beath, 1964; Cooper, 1967; Harr and Muth, 1972; Oehme, 1972; Thompson *et al.*, 1975). In man, early signs of acute Se poisoning are nervousness, fever, vomiting, and somnolence. The blood pressure falls, and labored breathing is accompanied by tetanic and clonic spasms; death is caused by respiratory failure from prolonged convulsions caused by the toxic action on the nervous system. Chronic intoxication produces depression, marked pallor, coated tongue, gastrointestinal disturbances, and a garlic odor of the breath. Continued intoxication causes kidney, liver, and spleen damage, hemolytic anemia, and loss of nails and hair. In animals blind staggers, ataxia, elevation of body temperature, labored respiration, rapid but weak pulse, and death from respiratory failure are the acute toxic symptoms. Ingestion of 2–30 ppm dietary Se over a prolonged period of time causes chronic toxicity; the common symptoms are alkali disease, loss of vitality, lameness, atrophy and cirrhosis of the liver, and degeneration and necrosis of the myocardium. The toxicity of Se is summarized in Table 6-2. Among common laboratory animals, cats are most susceptible, followed by rats and hamsters. Sodium selenite at 6–12 ppm in their drinking water had no effect on the survival of mice, but their growth was reduced at the 9 and 12 ppm levels (Schroeder 1967). Dietary levels of 400–800 ppm Se as selenite are fatal to sheep, hogs, and calves; Caravaggi *et al.* (1970) report deaths in lambs when 1.5–3.0 mg Se/kg body weight was consumed.

The toxicity of Se varies according to the chemical form and the species involved; the toxic dose varies from 0.4 to 6 mg Se/kg in several species, but injected doses of 200 μg Se/kg as selenites are lethal to cattle, horses, and pigs (Gabbedy and Dickson, 1969; Usov, 1969; Gleason *et al.*, 1969). Marked differences in the acute toxicity of inorganic compounds can be illustrated by the acute oral LD_{50} values in mice for selenium sulfide and sodium selenite, which were 3700 and 48 mg of the compounds/kg, respectively (Henschler and Kirschner, 1969). Earlier observations indicated that ingestion of elemental Se is relatively harmless, owing to its relatively low solubility. Intraperitoneal injection of powdered Se, 1000 mg/kg, is not toxic to rats (Hall, 1951), but inhalation of Se dust, 30 mg/m^3 for 6 hr produced mild pneumonitis, and intravenously administered colloidal Se, 30 mg/kg, is fatal to rats. Hydrogen selenide, 1–4 μg/liter, produces irritation of the respiratory tract and pulmonary edema, causing death in rabbits after 8 to 12 hr of inhalation (Rosenfeld and Beath, 1964).

The manifestations of selenosis in animals have been very well described (Moxon and Rhian, 1943; Rosenfeld and Beath, 1964; Underwood, 1971; Harr and Muth, 1972). Selenium salts and selenoamino acids cause subacute selenosis characterized by increased pancreatic weight and metabolic rate, and hemolysis. Chronic inorganic selenosis causes follicular skin rash, inflammation of the perivascular lymph channels, hemolytic anemia, serum bilirubin, and damage to spleen, pancreas, and liver. The

TABLE 6-2. Selenium Toxicity

| Compound | Animal | Route | Toxicity | Dosage/kg body weight | | | |
| | | | | Compound | Metal | | |
				mg	mg	mM	pT
Selenium	Rat	iv	LD_{50}		6	0.076	4.12
Selenium dioxide SeO_2	Rat	sc	LD_{50}	4	2.8	0.035	4.45
Selenium disulfide SeS_2	Mouse	oral	MLD	38	26.8	0.34	3.47
	Rat	oral	LD_{50}	370	260	3.29	2.48
Selenium chloride $SeCl_4$	Guinea pig	sc	LD_{50}	19	6.8	0.086	4.06
Selenium oxychloride $SeOCl_2$	Rabbit	sc	LD_{100}	7	3.3	0.042	4.38
Sodium selenide Na_2Se	Mouse	ip	LD_{50}	3.4	0.15	0.0019	5.72
Selenious acid H_2SeO_3	Rat	oral	MLD	25	15.3	0.193	3.71
	Rat	ip	MLD	10	6.12	0.077	4.11
Sodium selenite Na_2SeO_3	Mouse	oral	LD_{50}	7	3.2	0.041	4.39
	Rat	oral	LD_{50}	7	3.2	0.041	4.39
	Rat	ip	LD_{50}	7.5	3.4	0.043	4.40
	Rat	iv	LD_{50}	3.5	1.6	0.020	4.69
	Guinea pig	oral	LD_{50}	5.1	2.3	0.029	4.54
	Rabbit	oral	LD_{50}	2.25	1.0	0.0126	4.90
	Rabbit	ip	LD_{100}	4.0	1.8	0.023	4.64
	Rabbit	iv	MLD	2.0	0.9	0.011	4.94
	Cat	sc	MLD	4.0	1.8	0.023	4.64
	Dog	oral	MLD	4.0	1.8	0.023	4.64
	Dog	sc	LD_{100}	4.0	1.8	0.023	4.64
	Dog	iv	LD_{100}	3.5	1.6	0.020	4.69
Sodium selenate Na_2SeO_4	Rat	ip	MLD	9	3.7	0.047	4.33
	Rat	ip	LD_{50}	13.8	5.75	0.073	4.14
	Rat	iv	MLD	3	1.3	0.016	4.78
	Rat	iv	LD_{50}	8.35	3.5	0.044	4.35
	Rabbit	oral	MLD	7.0	2.9	0.036	4.35
	Rabbit	ip	LD_{100}	9.6	4.0	0.050	4.30
	Rabbit	iv	MLD	6.0	2.5	0.031	4.50
	Rabbit	iv	LD_{100}	2.0	0.84	0.006	4.97
	Dog	oral	MLD	4.0	1.7	0.021	4.67
Potassium selenocyanate KSeCN	Rat	ip	MLD	200	109	1.38	2.86
Selenoamino acids	Rat	oral	LD_{50}		13	0.16	3.77

specific symptoms in cattle include gastroenteritis and polyencephalomalacia; in sheep, myocardial degeneration and fibrosis, pulmonary congestion, and edema; in rats, hepatitis, nephritis, myocarditis, and other less striking symptoms. Acute organic selenosis was noted in cattle grazing on plants containing 100–10,000 ppm Se. The clinically acute syndrome is called blind staggers and is characterized by anorexia, emaciation, and collapse. A subacute organic selenosis, called alkali disease, is caused by consuming plant material containing 25–50 ppm Se. The symptoms are loss of weight, irritability, degeneration and fibrosis of the heart, liver, and kidneys, and impaired reproductive capacity with fetal resorption. Organic selenosis occurring in man living in seleniferous areas is characterized by chronic dermatitis, fatigue, anorexia, gastroenteritis, hepatic degeneration, and spleen enlargement. A review of the effects of Se in industrial workers revealed acute symptoms, such as local irritations, skin burns, and pulmonary edema, but no symptoms of chronic Se toxicity, death, or carcinogenicity were observed (Glover, 1970).

Selenium has significant toxic effects on reproduction, calcification, and teratogenesis. The reproductive capacity of mature male rats is impaired by a dietary level of 10 ppm Se as selenite (Wahlstrom and Olson, 1959), and 3 ppm selenite in drinking water reduced litter numbers and reduced the reproductive capacity of third generation mice (Schroeder and Mitchner, 1972). Selenium is concentrated in the testes and epididymis; it is difficult to evaluate its action. On the other hand, Se protects against the testicular necrosis, congenital malformations, and fetal resorption caused by Cd and Hg toxicity (Gunn and Gould, 1967; Mason and Young, 1967; Parizek *et al.*, 1969). The teratogenic effect of Se is seen in the abnormal development of embryos in rats (Rosenfeld and Beath, 1954), pigs (Wahlstrom and Olson, 1959), sheep (Rosenfeld and Beath, 1947), and cattle (Dinkel *et al.*, 1963), where many young are born with deformed hooves. There are also reports of malformations of chicken and sheep in seleniferous regions (Hunter, 1964). Several miscarriages among female laboratory technicians exposed to selenite powder and the birth of a deformed infant with bilateral clubfoot lend support to the teratogenic potential of Se (Robertson, 1970). However, Holmberg and Ferm (1969) failed to find teratogenic effects in hamsters given 2 mg/kg of intravenous Se as selenite on day 8 of gestation; Se protected against Cd- and As-induced malformations. A statistical survey showed that human neonatal death declined with increased environmental selenium in seleniferous areas (Shamberger, 1971).

The increased incidence of dental caries in children in seleniferous areas (Smith *et al.*, 1936; Tank and Storvick, 1960; Hadjimarkos, 1973) and in rats fed 5–10 ppm Se (Buttner, 1963) suggest that consumption of small amounts of Se during the developmental period of teeth may increase susceptibility to caries. The available evidence indicates changes in the protein components of enamel and bone matrix (Campo *et al.*, 1967; Hadjimarkos, 1969). Dietary Se increased the ratio of food-to-water intake; this delayed clearance of food particles, which was suggested to be caries-producing (Hadjimarkos, 1967*a,b*). Fluoride does not augment Se toxicity or interfere with Se excretion, and Se does not decrease fluoride deposition in bones and teeth; it is difficult to understand the biologic interaction between fluoride and selenium.

Reports on Se-caused carcinogenesis in different species are conflicting. Belief in the carcinogenicity of Se is based on the studies of Nelson *et al.* (1943), Tscherkes *et al.* (1961, 1963), and Volgarev and Tscherkes (1967). Nelson and co-workers induced nonmetastasizing hepatic tumors in rats by feeding them seleniferous grain containing 5–10 ppm Se; the latent period for cancer induction was about 18 months. In Tscherkes' studies, hepatic carcinomas, hepatic adenomas, and precancerous neoplasms were observed in 25% of the rats fed 4.3 ppm Se as selenate; these carcinomas metastasized to the lungs. Harr *et al.* (1967) reported that 8–16 ppm Se as dietary selenite did not induce neoplasms in rats, as the animals died of Se intoxication before tumors could form. Schroeder and Mitchner (1971*c*) produced data supporting carcinogenicity. At a level of 2 ppm in the drinking water of young rats, selenate was not toxic, but selenite caused growth depression, hepatic damage, and early death in the males. Subsequently selenate was substituted for selenite in the males after day 68, and the selenate level was increased to 3 ppm after one year. Growth and survival were unaffected by selenate in males, while selenite-fed females suffered from weight loss and high mortality. Selenate-fed rats lived longer than did control rats, but at autopsy selenate-fed rats showed a significantly increased incidence of benign and malignant liver tumors. The incidence of tumors was 42% compared with 17% in controls and 13% in selenite-fed females; control rats developed no tumors before two years, and 91% of the tumors occurred in selenate-fed males after two years.

Chronic ingestion of Se at subtoxic levels for prolonged periods may result in the recondite toxicity of Se manifesting itself in carcinogenesis, and selenate may undergo some sort of biotransformation to become carcinogenic.

However, reviews on selenium in carcinogenesis by Frost (1970, 1972), and Frost and Lish (1975) emphasize that there is no direct involvement of Se in carcinogenesis. Epidemiological studies reveal an inverse relationship between human cancer mortality and environmental Se and blood Se levels (Allaway *et al.*, 1968; Shamberger and Frost, 1969; Shamberger and Willis, 1971). When the mortality rates from several types of cancer are compared in high- and low-Se areas, there is seen a much lower death rate from cancer of the digestive organs and urinary organs in the high-Se areas. Selenium blood levels are lower in patients suffering from gastrointestinal cancer or liver metastases. Skin tumor formation was either prevented or decreased in mice fed diets supplemented with sodium selenide or selenite (Shamberger, 1970). Selenite also reduced carcinogen-induced chromosome breakage (Shamberger, 1974). The protective actions of Se against the induction of liver tumors in rats by N^{-1}-methyl-p-dimethylaminoazobenzene (Clayton and Baumann, 1949), against the cocarcinogenic effect of croton oil on mouse skin painted with 9,10-dimethyl-1,2-benzanthracene (Shamberger, 1970), and against the carcinogenic effect of N-2-fluorenylacetamide in vitamin-E-supplemented rats depleted in Se (Harr *et al.*, 1972) suggest that Se possesses anticarcinogenic capacity. Selenium, acting as an antioxidant, has been found to neutralize or eliminate hydroperoxide free radicals formed during radiobiological damage which trigger cancer. Malonaldehyde, a peroxide breakdown product of unsaturated fatty acids, is a carcinogen present in foods and human feces; dietary Se might topically protect the gastrointestinal lining against carcinogenesis by malonaldehyde, which may explain the lower mortality from cancer of the digestive organs in high-Se areas.

On the basis of the above evidence, the National Academy of Sciences Committee on Animal Nutrition (1972) suggested that the data on the carcinogenicity of Se are inconclusive. The U.S. Government also accepted that Se feed additives are not a cancer hazard for humans (Anon., Federal Register, 1974).

Interactions between Se and other metabolites either reduce or potentiate Se toxicity. Tocopherols, antioxidants, dietary protein, and organic sulfur in the form of S amino acids decrease Se toxicity. Linseed oil meal in the diet protects rats from the toxic effects of dietary Se at 10 ppm as measured by growth and survival, although kidney and liver Se levels remained high (Levander *et al.*, 1970). Lysine or fish meal also partially suppress the toxicity of dietary Se in rats (Chavez and Jafee, 1967). Interactions between Se and other minerals such as sulfate, arsenite, Hg, and other metal ions decrease Se toxicity in mammals; this could be caused by increased excretion of Se, competitive action, or the prevention of Se from reaching the sensitive receptor sites. Hemolytic anemia caused by dietary selenite in rats can be reduced by simultaneous administration of arsenite (Halverson *et al.*, 1970). Biliary excretion of Se is increased by As and Te; Hg and Tl reduce the urinary excretion of Se. Elimination of methyl selenide in expired air was depressed by As, Hg, and Tl (Levander and Argrett, 1969), and subcutaneous injection of Hg salts delayed the excretion of intravenously administered Se from mice. The excretion of Se in expired air and urine was decreased to varying degrees by increasing the molar ratios of Se/Zn, Se/Cd, Se/Te, and Se/As (McConnel and Carpenter, 1971). The fecal excretion of Se was decreased by Zn and Cd, but was increased by Te and As. Mercury, Tl, Cd, and Te increased the Se levels in the liver, spleen, and kidneys, of rats; conversely, Se decreases the toxicity of these matals. Dietary cobalt decreased the retention of Se in the heart and skeletal muscle and, to a lesser extent, in the liver and kidneys in rats (Gardiner and Nicol, 1971) and in ewes (Gardiner and Nairn, 1971). Cobalt deficiency may render sheep more susceptible to Se toxicity (Gabbedy, 1970); Co is presumed to affect the absorption of Se in cats.

Selenium or tocopherols have the capacity to detoxicate heavy metals by mutual interaction (National Research Council, 1971). It is not known whether Se binds to other metals as inert selenides or selenites. Illustrations of the ability of Se to detoxicate heavy metals include the counteraction of Cd testicular injury and protection against Cd-induced hypertension by Se, and protection against methyl Hg intoxication in rats by adding Se-containing tuna to an otherwise-Se deficient diet. Reciprocal detoxication between Cu and Se or Hg (Hill, 1974), and between Se and As (Holmberg and Ferm, 1969) have been established.

Selenium toxicity is generally attributed to its interference with sulfur metabolism and function. Selenium affects enzyme systems associated with cellular respiration and replaces thiol groups by SeH groups in some of the dehydrogenase enzyme systems, with subsequent inhibition of the dehydrogenases. Selenites react with sulfhydryl compounds such as cysteine or coenzyme A, forming stable selenotrisulfides and rendering these essential cofactors unavailable (Ganther, 1968). Frost (1972) postulated that the toxicity of Se is generally the result of excessive accumulation of selenite ion, an oxidant which may interfere with glutathione metabolism. Functional and structural proteins may undergo conformational changes if selenotrisulfide cross-links are formed between SH groups and selenites. Dietary selenomethionine is readily incorporated into protein by animals (Ochoa-Solano and Gittler, 1968). Selenite is converted into dimethyl selenide in animal tissues, but there is no conclusive evidence that selenite is converted into selenoamino acids in animals. Selenoamino acids bind to sulfhydryl groups of functional proteins forming complexes which prevent reversible redox reactions.

The major pathway of Se detoxication is methylation, but the major site of this detoxication is yet to be established. Trimethyl selenonium ion has been identified as the major excretory product of Se metabolism in rats fed selenate, selenite, selenocystine, selenomethionine, and seleniferous wheat (Palmer *et al.,* 1970). Vitamin E or fat-soluble antioxidants, such as butylated hydroxytoluene, are required for methionine reversal of Se toxicity in rats (Levander and Morris, 1970). It is presumed that vitamin E and other fat-soluble antioxidants render the methyl group of methionine more available for Se methylation and eventual detoxication. Arsenite inhibits the methylation of Se but increases the fecal excretion of selenite; dietary arsenite apparently binds Se in a form excreted via the bile.

Tellurium (Te)

Tellurium occurs to a very small extent in the free state; in general it occurs in the form of the silver and bismuth tellurides, hessite (Ag_2Te) and tetradymite (Bi_2Te_3). It is present in some seleniferous soils, but the extent of its presence in seawater and in the earth's crust is not well established. Tellurium is not essential for mammals, but it is stimulatory (Luckey, 1975b). There are no reports of the presence of Te in humans under normal conditions.

Tellurium is used in the steel industry in Cu and Pb alloys to provide increased resistance to corrosion and stress, in the glass industry as a coloring agent, and in the rubber industry as an improving agent. Bismuth telluride is used as a thermocouple material in refrigeration equipment. In the past Te salts were used therapeutically against syphilis. Tellurite solutions are used in the rapid diagnostic test for diphtheria, which reduces the tellurite to black elemental Te. Cases of tellurium poisoning are rare, and the toxicity of Tellurium is not as high as that of selenium in industrial operations.

Chemistry

Tellurium is the most electropositive element in subgroup VIA, and exhibits variable valence states from -2 to $+6$; the most common forms are -2, $+4$, and $+6$. Like S and Se, Te forms di- and trivalent oxides and water-soluble tellurite and tellurate anions. The cationic salts are unstable and are hydrolyzed by water to oxides and oxy acids. Tellurates readily undergo reduction to more stable tellurites. Telluric acid forms stable polymeric chains in aqueous solution.

Metabolism

The metabolism of Te in mammals has not been studied extensively. It is not an essential nutrient. The biochemistry of Te was reviewed by Amdur (1947, 1958) and Schroeder *et al.* (1967c), and the toxicology of Te was reviewed by Browning (1969) and Cooper (1971).

Metallic Te and TeO_2 are very poorly absorbed from the gastrointestinal tract, but about 25% of ingested water-soluble tellurites and tellurates are absorbed. Parenterally administered Te and TeO_2 remain at the site causing abscess and a bluish-black discoloration, while tellurites and tellurates are easily absorbed and distributed in the tissues. In the blood, Te is complexed to the plasma proteins, and very little enters the erythrocytes. The bones retain Te for a longer period than do the soft tissues in rats (Hollins, 1969). Retention is high in the kidneys, irrespective of mode of entry and chemical state of Te. The relative levels of retention in other organs are: heart > lungs > spleen > bone > liver. Metabolically active tissues evidently reduce tellurite to telluride, which is rapidly eliminated. The telluride is methylated to volatile dimethyl telluride, which has a garlic-like odor, and which is exhaled. Excretion of Te is mostly fecal and urinary, although a small amount is exhaled with the breath. The Te excretion pattern depends on the chemical nature, dose, and water-solubility of the Te compound, and the mode of administration. Excretion is greater in urine than in feces when Te is given parenterally, but orally ingested Te salts are excreted mainly in the feces. Intramuscularly administered [127,129]Te is excreted as NH_3TeO_6 in the urine within 24 hours (Durbin, 1960). Mammalian tissues accumulate both tellurite and elemental Te, as homeostatic mechanisms controlling the absorption and excretion of Te do not seem to exist in mammals.

Toxicity

The acute toxic effects of Te in animals are similar to those of arsenic: suppression of sweat, nausea, somnolence, inflammation of the gastric mucosa, intestinal hemorrhage and intense hyperemia of internal organs, and finally death from respiratory paralysis. Some of the chronic toxic effects are digestive disturbance, growth suppression, somnolence, and the more common garlic breath. The toxicity of Te is shown in Table 6-3.

Elemental Te is not toxic to rats at 1500 ppm of the diet, but intravenous elemental Te is toxic with a lethal dose of 500 mg/animal for dogs and 75 mg/animal for guinea pigs (Cerwenka and Cooper, 1961). TeO_2 is toxic; at 1500 ppm of diet it suppresses growth and causes renal and hepatic degeneration in guinea pigs and rats (Amdur, 1958). Tellurium tetrachloride at 500–1000 ppm of the diet markedly reduced growth and caused heavy mortality in Pekin ducks within two weeks; growth was not affected at 50–250 ppm, but 50–70% mortality occurred by the fourth week (Carlton and Kelley, 1967). Intramuscular injection of TeO_2 in guinea pigs caused hemorrhage and necrosis of the kidneys, with a dark-gray or bluish-black

TABLE 6-3. Tellurium Toxicity

| | | | | Dosage/kg body weight | | | |
| | | | | Compound | Metal | | |
Compound	Animal	Route	Toxicity	mg	mg	mM	pT
Tellurium metal	Guinea pig	iv	LD_{100}		75[a]	0.588	3.25
	Dog	sc	MLD		290	2.27	2.65
	Dog	iv	LD_{100}		500[a]	3.92	2.41
Sodium tellurite	Mouse	oral	LD_{50}	20	11.5	0.09	4.05
Na_2TeO_3	Rat	oral	LD_{50}	83	47.8	0.375	3.42
	Rat	ip	MLD	4.0	2.73	0.214	3.67
	Rat	ip	LD_{50}	4.3	2.47	0.019	3.72
	Rat	iv	LD_{50}	2.4	1.38	0.010	4.00
	Rabbit	oral	Ld_{50}	67	38.6	0.302	3.52
	Rabbit	oral	MLD	53.6	30.8	0.241	3.62
	Guinea pig	oral	LD_{50}	45	25.9	0.203	3.69
	Human	oral	MLD	29	16.7	0.131	3.89
Potassium tellurite	Dog	iv	MLD	35	17.6	0.138	3.86
K_2TeO_3							
Sodium tellurate	Mouse	oral	LD_{50}	165	88.6	0.694	3.16
Na_2TeO_4	Rat	oral	LD_{50}	385	206	1.61	2.80
	Rat	ip	MLD	37	19.8	0.155	3.81
	Rat	ip	LD_{50}	56.8	30.5	0.239	3.62
	Rat	iv	LD_{50}	55	29.5	0.231	3.64
	Rabbit	oral	LD_{50}	104	55.8	0.437	3.36

[a]Number of mg injected per animal.

discoloration owing to the trapping of elemental Te in the tissues. Tellurite and tellurate retard the growth of rats at 25–50 ppm Te in the diet (Franke and Moxon, 1937). Human workers exposed to excess tellurite during industrial operations show chronic toxicity symptoms such as nausea, somnolence, and loss of appetite (Cerwenka and Cooper, 1961).

Recent studies show that elemental Te can cross the blood-brain and placental barriers in rats; it also affects the nervous system. Weanling rats given 1% elemental Te in their diet suffered reversible paralysis in the hind limbs due to segmental demyelination of the sciatic nerves and spinal roots (Lampert *et al.*, 1970). Chronic intramuscular injection of elemental Te into rats and rabbits produced grossly visible "black brains," the accumulation of the metal in neuronal lysosomes in the form of fine needles (Cravioto *et al.*, 1970). The metal remains in the tissue for a long time, evidently bound to soluble proteins rendered inactive by a combination of protein binding and reduction by lysosomal enzymes (Agnew and Cheng, 1971). There is also evidence that continuous oxidation and reduction of tellurite to metallic Te occurs. When reduction exceeds oxidation, metallic Te is stored in brain cells. When fed to pregnant rats during days 10–15 of gestation at dietary levels of 3000 ppm, metallic Te induced hydrocephalus in fetuses and neonates (Agnew *et al.*, 1968; Duckett, 1972). Chronic consumption of metallic Te for five months by rats resulted in heavy deposits of Te in the neuronal lysosomes, and severely impaired their ability to learn a sequence of behavioral tasks (Dru *et al.*, 1972).

There are no specific mechanisms by which tellurite is detoxicated, except by immobilizing Te in an inert form in the tissues. If Po and element 106 were not radioactive, their toxicities should resemble that of Te, but their heavy atomic weights should make them less toxic chemically than Te, because of the poor solubility of their salts.

SUBGROUP VIB METALS

Chromium and molybdenum are essential trace metals in mammals, while tungsten is nonessential. Chromium is not an essential component of known metalloenzymes, but molybdenum is essential in enzymes such as ornithine oxidase. Subgroup VIB metals do not form stable hydrides; they readily form coordination complexes and polynucleate compounds, such as polyacids, which are capable of reacting with phosphates. Chelation is more pronounced with Mo and W than with Cr. Subgroup VIB metals differ from subgroup VIA metals in their lower toxicity, in the formation of toxic and volatile hydrides, in the greater absorption of anions of subgroup VIB metals from the alimentary tract in animals, in their inability to form stable coordination and polynucleate compounds, and in their decreased ability to form biologically active analogs of metabolically active sulfur compounds. The toxicity of subgroup VIB metals decreases directly with atomic weight and electropositivity on the basis of LD_{50} values of both cationic and anionic salts; the salts of the lighter metals are the most toxic in their highest state of oxidation. The metals of subgroup VIB are less toxic than the metals in subgroup VIA because at physiologic pH the salts of subgroup VIB metals are converted to insoluble polyacids through olation, while the anionic Se and Te readily complex with biologically active thiol compounds.

Chromium (Cr)

Chromium occurs in nature mostly as chrome iron ore ($FeO \cdot Cr_2O_3$). Chromium is present in small quantities in all soils and plants, at 1–2.5 ppb in sea water, and at about 200 ppm in the earth's crust. The normal human adult body contains about 6 mg Cr, with tissue concentrations of 0.02–0.04 ppm Cr on a dry weight basis. There is no high concentration in any particular issue. Chromium accumulates in the lungs with age, but the levels are usually harmless.

Industrially, chromium is used extensively in the manufacture of high-tensile structural steels and for ferritic disks in jet engines; chromium carbide is a substitute for tungsten carbide in the manufacture of cutting tools. Chromium salts and chromates are used in painting, lithography, textile printing, tanning, dyeing, photography, wallpaper, electric cells, matches, and rubber goods. Chromium is considered to be only a mild industrial hazard; Cr poisoning in humans is caused by accidental ingestion of chromates.

Chemistry

Chromium, a metal of the first transition series, exhibits five valence states from $+2$ to $+6$; Cr^{2+} is basic, Cr^{3+} amphoteric, and Cr^{6+} acidic. Trivalent Cr forms stable hexa- or octahedral coordinate complexes, and at pH levels above 7 these undergo hydrolysis forming insoluble hydroxo-aquo complexes, such as polynucleate bridge complexes, following olation. At physiologic pH Cr^{3+} salts and compounds undergo olation and become inactive. Trivalent Cr forms unstable inner complexes with α amino acids owing to its preferential coordination with oxygen and nitrogen atoms. Hexavalent chromium is linked to oxygen and is a powerful oxidizer, but it does not form coordination compounds, and is easily reduced to Cr^{3+}. It is very difficult to characterize the chemistry of naturally occurring biologically active chromium complexes such as the glucose tolerance factor, but these complexes are better utilized than inorganic Cr salts. Salts such as chromium perchlorate or salicylate possess a mild glucose-tolerance-factor activity.

Metabolism

Chromium is essential for most forms of animal life, as an essential component of the "glucose tolerance factor." A low-molecular-weight cofactor for insulin activity, it alleviates some symptoms of diabetes in humans. The Cr requirement of normal human adults is about 50 μg, and the average daily intake varies from 60 to 100 μg. This represents about 5% of the total body content, and there is a wide margin of safety.

Involvement of chromium in normal carbohydrate and lipid metabolism and its influence on glucose tolerance and lipogenesis have been studied extensively and reviewed (Schroeder *et al.*, 1962*a*; Schroeder, 1968; Mertz, 1967; Underwood, 1971; O'Dell and Campbell, 1971; National

Academy of Sciences, 1974). The chief essential function of chromium is in insulin activity and membrane transport of cell metabolites. Chromium does not serve as an essential component of any metalloenzyme nor as a specific enzyme activator, and it reacts strongly with nucleic acids, and is present in purified RNA preparations.

The gastrointestinal absorption of Cr^{3+} is very low, less than 3%, regardless of nutritional status and dosage (Visek *et al.*, 1963). The acid pH of the stomach keeps the Cr^{3+} in the trivalent form, and reduces much of the Cr^{6+} to Cr^{3+}; absorption of Cr^{3+} from the stomach is negligible. Oxalates enhance this absorption, while phytates inhibit it. The alkaline pH of the duodenum converts Cr salts into insoluble polynucleate bridge hydroxo-aquo complexes. Some chelated Cr complexes remain in solution and are absorbed into the blood stream (Hopkins and Schwarz, 1964; Nelson *et al.*, 1973). Intravenously administered Cr^{3+} salts and soluble forms of Cr absorbed from the alimentary tract into the blood bind progressively to siderophilin, transferrin, albumins, and γ globulins (Jet *et al.*, 1968). Transferrin, which has two distinct metal-binding sites, binds Cr^{3+} in a manner similar to iron and also complexes with Mn^{3+}, Co^{3+}, and Cu^{2+}. However, intravenously administered Cr^{6+} (chromates and dichromates) in physiologic doses permeates the erythrocyte membrane and is sequestered, although probably not as Cr^{6+}. Chromium bound to β globulins is distributed in the heart, pancreas, lungs, brain, spleen, liver, and testes (Baetjer *et al.*, 1959). In rat testes there is a dramatic quick uptake and slow release of absorbed Cr^{3+}. After transient storage in these tissues, 80% of the absorbed chromium is excreted via the kidneys. Part of the injected Cr^{3+} and all of the unabsorbed dietary Cr are excreted in the feces. Implanted and intramuscularly injected Cr salts remain at the site while being absorbed very slowly. Chromium deposited in the lungs seems to be largely insoluble, inert, and not generally available to the organism. In pregnant rats, Cr^{3+} and its simple complexes are not transferred to the fetus (Mertz, 1975) but Cr^{3+} incorporated into brewers yeast is efficiently transferred, indicating that the small, water-soluble metal complex of Cr^{3+} is the physiologically active form (glucose tolerance factor), the only one in which chromium behaves like an essential element in homeostatic regulation, placental transport, and general biologic activity.

Toxicity

Trivalent chromium is one of the least toxic of the trace metals (Schroeder *et al.*, 1962*a*; Mertz, 1969). The rat therapeutic index for Cr^{3+} is 1:10,000, but for Cr^{6+} the value is far less; hexavalent chromium (chromate) is more toxic than trivalent chromium. In physiologic doses, the animal's body alters the valence of soluble chromium salts to that required for absorption. There are no reports of oral toxicity of Cr^{3+}, and rats are unaffected when fed 100 ppm chromic lactate with a milk diet. Cats tolerate 1000 mg Cr^{3+} complexes. Life-term studies with mice and rats given 5 ppm Cr^{3+} show increased growth and decreased mortality over the controls. Tolerance of chromium trioxide in the diet is approximately 1% by rats and pigs, 0.5% by sheep and cattle, and 1–3% by chickens. Humans can tolerate 500 mg Cr_2O_3 daily. However, parenteral Cr^{3+} is toxic to experimental animals, while chrome alum and a Cr–niacin complex are quite toxic when injected, but not when fed (Mertz and Roginsky, 1975). The toxicity data of chromium are listed in Table 6-4.

Hexavalent chromium is topically corrosive, and oral ingestion of Cr^{6+} is toxic despite the organism's ability to reduce Cr^{6+} to the less toxic Cr^{3+}. Life-term studies with mice given 5 ppm Cr^{6+} in their drinking water showed a slight growth retardation and formation of malignant tumors (Schroeder and Mitchner, 1971a; Schroeder et $al.$, 1965). Laboratory animals can tolerate about 0.1% Cr^{6+} in their diet. In rabbits large doses of chromates were followed by albuminuria with desquamated cells; the kidneys showed hyperemia, fatty degeneration, and necrosis. In the past, treatment of warts by local application of chromic acid caused Cr^{6+} poisoning in humans, leading to nephritis, anuria, and extensive lesions in the kidneys. Accidental swallowing of dichromate causes gastrointestinal ulceration and symptoms affecting the central nervous system, but no other reports of the toxic effects of ingested Cr^{6+} were found.

Industrial inhalation of Cr^{6+} toxicity has been reviewed extensively (Browning, 1969). Air threshold-limit values for Cr^{6+} have been established: 0.1 mg/m³ for CrO_3 and 0.5 mg/m³ for other compounds; however these have to be revised. Chromate-contact dermatitis varies from a dry erythematous condition to eczema on the exposed limbs; the eczema is due to the direct necrotizing effect of chromate. A high incidence of this type of skin allergy occurs among workers in the cement, sulfite pulp, tanning, and electroplating industries. The injury to nasal mucosa caused by inhalation of Cr^{6+} compounds includes inflammation and ulceration, and the larynx is also affected.

Chromium compounds are not carcinogenic orally, but epidemiologic evidence involves fine particles of chromium salts in pulmonary cancer among workers in chromium refineries and related industries (Baetjer, 1956). Neoplasms such as squamous cell carcinomas, round cell carcinomas, and adenosarcomas are induced in the lungs. Chromium is a primary carcinogen, but does not induce cancer in other areas of the human body. Trivalent Cr is transformed into an active carcinogen during a latent period; a time lag of from four to twenty years is noticed between the initial exposure to Cr by inhalation and the onset of cancer. Intramuscular injection of a chromite suspension in lanolin, or subcutaneous injection of $CaCrO_4$ in arachis oil into rodents induced bronchial cancer (Hueper and Payne, 1959; Roe and Carter, 1969); however, when soluble chromates

TABLE 6-4. Chromium Toxicity

Compound	Animal	Route	Toxicity	Compound mg	Metal mg	Metal mM	pT
Chromium trioxide CrO_3	Dog	sc	MLD	330	172	3.30	2.48
Chromic acid H_2CrO_4	Rat	oral	LD_{100}	350	130	2.50	2.60
	Dog	sc	MLD	320	145	0.788	3.11
Chromium carbonyl $Cr(CO)_6$	Mouse	iv	LD_{50}	100	23	0.442	3.35
Sodium chromate Na_2CrO_4	Guinea pig	sc	LD_{100}	30	9.6	0.185	3.73
	Rabbit	sc	LD_{100}	243	78	1.50	2.82
	Rabbit	iv	LD_{100}	32	10.3	0.198	3.70
	Dog	iv	MLD	235	75	1.44	2.84
	Cow	oral	LD_{100}	700	224	4.30	2.37
Dipotassium chromate K_2CrO_4	Guinea pig	sc	MLD	70	18.5	0.355	3.45
	Rabbit	sc	MLD	12	3.2	0.062	4.21
	Dog	iv	MLD	4.0	1.0	0.020	4.66
	Man	oral	MLD	430	115	2.21	2.66
Ammonium dichromate $(NH_4)_2CrO_7$	Guinea pig	sc	LD_{100}	30	12.35	0.23	3.62
Sodium dichromate $Na_2Cr_2O_7$	Mouse	iv	MLD	26.2	10.4	0.20	3.70
	Rabbit	iv	MLD	18.4	7.2	0.138	3.86
	Guinea pig	sc	MLD	25.5	10.2	0.20	3.71
Potassium dichromate $K_2Cr_2O_7$	Mouse	sc	LD_{100}	100	35.3	0.679	3.17
	Rabbit	iv	LD_{100}	27.9	9.7	0.186	3.73
	Guinea pig	sc	LD_{100}	29.4	10.4	0.2	3.70
	Dog	oral	LD_{100}	2830	1000	19.2	1.72
Lead chromate $PbCrO_4$	Guinea pig	ip	LD_{50}	400	64.4	1.23	2.91
Chromium dichlorode $CrCl_2$	Rat	oral	LD_{50}	1870	791	15.2	1.82
Chromic trifluoride CrF_3	Guinea pig	oral	MLD	150	71.5	1.38	2.86
	Guinea pig	sc	MLD	120	57.2	1.10	2.96
Chromic trichloride $CrCl_3$	Mouse	ip	LD_{50}	140	45	0.865	3.07
	Mouse	iv	MLD	800	263	5.06	2.29
	Rat	oral	LD_{50}	1870	614	11.8	1.93
	Rabbit	iv	MLD	288	94.5	1.82	2.74
Chromyl chloride CrO_2Cl_2	Mouse	sc	MLD	5.45	2.32	0.045	4.35
Chromic nitrate $Cr(NO_3)_3 \cdot 9H_2O$	Rat	oral	LD_{50}	3250	422	8.11	2.09
Chromic sulfate $Cr_2(SO_4)_3$	Mouse	iv	MLD	247	65	1.25	2.90
	Rabbit	iv	MLD	217	57	1.09	2.97
Chrome alum $K\ Cr(SO_4)_2 \cdot 12H_2O$	Rat	iv	LD_{50}	112	10	0.192	3.89
Chromium hexaurea chloride	Rat	iv	LD_{50}	180	18	0.346	3.46

were deposited intratracheally in rodents, the incidence of cancer was low (Hueper and Payne, 1962). Both tri- and hexavalent Cr compounds are considered active in inducing bronchial carcinomas, but the mechanism of tumor formation is not known.

There are no specific or reported mechanisms by which Cr can be removed from living systems, but a possible detoxication of Cr could be the increased formation of ribonucleoprotein in the liver (Wacker and Vallee, 1959). The involvement of Cr in toxic doses in the reactions with these active molecules could be significant.

Molybdenum (Mo)

Molybdenum occurs as molybdenite sulfide (MoS_2) and as lead and iron molybdates (wulfenite and ochre) in nature. It is present at 12–16 ppb in seawater and at 1 ppm in the earth's crust and in many plants. The adult human body contains about 9.3 mg Mo, about 5 mg in the skeleton and 2.0 mg in liver. Molybdenum is essential for most living species. Industrially, Mo is used extensively in ferro- and manganese alloys, in pigments for paints, lacquers, inks, rubber, and leather, as coatings in glass and ceramics, and in vitreous enamel for increasing its adherence to steel, and in fertilizers for microbial fixation of nitrogen. Molybdenum has low toxicity when compared with other industrially important metals, and is not generally considered a serious industrial hazard. Molybdenum poisoning can occur in animals grazing in pastures rich in molybdenum or molybdenum-containing fertilizers.

Chemistry

Molybdenum, a metal of the second transition series, exhibits five valence states, from +2 to +6; the naturally occurring stable forms are the 5+ and 6+ states. In solution, Mo compounds disproportion to yield a mixture of different valences. Molybdenum complexes are found in a variety of stereoconfigurations with coordination numbers 4, 5, 6, and 8; minor differences in conditions result in shifts among coordination numbers. The preferred coordination numbers for Mo^{5+} are 5 and 6, and for Mo^{6+}, 6. Molybdenum is versatile in biologic reactions; the reduction potential for $Mo^{5+} \rightarrow Mo^{6+}$ is in the proper range for interaction with flavins. Its unique function is related to its ability to catalyze reactions that require the simultaneous exchange of two electrons and two protons. Among the simple cationic compounds, oxides and the sulfide of Mo are stable, while the others are not. Anionic molybdates are stable and common; Mo^{5+} complexes with oxygen, halogen, cyanide, and sulfocyanide ligands. Dimerization of Mo^{5+} complexes occurs through oxo bridges.

Pentavalent Mo complexes are unstable to oxidation, but hexavalent Mo complexes with oxygen as the preferred ligand and forms stable isopoly and heteropoly acids and salts:

$$(MoO_4)^{2-} \overset{H+}{\rightleftharpoons} Mo_2O_7 \overset{2e}{\rightleftharpoons} H_2(MoO_4)_6 \overset{10e}{\rightleftharpoons} H_2(Mo_2O_7)_6 \overset{10e}{\rightleftharpoons} (MoO_3)X$$

These polyacids can react with phosphates and tartrates in biologic fluids and tissues. Because of the preferred affinity of Mo^{6+} for oxygen ligands, molybdenum binds to carboxy groups, hydroxyl groups, and also sulfhydryl groups present in macromolecules. The single electron in the outermost orbital of Mo renders it more reactive than W.

Metabolism

The essentiality of Mo for animals is based upon its involvement in flavin-dependent metalloenzymes; Mo deficiency results in decreased xanthine oxidase in tissues. Molybdenum studies usually involve W antagonism, as manifestation of Mo deficiency alone is rare. Balance studies indicate that 0.1–0.3 mg Mo/day are adequate for man, but the exact requirements are not known, and the daily intake is not established, although the nutritional requirements of Mo and the nutritional and toxicologic interrelationships between Mo and Cu, and Mo and sulfate have been extensively reviewed (Nason, 1958; DeRenzo, 1962; Spence, 1965; Browning, 1969; Schroeder, *et al.,* 1970*a*; O'Dell and Campbell, 1971; Underwood, 1971).

Soluble hexavalent Mo compounds are readily and rapidly absorbed from the digestive tract; in pigs the absorption is very high. The sparingly soluble trioxide and $CaMoO_4$ are absorbed fairly well from the digestive tracts of rabbits and guinea pigs when fed in large doses (Gray and Daniel, 1954). Absorption of Mo compounds from sites of parenteral injection is rapid, and intravenously injected Mo salts are quickly distributed to the tissues with slightly preferential accumulation in the liver, kidneys, and bone. Most inhaled Mo is absorbed by lungs and distributed to other tissues, although compounds such as MoS_2 are poorly absorbed from lungs and the digestive tract. Excretion of soluble Mo salts is also rapid, mainly in the urine and partly in feces; the predominant form is a conjugated molybdate complex. Molybdenum salts are also secreted into bile and into the intestines in the enterohepatic circulation; much dietary Mo appears in the urine. Sulfate ions control Mo excretion and tissue Mo levels. In sheep, sulfate limits Mo retention by decreasing gastrointestinal absorption, increasing urinary excretion, and preventing Mo transport across tissue membranes (Dick, 1956). In physiologic doses, one-third of the injected Mo is rapidly excreted within 24 hr, one-third is retained in the liver, and the rest is distributed in the endocrine glands and other soft tissues. After temporary retention in the tissues, Mo is excreted completely, with no accumulation in mammals. Poorly characterized homeostatic excretory mechanisms maintain the level of Mo in tissues.

Toxicity

Monogastric animals tolerate Mo toxicity better than do ruminants, as shown by the following order of tolerance: horses > pigs > rats > rabbits > guinea pigs > sheep > cattle (Davis, 1950). The general symptoms of chronic Mo intoxication in animals are loss of body weight, anorexia, anemia, deficient lactation, male sterility owing to testicular degeneration, osteoporosis, and bone-joint abnormalities. Severe gastrointestinal irritation with diarrhea, coma, and death from cardiac failure are the symptoms of acute Mo toxicity. Rats subject to chronic Mo toxicity seem to develop a sensory ability to reject Mo-toxic diets by associating them with gastrointestinal disturbance (Monty and Click, 1961). Molybdenum intoxication depends upon the chemical form and dose, and is influenced by sulfate, Cu^{2+}, Zn^{2+}, Pb^{2+}, and protein intake, and the form and amount of copper and organic sulfur such as cystine and methionine in the body. The toxicity of Mo compounds is summarized in Table 6-5. The sparingly soluble MoS_2

TABLE 6-5. Molybdenum Toxicity

				Dosage/kg body weight			
				Compound	Metal		
Compound	Animal	Route	Toxicity	mg	mg	mM	pT
Molybdenum trioxide	Rat	inhal	LD_{75}		287^a		
MoO_3	Rat	oral	LD_{50}	188	125	1.30	2.89
	Guinea pig	ip	LD_{75}	400	267	2.78	2.56
Ammonium molybdate	Rat	oral	LD_{50}	680	370	3.86	2.41
$(NH_4)_6Mo_7O_{24} \cdot 4H_2O$	Rat	ip	MLD	203	110	1.15	2.94
	Guinea pig	oral	LD_{100}	2200	1200	12.5	1.90
	Guinea pig	sc	LD_{100}	1380	750	7.81	2.11
	Guinea pig	ip	LD_{100}	800	435	4.53	2.34
	Rabbit	oral	LD_{100}	1870	1020	10.6	1.97
	Rabbit	sc	LD_{100}	1600	870	9.07	2.04
	Cat	oral	LD_{100}	2400	1310	13.6	1.87
Disodium molybdate	Mouse	ip	LD_{50}	303	120	1.25	2.90
$Na_2MoO_4 \cdot 2H_2O$	Rat	ip	MLD	290	115	1.20	2.92
	Rat	ip	LD_{100}	385	153	1.59	2.80
	Rat	ip	LD_{50}	576	228	2.38	2.62
Calcium molybdate	Rat	ip	LD_{50}	208	99.8	1.04	2.98
$CaMoO_4$							

[a]Exposure for 1 hr daily for 30 days; number of mg/m³.

is nontoxic to rats even at 500 mg/kg diet for several weeks. Supplementary dietary Cu^{2+}, thiosulfate, methionine, and cysteine are effective in alleviating Mo toxicity in animals.

Teart syndrome, molybdenosis or scouring in cattle, is the major manifestation of Mo toxicity. Cattle die after grazing in "teart" pastures which may contain 20–100 ppm Mo, dry-weight basis (Lewis, 1943), while pigs show no ill effects when fed 1000 ppm Mo (Davis, 1950). Sterility in bulls results from damaged testicular interstitial cells and germinal epithelium, resulting in poor spermatogenesis; poor conception and deficient lactation are reported in cows (Thomas and Moss, 1951). In life-term experiments, feeding rats or mice 10 ppm Mo as Na_2MoO_4 impairs their reproduction (Schroeder and Mitchner, 1971b).

Inhalation of dusts of $CaMoO_4$ and MoS_2 is not toxic to rabbits, whereas the readily soluble MoO_3 is highly toxic (Fairhill $et\ al.$, 1945). Exposure to MoO_3 is irritating to the eyes and mucous membranes of the respiratory tract, and bronchial and alveolar exudates are present in moderate amounts in animals exposed to MoO_3 by inhalation (Maresch $et\ al.$, 1940).

The biological significance and involvement of Mo in the action of flavoproteins, xanthine, and aldehyde oxidase are well known. There are conflicting reports that these enzymes are inhibited $in\ vivo$ at higher Mo concentration, but high levels of molybdates inhibit the $in\ vivo$ activities of succinic acid oxidase (Miller and Engel, 1960), sulfite oxidase (Halverson, $et\ al.$, 1960), glutaminase (Capilna $et\ al.$, 1963), cholinesterase (Spiridonovaa and Suvorov, 1965), and cytochrome oxidase (Kolomiitseva and Stoliar, 1969). The inhibition of ceruloplasmin activity by excessive Mo has not been confirmed (Smith $et\ al.$, 1968), and Mo involvement in inhibiting ceruloplasmin appears to be indirect; in low doses it mobilizes hepatic Cu to increase ceruloplasmin activity, and in high doses it diminishes circulating Cu by increasing hepatic Cu storage and apparently decreasing ceruloplasmin activity. Liver sulfide oxidase activity is substantially reduced, with endogeneous production and accumulation of sulfide in the tissues during Mo intoxication and Cu deficiency; feeding of sulfides or sulfates in a low-Cu diet increases Mo toxicity in rats.

Excessive dietary molybdenum interferes with the metabolism of calcium and phosphorus, inducing osteoporosis, lameness, bone-joint abnormalities, and connective tissue changes in mandibular and maxillary exostosis. The mechanism for the inhibition of Mo toxicity by dietary supplements of Cu, sulfate, or S-containing amino acids is unknown, but hexavalent sulfate may displace the molybdate by simple mass action and increase the urinary excretion of molybdate (Schroeder $et\ al.$, 1970a). Tri- and hexavalent molybdenum is less toxic than chromium, on the basis of

LD_{50}, but is more toxic than tungsten. Although Mo has been implicated epidemiologically in certain human illnesses, there is a wide margin of safety in human foods.

Tungsten (W)

Tungsten occurs as tungstates in wolframite ($FeMnWO_4$) and scheelite ($CoWO_4$). It constitutes less than 0.5 ppm of the earth's crust, and traces are found in seawater. Mammalian bones and heart are reported to contain 0.25 ppb and 5 ppb, respectively (Wester, 1965). Tungsten is not a normal constituent of plant and animal tissues, and it is not essential for mammals.

Tungsten is used industrially for making hard tungsten-alloy tools; it is also used for plating material, for anticathodes in X-ray tubes as a source of ultraviolet radiation, and in the preparation of green and blue pigments. A W–Cu–Ni alloy is used as an effective substitute for lead as protection against ionizing radiation. Tungsten poisoning generally occurs after continued exposure to industrial dusts and vapors in the refining of tungsten, or after accidental ingestion of tungstates.

Chemistry

Tungsten forms stable divalent and hexavalent compounds. Tungsten oxides are stable, but the common cationic salts, such as halides, decompose in water. Hexavalent anionic salts are stable and water-soluble. Tungsten does not form many stabile chelate compounds; it forms coordination compounds such as $[W(CN)_8]^{3-}$ or $[W(OH)_3(CN)_5]^{4-}$, and it also forms complex halides and oxyhalides, e.g., $(W_2Cl_9)^{3-}$. In acid-medium tungstates readily form heteropolyhydrated tungstic acids:

$$(WO_4)^{2-} \underset{OH^-}{\overset{H^+}{\rightleftharpoons}} (W_2O_7) \overset{2e}{\rightleftharpoons} [H_2(WO_4)_6] \overset{10e}{\rightleftharpoons} (WO_3)_x$$

Among these polyacids, only phosphotungstic acid, $H_3[PO_4 \cdot W_{12}O_{18}(OH)_{36}]$, is stable. The existence of phosphotungstic acid in animal tissues has not been reported.

Metabolism

Tungsten is not an essential nutrient for animals, but it stimulates growth in plants (Bortels, 1936). The metabolism of W in sheep and swine (Bell and Sneed, 1970) has been studied, and studies of Mo metabolism in animals usually use W as as antagonist.

Indirect evidence indicates rapid gastrointestinal absorption of soluble tungstates but not of cationic or metallic W. Rats absorb a lethal dose of W from a diet which contained 2% W as tungstate (Kinard and Van de Erve, 1941). Rumen bacteria interfere with W absorption in sheep; tracer levels of tungstate adsorb to feed particles high in cellulose and hence are not absorbed from the digestive tract. Reports of the absorption of tungstate from sites of parenteral injection are not conclusive, but small doses are completely absorbed. The excretion of absorbed tungstate is mainly urinary, and about 2% of injected W is excreted fecally; the fecal excretion of about 25% of an orally ingested dose of W represents the unabsorbed portion. Tungstate can cross the placental barrier in rats. In laboratory animals, the retention of tungstate in the bones is high, about 18 mg W/100g. The spleen retains about 3 mg and liver retains < 1 mg W/100g; generally, W is poorly retained in soft tissues (Kinard and Aull, 1945). In sheep the levels of retention of tungstate in various tissues are: kidneys > liver > lymph nodes > bone > lungs > pancreas > muscle. In swine the retention of W in tissues is slightly different: kidneys > bone > brain > lymph nodes > liver > spleen > skin > muscle. The poor retention of tungstate in the liver and kidneys of laboratory animals illustrates the difference between W and other heavy metals. There are no specific reports of a homeostatic regulatory mechanism for W in mammals, but within physiologic limits, poorly understood homeostatic excretory mechanisms maintain the tissue level of W.

Toxicity

Acute tungstate poisoning by oral and parenteral routes causes nervous prostration, diarrhea, respiratory paralysis, and death. The symptoms of W toxicity depend on its chemical form and the route of administration. Diets containing about 0.5% W in the form of tungstate, tungstic oxide, or ammonium tungstate do not kill rats or rabbits, although their growth rate is decreased. With diets containing 0.5% W, both tungstic oxide and simple tungstate salts are toxic, while ammonium paratungstate is not. At higher levels, sodium tungstate (2% W), tungstic oxide (3.96% W), and ammonium paratungstate (5.0% W) in the diet are lethal to rabbits and rats (Kinard and Van de Erve, 1941; Kinard and Aull, 1945). Rabbits tolerate about 12 mg Na_2WO_4 daily in their diet for six weeks; data on the toxicity of injected Na_2WO_4 are summarized in Table 6-6.

Intratracheal innoculation of tungstate, tungsten metal, or tungsten carbide dust (about 150 mg) into guinea pigs produced pigmentation and interstitial cell proliferation near entrapped metal particles, and led to pneumonitis and bronchiolitis (Schepers, 1955*a,b*). Human pneumocon-

TABLE 6-6. Tungsten Toxicity

| | | | | Dosage/kg body weight | | | |
| | | | | Compound | Metal | | |
Compound	Animal	Route	Toxicity	mg	mg	mM	pT
Tungsten	Rat	ip	LD$_{50}$		5000	27.2	1.57
Tungsten trioxide WO$_3$	Rat	oral	MLD	840	666	3.62	2.44
Disodium tungstate Na$_2$WO$_4\cdot$2H$_2$O	Mouse	oral	LD$_{50}$	240	134	0.728	3.14
	Mouse	ip	LD$_{100}$	197	110	0.60	3.22
	Rat	oral	LD$_{50}$	1190	663	3.61	2.44
	Rat	sc	LD$_{50}$	240	134	0.728	3.14
	Rat	ip	LD$_{100}$	330	184	1.00	3.00
	Guinea pig	oral	LD$_{100}$	990	551	3.00	2.52
	Guinea pig	sc	LD$_{100}$	810	451	2.45	2.61
	Rabbit	oral	LD$_{50}$	875	487	2.65	2.57
	Rabbit	im	LD$_{50}$	105	58.5	0.318	2.50
	Rabbit	sc	MLD	78	43.5	0.237	3.63
	Dog	sc	LD$_{100}$	140	78	0.424	3.37

iosis, observed in workers in the tungsten carbide tool industry, was attributed formerly to tungsten toxicity, but later studies revealed that the cobalt and nickel present in tungsten carbide are probably the causative agents (Fredrick and Bradley, 1946).

Tungstate inhibits xanthine oxidase activity in lactating goats and cows (Bell and Sneed, 1970) but does not affect the milk yield. In rats dietary tungstate rapidly decreases the activities of xanthine oxidase and sulfite oxidase in the liver, lungs, kidneys, and intestines. The production of both these molybdenum-dependent enzymes in neonates is inhibited by feeding tungstate to pregnant rats (Johnson *et al.,* 1973).

There are no reports about W involvement in any other enzyme systems in mammals. Dietary tungstate at 5 ppm reduces the toxic effects of selenium in mammals (Moxon and Rhian, 1943).

7

TOXICITY OF GROUP VII METALS

INTRODUCTION

The highly reactive and electronegative halogens, fluorine (F), chlorine (Cl), bromine (Br), iodine (I), and astatine (At), and the moderatively active metals, manganese (Mn), technetium (Tc), and rhenium (Re), form group VII of the periodic table. Apart from their maximal heptavalence, the subgroup VIIB metals Mn, Tc, and Re differ from the nonmetallic halogens both in chemical and metabolic properties.

Fluorine, chlorine, and iodine, of the nonmetallic subgroup VIIA, are essential anions for mammals and other living organisms. Bromide can replace chloride in the growth of species of algae, but bromide has not been reported to perform any vital function in higher plants or animals. Astatine is too highly radioactive to be involved in mammalian nutrition, and there are no reports of nutrition studies on this halogen. The halogens in their elemental form are highly toxic. Fluorine is the most toxic halogen, as toxicity decreases with increasing atomic weight and decreasing electronegativity; soluble anionic fluorides are also toxic.

SUBGROUP VIIB METALS

Among the metals of subgroup VIIB, Mn is essential to mammals and Tc and Re are nonessential. No reports on the toxicity of Tc were found, but Mn and Re salts are toxic at high doses, the toxicity increasing with increasing molecular weight and electropositivity.

261

Manganese and rhenium are similar in electronic configuration, with two electrons in the outermost orbitals, and technetium has a single electron in the outermost orbital. All exhibit variable valence, Mn from $+2$ to $+7$, and Tc and Re from $+4$ to $+7$. All form water-soluble anions in the highest state of oxidation such as MnO_4^- and ReO_4^-, but the aqueous solubility of their cationic salts varies. All three metals form complexes with π-bonding ligands and carbonyls with carbon monoxide. Divalent Mn cations are stable; divalent cationic forms of Tc and Re do not exist. Divalent Mn coordinates with ligands to give an octahedral complex in biological systems; Tc and Re do not.

The gastrointestinal absorption of these metal ions depends upon their state of oxidation; the heptavalent anions are more readily absorbed than the cationic salts. Retention of these metal ions in mammalian tissues is transient. The thyroid gland appears to retain them for a longer time than do other tissues. While Mn excretion is fecal, Te and Re are excreted in the urine. Manganese is involved in metalloenzyme systems, but there is no indication of Re and Te involvement in mammalian metabolism.

Manganese (Mn)

Manganese occurs widespread in nature as the oxide (pyrolusite, braunite, manganite, and hausmannite), sulfide (hauserite), carbonate (manganesespat), and silicate (tephroite). Manganese is the 10th most

Group VII Metals

Atomic number	(Core) Active electrons
Name (Symbol)	
Atomic weight	**Usual valences**
Subgroup B	
25	(Ar) 5, 2
Manganese (Mn)	
59.94	$+2, +3, +4, +7$
43	(Kr) 6, 0, 1
Technetium (Tc)	
98.91	$+4, +6, +7$
75	(Xe) 5, 0, 2
Rhenium (Re)	
186.2	$+4, +6, +7$

abundant metal in the earth's crust, occurring at 1000 ppm or about 0.085%, and is the 19th most abundant metal in seawater at 1 ppb. Manganese is found in both plants and animals. The human adult body contains about 12 mg Mn, 43% of it in the bones and the rest in soft tissues, including the brain. Manganese is essential for mammals.

Manganese is used in industry in the manufacture of alloys, ferromanganese, silicospigel, manganese–bronze, and manganin; MnO_2 is used in dry-cell batteries as a depolarizer, in matches and fireworks, and in glass and ceramics as a pigment. Manganese salts are used in paints, varnishes, inks, and dyes. Manganous acetate is used in dyeing, in the tanning of leather, in fertilizers, and as a drier for linseed oil. Manganese sulfate is used as red glaze for pottery and is a fertilizer for vines and tobacco. Manganates and permanganates are powerful oxidizers used for disinfecting and bleaching; manganese carbonate is used in medicine as a hematinic; $MnSO_4$ is used in veterinary medicine to prevent perosis in poultry. Manganese cyclopentadienyltricarbonyl (MCT) is a good antiknock compound for internal combustion engines. Manganese ethylene-*bis*-dithiocarbamate (Maneb) is used as a chemical fertilizer.

Exposure to Mn metal and salts occurs through inhalation in industrial plants which make steel, alloys, glass and ceramics, varnishes, matches, and permanganates. Inhalation of gasoline vapor containing MCT or even contact for prolonged periods results in absorption of toxic MCT. In industry, Manganese salts in the air oxidize the toxic SO_2 to the more toxic SO_3. Food contamination occurs through Mn enrichment of deficient soils and through accidental poisoning.

Chemistry

Di- and trivalent Mn ions are basic; manganates 3^+, 4^+, and 5^+ are amphoteric; manganate 6^+ and permanganate 7^+ are acidic. Divalent cationic salts, and anionic manganates and permanganates are stable and water-soluble; Mn^{3+} is the biologically active form in mammals. It is more avid than Mn^{2+} toward chelation in biologic systems, owing to the smaller Mn^{3+} ionic radius. The near-alkaline state of mammalian biologic fluids is conducive to the retention of Mn^{3+}. Trivalent Mn can form stable and water-soluble halogen complexes such as $[MnF_5]^{3-}$; Mn^{2+} forms coordination complexes in octahedral geometry with a coordination number of 6, in aqueous solution Mn^{2+} exists as hexaquo ion $[Mn(H_2O)_6]^{2+}$. Manganese has a strong tendency to make coordination compounds with oxygen donors. Mn^{2+} coordination complexes are less stable than other transitional metal coordination complexes, because their ligand-field stabilization energy is weak or altogether lacking. Despite its variable valence, Mn is not directly involved in biological electron-exchange reactions. Manganese is firmly bound in enzyme systems such as pyruvate carboxylase.

Metabolism

Manganese is an essential metal for animals; birds require more Mn than do mammals, a fact attributed to their higher body temperature and higher oxygen consumption. The average daily dietary intake of Mn by man varies from 2 to 9 mg. Manganese is essential for the activity of many degradative enzymes such as pyruvate carboxylase, arginase, phosphatases, the biosynthetic enzyme of lipids, and mucopolysaccharides of cartilages. Manganese deficiency has not been clearly demonstrated in humans; although impaired blood clotting and lowered serum cholesterol have been reported (Doisy, 1972). In other animals, deficiency symptoms include mortality, decreased growth, impaired reproduction, poor bone formation, impaired glucose tolerance, slower blood clotting, and, in young animals born with Mn deficiency, ataxia (Cuthbertson, 1973).

The metabolism of Mn has been thoroughly studied and reviewed (Cotzias, 1958, 1964; Schroeder *et al.*, 1966a; Underwood, 1971; O'Dell and Campbell, 1971). Absorption of Mn from the gastrointestinal tract is poor under normal conditions, owing to the low solubility of cationic Mn. Ingested Mn^{2+} is converted into Mn^{3+} in the alkaline duodenal medium; absorption of Mn^{3+} depends upon the intestinal Mn concentration and the level and form of the Mn and organic chelates in the body. Manganese absorption is increased by iron deficiency (Pollack *et al.*, 1965; Diez-Ewald *et al.*, 1968). Excess dietary Ca or P reduces Mn absorption. Absorbed Mn is transported to the liver, conjugated with bile, and excreted into the intestine, where part of the Mn is reabsorbed in the enterohepatic circulation. The lower-valence oxides of Mn are poorly absorbed. Mn absorption from sites of parenteral administration is slow but continuous until it is completely cleared. Soluble Mn chelates are absorbed rapidly. Inhaled Mn salts are deposited in the lungs where slow but continuous absorption takes place, the rate depending on the body Mn reserves. Manganese is continually absorbed from such deposits. Some of the Mn absorbed from the lung is transferred to the gastrointestinal tract (probably by monocytes), and absorbed again. The gastrointestinal tract emerges as the effective route of entry and excretion whether Mn is swallowed or inhaled. Although absorption of cationic Mn by the skin is poor, organic Mn compounds are absorbed rapidly. Trivalent Mn is transported in the blood by a plasma β globulin, transmanganin, whose function is yet to be clearly established.

Turnover of Mn in mammalian tissues is rapid, depending upon exogenous supply. Adrenal corticoids act as ligands, accelerating the turnover of Mn. The absorbed Mn is retained initially by mitochondria in the liver, pancreas, and kidneys. About 40% of the total body content of Mn is retained in the bone marrow, and the rest is in a freely exchangeable, labile pool. Erythrocytes retain Mn in a porphyrin complex. In humans, the liver, pancreas, kidneys, and intestines are specific tissues for Mn retention. There is also a direct relationship between the Mn level and the darkness of hair and conjunctiva in man, dog and other mammals. The uniform Mn concentration in mammals and the lack of accumulation with age is owing to an efficient homeostasis which regulates Mn excretion more than its absorption. Manganese excretion is almost exclusively fecal and involves the liver, pancreas, and intestinal wall. Hepatic–pancreatic excretion is efficient, since one functions if the other is blocked. If Mn is administered as soluble and stable Mn chelates, about 6% is excreted in the urine. Under normal conditions other metal ions do not affect Mn metabolism, suggesting little or no interaction between Mn and other metal ions in biologic tissues. Manganese–vanadium antagonism is known in mammals; V inhibits while Cr and Mn stimulate the hepatic synthesis of cholesterol and fatty acids in rats.

Toxicity

Manganese is the least toxic of the essential metals. The daily intake of Mn by man is about 40% of the body pool; this ratio is much higher than that of most cationic trace metals. Mice, rats, and rabbits can tolerate about 1000 ppm Mn^{2+} in their diet or drinking water.

At concentrations greater than 1000 ppm, Mn is toxic to varying degrees, depending upon the type of Mn ion and its oxidation state. Growth retardation, nonspecific anemia, "metal fume fever," and psychic and neurological disorders associated with "manganism" are some of the symptoms of Mn intoxication. The toxicology and pharmacology of Mn has been reviewed by Stokinger (1963), Cotzias (1958), Fairhill and Neal (1943), and Von Oettingen (1935); the air pollution aspects of Mn and its compounds, and Mn toxicity following inhalation were reviewed by Sullivan (1969). The toxicity of some manganese salts is summarized in Table 7-1. Permanganate appears to be a highly toxic form of Mn because of its high solubility and powerful oxidizing action on the tissues; the permanganate ion is highly toxic when given intravenously, but is less toxic by the subcutaneous route. Divalent Mn^{2+} is about 2.5–3 times more toxic than is Mn^{3+}, but otherwise, higher oxides of Mn are more toxic than the lower

TABLE 7-1. Manganese Toxicity

Compound	Animal	Route	Toxicity	Compound mg	Metal mg	mM	pT
Manganese dioxide	Mouse	sc	LD_{100}	500	316	5.75	2.24
MnO_2	Rabbit	iv	LD_{100}	45	28	0.51	3.29
Manganese fluoride	Guinea pig	oral	MLD	200	118	2.15	2.67
MnF_2	Guinea pig	sc	MLD	700	414	7.54	2.12
Manganese chloride	Mouse	sc	LD_{100}	210	59	1.07	2.97
$MnCl_2$	Guinea pig	sc	LD_{100}	180	50	0.91	3.04
	Rabbit	sc	LD_{100}	180	50	0.91	3.04
	Rabbit	iv	MLD	65	18	0.33	3.48
	Dog	iv	LD_{50}	201	56	1.02	2.99
Manganese perchlorate	Rat	ip	LD_{50}	410	62	1.13	2.95
$Mn(ClO_4)_2 \cdot 6H_2O$							
Manganese sulfate	Mouse	ip	LD_{50}	120	44	0.80	3.10
$MnSO_4$	Mouse	ip	LD_{50}	120	44	0.80	3.10
Potassium permanganate	Mouse	sc	MLD	500	174	3.17	2.50
$KMnO_4$	Rat	oral	LD_{50}	1090	380	6.92	2.16
	Rabbit	iv	LD_{100}	70	24	0.44	3.36
Manganese acetate	Mouse	iv	LD_{50}	16	6.19	0.11	3.95
$Mn(C_2H_3O_2)_2 \cdot 4H_2O$	Rat	or	LD_{50}	3730	837	15.2	1.82

oxides (Levina and Rabachevsky, 1955). The anion of a Mn salt influences the overall Mn toxicity; the citrate is more lethal than the chloride or the sulfate by subcutaneous routes. Manganese acetate appears to be the least toxic among Mn salts when administered orally. Permanganate appears to be the most toxic among the Mn forms when given intravenously, and the MnO_4^- ion is less toxic by the subcutaneous route.

Massive feeding of Mn^{2+} to rats (1.75% of the dry weight of feed) retards growth and causes loss of Ca in the feces due to poor absorption, negative phosphorous balance, and severe ricketts. Anorexia, sexual impotence, and muscular fatigue are other symptoms. Excessive dietary Mn^{2+} (2000 ppm) interferes with the absorption and metabolism of Fe, and consequently reduces hemoglobin formation in calves (Cunningham *et al.,* 1966), lambs (Hartman *et al.,* 1955), rabbits, and pigs (Matrone *et al.,* 1959); Mn acts as an Fe antagonist. Excess Mn^{2+} causes erythrocyte agglutination and hemolysis.

Inhalation of Mn compounds in aerosols or fine dusts produces "metal fume fever". Chronic inhalation of manganese oxides (dust particles of about 3 μm in size) for a few months causes pulmonary pneumonitis in Mn industry workers. This pathologic condition is characterized by the onset of pneumonia, intense dyspnea, high body temperature, acute radiologic signs of lobar pneumonia, and hemorrhage in the lung with little expectoration; manganic pneumonia has a high mortality rate. Inhalation exposure of rabbits to MnO_2 dust at 10–20 mg MnO_2/m^3 levels (particles 5 μm in size) about 4 hours daily for 3 months did not produce penumonitis, but fibrotic changes in the lung occurred along with decreased hemoglobin and erythrocyte levels (Heine, 1944; Stokinger, 1963). Intratrachial injections of $MnCl_2$ and MnO_2 into rats produced characteristic changes in the lung (Davies and Harding, 1949).

Chronic Mn toxicity in humans following chronic exposure to Mn through inhalation, ingestion, or parenteral administration for periods of from six months to two years results in "manganism," a disease of the central nervous system involving psychic and neurological disorders. This disease is reversible if recognized very early and if exposure to dust of Mn oxides is eliminated. However, after the onset of the disease, the neurologic symptoms persist even after the exposure to Mn is removed and the tissue Mn levels revert to normal. A peculiar slapping gait, cramps or tremors of the body and extremities, slurred speech, hallucinations, insomnia, and mental confusion are some of the symptoms; these symptoms resemble those of Parkinson's disease (Cotzias, 1958). Divalent and trivalent Mn act as neurotoxins, inducing degenerative changes in the nervous system. Nephritis, cirrhosis of the liver, anorexia, muscular fatigue, sexual impotence, leukopenia, anemia, and monocytosis are other observed symptoms of manganism. The transitory psychiatric symptoms are followed by neurologic changes including disorders of the extrapyrimidal system. Experimental manganism was induced in monkeys by injection of $MnCl_2$ intraperitoneally for 18 months. The clinical symptoms observed in monkeys were similar to those of man; *viz.*, choreiform movements, rigidity of the muscles, disturbances in motility, and tremors. Respiratory symptoms also characterize Mn poisoning (Arena, 1974). Histologic examination reveals changes in corpus striatum and in the globus pallidus of the lenticular nucleus, degenerative changes in the caudate and striate nucleus, acute hepatitis, and liver necrosis (Mena *et al.*, 1967, Stokinger, 1963). Inadequate renal function is associated with decreased ATPase and increased DNAse and acid phosphatase activities, leading to renal degeneration (Jonek *et al.*, 1965). The efficient homeostatic mechanism fails, with increased retention of Mn (Wassarman and Mihail, 1964). Neuronal degeneration in the cerebral and cerebellar cortex of rats was observed after parenteral administration of $MnCl_2$, and succinic dehydrogenase activity is also decreased (Chandra and Srivastava, 1970). Manganese toxicity decreases the level of catecholamines and serotonin in the corpus striatum, and injures the dopaminergic neurons, for which the biogenic amines act as neurotransmitters.The derangement of amine metabolism is accompanied by a disturbance in functional reaction among these amines, cyclic AMP, and Mn^{2+}. In spite of the severity of the symptoms, chronic Mn intoxication or "manganism" is not a fatal disease, although it causes permanent disability. Intradermal or intramuscular injection of large doses of soluble Mn salts can result in calcified matrix, owing to the reaction of Mn with tissue proteins. Manganese is also a potential carcinogen. Chronic parenteral injections of $MnCl_2$ (0.1 ml of a 1% solution) induced an increased frequency of lymphosarcomas in female mice; tumors were found in tissues far away from the injection site. Leukemia and mammary adenosarcomas were also noted (Paola, 1964). However, manganese maleate inhibits new tissue growth in experimental cancer.

There are no specific detoxication mechanisms for Mn intoxication, unless the efficient Mn homeostasis can be considered to be one. This homeostatic mechanism and the severe gastrointestinal irritation caused by large doses of Mn salts may account for the lack of systemic toxicity following oral administration.

Technetium (Tc)

Technetium (formerly called masurium) does not occur in the earth's crust, but is found in some extraterrestrial material; Tc continues to be made artificially for industrial use. Technetium occurs in a number of radioactive forms, but ^{97}Tc and ^{98}Tc are the common forms; it is a radiation hazard, and must be handled carefully.

The anionic forms of Tc exhibit remarkable inhibition of steel corrosion; 0.5 ppm technate prevents the corrosion of soft iron and mild steel even under water. Technetium^{-95}, a high-γ-ray emitter, is used for scanning the brain, thyroid, and other organs. Chances of exposure to Tc are remote except by accidental mishandling of manmade Tc. It is not an essential nutrient.

Chemistry

Technetium exhibits variable valence of 0, +2, +4, +5, +6, and +7, of which valence states +4 and +7 are common. Technetium differs from manganese in electronic configuration, especially in the outer orbitals, but the two elements are similar in their chemistry and in the stability of their compounds.

Metabolism

Pertechnate containing ^{99}Tc has been used in technetium metabolism studies in rats. The gastrointestinal absorption of pertechnate in rats is very rapid and almost complete; 96% of the oral dose is absorbed in two days (Beasley *et al.*, 1966). However absorption from sites of intraperitoneal injection is not very rapid. Technetium is distributed mainly in the liver and thyroid glands. Technetium resembles rhenium in its transient deposition in the thyroid gland. Retention in other tissues was relatively poor; most of intravenously or intraperitoneally injected doses is excreted in the urine in two or three days.

No reports are available on the involvement of technetium in biochemical systems or on its toxicity. The unavailability of stable Tc isotopes and the high γ-radiation of Tc do not permit extensive studies on inhalation or oral toxicity. If chemically combined with various chelates or carriers, Tc might accumulate in the lungs, kidneys, liver, spleen, or brain.

Rhenium (Re)

Rhenium does not occur free in nature nor as a compound in a distinct mineral species, but it occurs in gadolinite, molybdenite, columbite, and rare earth minerals, and it is found as the sulfide in some molybdenum ores. Naturally occurring Re is a mixture of two stable isotopes. Although Re is widely spread throughout the earth's crust, it occurs only in the extent of about 0.001 ppm. There are no reports about its presence in any plant or animal tissues, and no indication that Re is an essential nutrient.

Rhenium is used in electron-tube and semiconductor applications. Rhenium is an additive in tungsten- and molybdenum-based alloys. Re-W alloy is used in thermocouples to measure temperatures up to 2200°C. Rhenium wire is used in photoflash lamps in photography and as filaments in mass spectrograph and ion gauges. Rhenium is also employed to make mirror backings and high-vacuum equipment. Rhenium catalysts are resistant to poisoning from N, S, and P and are used in the hydrogenation of fine chemicals, and in cracking and fractionating olefins. Rhenium compounds are not used therapeutically. Exposure to environmental Re compounds is minimal, and Re and its compounds are not considered industrial health hazards.

Chemistry

Rhenium is the most electropositive metal in this subgroup and exhibits valence ranging from $1-$ to $7+$; the more common forms are $4+$, $5+$, $6+$, and $7+$. Re forms a series of oxides, Re_2O_3, ReO_2, ReO_3, and ReO_7; the last dissolves in water, forming perrhenic acid ($Re_2O_3 + H_2O \rightarrow 2HReO_4$). The trihalides are stable and water-soluble, the tetrahalides dissolve with decomposition, and the pentahalides are hydrolyzed by water to form oxohalides such as $ReOF_3$; other oxohalides include anionic $ReOF^-_4$. Rhenium does not seem to form any coordination complexes with ligands in biological systems, but forms water-soluble compound salts such as K_2ReCl_6.

Metabolism

The metabolism of Re in mammals has not been investigated in detail, and Re involvement in enzymes is not known. Studies on Re absorption by mammalian tissues are rare, but the level of absorption in the digestive tract appears to be the same as, or greater than, Mn. The solubility and stability of either anionic or cationic Re in the gastrointestinal tract influences its absorption; perrhenates are absorbed better than the lower oxides. When given intraperitoneally to rats, K_2ReCl_6 and $ReCl_3$ are hydrolyzed to Re oxides. Distribution of stable Re in the tissues is little reported. Radioactive Re administered intravenously as $NaReO_4$ to rats accumulates temporarily in the thyroid (Baumann *et al.*, 1949). Transient thyroid accumulation of [183]Re is higher in iodine-deficient rats or in rats rendered goiterous with thiouracil than in controls (Shellabarger, 1956), but thyroid cannot bind the rhenium in an organic form. The bones and liver also retain Re to a small extent. The excretion of soluble Re salts is mainly urinary inrrespective of the mode of administration (Baumann *et al.*, 1949; Durbin, 1960), but the insoluble Re salts not absorbed in the digestive tract are excreted in the feces. The urinary excretion of Re is surprising, since the metabolically similar Mn is excreted in the feces. This could be due to the smaller (radioactive) quantities of Re administered, its slightly higher solubility, or the lack of any specific binding protein for Re.

Toxicity

Rhenium appears to be inert biologically, with low toxicity. The toxicity of $ReCl_3$ and $KReO_4$ is summarized in Table 7-2. Cationic Re is ten times more toxic than the ReO^-_4 anion. The greater toxicity of Re^{3+} was attributed to the free HCl liberated from $ReCl_3$ by hydrolysis. The symp-

TABLE 7-2. Rhenium Toxicity

| Compound | Animal | Route | Toxicity | Dosage/kg body weight | | | |
| | | | | Compound | Metal | | |
				mg	mg	mM	pT
Rhenium trichloride	Mouse	ip	LD$_{50}$	280	177	0.95	3.02
ReCl$_3$	Cat	iv	LD$_{100}$	20	12.6	0.07	4.17
Sodium rhenate	Rat	ip	LD$_{50}$	900	609	3.27	2.49
Na$_2$ReO$_4$							
Potassium rhenate	Mouse	ip	LD$_{50}$	2800	1790	9.61	2.02
K$_2$ReO$_4$	Cat	iv	LD$_{100}$	70	44.7	0.24	3.62
KReO$_4$							

toms of acute Re^{3+} toxicity in rat are sedation, abdominal irritation, and death from cardiovascular collapse. Rhenates in lethal doses cause intense sedation, severe ataxia, tonic convulsions, and cardiovascular collapse (Haley and Cartwright, 1968). In cats Re causes transient hypertension with tachycardia and transient auricular and ventricular fibrillations; lethal doses cause quick cardiovascular collapse. Unlike Mn, Co, and Ge, Re does not produce any erythropoietic effects in rats. Isolated rabbit ileum does not show any adverse effects when suspended in rhenate solutions. There are no reports about Re toxicity in humans.

There is no evidence of any homeostatic mechanisms for Re; the rapid urinary excretion of Re administered in very small doses does not necessarily support the presence of an efficient excretory mechanism for larger quantities.

8

TOXICITY OF GROUP VIII METALS

INTRODUCTION

Iron (Fe), cobalt (Co), and nickel (Ni); ruthenium (Ru), rhodium (Rh), and palladium (Pd); and osmium (Os), iridium (Ir), and platinum (Pt) form three triads of chemically similar metals which constitute Group VIII of the periodic table; the metals are not classified into subgroups A and B. Their valence states range from 2+ to 8+. The atomic weights of the metals in each triad differ only in small units. Iron and Co are considered essential, and Ni is stimulatory and is presumed by some to be essential to mammals, but the metals in the other two triads are not essential. Iron, Co, Ni and Pd are toxic at relatively high doses; Ru and Rh are slightly toxic. Data on Ir toxicity are scarce; inhalation of osmic acid vapor causes pulmonary irritation and congestion. Simple Pt salts are rarely toxic, but compound salts such as chloroplatinates are highly toxic. The carcinogenicity of Co, Ni, Rh, and Pd salts and iron–dextran in mammals have been established. Toxicity of metals within each triad increases with atomic weight.

The atomic structures of Group VIII metals are important factors in determining the stability and reactivity of their complexes, and have a dominant role in the biologic activity of these metal atoms; there are relatively minor differences in their electronic configuration. The electronic configuration of these metals differ in each triad, although the buildup of electrons in each triad is in $3d$, $4d$, and $5d$ orbitals. Ruthenium and Rh differ from Pd in that Pd possesses no electron in the $5s$ orbital. Osmium, Ir, and Pt differ in having 2, 0, and 1 electrons, respectively, in the $6s$ orbital. The second and third triads have empty $4f$ and $5f$ orbitals, respec-

Toxicity of Group VIII Metals

Atomic number	Atomic weight	Name (Symbol)	(Core) Active electrons	Valences
26	55.847	Iron (Fe)	(Ar) 6, 2	2+, 3+
27	58.933	Cobalt (Co)	(Ar) 7, 2	2+, 3+
28	58.71	Nickel (Ni)	(Ar) 8, 2	2+, 3+
44	101.07	Ruthenim (Ru)	(Kr) 7, 1	2+, 3+, 4+, 6+, 8+
45	102.95	Rhodium (Rh)	(Kr) 8, 1	2+, 3+, 4+
46	106.4	Palladium (Pd)	(Kr) 10, 0	2+, 4+
76	190.2	Osmium (Os)	(Xe) 14, 6, 2	2+, 3+, 4+, 6+, 8+
77	192.2	Iridium (Ir)	(Xe) 14, 7, 2	2+, 3+, 4+, 6+
78	195.09	Platinum (Pt)	(Xe) 14, 9, 1	2+, 4+

tively. Although the differences in the electronic structure of these metals are reflected in the formation, configuration, and stability of their coordination compounds, in biologic media these metals, with the exception of Pd and Os, form stable compounds with a coordination number of 6 and an octahedral configuration; the square planar configurations are highly labile. Simple salts such as chlorides, bromides, and sulfates, and complex hexamine and tetramine salts of these metals are water-soluble. Osmium forms no amine complexes. Chloroplatinate and chloroiridate are examples of the stable anions, and the oxy salts of Ru also are quite stable. The ferrous ion is capable of π or back-double-bonding in which the unused $3d$ electrons of Fe are contributed to the empty molecular orbitals in the ligand. This flow of electrons from the metal to the ligand confers extra stability to the bond. Cobalt and nickel are capable of limited π bonding. In these metal–ligand complexes the ligands change the reduction–oxidation potential of Fe and Co but not that of Ni. Ferrous and ferric ions preferentially combine with oxygen rather than nitrogen of ligands, whereas Co^{2+} and Ni^{2+} bind nonionically to sulfur more readily than to oxygen.

The salts of the first triad metals are absorbed better from the gastrointestinal tract than are salts of the second and third triad metals. Efficient homeostatic mechanisms control the absorption and retention of Fe, but the absorption of the salts of other metals is not controlled by homeostasis.

INDIVIDUAL METAL TOXICOLOGY

Iron (Fe)

Iron is the fourth most abundant element in the earth's crust, 5%, and occurs in seawater at about 3.5 ppb. A healthy adult human body contains about 4.5 g Fe, about 70% of which is in hemoglobin, 26% in storage proteins, ferritin, and hemosiderin, 3.5% in myoglobin, and most of the rest in enzymes throughout the body. Although there is no substantial increase in the total body content of Fe with age, some mammalian tissues tend to accumulate Fe. Iron is an essential nutrient for mammals. The average daily intake of Fe by healthy human adults is about 12–15 mg; the daily requirement for absorbed Fe in man depends upon age, sex, and physiologic condition: infants, 0.5–1.5 mg; children, 0.4–1.4 mg; adolescents, 1–2 mg; men, 0.5–1 mg; women, 0.7–2 mg; and pregnant women, 2.4–4.8 mg. Absorption is poor (10–20% of food Fe), but the liver efficiently recycles Fe from hemoglobin catabolism.

The industrial uses of Fe and its compounds are too numerous to be recounted here. Consequently, the chances for Fe intoxication are great. Inhalation of the vapor of Fe metal, alloy, or ore dust, or of compounds such as iron carbonyl causes industrial Fe intoxication. Accidental swallowing of vitamin pills fortified with high doses of Fe compounds is the main cause of Fe poisoning in children. The national clearing house for poison-control centers (U.S.A.) reports over one hundred hospitalizations of children and about ten deaths each year from accidental ingestion of mineral pills. The tendency to raise the recommended daily allowance for iron to 10–20 times the actual requirement and American reluctance to support regulation of mineral pills portends increased morbidity and mortality from ingested iron.

Chemistry

Iron, in both the di- and trivalent states, is stable and important physiologically; under biological conditions Fe exists completely in a non-ionized state. The ferrous ion coordinates very strongly with N- and O-containing ligands and forms stable salts in octahedral coordination complexes with anions. However, bulky groups on the ligands can prevent Fe–N chelation. The availability of Fe in a coordination complex rather than as an ion determines its absorption from the digestive tract and transfer across the mucosal cell. Hemoglobin contains divalent iron coordinated to four heme nitrogens, one to the imidazole nitrogen of histidine, and one to either water or molecular oxygen, but oxygen will not coordinate with the iron of hemoglobin if it is in the trivalent state. However, the free trivalent iron ion has a greater affinity than the divalent ion toward oxygen-containing ligands. Divalent iron can also coordinate with carbon monoxide or cyanide ions. Both molecular oxygen and strong electron receptors can convert Fe^{2+} into Fe^{3+} at both acidic and alkaline pH.

Metabolism

Iron is an essential nutrient for mammals, since Fe is involved in a number of physiologic reactions. Iron deficiency is characterized by anemia, stunted growth, fatigue, lowered resistance to infection, anorexia, and death. Iron is involved in oxygen transport from the lungs to tissues by hemoglobin and in oxygen storage in myoglobin; divalent Fe is a cofactor in heme enzymes such as catalase and cytochrome *c,* and in nonheme enzymes such as aldolase and tryptophan oxygenase. The biochemistry and metabolism of Fe have been extensively reviewed (Moore and Dubach, 1962; Chance and Estabrook, 1966; O'Dell and Campbell, 1971; Underwood, 1971).

Iron absorption by mammals has been studied extensively and excellent reviews are available (Pollack, 1968, Forth and Rummel, 1971, 1973). Iron absorption from the mammalian digestive tract is a mediated transfer across the mucosal epithelium, the rate and amount of absorption depending on the release of iron from mucosal cells into the blood. Absorption is biphasic; an initial rapid phase in the duodenum and jejunum involving 75% of the absorbed Fe, which lasts about 2 hours, is followed by a slow phase in the ileum involving 25% of the absorbed Fe, which continues for many hours. The receptor sites in the mucosa have good specificity for Fe, but they also bind chemically related metals, such as Mn, Co, Ni, Cr, and Zn, which compete with Fe. The Fe-specific acceptors in the intestinal lumen are two Fe-binding proteins of the mucosal cells. One protein, present in the ileum, is similar to ferritin and is heat-stable, with a molecular weight of 150,000 daltons. Another protein, present in the duodenum and jejunum, similar to transferrin, is heat-labile and has a molecular weight of 80,000 (Worwood and Jacobs, 1971; Sheehan and Frenkel, 1972). The amount of intestinal absorption of dietary iron is about 8–10% in normal adults and about 15–20% in iron-deficient subjects. The degree of absorption is influenced by a number of dietary factors, and depends upon the solubility and stability of the compound or chelate. Sulfates, phosphates, phytates, and long-chain fatty acids convert soluble Fe compounds into insoluble forms; alkaline digests convert Fe salts into insoluble hydroxides, and excessive dietary intake of metal ions such as Ca, Zn, Mn, Co, or Mg compete with Fe absorption; decreasing the level of Fe absorption. Short-chain fatty acids, amino acids, citrates, fructose, and ascorbates, which act as low-molecular-weight ligands for Fe, increase Fe absorption.

The Bantus, who cook and brew in iron pots, have a daily intake of 100–200 mg Fe (Bothwell and Finch, 1962), and suffer a high incidence of Fe intoxication; the daily Fe intake of Ethiopians (Hofoander, 1968), who consume Fe-containing cereals, is double that of the Bantus. The Bantus' higher level of Fe intoxication is attributed to the greater absorption of solubilized Fe, since only a small part of the Fe intake of the Ethiopians is absorbed. The effects of proteins on Fe absorption can be either positive or negative; egg proteins retard Fe absorption, as conalbumin forms a stable complex with Fe and is not readily digested, but gastroferrin, an Fe-binding protein first detected in human gastric juice, is associated with the solubilization and absorption of Fe (Davis *et al.*, 1967). The role of the intestinal microflora in the intestinal absorption of Fe in the host mammals is yet to be evaluated fully; some microflora synthesize and excrete suitable chelates into the intestinal tracts of the hosts to bind Fe for eventual absorption by themselves, but it is not known whether these chelates can be utilized by the mammalian host.

Absorption of Fe from sites of parenteral administration depends upon the solubility of the compound, and is fairly rapid. The lungs retain inhaled insoluble Fe compounds, while soluble compounds are readily absorbed into the blood. The Fe-binding protein, transferrin, transports Fe in the blood. Iron is stored intracellularly as ferritin and hemosiderin in the liver, bone marrow, and spleen.

Iron is excreted mostly in feces, and smaller amounts are excreted in urine, sweat, and sloughed cells. Bleeding is another form of Fe excretion. Iron excreted in the feces comes from blood loss into the alimentary tract, unabsorbed ingested Fe, biliary Fe, and desquamated intestinal mucosal cells. The dermal loss of Fe can be appreciable in hot or moist atmospheric conditions. However, the overall rate of excretion of absorbed Fe is very slow and the amount excreted is very low. Mammals retain Fe tenaciously; of 25 mg Fe released by hemoglobin catabolism and 1 mg absorbed dietary iron daily, only a very small amount (1 mg, or 0.02% of the total body content) is excreted.

Homeostasis of Fe metabolism depends mostly on absorption, since very little iron is excreted (Bothwell and Finch, 1962). The regulated absorption is governed by intraluminal, humoral, and mucosal factors which ensure maximal absorptive utilization of dietary iron. The efficiency of this homeostasis increases with the degree of saturation of the mucosa. When excessive soluble Fe is available, absorption increases in direct proportion to the available Fe, resulting in an overload of nontransferrin-bound Fe in the blood.

Toxicity

Iron overload in mammals occurs when the regulatory homeostatic mechanism fails and alters the absorptive capacity of the digestive tract, or when repeated blood transfusions are given (Al-Rashid, 1971). Excessive Fe absorption occurs in (1) "idiopathic" hemachromatosis, (2) excessive intake of soluble dietary Fe or prolonged therapeutic administration, (3) chronic alcoholism with liver cirrhosis and pancreatic insufficiency, and (4) refractory anemia with ineffective erythropoiesis and increased hemolysis. Excess Fe absorbed from the gastrointestinal tract accumulates in parenchymal cells, whereas excess Fe from parenteral administration accumulates in reticuloendothelial cells (Nath *et al.*, 1972); iron accumulation (i.e., hemosiderosis and hemochromatosis) in tissues and serum causes a series of heterogeneous disorders. Spencer (1951), Hoppe *et al.* (1955), Aldrich (1958), Rigby (1971), and Bothwell and Finch (1962) have reviewed the symptoms of Fe intoxication in humans.

Whitten and Brough (1971) reviewed the pathophysiology of acute Fe poisoning. Acute Fe toxicity data are summarized in Table 8-1. In humans, the ingestion of ferrous sulfate in doses ranging from 40 to 1600 mg Fe/kg body weight has resulted in death. Hoppe *et al.* (1955) estimated the fatal dose for a two-year-old or younger child to be 900 mg Fe/kg. The average human lethal dose of ferrous sulfate·has been computed to be about 200–500 mg Fe/kg (Arena, 1970); however, the minimum lethal dose is much lower, since fatalities have occurred at the 100 mg Fe/kg level.

Acute Fe poisoning causes a biphasic shock; initial but rapid increases in the respiration and pulse rates, with congestion of blood vessels, lead to hypotension, pallor, and drowsiness in 6–8 hr. This is followed by prostration, coma, and finally death from peripheral cardiac failure in 36 hr. Chronic Fe poisoning causes hemorrhagic necrosis of the gastrointestinal tract, hepatotoxicity, metabolic acidosis, a greatly prolonged blood-clotting time, and elevation of plasma levels of serotonin and histamine. Dilute solutions of Fe salts are mildly astringent, but at high concentrations there is a marked corrosive action which causes intense mucosal ulceration and necrotizing gastritis at the pyloric end of the stomach.

Dietary ingestion of Fe in toxic doses results in hepatic damage, which causes jaundice by elevating the bilirubin level and inhibiting hepatic cellular enzyme activity; a study in the enzyme histochemistry of the liver suggests that there is inhibition of glucose-6-phosphatase, succinic acid dehydrogenase, and other oxidative enzymes (Witzleben and Chaffey, 1966). Inactivation of Krebs cycle enzymes leads to accumulation of lactic and other acids in blood and tissues, and plasma volume decreases owing to increased capillary permeability.

Intravenous administration of 700 mg Fe/kg as $FeSO_4$ to rabbits causes death within five minutes from thrombus formation with thrombocytopenia (Wilson *et al.*, 1958); at sublethal doses, blood coagulation ceases and a breakdown in the cellular oxidative metabolism of the liver occurs. Metabolic acidosis is induced in dogs and rabbits by an intravenous dose of 10 mg Fe/kg, or by about 150–250 mg Fe/kg through other routes (Reissman and Coleman, 1955). After absorption exclusively in the ferrous form, Fe is oxidized to the ferric state in the blood. When the Fe^{3+} concentration exceeds the binding capacity of transferrin, Fe^{3+} is precipitated in the form of $Fe(OH)_3$, releasing an excess of H^+ ions which can lower the blood pH to 6.7.

$$Fe^{3+} + 3H_2O \rightarrow Fe(OH)_3 + 3H$$

The $Fe(OH)_3$ chelates with nitrogen-containing ligands and is retained in the tissues.

Transferrin, lactoferrin, and conalbumin prevent the uptake of Fe by bacteria, and thereby reduce the virulence and lethal capacity of the

TABLE 8-1. Iron Toxicity

Compound	Animal	Route	Toxicity	Compound mg	Metal mg	Metal mM	pT	Time until death
Ferrous chloride	Rat	oral	LD$_{50}$	984	276	4.94	2.31	6 hr
FeCl$_2$·4H$_2$O	Rat	rectal	LD$_{100}$	591	208	3.72	2.43	48 hr
	Rabbit	oral	LD$_{100}$	890	250	4.48	2.35	2 hr
	Rabbit	rectal	LD$_{100}$	984	276	4.94	2.31	6 hr
	Rabbit	sc	LD$_{100}$	188	53	0.95	3.02	12 hr
Ferric chloride	Mouse	ip	LD$_{100}$	260	53.7	0.96	3.02	
FeCl$_3$·6H$_2$O								
Ferric chloride	Mouse	ip	LD$_{50}$	68	23.4	0.42	3.38	
FeCl$_3$	Rat	oral	LD$_{50}$	900	310	5.55	2.26	
	Rabbit	iv	LD$_{50}$	7.2	2.4	0.04	4.37	Instant
Ferrous nitrate	Rabbit	sc	LD$_{100}$	428	83	1.49	2.83	24 hr
Fe(NO$_3$)$_2$·6H$_2$O								
Ferric nitrate	Rat	oral	LD$_{50}$	3250	449	8.04	2.09	
Fe(NO$_3$)$_3$·9H$_2$O								
Ferrous sulfate	Mouse	oral	LD$_{50}$	1170	430	7.70	2.11	
(anhydrous)	Mouse	ip	LD$_{50}$	100	36.7	0.66	3.18	
FeSO$_4$	Mouse	iv	LD$_{50}$	81	29.8	0.53	3.27	
	Rat	oral	LD$_{50}$	1480	544	9.74	2.01	
Ferrous sulfate	Rat	oral	LD$_{100}$	2130	418	7.48	2.13	28 hr
FeSO$_4$·7H$_2$O	Rat	rectal	LD$_{100}$	1040	209	3.74	2.43	4 hr
	Rabbit	oral	LD$_{100}$	2780	558	9.99	2.00	8 hr
	Rabbit	sc	LD$_{100}$	278	55.8	1.00	3.00	8 hr
	Rabbit	iv	LD$_{100}$	99	19.8	0.35	3.45	8 hr

Compound	Animal	Route						
Ferric sulfate $Fe_2(SO_4)_3 \cdot 6H_2O$	Rabbit	sc	LD_{100}	1070	235	4.21	2.38	Several days
Ferrous ammonium sulfate $FeSO_4(NH_4)_2SO_4 \cdot 6H_2O$	Rat	oral	LD_{50}	3250	639	11.4	1.94	
Iron carbonyl	Mouse	inhal	LD_{100}	7^a	2^a			
$Fe(CO)_5$	Guinea pig	oral	MLD	36	10.3	0.18	3.74	
	Rabbit	oral	MLD	18	5.13	0.092	4.04	
	Rabbit	iv	MLD	17	4.84	0.088	4.06	
Ferrous acetate	Rabbit	sc	LD_{100}	492	158	2.83	2.55	18 hr
Ferrous gluconate	Mouse	sc	MLD	2600	650	11.6	1.93	
$(Fe_2C_{12}H_{22}O_{14})$	Rat	oral	LD_{50}	4600	1150	20.6	1.69	
Ferrous lactate	Mouse	sc	MLD	4200	1000	17.9	1.75	
$Fe(C_3H_5O_3)_2$	Rabbit	sc	MLD	578	138	2.47	2.61	48 hr
	Rabbit	iv	MLD	287	68.5	1.23	2.91	8 hr
Potassium ferricyanide $K_3Fe(CN)_6$	Rat	oral	MLD	1600	271	4.85	2.31	
Sodium nitroprusside or nitro ferricyanide	Mouse	sc	MLD	10	1.87	0.03	4.48	
	Dog	oral	MLD	200	37.5	0.67	3.17	
$Na_3Fe(CN)_6NO \cdot 2H_2O$	Rabbit	sc	MLD	4.8	0.9	0.016	4.79	
	Pigeon	sc	MLD	10	1.87	0.033	4.48	
$Na_3Fe(CN)_6NO$	Mouse	sc	MLD	9.5	2.03	0.036	4.44	
	Rat	oral	MLD	20	4.26	0.076	4.12	
	Cat	iv	MLD	1	0.21	0.0038	5.42	
	Rabbit	oral	MLD	40	8.53	0.15	3.82	
	Dog	iv	MLD	1	0.21	0.0038	5.42	

(Cont'd)

TABLE 8-1. (Cont'd)

| Compound | Animal | Route | Toxicity | Dosage/kg body weight | | | | | Time until death |
| | | | | Compound | Metal | | | | |
				mg	mg	mM	pT		
Sodium nitroprusside or	Mouse	oral	LD$_{50}$	61	13.15	0.24	3.63		
nitro ferrocyanide	Mouse	ip	LD$_{50}$	9.4	2.03	0.036	4.44		
Na$_2$Fe(CN)$_5$NO	Rat	oral	LD$_{50}$	51	11	0.20	3.71		
	Rat	ip	LD$_{50}$	9.4	2.03	0.036	4.44		
	Rabbit	iv	**MLD**	3.04	0.65	0.012	4.93		
Cobalt nitroprusside	Mouse	oral	LD$_{50}$	74	15.1	0.26	3.59		
CO Fe(CN)$_5$NO	Mouse	ip	LD$_{50}$	10.7	2.18	0.037	4.43		
	Rat	oral	LD$_{50}$	147	30	0.51	3.29		
	Rat	ip	LD$_{50}$	15	3.05	0.051	4.29		
	Rabbit	iv	MLD	9.4	1.91	0.032	4.49		
Ferrocene (iron	Mouse	ip	LD$_{50}$	335	100	1.79	2.75		
dicylopenta dienyl)	Rat	oral	LD$_{50}$	1180	354	6.34	2.20		
FeC$_{10}$H$_{10}$	Rat	ip	LD$_{50}$	500	150	2.69	2.57		

[a]Exposure for 4 hr; number of mg/m^3.

bacteria, but ferric ions in toxic doses increase bacterial virulence by saturating serum transferrin, which otherwise enhances serum-β2- and γ-globulin interference with bacterial Fe metabolism (Bullen and Rogers, 1969). Iron accumulates in polymorphonuclear leukocyte lysosomes because of its affinity for the acidic glycoproteins of the lysosomes, for lactoferrin (Baggiolini *et al.*, 1970), and for basic proteins (Zeya and Spitznagel, 1968); the binding of Fe^{3+} to these proteins destroys their bactericidal properties. It should be remembered that although Fe is required for toxin production by some bacteria, excess Fe in culture media inhibits the production of toxins in some bacterial species (Van Heyningen and Gladstone, 1953).

Some Fe compounds are reported to be cocarcinogens. Parenteral administration of iron–dextran complexes induce malignant tumors in rats at the site of injection (Richmond, 1959; Haddow and Horning, 1960). Long-term inhalation of iron oxide results in a mottling of the lungs called siderosis; iron oxide penetrates the bronchial and alveolar walls and enters into lung tissue without extensive damage to the ciliary or mucous barrier. Iron oxides enhance the carcinogenic action of organic carcinogens such as benzpyrene by acting as an inert carrier to transport the carcinogen in high concentration to healthy cells (Stokinger and Coffin, 1968; Saffiotti *et al.*, 1968). The only known specific detoxication mechanism for Fe is the accumulation of particulate Fe compounds in lysosomes.

Cobalt (Co)

Cobalt occurs in the earth's crust at 23 ppm and in seawater at 0.1 ppb; soil may contain up to 100 ppm. Cobalt is also present in plant and animal tissues, and the human adult body contains about 1.2 mg of cobalt; 14% of it held in bone, 43% in muscle, and the remainder in other soft tissues. About 100 μg is in the form of vitamin B_{12}. Traces of Co have been detected in newborn infants, and the body content of Co does not increase with age. Cobalt is an essential nutrient.

Cobalt is used extensively in making high-speed tools; high-resistance Co alloys are used in jet engines, rocket nozzles, and gas turbines. Its magnetic properties are useful in making permanent magnets for high-fidelity speakers. Industrially, Co compounds are used as pigments in glass and porcelain, as catalysts, and as binders in tungsten carbide tools. Poisoning can occur from the inhalation of Co-alloy vapors in industry and from the therapeutic use of Co compounds in treatment of anemia and of radioactive Co in cancer.

Chemistry

Cobalt has two normal valence states; in solution Co^{2+} is the more stable, while Co^{3+} is the more powerful oxidizing catalyst. The coordination numbers of Co complexes involving nitrites, cyanides, and halides with planar or octahedral geometry are 4 and 6 for Co^{2+}, and 6 for Co^{3+}. Biologically active Co chelates form the corrinoid structure of cyanocobalamine. Both Co^{2+} and Co^{3+} have a special affinity for ligands in which oxygen and nitrogen are donor atoms; but the relative affinity of chelated cobalt toward ligands changes depending on the variations in reduction–oxidation potential of Co.

Metabolism

Cobalt is an essential micronutrient for mammals, but except for its role in vitamin B_{12}, the function of cobalt is unknown. However, glycylglycine dipeptidase contains Co in a bound form. Unlike other transition metals, Co is not required for mammals either as ions or in loosely bound organic forms. Cobalt deficiency leads to stunted growth, anemia, and death. Deficiency diseases of cattle and sheep in Co-deficient areas of the world are well documented, but it is very difficult to develop Co-deficiency symptoms in laboratory animals since Co is readily stored, is ubiquitous in nature, and is required in such small quantities (1 μg vitamin B_{12}/day for laboratory rodents). A daily intake of about 50 μg cobalt (of which 40 μg is in the form of vitamin B_{12}) maintains cobalt equilibrium in the human. Ruminants utilize Co directly, since ruminant bacteria use Co to synthesize vitamin B_{12}.

The biochemistry and metabolism of Co has been extensively reviewed (Grant and Root, 1952; Schroeder *et al.,* 1967a; Underwood, 1971, 1975; O'Dell and Campbell, 1971). Divalent Co ions activate a number of enzymes *in vitro,* and Co can replace Zn in bovine carbonic anhydrase to form an active enzyme. Corrinoid compounds lacking cobalt are metabolic inhibitors of vitamin B_{12}.

The efficiency of absorption of small quantities of Co from the gastrointestinal tract is greater than that for other metals of Group VIII, and depends on the amount given; very small doses, such as 30 or 50 μg/kg are readily absorbed, while larger doses are poorly absorbed. Cobalt is not sequestered; it is stored in the intestinal mucosa and subsequently lost through normal desquamation of the epithelium. Since Co forms no insoluble complexes in neutral and alkaline media, absorption occurs both in proximal and distal regions of the intestine. The major, if not only form of Co absorbed by ruminants, is vitamin B_{12}. Iron and cobalt seem to share at least part of the same mediated transport across rat intestinal mucosa (Valberg, 1971); increased Co absorption in Fe deficiency and mutual inhibition of the absorption of Fe, Co, Mn, and Zn in anemic rats are reported (Schade *et al.*, 1970). The preferential binding of the Fe-binding proteins of rat intestinal wall is Fe > Co > Ni > Mn > Zn (Forth and Rummel, 1971).

Small doses of parenterally injected Co salts are absorbed slowly from injection sites, while intravenously injected soluble Co salts are bound to α globulins and are distributed to the tissues quickly. Inhaled metallic powder and salts are retained in the lungs and subsequently absorbed slowly. Cobalt salts are retained primarily in the bone marrow, pancreas, liver, spleen, kidneys, and heart.

The excretion of Co is mainly urinary; chelated and coordinate Co complexes and injected Co salts are excreted principally in the urine, but about 20% of injected Co is excreted in the feces. Fecal Co is mainly the unabsorbed portion from the digestive tract plus the Co which passes into the intestines by desquamation and via the enterohepatic pathway. When the level of Co absorption exceeds that of its excretion, its toxicity is cumulative. Homeostatic regulation of Co is effective in both absorption and excretion, and excess Co is eliminated from mammals as a Co-histamine complex (Hatem 1958); this excretion can serve for both detoxication and homeostasis.

Toxicity

Acute Co poisoning by oral ingestion in rats causes diarrhea, paralysis of the hind limbs, and lowering of blood pressure and body temperature prior to death. Also observed are congestion of all organs, with small focal hemorrhages on serosal surfaces, large hemorrhages in the liver and adrenals, alveolar thickening in the lungs, and renal and pancreatic degeneration. Inhalation of metallic Co or intratracheal injection of Co salts causes acute lung irritation, edema, and hemorrhage, hyperplasia of bone marrow, massive pericardial effusion, and extensive loss of fluids from the capillaries into the peritoneal cavity. Chronic Co poisoning produces polycythemia, hyperplasia of the thyroid gland, pericardial effusion, damage to α cells of the pancreas, and cancer in rats and humans. The toxicity of Co is summarized in Table 8-2.

The gastrointestinal absorption of Co evidently influences its toxicity. Mammals tolerate low levels of ingested water-soluble Co salts without any toxic symptoms; safe dosage levels include: in man, 2–7; in rats, 25–200; in sheep, 3–10, and in cattle, 1–20 mg Co/kg body weight per day (Schroeder *et al.*, 1967*a*). Children are more susceptible than adults; cobalt toxicity is

TABLE 8-2. Cobalt Toxicity

Compound	Animal	Route	Toxicity	Compound (mg)	Metal (mg)	Metal (mM)	pT
Cobalt metal	Mouse	ip	LD_{100}		150	2.54	2.59
	Rat	oral	MLD		1500	25.5	1.59
	Rat	im	MLD		112	1.90	2.72
	Rat	ip	LD_{100}		375	6.36	2.20
	Rat	it[a]	MLD		25	0.42	3.37
	Rat	iv	LD_{100}		100	1.70	2.77
	Rabbit	oral	MLD		700	11.9	1.93
	Rabbit	iv	LD_{100}		20	0.34	3.47
Cobaltous oxide	Mouse	im	MLD	800	629	10.7	1.97
CoO	Rat	oral	LD_{50}	1700	1337	22.7	1.64
Cobaltic oxide	Rat	ip	LD_{100}	703	500	8.48	2.07
Co_2O_3							
Cobalt chloride	Mouse	oral	LD_{50}	80	36.3	0.62	3.21
$CoCl_3$	Mouse	sc	MLD	54.5	24.7	0.42	3.38
	Rat	oral	LD_{50}	180	81.7	1.39	2.86
	Rat	sc	LD_{100}	66	30	0.51	3.29
	Rat	iv	LD_{50}	20	9.08	0.15	3.81
	Rabbit	sc	MLD	109	49.5	0.84	3.08
	Rabbit	iv	LD_{100}	13.8	6.26	0.11	3.97
	Guinea pig	oral	LD_{50}	55	25	0.42	3.37
	Cat	iv	LD_{50}	27	12.3	0.21	3.68
	Dog	iv	LD_{100}	16.5	7.5	0.13	3.90
Cobalt nitrate	Rabbit	oral	MLD	250	80.5	0.37	2.86
$Co(NO_3)_2$	Rabbit	oral	LD_{100}	400	129	2.19	2.66
	Rabbit	sc	LD_{100}	75	24.1	0.41	3.39
Cobalt sulfate	Mouse	ip	LD_{50}	54	20.5	0.35	3.46
$CoSO_4$							
Cobalt sulfate	Dog	iv	LD_{100}	16.2	3.4	0.058	4.24
$CoSO_4.7H_2O$							
Cobalt carbonyl	Rat	inhal	LD_{100}	1400[b]			
Cobalt fluoborate	Rat	oral	MLD	500	126	2.14	2.67
CoB_2F_8							
Cobalt acetate	Mouse	iv	LD_{50}	28	9.3	0.16	3.80
$(CH_3CO_2)_2Co$							
Cobalt acetate	Rat	oral	LD_{50}	821	194	3.29	2.48
tetrahydrate							
Cobalt nitroprusside	Mouse	oral	LD_{50}	74	16	0.27	3.57
$CoFe(CN)_5NO$	Mouse	ip	LD_{50}	10.7	2.3	0.039	4.41
	Rat	oral	LD_{50}	147	31.6	0.54	3.27
	Rat	ip	LD_{50}	15	3.2	0.054	4.27
	Rabbit	iv	MLD	9.4	2.02	0.034	4.47

(Cont'd)

TABLE 8-2. (Cont'd)

| Compound | Animal | Route | Toxicity | Dosage/kg body weight | | | |
| | | | | Compound | Metal | | |
				mg	mg	mM	pT
Cobaltocene	Mouse	ip	LD_{50}	80	24.9	0.42	3.37
$CoC_{10}H_{10}$	Rat	ip	LD_{50}	125	38.8	0.66	3.18
Dicobalt EDTA	Mouse	iv	LD_{50}	50	14.5	0.25	3.61
$Co_2C_{10}H_{12}N_2O_8$	Rat	ip	LD_{50}	41	11.9	0.20	3.70
	Rat	iv	LD_{50}	43	12.5	0.21	3.67
Cobalt sodium EDTA $CoC_{10}H_{12}N_2O_8Na_2$	Mouse	ip	LD_{50}	1950	291	4.94	0.231

[a]Intratracheal.

[b]Exposure of 4 hr; number of mg/m^3.

observed in children given <1 mg Co/kg body weight per day for anemia therapy (Little and Sunica, 1958; Caplan and Curtis, 1961). A single oral dose of 150 mg $CoCl_2$ in man is harmless except for polycythemia; however, 500 mg $CoCl_3$ causes serious toxic manifestations. Young rats are unable to survive repeated 30-mg daily doses of Co metal powder in the diet for a month (total dosage of about 900 mg), whereas they tolerate 1250 mg of the metal in a single dose. Tolerance to Co toxicity can be developed experimentally in rats; rats given subcutaneous injections of 2.5 mg Co (as $CoCl_2 \cdot 6H_2O$) daily for two weeks, can tolerate a normally lethal dose of 250 mg Co. Sheep tolerate daily doses of 6 mg Co/kg as a soluble salt in their diet for eight weeks, but a single dose of 300 mg Co/kg is lethal, and occasionally morbidity is observed with 40–60-mg doses. After absorption, Co may undergo biotransformation to a toxic form, which could explain the toxicity of small doses of ingested Co compounds. When this biotransformation mechanism becomes inoperative, the animals develop Co tolerance.

At toxic levels, Co salts produce polycythemia in all species, which is noted with oral doses of about 10 mg Co (as $CoCl_2$)/kg daily for two weeks, the exact dosage depending on the species and animal conditioning. In cattle polycythemia occurs in calves prior to rumen development but not in mature animals. Polycythemia is primarily due to the stimulatory action of Co on red bone marrow and on extramedullary hemopoietic tissue in other organs. Hyperplasia in bone marrow results in poorly developed reticulocytes with impaired respiratory function and the release of immature erythrocytes. In the liver, Co is reported to interfere with the formation of metabolic precursors which are later required for hemoglobin formation (Warren *et al.*, 1944). Excess Co evidently inhibits Fe absorption and blocks carriers involved in Fe transport. Mild Co toxicity causes hyperglycemia in dogs because of transient and reversible damage to the α cells of the pancreas (Lazarus *et al.*, 1953). A goitrogenic effect in humans is observed after the therapeutic administration of 3–4 mg Co as $CoCl_2$ daily for three weeks. Aerosols containing ionizable cobalt compounds such as $CoCl_2$, Co(Co–EDTA), or Na_2(Co–EDTA) cause pulmonary edema in rats; the further distribution of Co compounds to tissues depends upon its ionization (Hobel *et al.*, 1972). In rabbits, Co salts cause cardiotoxicity (Hall and Smith, 1968).

Fatal heart disease among heavy beer drinkers was attributed to the cardiotoxic action of Co salts which were formerly used as additives to improve foaming; these levels of Co beer-additives were much lower than doses tolerated in other conditions, such as the therapeutic intake of 300 mg Co/day in the treatment of anemia. Bottled beer contains sulfur dioxide which destroys thiamine. Thiamin and protein deficiencies aggravate the cardiotoxicity of Co. Thiamine deficiency was found to be more potent than protein deficiency in aggravating Co-induced heart lesions in rats (Grice *et al.*, 1969*a,b*). However, oral ingestion of proteins such as casein (rich in S) or dried-pea protein (low in S) reduced the toxic effects and cardiac levels of Co; wheat gluten, a protein low in titrable amino groups and adequate in cysteine and methionine, did not reduce the cardiotoxicity of Co. Thus, the binding capacity of the protein with Co depends on the availability of free amino groups rather than thiol groups (Wiberg *et al.*, 1967, 1969).

Ethanol sensitizes animals to Co intoxication. Cobalt and alcohol are synergistic in their toxicity in rats (Alexander, 1969; Derr *et al.*, 1970); the symptoms include growth retardation and reduction in heart weight following a 35-day ingestion of water which contained 1 ml ethanol and 1 mg Co per 10 ml water. Guinea pigs also show this Co–ethanol synergism by growth retardation and myocardial fatty degeneration (Desselberger and Wegener, 1971).

Cobalt ions depress oxygen uptake in heart mitochondria by inhibiting the enzymes α-ketoglutarate dehydrogenase and pyruvate dehydrogenase. Cobalt–amine complexes activate the serum inhibitor of hyaluronidase. The heart is susceptible to Co toxicity owing to its ability to accumulate Co ions, combined with its dependence on oxidative metabolism for its high energy requirements (Wiberg, 1968). The cardiotoxicity of Co is confirmed by structural changes in rat heart myofibrils and mitochondria produced by intraperitoneal injections of 5 mg Co/kg (in salt form) daily for 11 days (Grice *et al.*, 1969*a,b*; Lin and Duffy, 1970).

The toxic action of Co in tissues is attributed to the inactivation of thiol groups. For example, the Co inhibition of α-ketoglutarate dehydrogenase is caused by a complexing of Co with thiol groups of α-lipoic acid, a cofactor in these enzyme systems (Heggtveit *et al.*, 1970). Oral administration of sulfur-containing amino acids and histidine reduces cobalt toxicity by forming non-ionized, soluble chelates which protect thiol-group-containing cofactors and enzymes (Orten and Bucciero, 1948; Grant and Root, 1952; Derr *et al.*, 1970). Note that this observation about sulfur-containing amino acids is at variance with the observation that amino acids with a greater number of free amino groups reduced cobalt-induced lesions in rats. Excess Co is reported to block carriers involved in iron utilization, and thereby reduce erythrocyte formation (Goodman and Gilman, 1970). Plasma kininogen, which serves as either substrate or precursor for the production of erythropoietin, is significantly elevated following subcutaneous injection of cobaltous chloride in rats (Smith and Cantrera, 1972).

The carcinogenicity of Co in rats has been verified; there is no evidence for the involvement of dietary Co in carcinogenesis in mammals, but parenteral injection of oxides and sulfides of cobalt causes cancer in experimental animals (Gilman and Heath, 1956; Ruckerbauer, 1962; Sunderman, 1972). Fibrosarcomas appeared in rats six months after an injection of a dust commonly present in nickel refineries, containing cobaltous oxide or sulfide. Tumors were produced in the thyroid gland and at the site of injection of Co-metal powder (Weaver *et al.*, 1956). Transplantable liposarcomas were produced by injection of Co metal dust in rabbits, and hyperplasia of adipose tissue was also observed (Thomas and Thiery, 1953).

No direct evidence for Co homeostasis is available, but the presence of Co-histamine complexes in urine suggests a possible detoxication mechanism. When intravenously injected, insoluble Co compounds are phagocytized by macrophages, and excreted in residual vacuoles.

Nickel (Ni)

Nickel occurs widespread in nature, and is present in the earth's crust at 80 ppm, and in seawater at 2–5 ppb. The earth's core contains 8.5% Ni. Nickel occurs in very minor quantities in all types of coal; the Ni content of urban atmosphere (0.15–0.16 $\mu g/m^3$) is twice that in nonurban atmospheres. The level of occurrence of Ni in some food stuffs is: fresh oysters, 1.50 ppm; wheat or buckwheat, 3–5 ppm; rye, 2.70 ppm; gelatin, 4.5 ppm; cocoa, 5 ppm; tea, 7.6 ppm; baking powder, 13 ppm (all on a fresh-weight basis); and cabbage, mushrooms, and kidney beans, about 3 ppm (dry-weight basis). The human adult body contains about 10 mg Ni, about 18% present in the skin and the rest distributed in other tissues. Human fetuses contain traces of Ni. Nickel does not accumulate with age in humans (Tipton and Cook, 1963). It is considered to be an essential nutrient by some nutritionists (Nielsen, 1971; Nielsen *et al.,* 1975), and a hormetic by others (Luckey, 1975*b*).

Nickel is extensively used in electroplating, in the manufacture of steel and other alloys, and in the manufacture of electronic devices. Colored-glass and ceramic industries utilize large quantities of Ni. Finely ground Ni is used as catalyst in the hydrogenation of oils. Nickel salts are considered an industrial health hazard because many Ni compounds are released into the atmosphere during mining, smelting, and refining operations. It is not profitable to recover Ni; hence, coal-fired power plants release Ni compounds into the atmosphere. Humans and animals are exposed to an environmental health hazard as Ni compounds from the soil, water, and atmosphere find their way into human food from plants. The toxicity of nickel is considerably less serious than its carcinogenicity; nickel toxicity for mammals is low and is similar to chromium. Nickel poisoning can occur industrially following inhalation of gaseous nickel carbonyl and nickel oxide or sulfide dusts.

Chemistry

Most simple salts of Ni are water-soluble. Divalent Ni salts are more stable than are trivalent salts. Nickel forms square planar and tetrahedral complexes with coordination numbers 4 and 6, common to essential metals. Nickel complexes involving nitrogen and oxygen ligands are more stable than are similar complexes of all other transitional essential metals except copper. Nickel has little affinity for sulfur ligands, but it reacts more strongly with amino acids and serum albumin than does copper.

Nickel is found in normal human and rabbit serum in three forms: ultrafiltrable Ni, albumin-bound Ni, and a Ni metalloprotein, "nickeloplasmin." The presence of Ni in RNA from different sources indicates a possible structural role for Ni, perhaps in the maintenance of the configuration of RNA. Nickel is present in human tissues in a strongly chelated form; repeated injections of EDTA do not change Ni excretion.

Nickel can bind to the pyrophosphate and the pyrimidine base of thiamine pyrophosphate, and divalent Ni reacts with all active forms of vitamin B_6. Nickel forms paramagnetic octahedral complexes and diamagnetic square planar complexes with triglycine and complex peptides. Nickel can also complex with porphyrins and related compounds.

Metabolism

There is some evidence that Ni is an essential micronutrient for animals, particularly for chickens (Nielsen 1971). Nickel satisfies several of the criteria established for essential trace metals and also occupies a position (atomic number 28) amidst a row of essential metals in the periodic chart; Nielsen *et al.* (1975) report specific Ni-deficiency symptoms in mammals. Ni may be either an essential nutrient or a hormetic (Luckey, 1975*b*). It is involved in *in vitro* activation and inhibition of several enzyme systems; urease contains firmly bound Ni (Dixon *et al.*, 1975). The involvement of Ni in enzyme activation, hormone action, structural stability of biological macromolecules, and general metabolism has been well reviewed (Schroeder *et al.*, 1962; Underwood, 1971, National Academy of Sciences, 1975). The daily intake of Ni by humans is 0.3–0.6 mg/day, most coming from plant sources as the Ni content of meat is very low.

The level of absorption of Ni salts from the gastrointestinal tract is about 3% of the dietary Ni. This is less than that of Co and Fe, although all three metals seems to share the same transport mechanism across the intestinal wall. Nickel in drinking water is absorbed better from the digestive tract than is the firmly bound Ni in foods. Most Ni found in ingested food is not absorbed and is excreted in the feces. The removal of injected soluble Ni salts from the peritoneal cavity and intramuscular sites is rapid, but insoluble salts are cleared slowly. The lungs retain inhaled Ni salts, especially those from cigarette smoke, and their removal from the pulmonary tissues is low. Approximately 50% of inhaled Ni dust is deposited on the bronchial mucosa and swept upward in mucous to be swallowed, about 25% is exhaled, and the rest is deposited in the pulmonary parenchyma. Gaseous Ni compounds, such as nickel carbonyl, are deposited in much larger proportions. Intravenously injected Ni salts disappear quickly from the blood, indicating widespread distribution in the tissues (Smith and Hackley, 1968). In spite of the presence of firmly chelated Ni in human tissues, retention of newly acquired Ni in the tissues is transient and poor. However, under certain pathological conditions, such as gastric ulcers, thyrotoxicosis, and cancer, increased amounts of Ni are found in the blood and neoplastic tissues.

Excretion of ingested Ni compounds is mainly fecal, with only about 10% in the urine; this pattern of excretion is noted both in dogs (Tedeschi and Sunderman, 1957) and in humans (Horak and Sunderman, 1973). Excretion of absorbed Ni and intravenously administered Ni compounds is primarily urinary (about 60%) and the rest fecal, presumably from the bile, which indicates an enterohepatic transfer. Inhaled nickel carbonyl is excreted primarily in the urine (Sunderman and Selin, 1968).

The existence of an efficient homeostatic mechanism for Ni metabolism is suggested by the limitation in the intestinal absorption of Ni from the large amounts of Ni present in food. Serum Ni levels are maintained within relatively narrow ranges in 16 different animal species (Nomoto *et al.,* 1971), which suggests the existence of an active excretory mechanism for Ni in the kidneys (Mertz *et al.,* 1970), and substantiates Ni homeostasis.

Toxicity

Nickel is relatively nontoxic, ranking with iron, cobalt, copper, chromium, and zinc. The content of Ni in the normal human body is about 1% of the toxic oral dose of inorganic salts. Nickel metal is less toxic than its soluble salts. Acute Ni toxicity causes severe gastroenteritis following both oral and parenteral administration. Nervous symptoms such as tremor, chorea-like movements, and paralysis occur prior to death, which occurs mostly from heart failure. Toxicity data for Ni are summarized in Table 8-3. Chronic toxicity results in degenerative changes in heart muscle, brain, lung, liver, and kidney tissues, and can result in cancer. Inhalation of nickel carbonyl causes respiratory tract neoplasia and myocardial infarction; intraalveolar hemorrhage, edema, and exudates are noticed prior to neoplasia. Perivascular leukocytosis and neuronal degeneration occur in the brain.

Exposure to industrial Ni dusts causes Ni dermatitis. Sensitivity to Ni may be exhibited from skin contact and even internal prostheses (Symeonides *et al.,* 1973). The degree of one's sweating exerts a profound influence on the degree of dermatitis in Ni-sensitive persons because Ni is leached from Ni-plated objects worn by the individuals. Divalent Ni ions can penetrate the skin at sweat-duct and hair-follicle ostia, and bind with keratin. Contact dermatitis thus results from the permeability of the human dermis and epidermis to Ni (Matten *et al.,* 1969). Eczematous dermatitis occurs from internal exposure to Ni from stainless steel screws in the patella (Barranco and Soloman, 1972), and urticaria can occur after fixation of a humeral fracture using Vitallium plates (Symeonides *et al.,* 1973).

Earlier studies gave conflicting evidence about the oral toxicity of Ni. Nickel metal fed daily to cats and dogs at 4–12 mg/kg body for 200 days did not affect their health, and dogs could tolerate 1–3 g Ni/kg without any untoward effect. Rodents can tolerate inorganic Ni salts. Ingestion of nickel (as Ni_2SO_4, or nickel soap), up to 1000 ppm, did not affect growth or reproduction in the mouse, rat, chicken, or monkey (Phatak and Patwardhan, 1950, 1952), but growth retardation in mice was reported when they were fed more than 1000 ppm nickel (as nickel acetate). Growth retardation was attributed to the reduced activities of liver cytochrome oxidase and

TABLE 8-3. Nickel Toxicity

| Compound | Animal | Route | Toxicity | Dosage/kg body weight | | | | | Time until death |
| | | | | Compound | Metal | | | | |
				mg	mg	mM	pT		
Nickel (spongy or	Rat	im	MLD		25	0.43	3.37		—
colloidal	Rat	it	MLD		12	0.20	3.69		—
	Guinea pig	oral	MLD		5	0.086	4.07		—
	Dog	iv	MLD		10	0.17	3.77		—
Nickel (refinery dust)	Mouse	im	MLD		800	13.6	1.87		—
Nickel oxide	Rat	im	MLD	180	141	2.40	2.62		—
NiO	Cat	iv	MLD	10	7.8	0.13	3.88		—
	Dog	iv	MLD	7	5.5	0.094	4.03		—
Nickel fluoride NiF_2	Mouse	iv	LD_{50}	130	112	1.91	2.72		—
Nickel chloride $NiCl_2 \cdot 6H_2O$	Dog	iv	LD_{50}	40	10.4	0.18	3.75		—
Nickel chloride	Rat	ip	LD_{50}	11	5.0	0.086	4.07		—
(anhydrous) $NiCl_2$	Dog	iv	MLD	10	4.5	0.076	4.12		—
Nickel perchlorate $Ni(ClO_4)_2 \cdot 6H_2O$	Mouse	ip	MLD	100	16.05	0.27	3.56		—
Nickel nitrate $Ni(NO_3)_2 \cdot 6H_2O$	Rat	oral	LD_{50}	1620	242	4.12	2.38		—
Nickel sulfate	Mouse	ip	LD_{50}	34	8	0.14	3.87		—
$NiSO_4 \cdot 6H_2O$	Guinea pig	sc	LD_{100}	62	13.8	0.24	3.63		—
	Rabbit	sc	LD_{100}	500	112	1.91	2.72		—
	Rabbit	iv	MLD	35.8	8	0.14	3.87		—
	Cat	iv	MLD	71.6	16	0.27	3.56		—
	Dog, rat	sc	LD_{100}	500	112	1.90	2.72		—
	Dog, rat	iv	MLD	89.5	20	0.34	3.47		—

(Cont'd)

TABLE 8-3. (Cont'd)

| Compound | Animal | Route | Toxicity | Dosage/kg body weight | | | | | Time until death |
| | | | | Compound | Metal | | | | |
				mg	mg	mM	pT		
Nickel fluoborate	Mouse	inhal	MLD	0.53[a]					—
NiB_2F_8	Rat	oral	MLD	500	126	2.15	2.67		—
Nickel carbonyl	Mouse	inhal	LD_{50}	0.07[a]	0.024[a]				30 min
$Ni(CO)_4$	Rat	inhal	LD_{50}	0.24[a]	0.083[a]				—
	Rat	ip	LD_{50}	39	13.4	0.23	3.64		—
	Rat	sc	LD_{50}	63	21.6	0.37	3.43		—
	Rat	iv	LD_{50}	66	22.6	0.38	3.41		—
	Cat	inhal	LD_{50}	1.89[a]	0.65[a]	0.011	4.96		—
Nickel hexafluoro-	Rat	oral	MLD	50	14.6	0.25	3.60		—
silicate	Rat	oral	LD_{50}	500	155	2.64	2.58		—
$NiSiF_6$									
Nickelocene	Rat	ip	LD_{50}	50	15.5	0.26	3.58		—
Nickel acetate	Mouse	oral	LD_{50}	410	136	2.32	2.64		—
$Ni(C_2H_3O_2)_2$	Mouse	ip	LD_{50}	32	10.6	0.18	3.74		—
	Rat	oral	LD_{50}	350	116	1.98	2.70		—
	Rat	ip	LD_{50}	23	7.64	0.13	3.89		—
	Rat	im	MLD	50	16.6	0.28	3.55		—

[a]Exposure of 4 hr; number of mg/m^3.

isocitrate dehydrogenase, kidney malic acid dehydrogenase, and heart cytochrome oxidase and malic acid dehydrogenase (Weber and Reid, 1969*b*). Calves fed 1000 ppm nickel show lowered retention of total nitrogen, calcium, and phosphorus, and decreased food intake and growth (O'Dell *et al.*, 1970*a,b,* 1971). Hyperglycemia, decreased liver glycogen, and increased muscle glycogen were found in rabbits fed 0.5 mg Ni as $NiCl_3$ per day for 5 months (Gordynya, 1969). The growth rate is reduced in chickens fed 900–1300 ppm Ni as either the acetate or sulfate.

The reproductive capacity of male rats is impaired by reduced spermatogenesis and sperm motility following an oral intake of 23 mg $NiSO_4$/kg daily for three months (Waltschewa *et al.*, 1972), and other changes noticed were smaller testes, proliferation of connective tissue in the interstitial region, and increased interstitial cells. The number of litters and the survival of the pups decreased with a dietary intake of 1600 ppm Ni (Weber and Reid, 1969*b*); similar effects on rat reproductive processes were observed upon subcutaneous administration of $NiSO_4$ (Hoey, 1966). Shrinkage of the central tubules, hyperemia of intertubular capillaries, and disintegration of spermatozoa were observed 18 hr after a single dose of 2.3 mg $NiSO_4$/kg, but these effects are reversible when the intake is reduced to physiologic levels.

Nickel salts are highly toxic when administered intravenously or subcutaneously. Gastroenteritis, tremors, and paralysis occur in dogs and guinea pigs immediately following intravenous administration of sublethal doses; hyperglycemia develops, but can be prevented by administration of insulin. The localization of injected Ni salts in the pituitary gland and the subsequent decreased serum level of prolactin suggest that divalent Ni exerts a direct specific inhibitory action on prolactin-secreting cells in the anterior pituitary (LaBella *et al.*, 1973). Injected $NiSO_4$ decreases the activity of several enzymes, and changes the distribution of serum LDH isoenzymes.

Intoxication following inhalation of nickel compounds varies according to the nature of the compounds; nickel carbonyl is the most toxic, while dusts of nickel oxide, chloride, and sulfate are less toxic. Inhalation of the less toxic nickel oxides and chlorides produces pneumonitis and increases the number of alveolar macrophages and the viscosity of pulmonary washings (Bingham *et al.*, 1972).

Nickel carbonyl causes the same toxic symptoms whether administered by inhalation, intravenously, or parenterally. Immediate lethal toxicity occurs at 30 ppm in the air. An 8-hr exposure to 0.001 ppm $Ni(CO)_4$ in the air is also toxic, producing severe pneumonitis. Chronic exposure to very low levels of $Ni(CO)_4$ produces squamous metaplasia of the bronchial epithelium and carcinoma of the upper respiratory tract and lungs (Sunderman *et al.,* 1959). Nickel carcinogenicity from tobacco smoke is attributed to Ni(CO)₄ (Sunderman and Sunderman, 1961).

Epidemiologic studies among nickel refinery workers (National Academy of Sciences, 1975) and experimental studies in laboratory rodents have established Ni carcinogenesis in man and animals. Respirable particles of Ni, Ni subsulfide, nickel oxide, and vapors of nickel carbonyl are primarily responsible for Ni carcinogenicity. The highest risk of mortality from cancer of the respiratory tract is found among nickel mine workers involved in roasting, smelting, and electrolysis. Nickel carcinogenesis has been established following parenteral administration or inhalation in several species of animals. When administered parenterally, finely ground or colloidal Ni metal causes cancer of the bone, connective tissue, nerve tissue, and muscle in rats, rabbits, and guinea pigs. Malignant tumors are also induced at the site of injection (Heath and Daniel, 1964*b*). Nickel sulfide causes rhabdomyosarcomas and adenocarcinomas in rats and cats following intramuscular injection or after implantation of nickel sulfide disks. Pulmonary carcinomas have been induced in rodents following the inhalation of nickel dust (particle size 4 μm) (Hueper, 1958*a,b*), and following nickel carbonyl inhalation (Sunderman *et al.,* 1959). Multiple intravenous injections of nickel carbonyl to rats causes carcinomas and sarcomas in various organs, including the liver and kidneys (Lau *et al.,* 1972). Carcinogenic synergism between Ni compounds and polycyclic aromatic hydrocarbons, such as benzpyrene, has been established in rats (Maenza *et al.,* 1971). Gilman (1966) developed a simple, convenient, and reproducible method of Ni carcinogenesis in rodents, establishing unequivocally that Ni is carcinogenic.

The latent period for induction of lung cancer in laboratory animals by inhalation of nickel carbonyl is about 2 years. Nickel carbonyl is able to cross cell membranes without hydrolysis because it is lipid-soluble; this ability to penetrate cellular barriers and to release Ni ions is responsible for its high toxicity and carcinogenicity. Other Ni compounds form complexes with serum proteins or ultrafiltrable molecules, such as histidine or other amino acids. These metal–protein (nickelocene) or amino acid complexes adsorb to the surface of a cell and enter the cell by endocytosis. Within the cell, lysosomal proteinases hydrolyze the carrier protein to release the electrophilic metal ion. Nickel ions then bind with nucleic acids and other cellular constituents (Hatem-champy, 1961).

Nickel carbonyl inhibits the induction of several enzymes in the lungs and liver (Sunderman, 1967). Following the injection of a sublethal dose of nickel carbonyl to rats, DNA-dependent RNA polymerase activity in liver nuclei was reduced by 60% (Heath and Webb, 1967; Kasprzak and Sunderman, 1969; Witschi, 1972).

The mechanism of Ni carcinogenesis has not been established, but a few interesting hypotheses have been put forward: (1) alteration of DNA polymerase to temporarily decrease the fidelity of DNA replication, causing mutation to the genome, (2) chemical modification of RNA or nucleoproteins that regulate DNA template activity, causing expression of normally repressed portions of the genome, (3) metabolic block at the level of m-RNA, (4) inhibition of the induction of enzymes involved in the metabolic degradation of organic carcinogens such as benzpyrene, and (5) absorption by myoblasts involved in the repair of muscle injury of the diffusible Ni–protein complexes, resulting in the myoblasts' neoplastic transformation.

There is no direct evidence that Ni can be bound safely in proteins such as metallothionein, nor is there evidence for any detoxication mechanism for Ni in mammals. However, the efficient renal excretion suggests a subtle detoxication mechanism.

Ruthenium (Ru)

Ruthenium occurs in minerals with the less abundant Group VIII metals in trace amounts, and is not found in seawater. It has not been reported to be present in mammals. Ruthenium is used as a hardener in Pt and Pd alloys for electrical contacts and jewelry. Ruthenium is not an essential nutrient, and there are no reports about stimulatory effects or dietary requirements of it in mammals; Ru poisoning is rare.

Chemistry

Trivalence is the most common state for Ru; however, it can exhibit valence states from $2+$ to $8+$. Ruthenium and rhodium are isoelectronic, with a single electron in their $5s$ orbital. Ruthenium readily forms stable water-soluble coordination compounds such as $Ru(NH_3)_4OHCl$, but ruthenium coordination complexes are not reported in mammals. $Ru(OH)_2$, $RuCl_4$, and RuO_2 are stable and water-soluble, but in general the trivalent salts are not soluble. Tetroxides of Ru sublime upon exposure to air.

Metabolism

The biochemistry and metabolism of stable Ru and its salts have not been investigated in detail, but reports are available on radioactive [103]Ru and [106]Ru obtained from nuclear fission products (Burykina, 1962; Stara *et al.*, 1971). Studies with minute doses of [103]Ru indicate that the gastrointestinal absorption of Ru depends on its chemical form. Absorption in rats and cats is very low for ruthenium oxides, and is a little higher with ruthenate ions, ruthenium nitrosochloride, or ruthenium nitrosonitrate. Absorption of orally ingested ^{103}Ru-phenanthroline complex (Tan *et al.*, 1968–1969) and ^{103}RuCl$_3$ in adult rats (Thompson and Hollis, 1958) and in newborn rats (Sikov *et al.*, 1969) was negligible. The absorption of Ru from sites of parenteral injections is similar to that from the gastrointestinal tract. Absorption from inhalation is greater; the lungs retain half of the inhaled dose, and the removal from the lung is slow. Following intravenous injection, soluble Ru salts such as Na$_2$RuCl$_5$OH are quickly distributed in the bone and soft tissues; about 20% of the dose is excreted in the urine within twenty-four hours, with continued excretion extended over a period of time (Durbin, 1960). However, intravenous administration of ^{103}Ru-chelates, such as Tris 1,10 phenanthraline ruthenium-perchlorides, result in rapid urinary excretion of about 90% of the dose injected. The levels of retention of Ru in various tissues are: kidneys > muscle > liver > bone (Ramasastry, 1966). Skeletally deposited Ru is retained for a long time, and this is a health hazard if it is ^{103}Ru or ^{106}Ru, since these emit strong γ radiations.

Toxicity

Reports on Ru toxicity are scarce. Ruthenium salts are considered slightly toxic; the LD$_{50}$ of RuCl$_3$ in rats is 360 mg Ru/kg when injected intraperitoneally. Symptoms of morbid Ru intoxication following parenteral administration are not available. Ruthenium tetroxide fumes are highly injurious to lungs and eyes, causing severe irritation and discoloration in the mucous membranes of the lungs.

Rhodium (Rh)

Rhodium occurs in trace amounts with the other rare metals of Group VIII, and it has been reported to be present in mammals in detectable quantities. Rhodium is used to harden platinum and palladium alloys for use in high-resistance furnaces, thermocouple elements, and electrical contact material, and is also used for electroplating scientific and optical instruments. Industrial Rh poisoning is rare. Rhodium is not an essential nutrient for mammals; rhodium chloride has therapeutic activity against viral infection.

Chemistry

Trivalent Rh forms salts such as chlorides, nitrates, sulfates, and rhodates, and soluble complexes such as hexachlororhodates with coordination numbers 4 and 6. Although Rh can form stable complexes with S-containing ligands, no such complexes are reported in biological systems. Rhodium oxides are insoluble in water, but hydrated trichlorides are soluble.

Metabolism

Studies on the metabolism of Rh salts are mostly on [103]Rh and [103]Ru studied goether (Ramasastry, 1966); there are no reports about Rh requirements in mammals or about Rh involvement in biochemical systems. The gastrointestinal absorption of small doses of [103]Rh compounds is similar to that of Ru, as are the retention of Rh in the lungs after inhalation and the distribution of Rh to soft tissues after intravenous injections. Excretion of [103]Rh is mainly urinary and biphasic; a rapid initial excretion of about 45% of the parenteral dose within one day is followed by a gradual and extended excretion (Durbin, 1960). There are no extensive reports about retention of Rh in animal tissues or about Rh homeostasis.

Toxicity

The few studies on Rh toxicity indicate that Rh salts are more toxic than Ru salts. In general, Rh salts are slightly toxic (Table 8-4). Landolt *et al.* (1972) attribute Rh toxicity to its effects on the central nervous system. $RhCl_3$ has chemotherapeutic activity against viral infection in mice, and inhibits the formation of viral phospholipid by acting as a Co antagonist (Bauer, 1958).

When fed to mice at 5 ppm in life-term studies, $RhCl_3$ produced tumors (Schroeder and Mitchner, 1971*a*), both lung adenocarcinoma and lymphoma–leukemia, indicating that rhodium at these dosage levels is slightly carcinogenic. Deposits resembling amyloid were found in the kidneys, liver, adrenals, spleen, and heart, although growth rates and longevity were not affected. Thus, significant absorption does occur.

Palladium (Pd)

Palladium occurs at 0.2 ppm in the earth's crust in association with the rare metals of the Group VIII. Palladium is detectable in mammalian tissues, but is not essential to mammals. Palladium is used extensively in industry in alloys, as a catalyst in hydrogen purification, and in dental inlays with platinum and gold. A colloidal form of Pd has been used therapeutically for tuberculosis, gout, and obesity. There are no reports of

TABLE 8.4. Toxicity of Other Group VIII Metals

| | | | | Dosage/kg body weight | | | | |
| | | | | Compound | Metal | | | Time until |
Compound	Animal	Route	Toxicity	mg	mg	mM	pT	death
Rhodium trichloride	Rat	ip	LD_{50}	280	138	1.34	2.87	24 hr
RhCl$_3$	Rat	iv	LD_{50}	198	97.5	0.95	3.03	48 hr
	Rabbit	iv	LD_{50}	215	106	7.03	2.99	12 hr
Palladium dichloride	Mouse	ip	LD_{50}	104	62.4	0.59	3.23	
PdCl$_2$	Rabbit	iv	MLD	19	11.4	0.11	3.97	
Osmium tetroxide	Rabbit	inhal	LD_{100}	1.316[a]				30 min
OsO$_4$	Human	inhal		0.1[a]				
Platinous chloride	Rabbit	iv	MLD	23	16.9	0.089	4.06	
PtCl$_2$								
Platinum diamino-dichloride	Rat	ip	LD_{50}	12	7.8	0.040	4.40	
H$_2$PtCl$_2$(NH$_2$)$_2$								
Ammonium chloro-platinate	Human	inhal		0.9[b]				
(NH$_4$)$_2$PtCl$_6$								

[a]Exposure for 4 hr; number of mg/m^3.
[b]Exposure for 4 hr; number of μg/m^3.

a toxic effect from exposure to Pd in industrial processes, but Pd poisoning could occur by continued therapeutic use.

Palladium forms di- and tetravalent salts; hexa- and tetra- chloropalladates with coordination numbers 6 and 4 are stable. Palladium salts are water-soluble; the nitrate is partially hydrolyzed, indicating that palladium nitrate could be hydrolyzed at tissue pH to palladium oxy salt or oxides. Palladium coordination complexes are not reported to be present in biologic systems.

Metabolism

While Pd metabolism has been studied more extensively than either Ru or Rh, there are no reports about Pd involvement in any biochemical systems *in vivo*. Its diuretic activity and oxidative capacity are said to be involved in the affect of Pd in decreasing obesity (Dodds *et al.*, 1937; Kaufman, 1913). Absorption of Pd salts from the digestive tract and the peritoneal cavity is poor. Subcutaneous injection of soluble Pd salts into rabbits caused local black discoloration owing to deposition of palladium oxides, which is similar to the reaction seen with osmium salts. The lungs retain most of the inhaled Pd salts.

Excretion of Pd depends upon its mode of administration and its form. Orally ingested Pd salts are excreted in feces, and intravenously injected stable Pd salts are excreted in the urine after a transient retention in the kidneys and liver; intravenously injected Na_2PdCl_4 (minute quantities) containing ^{103}Pd is excreted very rapidly in the urine (Durbin, 1960). Palladium is not retained in the blood to the same extent as Ru and Rh.

Toxicity

In life-term studies, mice given 5 ppm Pd as $PdCl_2$ in their drinking water exhibited decreased growth, increased longevity, and malignant lung tumors (Schroeder and Mitchner, 1971*a*). Amyloid-type deposits are seen in the kidneys, liver, and spleen. Rats and rabbits tolerate about 25 mg Pd/kg as soluble Pd salts injected subcutaneously (Meek *et al.*, 1943), but higher doses caused necrosis at the site. Rapid intravenous administration of Pd salts is immediately lethal to rabbits at a level of 0.6 mg Pd/kg; when Pd salt was given slowly, at intervals, or in small doses, the lethal dose was 16.7 mg Pd/kg, with a survival period of two weeks. Since colloidal $Pd(OH)_2$ even at 1:25,000 dilution is reported to cause hemolysis, the lethal effect could be caused by this. Liver and kidney cells also suffer acute damage. Contact dermatitis has been associated with metallic Pd, especially with Pd alloys (Sheard, 1955).

Palladium is not known to accumulate in mammals following oral ingestion or parenteral injection, and there are no specific detoxication or homeostatic mechanisms for Pd salts.

Osmium (Os)

Osmium, the heaviest substance known, occurs with iridium in trace amounts with the rare metals of Group VIII. Osmium is not normally present in mammals and other organisms; it is not required by animals, but it stimulates growth. Osmium–iridium alloy is very hard, and is used in instrument pivots, electrical contacts, and fountain-pen tips. Osmium tetraoxide is used to detect fingerprints and to stain fatty tissue for microscopy. Osmium poisoning occurs through inhalation of OsO_4 which readily vaporizes from aqueous solutions even at room temperature; even the Os–Ir alloy readily releases OsO_4 vapors at the heat required for annealing.

Chemistry

Osmium forms stable tri-, tetra-, and octavalent compounds. Osmium coordinates readily with atoms and radicals to form hexahalo-, cyano-, nitro- and carbonyl coordination complexes with a tetrahedral configuration. These complexes are not stable and there are no reports of the presence of similar Os complexes in mammals. The tri- and tetrahalides and the tetroxides are water-soluble, and the osmiates (OsO_4^{2-}) are slightly soluble. The rapid electron exchange between coordinately saturated Os^{2+} and Os^{3+} complexes could involve them in oxidation and reduction processes of living organisms.

Metabolism

Reports on Os metabolism are few and conflicting. Osmium as osmic acid at 1 ppm in the rat diet may be a growth stimulant (Bunyan *et al.*, 1958), but it has no status as an essential dietary factor (Schwartz *et al.*, 1959). Reports on gastrointestinal absorption of Os compounds are not available. While ingested insoluble Os salts are excreted in the feces, parenterally injected Os salts remain at the sites, where they are reduced to the metal; the absorption of Os from these sites is slow. Minute doses of intravenously injected osmiates are distributed in soft tissues such as the liver, kidneys, and muscles, and about 80% of the dose is excreted in the urine in about a week (Durbin, 1960). The lungs and the respiratory tract retain most of the inhaled OsO_4 vapors. Severe irritation and black discoloration occur in the mucous membranes of the lungs and gastrointestinal tract when OsO_4 comes in contact with tissues; OsO_4 is reduced to lower oxides and metallic Os by organic matter (McLaughlin *et al.*, 1946). There are no reports about Os accumulation in animal tissues.

Toxicity

Data on Os toxicity in laboratory animals by oral, subcutaneous, or intravenous administrations are not available. Inhalation of OsO_4 vapor or its contact with eyes produces acutely toxic symptoms; a concentration of 400 mg Os/m^3 of air is lethal to most laboratory animals (Brunot, 1933). Rabbits which survived four days after exposure to 250 mg Os/m^3 showed degenerative changes in the lungs in the form of a purulent bronchopneumonia, with discoloration of the epithelial lining of the bronchi and bronchioli. The corneas and sclera became opaque and ulcerated, and were covered with a brown film. Chronic toxicity through inhalation of small doses of OsO_4 for a prolonged time reduced erythrocyte and leukocyte counts in the blood of guinea pigs (Mazturzo, 1951). The toxic action of Os seems to be in the bone marrow, affecting the maturation of reticulocytes.

The renal tubular epithelium undergoes degeneration. Mild industrial exposure affects human eyes more than the lungs, causing acute conjuctivitis; dermatitis with painful skin eruption develops following OsO_4 contact with skin.

Iridium (Ir)

Iridium occurs in trace amounts in alluvial deposits with Os and the other rare metals of Group VIII. Iridium has not been reported to be present in mammals and other organisms, and is not an essential nutrient. It is used as a hardening agent in a number of Pt and Os alloys. Iridium poisoning is rare and iridium oxides do not vaporize as does OsO_4.

Iridium forms di-, tri-, and tetravalent compounds. Iridium forms stable coordination complexes with carbonyl, thioethers, thiourea, and ammonia. With thiourea the coordination of Ir is with S atoms and not with N. The anionic complexes hexachloroiridate and hexaoxaloiridate are stable and water-soluble, and divalent Ir salts, such as halides and sulfates, are water-soluble. Although Ir readily forms complexes *in vitro* with S-containing ligands, such Ir complexes are not reported to be present in tissues.

There are very few reports about Ir metabolism, absorption, or excretion, and Ir involvement in biochemical systems *in vivo* is unknown. Radioactive Ir, injected intravenously in very minute doses as soluble hexachloroiridate, is distributed mostly in the liver and blood, with less in kidney, bone, and muscle; half of the injected dose is excreted in the urine (Durbin, 1960). Iridium dioxide is used as a nutritional and fecal marker in balance study experiments on animals, since it has low solubility and negligible absorption from the gastrointestinal tract (Luckey *et al.*, 1975).

There are no reports about Ir toxicity, but high toxicity could be predicted for soluble Ir salts, since Ir is capable of coordinating with S- and N-containing ligands and since Os and Pt, the other two members of this triad, are toxic.

Platinum (Pt)

Platinum occurs in nature with other rare metals of Group VIII, and sometimes with Cu. It is reported to be present in normal mammalian tissues, but it is not an essential nutrient. Platinum is used extensively in industrial and chemical laboratories. Platinum and Pt-rich alloys are used as contacts in electronic and telecommunication equipment and as high-resistance wire in electric furnaces. Platinum catalysts are used in the manufacture of sulfuric acid, persulfuric acid, nitric acid, organic and vitamin products, in the cracking of petroleum and the manufacture of high-octane fuel, and in the manufacture of fiber glass. Tetra- and hexachloroplatinates are used in photography, refining, and analysis. Platinum poisoning by inhalation of, or contact with, vapor and dusts of platinum complex salts occurs most frequently from industrial exposure. The therapeutic use of Pt was known as early as 1841, when it was used against syphilis and rheumatism. Since Pt compounds have now been found to be potent antitumor agents (Rosenberg *et al.*, 1969), Pt poisoning could occur owing to excessive therapeutic use.

Chemistry

Platinum forms di-, tri-, tetra-, and hexavalent salts. Chloroplatinic acid, obtained during platinum extractions, and its Na, K, and NH_4 salts, used in photography, are stable and water-soluble. The coordination chemistry of Pt compounds has been studied in detail; Pt readily complexes *in vitro* with O-, N-, and S-containing ligands such as ethers, or thiocarbamide, and thus Pt ions can form complexes and chelates with biological ligands, although there are no reports about the presence of such Pt complexes in mammalian tissues. Platinum ions would never be expected to occur free in living organisms.

Metabolism

The biochemistry of Pt and its salts has not been well investigated. Platinum metal is biologically inert and is used as support for fractured bones in humans. *In vivo* involvement of Pt in biochemical enzyme systems has not been reported.

Data on the gastrointestinal absorption of Pt and its cationic salts are rare. Studies on the tissue distribution of a single oral dose of cationic ^{191}Pt in rats revealed practically no tissue retention or absorption and almost total excretion in the feces (Moore *et al.*, 1975). Intravenous administration to rats of minute doses of nontoxic Na_2PtCl_4 labeled with ^{191,193}Pt indicated higher Pt retention in the liver, kidneys, spleen, and muscle than in the bones; about 35% of the dose was excreted with 24 hr in urine and feces (Durbin, 1960; Moore *et al.*, 1975). The low fecal excretion suggests an enterohepatic excretion pathway. The appearance of ^{191}Pt in the fetus of pregnant rats indicates that Pt can cross the placental barrier.

Parenterally injected Pt salts are retained at the injection sites, until later they are partially removed. There are no reports about accumulation of Pt in animal tissues.

Toxicity

There are no reports on the oral toxicity of Pt compounds. Acute toxic symptoms of parenterally injected Pt salts are epileptiform convulsions, coma, and death in animals. Chronic toxicity causes respiratory and skin disorders. Industrial workers exposed to complex Pt salts suffer from rhinorrhea, sneezing, coughing, wheezing, shortness of breath, and cyanosis. The symptoms are similar to a combination of hay fever and asthma. Strong solutions of Pt salts cause immediate urticaria if splashed on the face (Hunter *et al.*, 1945). The toxic effects of soluble simple salts are quite different from those of complex salts; simple salts generally cause vomiting and severe diarrhea, while complex salts act on the nervous system.

Platinum oxides and soluble platinum salts cause contact dermatitis (Schwartz *et al.*, 1947). Finely powdered elemental Pt also causes allergic dermatitis (Sheard, 1955), a loose term describing erythematus dermatitis, urticarial rash, cracking of the skin, eczematous patches, and edematous scaling. The acute toxicity of chloroplatinate (20 mg/kg by intravenous injection) in guinea pigs is exhibited as a violent attack of asthma and death in three min; this effect is due to the capacity of chloroplatinate to suddenly release histamine (Parrot *et al.*, 1969). Intense bronchiospasms can be produced by injecting either 1–2 mg chloroplatinate/kg or 3 μg histamine/kg. The action of chloroplatinate is lessened by repeated small doses, and thereafter the animal is able to resist a lethal dose. Platinum–ammonium complexes are toxic to rabbits by subcutaneous administration; platinum diammonium chloride is more toxic than platinum tetrammonium chloride.

Platinosis, the syndrome of respiratory and skin disorder caused by exposure to Pt salts in industrial processes, has been studied extensively (Roberts, 1951). Irritation of nasal and upper-respiratory-tract mucous membrane and high lachrymation are followed by severe asthmatic symptoms; platinosis is considered to be an allergic response to complex Pt salts. Levene (1971) indicated that the mechanism of histamine liberation by Pt salts in guinea pigs is not applicable to humans. The hypersensitivity hypothesis is supported by the successful hyposensitization produced by a course of intradermal injections of ammonium hexachloroplatinate in increasing amounts. Levene proposes that reaginic (IgE) antibody mediates the original symptoms and that injections of Pt salts stimulate the production of IgE (blocking) antibody, which combines preferentially with the antigen to prevent further anaphylactic reactions.

9

CHEMICAL TOXICITY OF METALS IN MAMMALS

Toxicology is the study of the adverse effects of chemical agents upon biological systems. Toxicity is the capacity of a chemical agent to adversely effect the activity of a living organism, its growth, health, life span, and reproductive capacity. Early mortality, growth retardation, impaired reproduction with mortality of offspring, neoplasms, and chronic disease symptoms are some of the common criteria for metal toxicity in mammals. Adverse effects also include behavioral changes in individual organisms and ecological changes that affect collective populations. High doses of nutritionally essential metals can also cause adverse effects. Under pathological conditions, such as renal disorders, biliary obstruction, breakdown of homeostasis, or physiologic stress, metals accumulate more readily than in healthy individuals; these conditions accentuate toxicities and accelerate degenerative processes, leading occasionally to death. Industrial activities have disturbed geologic caches of heavy metals, such as Cd, Hg, Tl, and Pb, and released high concentrations of these metals into the environment from the world's reservoir of immobilized ores and minerals. Consequently man and animals are increasingly exposed to uncommon metals at an alarming rate. Metals and their salts can have toxic effects on the organism at the tissue, cellular, subcellular, and molecular levels. Radiation effects of radioactive metals are excluded from this treatise. Toxicity at the cellular level causes deranged reproduction, differentiation, and maturation, as exemplified in teratogenesis. Metals may affect the permeability of cell membranes or act as antimetabolites; some metals affect the mitochondrial membrane and disturb energy metabolism; others decrease the stability of lysosomal membranes, leading to the disruption of cell functions by the

307

release of acid hydrolases. At the molecular level, some metals interact with proteins, leading to denaturation, precipitation, allosteric effects, and/ or enzyme inhibition; other metals bind to nucleic acids, leading to irreversible conformational changes. Interaction with DNA can cause mutation or carcinogenesis. Osmotic pressure is important in the toxicity of some metal salt solutions.

Metals cause acute, chronic, latent, and recondite toxicity symptoms, depending on the susceptibility of the animal, and the dose, mode of administration, and duration of exposure. Acute or chronic toxicity symptoms can be produced in experimental animals and are readily detected in man, whereas symptoms associated with recondite toxicity are difficult to produce experimentally and are difficult to diagnose in man. Delayed, or latent, toxicity is caused by metals such as Cr and Be, in which clinical symptoms are observable only months or years following exposure; latent toxicity is based on epidemiologic evidence and is yet to be proved experimentally. The lethal doses of a metal compound for acute, chronic, recondite, and latent toxicity vary and depend upon biological and environmental variations. This includes diurnal variations, or chronotoxicity.

The inherent toxicity of a metal depends on its capacity to disturb the dynamic equilibrium of the life processes in biologic systems by combining with cell organelles, macromolecules, and/or metabolites. The inherent toxicity of a metal is an expression of its electrochemical character, the solubility, stability, and reactivity of the metal compounds in body fluids and tissues, its ability to chelate with ligands of biologic macromolecules and the stability of these metal chelates, and finally, its physical form (ionic, colloidial, or radiocolloidal) in susceptible target tissues. This inherent toxicity is influenced by the rate of the metal's absorption from the alimentary and/or respiratory tracts, the rate of its absorption into the blood from the sites of parenteral injections, and the rate of its distribution from the blood to the various susceptible tissues. The inherent toxicity of a metal in a mammal is attenuated by the efficiency and capacity of homeostatic mechanisms which control the absorption, distribution, retention, and excretion of the metal ion.

The complete biologic activity spectrum of any material is indicated in Fig. 9-1. As discussed in Chapter 1 of Volume 1, if stimulation is encountered, a zero-equivalent point (ZEP) is discernable. The specific response to a given dose would change for each compound, each species, each parameter, and each condition, as is expected in biologic experimentation. The patterns represent different responses from acute toxicity studies. There is a great need for systematic appraisal of harm from chronic and

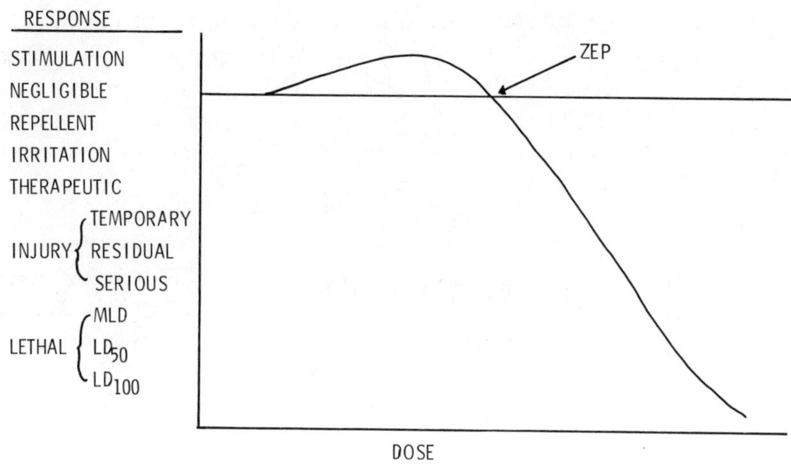

FIG. 9-1. Dose–response curve showing the complete spectrum of biologic activity as the dose is increased from zero to lethality.

multiple toxic materials, but society has yet to support such work. The paucity of information makes the following generalizations tentative.

A comparison of acute LD_{50} toxicity values of metallic salts with that of other compounds (Table 9-1) shows that metal toxicity ranges from negligibly toxic, pT 0–1, to mildly toxic, pT 4–5 (Luckey and Venugopal, 1977a,b). Since each pT whole number indicates a tenfold greater than the next lower number, metals are far less toxic than the natural toxins when taken parenterally, but many protein toxins are digested when taken orally, which would cause the toxicity from these toxins, some with pT values above 10, to lessen. Few pT values above 6 would be found in oral toxicity studies with healthy individuals when tested with compounds from urban environments. Thus, metal toxicity assumes great relative importance, since most harmful materials enter through the mouth or lungs, and not from injections. This and the widespread disperson of unusual metallic compounds by industrial and technical societies make metal toxicity a major environmental hazard of the world.

Published data for MLD and LD_{100} is less extensive and less reliable than LD_{50} data. Comparisons of pT data for these three levels of toxicity in

TABLE 9-1. Mouse LD$_{50/ip}$ Data Classified by pT

Toxin or toxicant	Molecular weight	Toxicity			Reference
		Mg/kg	Mol/kg	pT$_{50}$	
Botulinal D	1,000,000	3.2×10^{-7}	3.20×10^{-16}	15.49	LaManna and Sakaguchi, 1971
Botulinal A	900,000	1.14×10^{-6}	1.27×10^{-15}	14.90	LaManna and Sakaguchi, 1971; van Heynigen and Mellanby, 1971
Botulinal B	165,000	8.08×10^{-7}	4.90×10^{-15}	14.31	LaManna and Sakaguchi, 1971; van Heynigen and Mellanby, 1971
Botulinal Eδ[a]	350,000	5.68×10^{-6}	1.62×10^{-14}	13.79	LaManna and Sakaguchi, 1971; van Heynigen and Mellanby, 1971
Tetanus	66,000	1.67×10^{-6}	2.53×10^{-14}	13.60	Raskova and Masek, 1971
Botulinal E	350,000	2.50×10^{-3}	7.14×10^{-12}	11.15	LaManna and Sakaguchi, 1971; van Heynigen and Mellanby, 1971
Shigella	82,000	1.35×10^{-3}	1.65×10^{-11}	10.78	Montie and Ajl, 1970
Palytoxin	3,300	1.5×10^{-4}	4.55×10^{-11}	10.34	van Heynigen, 1970
Perfringens ϵ	40,500	3.2×10^{-3}	7.90×10^{-11}	10.10	LaManna and Sakaguchi, 1971
Pestis	120,000	4.0×10^{-2}	3.33×10^{-10}	9.48	Arbuthnott, 1970
Perfringens Cθ	74,000	8.1×10^{-3}	1.09×10^{-10}	9.96	Altman and Dittmer, 1973
Streptococcal o	80,000	1.0×10^{-1}	1.25×10^{-9}	8.90	LaManna and Sakaguchi, 1971
Perfringens A	36,000	1.25×10^{-1}	3.47×10^{-9}	8.46	Christensen and Luginbyhl, 1974
Staphylococcae α	21,000	4.0×10^{-2}	1.90×10^{-9}	8.72	LaManna and Sakaguchi, 1971; Wieland and Wieland, 1972
Saxitoxin	372	3.4×10^{-3}	9.14×10^{-9}	8.04	Barnes and Eltherington, 1964
Tetrodotoxin	319.3	1×10^{-2}	3.13×10^{-8}	7.50	Altman and Dittmer, 1968
α Amanitin	916	0.3	3.28×10^{-7}	6.48	Karasek *et al.*, 1948, 1949
Actinomycin D	1,255	0.7	5.58×10^{-7}	6.25	Altman and Dittmer, 1968

Strychnine	334.4	0.98	2.93×10^{-6}	5.53	Altman and Dittmer, 1968
Rotenone	394.5	2.8	7.10×10^{-6}	5.15	Altman and Dittmer, 1968
$HgCl_2$	271.5	5	1.84×10^{-5}	4.74	Altman and Dittmer, 1968
Parathion	291.3	5.5	1.89×10^{-5}	4.72	Spector, 1956
Epinephrine	183.3	4	2.18×10^{-5}	4.66	Altman and Dittmer, 1968
NaH_2AsO_4	163.9	9	5.49×10^{-5}	4.26	Altman and Dittmer, 1968
TlCl	239.8	24	1.00×10^{-4}	4.00	Altman and Dittmer, 1968
HCN	27.0	3	1.11×10^{-4}	3.95	Altman and Dittmer, 1968
$BeCl_2$	79.9	12	1.50×10^{-4}	3.82	Altman and Dittmer, 1968
Sodium fluoroacetate	100.1	18	1.80×10^{-4}	3.74	Altman and Dittmer, 1968
Benedryl	255.4	84	3.29×10^{-4}	3.48	Barnes and Eltherington, 1964
Codeine	299.4	130	4.34×10^{-4}	3.36	Barnes and Eltherington, 1964
Morphine	285.4	285	9.99×10^{-4}	3.00	Altman and Dittmer, 1968
Streptomycin	581.6	610	1.05×10^{-3}	2.98	Barnes and Eltherington, 1964
Caffeine[b]	194.2	250	1.29×10^{-3}	2.89	Barnes and Eltherington, 1964
Tetracycline HCl	480.9	650	1.35×10^{-3}	2.87	Altman and Dittmer, 1968
$BaCl_2$	208.3	500	2.40×10^{-3}	2.62	Barnes and Eltherington, 1964
Aspirin	180.2	495	2.75×10^{-3}	2.56	Altman and Dittmer, 1968
NaF[b]	42	125	2.98×10^{-3}	2.53	Barnes and Eltherington, 1964
Plutonium citrate	435.1	1750	4.02×10^{-3}	2.40	Altman and Dittmer, 1968
Pantothenic acid	219.2	900	4.11×10^{-3}	2.39	Spector, 1956
$CdCl_2$	183.3	1350	7.36×10^{-3}	2.13	Altman and Dittmer, 1968
Nicotinic acid	123.1	1860	1.51×10^{-2}	1.82	Karasck *et al.*, 1948, 1949
CCl_4	153.8	4620	3.00×10^{-2}	1.52	Altman and Dittmer, 1968
NaCl	58.4	2600	4.45×10^{-2}	1.35	Altman and Dittmer, 1968

[a]δ = activated toxin.
[b]Intraperitoneal?

TABLE 9-2. Comparison of pT Values for Rat MLD, LD$_{50}$, and LD$_{100}$

Salt	pT$_M$	Difference	pT$_{50}$	Difference	pT$_{100}$
Oral Administration					
$BaCO_3$			2.50	.03	2.47
$BaCl_2$			3.12	.38	2.74
$K_2Nb_6O_6$	3.07	.10	2.97	.28	2.69
$Al_2(SO_4)_3$			1.95	.09	1.86
Thallium acetate	3.85	.05	3.80		
ip Administration					
$K_2Nb_6O_6$	4.82	.18	4.64		
	4.00	.19	3.81		
Thallium acetate	4.01	.06	3.95		
$LaCl_3$	3.12	.27	2.85		
iv Administration					
Strontium acetate			2.31	.35	2.96
$SrCl_2$	2.86	.22	3.08		
$P(NO_3)_3$	4.61	.86	3.75		
$Dr(NO_3)_3$			3.95	.20	3.75
			4.11		4.11
$ThCl_4$			4.12	.08	4.04
sc Administration					
$In_2(SO_4)_3$			4.06	.10	3.96
$Tl(NO_3)_3$			4.13	.09	4.04

rats (Table 9-2) indicate that lethality depends upon both the inherent chemical character of the metal and the concentration. When the high and low values for each set are deleted, the difference between pT$_M$ and pT$_{50}$ is 0.16 and the difference between pT$_{50}$ and pT$_{100}$ is 0.15. Thus only about 0.8 mol of compound per kg body weight separates MLD from LD$_{50}$ and LD$_{50}$ from LD$_{100}$ for an average metallic salt.

INTERSPECIES COMPARISONS

Within certain limitations, interspecies comparisons can be made from available data for acute chemical LD$_{50}$ toxicity of metal compounds (Table 9-3). The toxicity data for the rat and mouse are summarized in Figs. 9-2–9-5 (pp. 316–319). Only lethality was considered, and other symptoms and epidemiologic evidence were excluded. This compilation includes all data from the literature that were judged to be dependable and pertinent for each metal except that the lanthanides were treated as a group. It includes many variables: different strains, animal conditions, physical conditions, food, and times of observation. Diurnal variations are less than might be suspected from chronotoxicology considerations, because most studies are started during late morning, the low ebb of rodent activity. Very few data are available for mammals other than those listed. This paucity of information suggests areas for future investigation. Most comparisons have been made between rodents, and few of these are based on adequate numbers to make the observations more than tentative.

The mouse shows less toxicity to oral Sn than the other animals, but is more sensitive than the rat to oral toxicity from the subgroup VB compounds, Nb and Ta, and the first triad of Group VIII, Fe, Co, and Ni. Mice are quite sensitive to orally administered Se compounds, while the rat is most sensitive to orally administered W compounds. Rats are less susceptible than mice to subcutaneous administration of compounds of subgroup IIIA, Al and Ga, and subgroup VA, as illustrated by Sn. Rats are the least susceptible to oral administration of subgroup VIA metals, Se and Te, and rabbits show the most sensitivity to oral administration of subgroup VIA compounds, particularly those of Se.

Intraperitoneal pT_{50} toxicity studies indicate that the mouse is often more susceptible to harm from metals than the rat. This is noted for Hg, Al, and Tl, and it is especially evident for Ba, Sb, and Se. The rat shows somewhat greater toxicity than the mouse for Co, Cu, and Sn. The few data available indicate that the guinea pig may be slightly more susceptible than the mouse to metal toxicity.

The rat and mouse show similar susceptibility to toxicity to most metals administered intravenously. Exceptions include Ca, Ni, and the lanthanides, where the rat is more susceptible than the mouse. There are no iv studies of metal toxicity in guinea pigs. The rabbit seems especially susceptible to iv Na, V, and Fe.

Subcutaneous administration of Ga and In shows the rat to be more susceptible than the mouse to these metals. Metals other than these two subgroup IIIA metals have similar toxicities for both rodents. The guinea pig is quite susceptible to V and Fe by the sc route. The few data for rabbits indicates their susceptibility to sc metal toxicity is similar to that of rodents. Another view is presented in Table 9-4 from the limited data on Be from Stokinger (1972). The rat is less susceptible than the mouse, dog, or monkey to iv or intratracheal $BeSO_4$.

COMPARISON BY MODE OF ADMINISTRATION

In Chapter 2 of Volume 1 rationale was developed for the translation of inhalation data to mg/kg by using standard values for alveolar air volumes and animal weights, with stated time of exposure and concentration for

TABLE 9-3. pT_{50} Summation of Metal Toxicity in Mammals[a]

Atom	Mouse				Rat				Guinea pig				Rabbit			
	ip	iv	oral	sc	ip	iv	oral	sc	ip	iv	oral	sc	ip	iv	oral	sc
Group I																
Li	(4) 1.75						(1) 1.75									
Na	(7) 2.66	(3) 2.56		(1) 4.11	(5) 2.25		(8) 2.32						(1) 2.37	(1) 3.04	(1) 1.90	
K	(3) 3.08	(1) 3.70	(4) 2.71	(2) 3.90	(4) 2.53	(3) 2.70	(13) 2.55				(1) 1.47					
Rb and Cs	(2) 2.01	(1) 2.24	(1) 2.06		(6) 2.32		(3) 1.28		(1) 2.15							
Ave	(16) 2.43	(5) 2.72	(5) 2.58	(3) 3.97	(15) 2.35		(25) 2.29									
Cu	(2) 3.70				(1) 4.01		(9) 2.70					(1) 3.13				
Ag	(1) 3.53		(2) 3.08			(1) 3.72	(2) 3.51									
Au																
Ave	(3) 3.64						(11) 2.84									
Group II																
Be		(4) 4.52	(2) 2.89	(1) 4.84	(3) 3.14	(6) 4.37	(5) 2.92	(1) 4.84	(4) 3.38		(1) 3.14					(1) 4.78
Mg	(2) 2.31	(1) 3.84	(1) 2.57		(1) 3.10		(1) 1.53									
Ca	(1) 2.60	(2) 1.71	(1) 2.70		(2) 2.65	(1) 2.81	(6) 2.41								(1) 1.76	
Sr	(1) 2.24	(1) 3.03			(6) 2.67	(1) 3.29								(1) 2.17		
Ba	(1) 3.59	(1) 4.24			(2) 2.35		(5) 2.53	(1) 3.49								
Ave	(5) 2.61	(9) 3.63	(4) 2.76		(14) 2.76	(8) 4.04	(17) 2.55	(2) 4.16								
Zn	(1) 3.74		(1) 2.59		(1) 3.76		(3) 2.61				(1) 2.74					
Cd	(2) 3.21			(1) 4.32			(2) 2.93				(1) 3.00					
Hg	(9) 4.31	(1) 4.55	(9) 3.79	(1) 4.07	(2) 3.53		(12) 3.69	(1) 3.77	(1) 4.56		(1) 4.08					
Ave	(12) 4.07		(10) 3.67	(2) 4.20	(3) 3.61		(17) 3.41				(3) 3.27					
Group III																
Al	(2) 2.98		(2) 1.62		(2) 3.36	(2) 4.34	(4) 2.16	(2) 2.93								
Ga	(2) 3.59			(1) 2.07	(1) 3.79	(2) 3.39	(2) 1.53	(1) 4.06			(1) 1.80			(2) 3.43		(2) 2.87
In	(3) 3.80	(1) 5.55	(1) 2.07	(1) 2.57	(2) 3.51	(1) 4.66	(2) 1.59	(1) 4.04								
Tl	(1) 4.00		(4) 3.93		(2) 3.70		(3) 3.91	(4) 3.48						(1) 4.28		
Ave	(8) 3.57		(7) 3.01	(2) 2.32	(7) 3.56	(5) 4.02	(11) 2.42							(3) 3.71		
Sc	(2) 2.76		(1) 1.58													
Y	(4) 3.07				(4) 2.84				(2) 3.58				(1) 2.85			
Lanthanides	(49) 3.14	(2) 2.26	(12) 1.85	(1) 2.17	(23) 3.05	(13) 4.25	(15) 2.08		(26) 3.55			(2) 2.31		(1) 3.85		
Ave	(55) 3.13		(13) 1.90		(27) 3.04				(28) 3.55							

Group IV														
Ge					(1) 2.14	(1) 3.56	(1) 2.86							(1) 2.09
Sn	(2) 3.71		(1) 2.20		(3) 4.02	(1) 3.74	(2) 3.30			(1) 3.80			(1) 4.53	
Pb	(1) 3.40				(5) 3.44	(1) 3.44	(2) 3.76	(6) 2.87		(2) 2.13			(1) 3.45	
Ave	(3) 3.61				(9) 3.19	(3) 3.58	(5) 3.39			(3) 2.69			(2) 3.99	
Ze					(13) 2.91		(9) 1.91				(1) 3.37			
Ti and Ht	(3) 3.13	(1) 3.23	(1) 2.55											
Ave	(3) 3.13				(13) 2.91		(9) 1.91							
Group V														
As	(1) 4.31		(1) 3.36	(1) 3.92			(3) 3.76			(1) 3.82				
Sb	(2) 3.75	(2) 4.10	(1) 2.74	(2) 3.78	(6) 2.56		(4) 2.57	(1) 3.94	(2) 3.63			(1) 4.19	(1) 3.46	
Bi														
Ave	(3) 3.94		(2) 3.05	(3) 3.83	(6) 2.56		(7) 3.08							
V				(1) 3.08			(5) 3.01	(1) 3.22		(1) 4.66		(1) 4.48		
Nb	(2) 3.75		(2) 2.91		(2) 3.43		(1) 2.11							
Ta	(1) 3.60	(1) 3.40	(1) 3.55		(2) 3.35		(3) 1.97							
Ave	(3) 3.70		(3) 3.12		(4) 3.39		(9) 2.56							
Group VI														
Se	(1) 5.72		(1) 4.39		(2) 4.25	(3) 4.39	(3) 3.55	(1) 4.45		(1) 4.54	(1) 4.06		(1) 4.90	
Te			(2) 3.60		(2) 4.17	(2) 4.30	(2) 3.11			(1) 3.69			(2) 3.44	
Ave			(3) 3.87		(4) 4.21	(5) 4.35	(5) 3.37			(2) 4.11			(3) 3.93	
Cr	(1) 3.06	(1) 3.35				(2) 3.58	(3) 1.95		(1) 2.71					
Mo	(1) 2.90				(2) 2.75		(2) 2.65							
Wo			(1) 2.14		(1) 1.57		(1) 3.44	(1) 3.14					(1) 2.58	
Ave	(2) 2.98	(1) 3.35	(1) 2.14		(3) 2.39	(2) 3.68	(6) 2.26						(1) 2.58	
Group VII														
Mn	(1) 3.09	(1) 3.95			(1) 2.95		(2) 1.99							
Re	(2) 2.52				(1) 2.49									
Ave	(3) 2.71				(2) 2.72		(2) 1.99							
Group VIII														
Fe	(5) 3.64	(1) 3.27	(3) 3.11		(3) 3.77		(9) 2.39					(1) 4.37		
Co	(4) 3.39	(2) 3.71	(2) 3.39		(3) 3.71	(2) 3.74	(4) 2.56			(1) 3.37				
Ni	(2) 3.80	(1) 2.72	(1) 2.64		(4) 3.79	(1) 3.41	(3) 2.56	(1) 3.43						
Ave	(11) 3.58	(4) 3.35	(6) 3.13		(10) 3.76	(3) 3.63		(1) 3.43		(1) 3.37		(1) 4.37		
Others	(1) 3.23				(2) 3.64	(1) 3.02						(1) 2.99		
Ave	(12) 3.55	(4) 3.35	(6) 3.17		(12) 3.74	(4) 3.48	(16) 2.46	(12) 3.43		(1) 3.37		(2) 3.68		

[a]Number of experiments is indicated in parentheses preceding pT_{50} values.

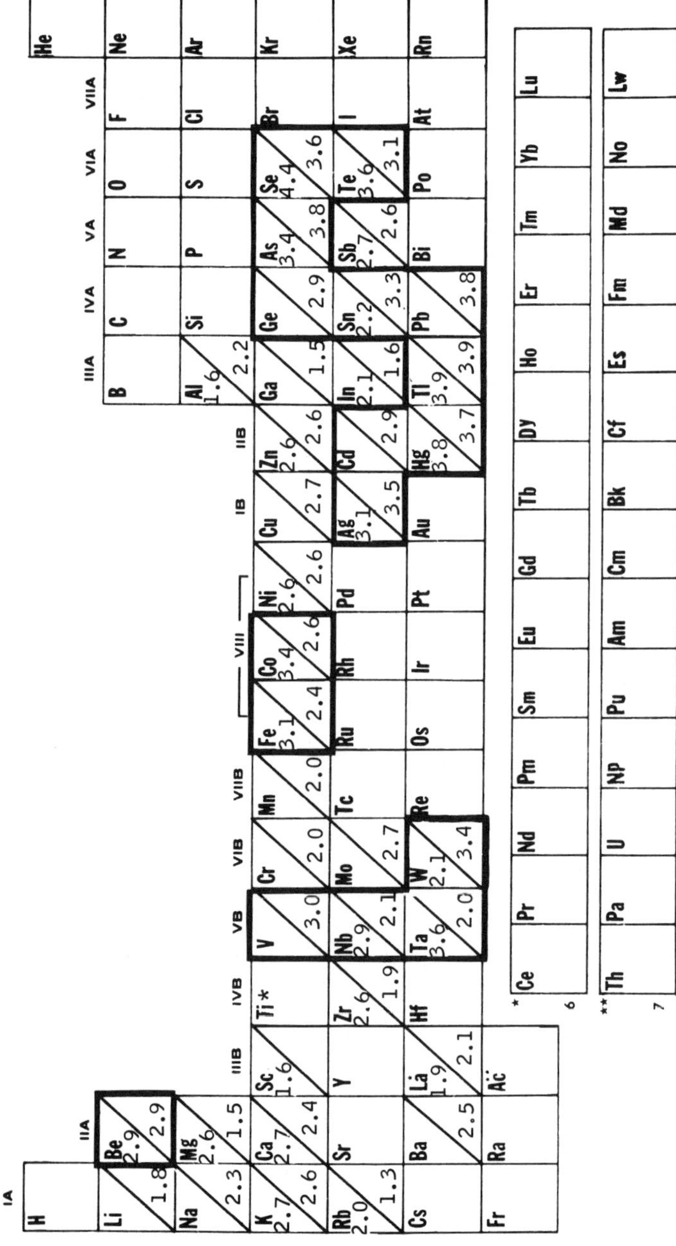

FIG. 9-2. The oral pT_{50} toxicity of metals for mouse/rat. Toxicities with $pT \geq 2.9$ are outlined in bold.

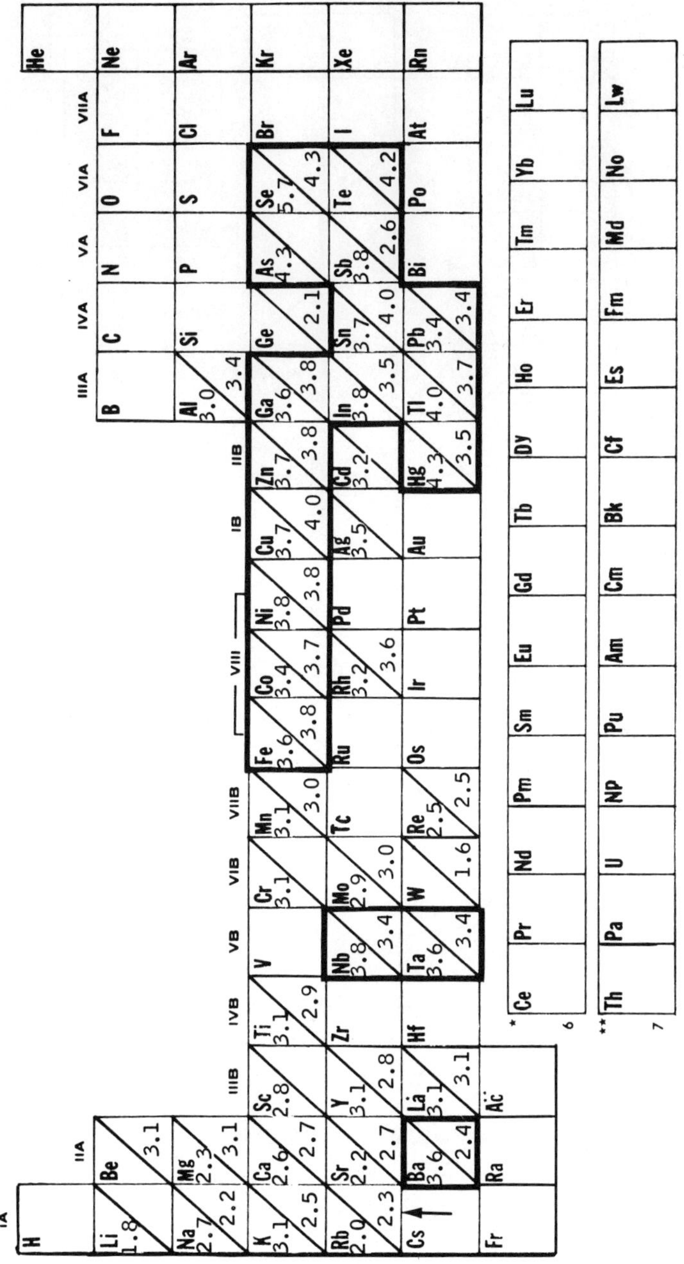

FIG. 9-3. The ip pT_{50} of metals for mouse/rat. Toxicities with $pT \geq 3.6$ are outlined in bold.

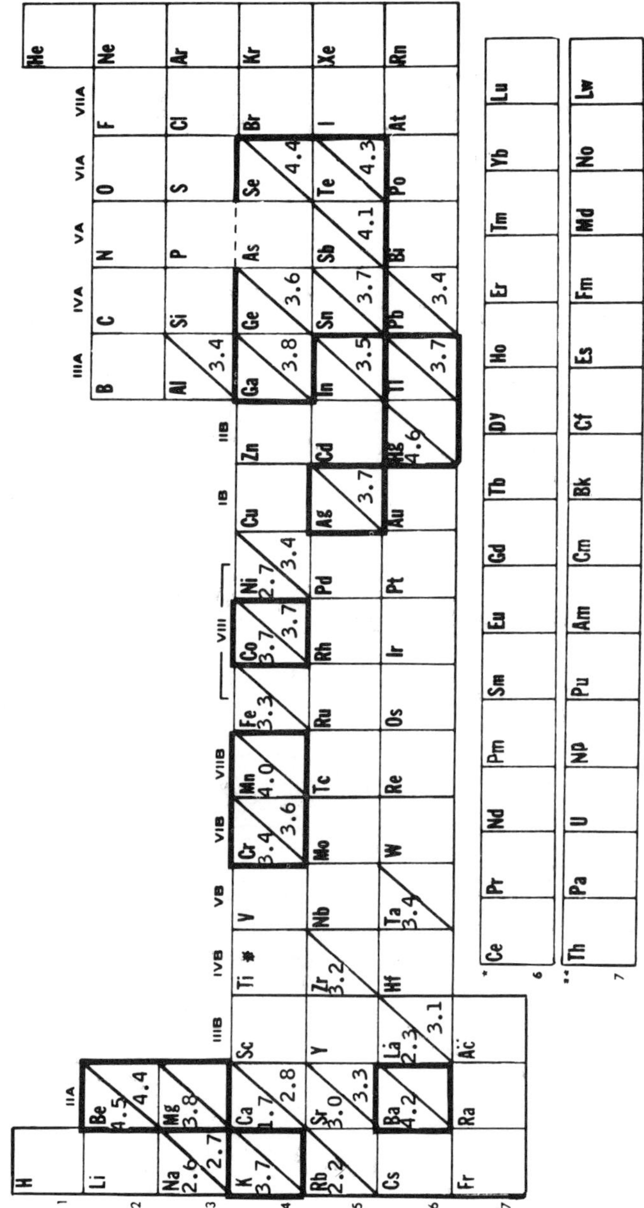

FIG. 9-4. The iv pT_{50} of metals for mouse/rat. Toxicities with $pT \geq 3.6$ are outlined in bold.

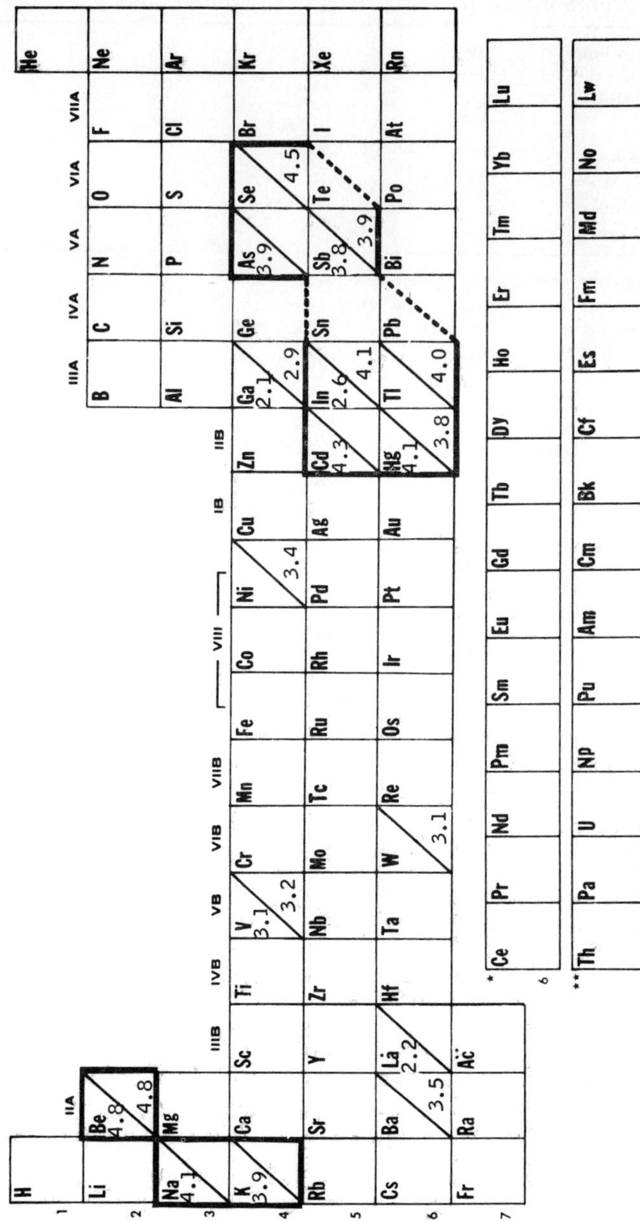

FIG. 9-5. The sc pT_{50} of metals for mouse/rat. Toxicities with $pT \geq 3.9$ are outlined in bold.

TABLE 9-4. Comparative Acute Toxicity of Beryllium Salts

Salt	Animal	iv	ip	it[a]	ih[a]	sc	po[a]
$BeSO_4$	Mouse	0.5				1.5	
	Rat	7.2	18[b]	10[b]	3.6[b]	1.5	3,750[c]
	Rabbit					1.5	
	Dog	0.6[b]		1.0[b]			
	Monkey	0.6[b]		1.0[b]			
BeF_2	Mouse	1.8				20	100

All data are mg compound/kg for LD_{50} from Tabershaw, 1972.

[a] it = intratracheal; ih = inhalation; po = *per os*.

[b] Estimated to be in excess by 1.5–2 times.

[c] Cumulative from 172 days feeding.

each toxicant. These estimations are acceptable for healthy individuals within usual limits of biologic variation. The translation of LD_{50} inhalation data for minerals is illustrated in Table 9-5. The long exposure for $BeSO_4$ prevents these values from being compared to the acute toxicity data given in Table 9-4. Apparently Be is much less dangerous in chronic than in acute conditions, a concept reinforced by comparing low-level feeding of $BeSO_4$ to other modes of administration (Table 9-4). Except in the rat, Be toxicity varies with the mode of administration as: iv ~ it > sc > *per os*.

A more general concept of the effect of mode of administration upon acute mineral LD_{50} toxicity is taken from Table 9-3. For many of the metals studied, subcutaneous injection is the most sensitive mode of administration and oral administration is the least sensitive. The intraperitoneal and intravenous modes are close to subcutaneous administration in their sensitivity. When individual metals are considered, pT values over 4 are not unusual for the most toxic metals administered parenterally. A pT value over 3.0 indicates considerable toxicity for metals given orally. A comparable dividing line for metals administered intravenously or intraperitoneally is pT 3.6. The few LD_{50} data available from other modes of administration and data from chronic toxicity studies do not allow comparison. The most toxic metals for each mode of administration are listed in Table 9-6.

The comparison of metal toxicity for all species by different modes of administration (Tables 9-3 and 9-7) is typified by subgroup IA metals. The acute pT_{50} for this subgroup for all species is: oral, 2.3; ip, 2.4; iv, 2.8; and sc, 4.0. The high pT_{50} values for Na and K by the sc route are evident in the mouse. Metals of subgroup IB show less variety, and the pT average for all modes is 3.4. Metals of Group II follow the general pattern: pT_{50} is high for sc administration, and low for oral administration (except Hg), with iv toxicity somewhat greater than ip. Beryllium makes the toxicity of

TABLE 9-5. *Translation of LD$_{50}$ Inhalation Toxicity Data*

Compound	Animal	kg[a]	Time			Compound			Metal		
			Air[b]	min	m^3	mg/m^3	mg	mg/kg	mg/kg	mM	pT
Ni(CO)$_4$	Mouse	0.03	0.02	30	0.6	67	40.2	1340	461	7.85	2.11
	Rat	0.3	0.05	30	1.5	240	360	1200	413	7.04	2.15
	Cat	3	0.7	30	21	1900	39,900	13,300	4580	78.0	1.11
BeSO$_4$	Rat	0.03	0.05	1800	90	10	900	3,000	255	28.3	1.55
	Guinea pig	0.5	0.11			47			45.8		

[a]Standard values, see Table 2.1, Volume 1.

[b]Alveolar air volume in meters per minute, see Table 2.1 and discussion in Chapter 2, Volume 1.

TABLE 9-6. Most Toxic Metals for Rodents by Different Modes of Administration[a]

Mode	Metals
po	Ag, Hg, Tl, Sn, Pb, As, V, Ta, Se, Te, Fe, Co, W
ip	Cu, Hg, Tl, Sn, As, Se
iv	Be, Ba, Hg, In, Al, Sb, Se, lanthanides
sc	Na, K, Be, Cd, Hg, In, Tl, Sb, Se

[a]Arranged in order of decreasing toxicity.

subgroup IIA somewhat variable, since its toxicity is quite high by the sc mode and surprisingly low by oral administration. The oral toxicity of subgroup IIB metals is high because of the value for Hg, $pT_{50} = 3.8$. The oral toxicity of Group III metals is low except for that of Tl. Toxicity of these metals following sc administration is low in the mouse and variable in the rat (see Figs. 9-2 and 9-3). The hydrolysis of subgroup IIIA metals to acids and acid salts makes their ip and iv toxicity high, and ip > iv. The toxicity of subgroup IIIB by the iv route is high, $pT = 4.1$, owing to a special preparation of In stabilized with gelatin.

In Group IV the toxicity of Ge may be greater by the iv than the ip mode. Otherwise the other metals of subgroup IVA (Sn and Pb) resemble

TABLE 9-7. Comparison of Toxicity (pT_{50}) for all Species by Mode of Administration[a]

Group	ip	iv	oral	sc	im	it
IA	34/2.39 (.709)	10/2.76 (.428)	33/2.28 (.935)	3/3.97 (.179)	1/2.74	
IB	4/3.73 (.250)	1/3.72	13/2.88 (.593)	1/3.13		
IIA	23/2.83 (.550)	18/3.73 (1.26)	23/2.57 (.657)	4/4.49 (.666)	1/3.84	
IIB	16/4.02 (.614)	1/4.55	30/3.48 (.603)	3/4.05 (.276)	1/3.87	
IIIA	15/3.56 (.893)	8/3.97 (.810)	19/2.60 (1.096)	9/3.12 (.682)		1/2.36
IIIB	120/3.20 (.370)	25/4.10 (.810)	32/1.99 (.339)	5/2.66 (1.082)		
IVA	18/3.15 (.716)	4/3.67 (.218)	11/3.20 (.886)	1/2.09		
IVB	17/2.98 (.533)	1/3.23	10/1.97 (.296)		1/3.00	
VA	11/3.13 (1.074)	3/4.13 (.221)	10/3.11 (.913)	4/3.83 (.077)		
VB	7/3.52 (.351)	3/4.05 (.573)	12/270 (.629)	3/3.65 (.873)		
VIA	5/4.51 (.784)	5/4.35 (.515)	13/3.73 (.709)	2/4.26 (.272)		
VIB	5/2.68 (.738)	3/3.51 (.416)	8/2.54 (.634)	2/3.27 (.210)	1/3.50	
VII	5/2.71	2/3.47	2/1.99			
VIII	24/3.65	12/3.5	23/2.68	1/3.43		

[a]Number of experiments considered/mean pT_{50} (variation between highest and lowest pT_{50} values found).

those of subgroup IB in that $pT_{50} = 3.2–3.7$ for all modes except sc ($pT_{50} = 2.1$). No data are available for sc administration on subgroup IVB. The oral toxicity of subgroup IVB metals is lower than the toxicity by other modes. The high toxicity of Group V iv > sc > ip >oral, with the high oral toxicity of As making the average oral toxicity of subgroup VA considerably higher than that for subgroup VB metals. All modes of administration give about the same toxicity for Group VI metals, but those of subgroup VIA show relatively high toxicity and those of subgroup VIB show low toxicity when administered orally. Oral administration of Se, Te, and V show high toxicity values, while Mo and Cr are less toxic by this mode. There is inadequate data for comparison with Group VII metals. Amoung Group VIII metals toxicity is: ip > iv > oral, and there is little sc data.

Oral administration is the least harmful route for all subgroup metals except IVA. Intravenous administration was either the most harmful or at least as harmful as any other route of administration for metals of subgroups IB, IIB, IIIA, IIIB, IVA, IVB, VA, VB, VIB, and Group VII. Subcutaneous administration was the most harmful route for metals of subgroups IA and IIA. Intraperitoneal administration was either the most harmful or as harmful as any other route for metals of subgroups IB, IVA, VIA, and Group VIII.

Intravenous and intraperitoneal routes of administration were the most harmful for metals of subgroups IB and IVA. Harm from im administration paralleled that from iv administration. These tentative conclusions resemble those drawn from consideration of the mouse and rat data, as these two species provide the bulk of the information available. The relative pT_{50} toxicity of metal subgroups for each route of administration is summarized in Table 9-8. Less-well-documented averages for pT_M and pT_{100} (Tables 9-9 and 9-10) corroborate the trends established by the pT_{50} data.

TABLE 9-8. Toxicity of Metal Subgroups from Different Routes[a]

ip	iv	sc	oral
VIA	IIB	IIA, VIA	VIA, IIB
IIB	VIA	IIB, IA, VA	IVA, VA
IB, IIIA, VIII, VB	IIIB, VA, VB, IIIA, IVA, VII	VB	IB
VIB	IB, IIA	VIII	VB, IIA, IIIA, VIB, VIII
IIIB, IVA, VA	VIB, VIII	VIB	IA
IVB, IIA, VII	IVB	IB, IIIA	IIIB, IVB, VII
IA	IA	IIIB	
IA	IA	IIIB	
		IVB	

[a]Vertical position indicates relative toxicity for combined species.

TABLE 9-9. pT_M Toxicity for All Species[a]

Group	ip	iv	oral	sc	im
IA	(11) 2.99	(13) 2.64	(21) 2.58	(15) 2.74	
IB	(2) 3.81	(3) 4.64	(4) 2.51	(2) 2.60	(3) 5.44
IIA	(6) 2.81	(12) 3.18	(13) 2.11	(13) 2.69	
IIB	(7) 4.15	(3) 4.34	(8) 3.20	(5) 3.46	(1) 4.94
IIIA	(9) 3.96	(5) 4.40	(17) 3.42	(8) 3.43	
IIIB	(11) 3.12	(13) 3.78	(3) 2.01	(10) 2.18	
IVA	(8) 3.33	(2) 3.83	(10) 3.12	(1) 3.01	
IVB			(1) 3.08	(3) 2.68	(3) 3.47
VA	(5) 3.51	(2) 4.40	(5) 3.98	(4) 4.32	(1) 3.48
VB	(6) 3.47		(1) 2.20	(1) 3.89	
VIA	(5) 3.96	(4) 4.77	(7) 4.06	(2) 3.64	
VIB	(2) 2.93	(8) 3.24	(3) 2.65	(8) 3.42	
VII	(1) 2.64	(1) 3.48	(1) 2.67	(2) 2.31	
VIII	(1) 3.56	(17) 4.11	(13) 3.12	(9) 3.44	(6) 2.8

[a]Values in parentheses indicate the number of experiments considered.

COMPARISON OF TOXICITY BY GROUPS OF THE ATOMIC TABLE

A review of the inherent toxicity of metals, relevant to their position in the vertical groups of the periodic chart, and of their physiological behavior (the solubility of the metallic salts in tissues and fluids, their absorption, transport, and distribution in tissues, their excretion, and their retention in mammals) reveals interesting relationships between inherent toxicity, electropositivity, and solubility. Inherent toxicity increases with the increased electropositivity and solubility of the metallic cations in water and lipids. Greater toxicity is found when electropositivity is associated with the formation of covalent and coordinate covalent compounds than with electrovalent compounds. The most highly electropositive of the metals are the alkali-group metals, which form strong polar and electrovalent compounds; their salts ionize in aqueous solution. The alkali metals do not normally form stable coordinate or chelate compounds with biologic macromolecules. Metals of Groups II and III, which are less electropositive than subgroup IA metals, form stable polar and covalent bonds with biologic macromolecules, the stability of the compounds depending on the electropositivity of the metals. The greater the stability, the greater is the inherent toxicity, and the higher the solubility of the heavy metal salts in biologic fluids, the greater their inherent toxicity; this generalization does not apply to the alkali metals. Lead, thallium, and mercury ions are highly toxic due to their high solubility, allowing transport to tissues where chelation and/or complexation with macromolecules can cause damage. Salts of Group III metals possess high inherent toxicity, but these salts are hydrolyzed to insoluble hydroxides or oxides at the tissue pH. The sparingly soluble nature of these hydroxides immobilizes the metals at the site of administration and renders them less toxic; when the cations of subgroup IIIA metals are kept hydrated in a solution by suitable buffers or chemical agents, these metals exhibit high inherent toxicity.

Within each vertical group of the periodic chart, increased electropositivity and inherent toxicity are associated with increased atomic number or weight. Exceptions to the above generalization are Li and Be, the lightest metals of subgroups IA and IIA respectively. The heavier metals in each group have the capacity to form irreversible and stable complexes with biologic macromolecules, which changes their conformation and biological function; hence these heavy metals are toxic The lighter metals form reversible complexes with the macromolecules, which allows essential functions to proceed. Essential metals are not toxic in physiologic doses, but are toxic at higher levels. Barium and Mg of subgroup IIA or Zn of subgroup IIB are suitable illustrations. The inherent toxicity of heavy metals tends to be greater the higher the solubility of the salts of these metals; however, the aqueous solubility of the simple salts of the metals in a group generally progressively decrease with increases in atomic weight. Hence, the heavier metals do not exhibit their potential inherent toxicity owing to their poor solubility. Lead, Tl and Hg salts are exceptions to these

TABLE 9-10. pT_{100} Toxicity for All Species[a]

Group	ip	iv	oral	sc	im
IA	(3) 1.65	(1) 3.82	(3) 1.52	(4) 1.90	
IB	(3) 4.18	(3) 4.18	(3) 3.16	(1) 2.35	
IIA	(1) 3.16	(9) 3.04	(12) 2.75	(8) 3.37	
IIB		(7) 4.10	(6) 3.15	(8) 3.67	
IIIA	(5) 3.17	(3) 3.68	(3) 3.23	(9) 3.67	
IIIB	(2) 2.89	(9) 4.42	(3) 2.38	(9) 3.67	
IVA	(5) 3.01	(4) 3.52		(4) 2.90	
IVB					
VA		(9) 4.55	(3) 2.87	(3) 4.08	(6) 3.52
VB	(3) 3.35	(9) 4.08	(3) 2.78	(22) 3.54	
VIA	(2) 4.47	(4) 3.83		(2) 4.51	
VIB	(5) 2.80	(2) 3.72	(7) 2.14	(9) 3.02	
VII	— —	(4) 3.61	— —	(4) 2.82	
VIII	(4) 2.47	(6) 3.63	(4) 2.28	(11) 2.93	

[a]Values in parentheses indicate the number of experiments considered.

generalizations owing to the greater solubility of their simple salts. The inherent toxicity of metals of subgroup IIA differs from that of the subgroup IIB metals owing to the difference in their electronic configurations, especially in the outermost orbitals; however, in some groups metals of subgroup B are more toxic than metals of subgroup A.

Metals of subgroup IA form water-soluble and polar compounds; their toxicity generally increases with increased atomic weight and electropositivity. Subgroup IA metals are considerably more toxic by subcutaneous than other routes (Table 9-7). Subgroup IB metals are more toxic than subgroup IA metals by all routes, and intravenous and intraperitoneal administration show more toxicity than subcutaneous injection. Rubidium and cesium toxicity results from their competition with and displacement of Na and K in cell reactions. Lithium occupies a unique place in the toxicity scale in this subgroup owing to its small ionic size and high charge/mass ratio. The caustic action of strong solutions of the alkali hydroxides of subgroup IA metals results in acute toxicity from tissue dissolution. In contrast, metals of subgroup IB are less electropositive than subgroup IA metals, and are more toxic. Within the subgroup IB, Au has the highest inherent toxicity, but the poor solubility of Au salts results in negligible absorption from the alimentary tract, rendering Au less toxic than Cu or Ag. Gold salts tend to decompose into elemental Au owing to their instability in tissues. Intravenous Cu is particularly toxic in rats; the greater solubility of Cu salts in tissue fluids renders Cu more toxic than Ag when administered intravenously. Divalent Cu is more toxic than monovalent Cu. Silver is equally toxic by all routes of administration. Silver is immobilized by reacting with thiol groups of tissue proteins following parenteral administration, while Cu ions are rapidly absorbed.

Metals of subgroup IIA are less toxic than are metals of subgroup IIB and somewhat more toxic than subgroup IA metals. Within the subgroup IIA, the lightest metal, Be, is the most toxic irrespective of the nature of its salts or the mode of administration. The great inherent toxicity of Be is owing to its small ionic size and high charge/mass ratio which allows ready penetration into tissues and cells. The lack of a homeostatic mechanism in mammals for Be enhances this inherent toxicity. Magnesium, an essential metal for mammals, also possesses high inherent toxicity and acts by disturbing ionic and osmotic equilibria in biologic tissues and fluids. However, this inherent toxicity is manifested only from parenteral administration. Magnesium salts are practically nontoxic by oral administration, since effective homeostasis operates in intestinal absorption. The inherent toxicity pattern of Ca, Sr, and Ba are similar to Mg; the toxicity increases with increase in electropositivity and atomic weight. The poor solubility of a heavy metal salt, such as $BaSO_4$, renders it nontoxic while the soluble

BaCl$_2$ is highly toxic. These metals exhibit considerable variability in toxicity when given intravenously.

Metals of subgroup IIB exhibit a similar toxicity pattern to those of subgroup IIA. The high solubility of mercury and its inorganic and organic salts in lipids and in water enhances its great inherent toxicity. Mercury is the most lethal of subgroup IIB metals, $pT_{50} = 4.1$ for guinea pigs *per os*. Cadmium has relatively high lethality when given orally to guinea pigs or subcutaneously to mice.

Metals of Group III possess the highest inherent toxicity among the metals, but this toxicity is attenuated by the behavior of the soluble salts of these metals at physiologic pH; the salts are hydrolyzed to the corresponding hydroxides which are insoluble in tissue and fluids and separate as stable precipitates or colloidal or radiocolloidal particles. Thallium is the only exception to this behavior; it exists as stable monovalent ion and behaves metabolically like K^+ in its absorption from the alimentary tract and subsequent distribution to tissues, Thallium is the most toxic of all nonradioactive metal ions, exhibiting high toxicity by all routes of administration in all species tested. The high intravenous toxicity of In may be owing to an anomaly of protein activation as much as to activated In. Subcutaneous In is relatively lethal when given to rats, but not to mice. The inherent toxicity of the other metals in this group can be demonstrated by preparing their stable hydrated ionic forms and administering these ions parenterally into experimental animals. The hydrolysis of the soluble salts of the metals of Group III at the tissue pH levels is more pronounced in subgroup B (lanthanides) than in subgroup A. The lanthanide hydroxides are not soluble at the pH of the intestinal fluids and are scarcely absorbed, and hence are used as unabsorbable nutritional markers. These oxides and hydroxides of lanthanides behave as negligibly toxic compounds when administered orally. However, soluble salts of lanthanides are toxic when given intravenously, because the hydrolyzed products (i.e., the hydroxides) accumulate in the reticuloendothelial system and block its normal functioning. Gallium and indium salts are effective toxicants by this route.

The inherent toxicity of Group-IV metals can be partially masked by the hydrolysis of their soluble salts at tissue pH levels and subsequent olation. The solubility of the cationic salts of Group IV metals is generally poor, but wherever the solubility is high, the metal salt exhibits its inherent toxicity (e.g., lead acetate or carbonate). Group IV metals exhibit electroneutral and amphoteric character with variable valence. Tetravalence is more common with the anions of the lighter metals such as Ge and Ti. The higher the valence the lower is the toxicity of these two values, owing to the rapid excretion of salts having higher valence. Within subgroup IVA the comparative electropositivity, stability, solubility, and toxicity of simple cationic salts increase as the atomic weight increases. Group IV metals all have about the same toxicity for mice; however, subgroup IVA metals are definitely more toxic than subgroup IVB metals for rats. Tin is surprisingly toxic *per os,* particularly in rabbits. Soluble Sn^{4+} salts are inherently more toxic than Pb salts; the diffusibility of Sn salts across biologic membranes is lower than that of Pb salts; the number of soluble stannic salts is very limited whereas the number of soluble Pb^{2+} salts are numerous. The toxicity of the Sn^{4+} salts depends somewhat upon the anionic part of the salt. It is very difficult to evaluate the inherent toxicity of subgroup IVB metals, The salts of these metals are very slightly toxic (Table 9-7); they undergo olation and readily form insoluble and hydrated polymeric basic salts. Among Group IV metals, Pb appears to be highly toxic due to the solubility of its salts and their high diffusibility across biologic membranes, and toxicity data on inorganic Pb compounds are far more numerous than are data on the salts of any other metal in this group.

Subgroup VA metals may be somewhat more toxic than subgroup VB metals for mice; this is not the case in rats and may be the opposite in rabbits and guinea pigs. The average for all species (Table 9-7) shows a similarity for both subgroups for all species and all modes of administration, except oral, where subgroup VB is less toxic than VA. The data are inadequate for many comparisons. Arsenic, the lightest metalloid of subgroup VA, is the most toxic for all species and all modes tested. Trivalent anionic As, in the form of soluble arsenites, and As_2O_3 are highly toxic. The poor solubility of Sb and Bi salts mask their inherent toxicity. Antimony and bismuth do not form soluble stable anions, and the solubility of their cationic salts decreases with increased atomic weight; the soluble cationic salts of Bi and Sb hydrolyze at tissue pH to form insoluble SbO^+ and BiO^+ cationic salts. In general the anionic forms of the metals of subgroup VA are more toxic than are the cationic salts because the solubility of the metals is higher. Organic Sb compounds are more toxic than are similar As and Pb compounds. Metals of subgroup VB are less toxic than their counterparts in subgroup VA. The difference in the electronic structure of V accounts for the greater solubility and variety of its salts than is found in Nb and Ta. Tantalum with a valence of 5^+ forms stable but insoluble cationic salts and is relatively nontoxic; Nb forms stable cations and also stable niobates. Cationic Nb is more toxic than anionic Nb, but the solubility of cationic Nb salts is poor. Cationic V salts are not stable, but V forms a variety of soluble oxy acids and anionic salts which are more toxic than are Nb salts. It can be speculated that if cationic V salts were stable they would be less toxic than comparable cationic Nb salts.

Among the metals of Group VI, toxicity is associated more with anions than cations. The anions of the lighter metals, which are essential in the nutrition of mammals, are more toxic than are anions of the heavier members. If the toxicity of the cations alone were to be compared, the increase in the inherent toxicity could be related directly to increased electopositivity and atomic weight. The variable valence states exhibited by these lighter metals seems to increase their inherent toxicity. Among anions, those with lower valence states are more toxic than are those with higher valence states (eg., selenite is more toxic than selenate). However, anionic hexavalent Cr salts are more toxic than are cationic trivalent Cr salts. The anionic salts of metals of subgroup VIA are more toxic than are the corresponding salts of the metals of subgroup VIB. For each route of administration, the toxicity of subgroup VIA and subgroup VA metals is similar, and that of subgroup VIB and VB metals are similar. Selenium shows the highest overall lethality for all species and all modes of administration for all metals. Tellurium also shows high lethality, particularily *per os*. Anionic Cr salts are more toxic than either Mo or W anionic salts. Tungsten toxicity is highly variable according to the mode of administration: in rats the oral pT_{50} is high, 3.44, and the ip pT_{50} is only 1.6, compared with the sc pT_{50} of 3.1.

Group VII metals exhibit no unusual toxicity characteristics. Among the metals of subgroup VIIB, anionic Mn, particularly when given intravenously, shows high toxicity in mice ($pT_{50} = 3.95$), but the other two metals have not been extensively investigated for their toxicity. Cationic Re is somewhat more toxic than cationic Mn. Again, the general pattern of increased inherent toxicity with increased electropositivity and atomic weight is observable. However, at physiologic pH the solubility of cationic Re salts is decreased and the degree of hydrolysis to insoluble compounds is increased. Another interesting feature is the increased toxicity of the oxy acid anions; the higher the valence state, the greater is the toxicity of the oxy acids in Group VII metals.

Group VIII differs from the other groups in that it is not subdivided into subgroups A and B but is subdivided into sets of horizontal triads. Group-VIII metals all show consistently high toxicity for all modes of administration in most species. The first triad, Fe, Co, and Ni, does not have high oral toxicity values in rats ($pT_{50} = 2.4–2.6$), but Fe and Co have high oral pT values for mice and guinea pigs. Most pT_{50} values for Group VIII metals are above 3. If these metals are divided in an unconventional manner into three vertical subgroups (Fe, Ru, Os; Co, Rh, Ir; Ni, Pd, Pt), the metals of the first and third subgroups are more toxic than are metals of the second subgroup. The lighter metals of Group VIII are essential; the soluble cationic salts and volatile compounds of these metals are more toxic than are the heavier metals. However, the anionic complex soluble salts of the heavier group (e.g., chloroplatinates, osmic acid) are more toxic than are the anionic complexes of the lighter metals. Thus, increased inherent toxicity concomitant with increased electropositivity and atomic weight can be speculated for the Group VIII metals.

COMPARISON OF TOXICITY BY PERIODS OF THE ATOMIC TABLE

A review of the inherent toxicity of metals, on the basis of horizontal periods in the periodic chart, reveals a pattern of toxicity which increases with successive periods. Within a given period there is a general progressive decrease in the inherent toxicity of the metal cation from Groups I to IV, followed by a progessive increase in the toxicity of the metal anion (oxy acid exhibiting the maximum valence state) from Groups V to VII. Metals of Group VIII exhibit a similar inherent toxicity associated with both types of ions. The increased solubility of the alkali salts of the oxy acids of metals of Groups V to VII enhances this inherent toxicity. The above generalization holds more for horizontal periods 3 and 4 than for periods 5 and 6 or the incomplete period 7. The electropositivity of metals increases progressively with each successive horizontal period.

The first horizontal period or series comprises H and He, and these gaseous nonmetals are not toxic. Lithium and beryllium, the lighter members of subgroups IA and IIA, respectively, are the two metals in the second period, and these two metals possess greater inherent toxicity than do metals which are essential nutrients for mammals.

The metals Na, Mg, and Al of the third period are comparatively nontoxic by oral administration, although cationic Mg and Al salts are toxic when introduced parenterally into experimental animals. The metals K to Se of the fourth horizontal period contain the first transitional series of elements. These metals generally form soluble salts, both anionic and cationic, and are nontoxic at physiologic levels. Some salts of these metals in their lower valence states are practically nontoxic (values of less than 1 on the pT_{50} scale). Some metals, such as Se, are more toxic in lower valence than in higher valence states. The inherent toxicity of metals belonging to Groups III to V in this horizontal series is not apparent because the soluble simple cationic salts of these metals are hydrolyzed to insoluble metal hydroxides or oxides; hence, they are not easily absorbed from the digestive tract. These salts exhibit their inherent toxicity when introduced parenterally. Citrates of these metals can be toxic because they are stable enough for absorption. Subsequent metabolic breakdown of the citrates in the tissues releases the metal ions within the tissues.

Metals of the 5th, 6th, and 7th horizontal periods are more electropositive than are the lower series, and their salts are progressively less soluble than are the metal salts of the earlier series. The greater electopositivity of the metals of the 5th horizontal period, extending from Rb to Te, renders them more toxic than the metals of the 4th series. Within the 5th period, metals of subgroups IIIB, IVB, VB, VIB, VIIB, and Group VIII are of low inherent toxicity, when compared the metals of subgroups IB, IIB, IIIA, IVA, VA, and VIA, owing mainly to the insoluble nature of the simple salts and the hydrolysis of soluble salts to insoluble oxy salts of the former groups. The 6th horizontal series, Cs to Po, comprises 30 metals (excluding At and Rn), including the lanthanides or rare-earth metals. These metals are the most electropositive of the nonradioactive metals of all horizontal series, and hence possess the greatest inherent chemical toxicity. These metals are capable of forming cationic and anionic salts which form stable complexes with cellular macromolecules and active metabolites. Simple salts of metals such as Cs, Ba, Os, Hg, Tl, and Pb are more soluble than are the corresponding salts of La (and the lanthanides), Hf, Ta, W, Re, Ir, Pt, or Bi. Some of the anionic oxy salts of Hf and W and salts such as aurothiomalates or chloroplatinates are somewhat soluble in water and biologic fluids, but the most soluble metal salts in this series are those of Cs and Tl. The soluble salts of the metals in this series are the most toxic of all metal salts, with Tl salts at the top of the toxicity scale. Analogous to Cs, Hg, Tl, and Pb salts, the salts of other metals in this series should also be toxic; their insoluble nature and poor absorption from the digestive tract mask their potential inherent toxicity. However, salts of these metals are highly toxic when administered parenterally, although few experimental data are available to substantiate this observation.

The last of the horizontal series of metals, the 7th period, extends from Fr to element No. 104. The series includes all the highly radioactive, unstable, and artifically made metals; the metals of this series should possess very high inherent toxicity. It is impossible to assess their chemical toxicity because of the nonavailability of these metals in larger quantities, their high radioactivity, and their short half-lives. The retention of these metals, especially the actinides, in animal tissues renders them the most dangerous metals when considering both chemical and radiation toxicity.

An overall review of the toxicity of metals in horizontal periods 4, 5, and 6 reveals the following pattern of inherent toxicity. The toxicity associated with metal cations progressively decreases from subgroup IA to subgroup VIIB; it progressively increases from Group VIII to subgroup IVA, and decreases from subgroups VA to VIA. The toxicity of anionic oxy salts of metals increases from subgroups IVB to VIIB and from subgroups VA to VIA. The metals of Groups I, II, and III do not form anionic oxy salts.

GENERAL COMPARISONS

The pattern of metal toxicity for mice and rats is outlined in Figs. 9-2 through 9-6. There are too few data from other species for a good general discussion. Metals most toxic *per os* are outlined with heavy lines when oral $pT_{50} \geqslant 2.9$. Selenium is the most toxic of the metalloids, and Tl, Hg, and Pb are the most toxic metals. Surprisingly, metals of subgroup VB are toxic by this criterion. Note that Be toxicity just falls into the toxic category. Selenium, Co, and Fe are the most toxic essential metals when taken orally.

Metals with $pT_{50} \geqslant 3.6$ are designated as toxic in both ip and iv routes. The pattern of ip toxicity (Fig. 9-3) resembles that of *per os* toxicity; Se, As, and Te are very toxic metalloids, and Hg, Tl, and Sn are the most toxic metals. All the heavy transitional metals which are essential nutrients are considered toxic by this criterion. All Group V metals tested and Ba fall within the toxic limit.

When administered intravenously Fe, Ni, Cu, and Zn are not excessively toxic, $pT_{50} \geqslant 3.6$ (Fig. 9-4). Selenium and Te are very toxic metalloids (As has not been tested), and Be, Hg, and Mn are the most toxic metals by this mode of administration. Essential metals which are toxic by this limit are Co, Mg, K, and Cr.

Subcutaneous injection shows most metals to be harmful in relatively small quantities; except for La and Ga, all tested have $pT_{50} > 3$. Those with $pT_{50} \geqslant 3.9$ are designated as toxic relative to the others. Beryllium is the most toxic, and Cd, and Hg form the corners of a cabal which shows toxicity by all routes. Surprisingly, Na, and K are quite toxic when administered subcutaneously (Fig. 9-5).

The above patterns focus when both rat and mouse data are combined for comparisons. (Fig. 9-6). The averages used are not weighted for number of observations, and too much reliance cannot be given in all areas except those having more than 2–3 data points (see Table 9-3). Oral and ip administration show similar patterns with metals administered intraperitoneally being about 0.5 pT unit (0.3 M/kg) more toxic than when given orally. Intravenous administration shows a similar pattern. The alkali metals are quite toxic. The few data for sc administration provide little information for comparisons. In general, the average toxicity for all modes of administration increase from left to right in the periodic table of elements. The four peaks of activity suggest these data would conform better to a conical representation. These perturbances in periodicity correlate well with lattice energies for the metals of the fourth and fifth periods; this relates them to the inverse of interatomic distances (Phillips and Williams, 1965). A good correlation also exists between toxicity of metals and

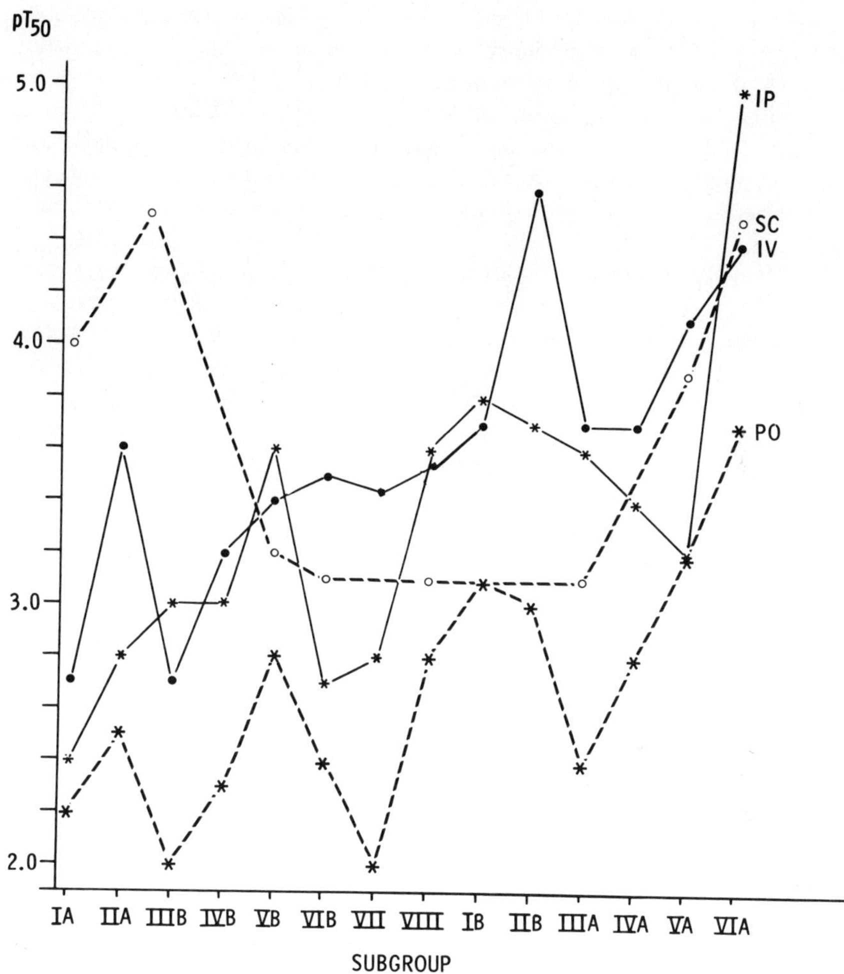

FIG. 9-6. *Summary of pT₅₀ toxicity of metals for mice and rats. The points represent the unweighted means of combined values for rats and mice.*

electronegativity for metals of periods 2, 3, and 4. Metals in periods 4 and 5 correlate less well with Pauling's electronegativity data than with those of Radner and Alfred (Mackay and Mackay, 1969). Toxicity by iv administration correlates with energy of ionization. Other parameters examined showed no direct correlation. The atomic characters contributing to toxicity are limited by solubility. This depresses the potential toxicity and probably accounts for the less-good correlations noted for metals of period

5 when toxicity is compared to electronegativity. This satisfying correlation of toxicity with the periodicity of atomic structure and characteristics provides for predicting metal toxicity in mammals.

When toxicity data from all species by mode of administration are pooled (Table 9-7), the averages may be biased from the vagaries of different numbers of observations in different subgroups. However, trends should be indicated and the compilation shows where more information is desirable.

When data from all species and modes of administration are pooled (Table 9-11), the number of observations is increased, but the differences already noted in number of observations for oral and parenteral administration make for a factor of uncertainty. This is a possible explanation for those cases in which the pT_{100} is greater than the pT_{50}. Since more toxicant should be needed for a pT_{100} than a pT_{50}, pT_{100} values should always be smaller than pT_{50} values. The relative pT_{50} toxicity of different subgroups of the metals for all species and all modes of administration is: VIA > IIB > IB, IIA, IIIA, IIIB, IVA, VA, VB, VIII > IA, IVB, VII.

When the pT_{50} toxicities of all metals are averaged (Table 9-12), the rabbit appears to be the least susceptible to intraperitoneal administration and the most susceptible to oral administration. Both the rat and rabbit have relatively high susceptibility to intravenously administered metals, when compared with other mammals. All animals seemed to be uniformly susceptible to subcutaneously and intramuscularly administered metals.

TABLE 9-11. Toxicity Comparison: All Species and All Modes[a]

Group	pT_M	pT_{50}	pT_{100}
IA	(61) 2.70	(81) 2.45	(11) 1.90
IB	(14) 3.79	(20) 3.21	(7) 3.48
IIA	(58) 2.60	(73) 3.13	(31) 3.02
IIB	(26) 3.72	(51) 3.71	(24) 3.56
IIIA	(40) 3.58	(51) 3.17	(21) 3.45
IIIB	(37) 3.41	(14) 2.91	(21) 3.48
IVA	(21) 3.26	(34) 3.20	(14) 3.06
IVB	(7) 3.07	(29) 2.64	
VA	(24) 4.07	(28) 3.33	(22) 3.88
VB	(8) 3.37	(25) 3.21	(38) 3.58
VIA	(18) 4.14	(25) 4.05	(8) 4.16
VIB	(21) 3.20	(18) 2.83	(23) 2.76
VII	(5) 2.68	(9) 2.72	(8) 3.22
VIII	(51) 3.30	(63) 3.36	(31) 2.73

[a]Values in parentheses indicate number of experiments considered.

TABLE 9-12. pT_{50} Toxicity of All Metals[a]

	ip	iv	oral	sc	im
Mouse	(132) 3.24	(30) 3.48	(60) 2.83	(13) 3.43	(2) 3.42
Rat	(125) 3.05	(47) 3.94	(157) 2.58	(12) 3.68	(2) 3.30
Hamster		(1) 2.87			
Guinea pig	(43) 3.42	(1) 2.56	(12) 2.99	(6) 3.38	
Rabbit	(3) 2.87	(12) 3.87	(9) 3.27	(4) 3.15	(1) 3.50
Dog	(1) 2.46	(4) 3.58	(1) 1.87	(1) 4.56	

[a]Values in parentheses indicate number of experiments considered.

TABLE 9-13. Toxicity of All Metals for All Animals[a]

Mode	pT_M	pT_{50}	pT_{100}
Intraperitoneal	(74) 3.42	(304) 3.18	(30) 3.01
Intravenous	(83) 3.68	(96) 3.76	(70) 3.91
Per os	(107) 2.97	(239) 2.69	(47) 2.63
Subcutaneous	(83) 3.01	(37) 3.51	(85) 3.30
Intramuscular	(14) 3.66	(5) 3.39	(6) 3.52
Intratracheal	(4) 2.93	(1) 2.37	
Rectal			(1) 2.39

[a]Values in parentheses indicate number of experiments considered.

Hopefully, these emerging patterns of toxicity will be reinforced as new data become available. The information gathered here will serve as a standard to compare later data and to stimulate more systematic work.

When all metals and all modes of administration for all species are grouped (Table 9-13), the numbers increase to allow a general view of the whole. Oral administration is less harmful than parenteral administration. When the minimal intratracheal and intrarectal data are ignored, intraperitoneal administration is generally the least harmful of the parenteral routes. Intravenous administration is most toxic for many metals; there is often a direct correlation between ionization energy and metal toxicity.

APPENDIX

1969 International Atomic Weights, based on the assigned relative atomic mass of $^{12}C = 12$: The following values apply to elements as they exist in materials of terrestrial origin and to certain artificial elements.

Element	Symbol	Atomic number	Atomic weight	Element	Symbol	Atomic number	Atomic weight
actinium	Ac	89	(227)	fluorine	F	9	18.9984
aluminum	Al	13	26.9815	francium	Fr	87	(223)
americium	Am	95	(243)	gadolinium	Gd	64	157.2_5
antimony	Sb	51	121.7_5	gallium	Ga	31	69.72
argon	Ar	18	39.94_8	germanium	Ge	32	72.5_9
arsenic	As	33	74.9216	gold	Au	79	196.9665
astatine	At	85	(210)	hafnium	Hf	72	178.4_9
barium	Ba	56	137.33	hahnium	Ha	105	(260)
berkelium	Bk	97	(247)	helium	He	2	4.00260
beryllium	Be	4	9.01218	holmium	Ho	67	164.9303
bismuth	Bi	83	208.9806	hydrogen	H	1	1.008_0
boron	B	5	10.81	indium	In	49	114.82
bromine	Br	35	79.904	iodine	I	53	126.9045
cadmium	Cd	48	112.41	iridium	Ir	77	192.2_2
calcium	Ca	20	40.08	iron	Fe	26	55.84_7
californium	Cf	98	(249)	krypton	Kr	36	83.80
carbon	C	6	12.011	kurchatovium	Ku	104	(261)
cerium	Ce	58	140.12	lanthanum	La	57	138.905_5
cesium	Cs	55	132.9055	lawrencium	Lr	103	(257)
chlorine	Cl	17	35.453	lead	Pb	82	207.2
chromium	Cr	24	51.996	lithium	Li	3	6.94_1
cobalt	Co	27	58.9332	lutetium	Lu	71	174.97
copper	Cu	29	63.54_6	magnesium	Mg	12	24.305
curium	Cm	96	(245)	manganese	Mn	25	54.9380
dysprosium	Dy	66	162.5_0	mendelevium	Md	101	(256)
einsteinium	Es	99	(254)	mercury	Hg	80	200.5_9
erbium	Er	68	167.2_6	molybdenum	Mo	42	95.9_4
europium	Eu	63	151.96	neodymium	Nd	60	144.2_4
fermium	Fm	100	(255)	neon	Ne	10	20.17_9

(Cont'd)

1969 International Atomic Weights (*Cont'd*)

Element	Symbol	Atomic number	Atomic weight	Element	Symbol	Atomic number	Atomic weight
neptunium	Np	93	237.0482	selenium	Se	34	78.9_6
nickel	Ni	28	58.7_1	silicon	Si	14	28.0855
niobium	Nb	41	92.9064	silver	Ag	47	107.868
nitrogen	N	7	14.0067	sodium	Na	11	22.9898
nobelium	No	102	(254)	strontium	Sr	38	87.62
osmium	Os	76	190.2	sulfur	S	16	32.06
oxygen	O	8	15.999_4	tantalum	Ta	73	180.947_9
palladium	Pd	46	106.4	technetium	Tc	43	98.9062
phosphorus	P	15	30.9738	tellurium	Te	52	127.6_0
platinum	Pt	78	195.0_9	terbium	Tb	65	158.9254
plutonium	Pu	94	(244)	thallium	Tl	81	204.3_7
polonium	Po	84	(210)	thorium	Th	90	232.0381
potassium	K	19	39.0983	thulium	Tm	69	168.9342
praseodymium	Pr	59	140.9077	tin	Sn	50	118.6_9
promethium	Pm	61	(147)	titanium	Ti	22	47.9_0
protactinium	Pa	91	231.0359	tungsten	W	74	183.8_5
radium	Ra	88	226.0254	uranium	U	92	238.029
radon	Rn	86	(222)	vanadium	V	23	50.941_4
rhenium	Re	75	186.2	wolfram	W	74	183.8_5
rhodium	Rh	45	102.9055	xenon	Xe	54	131.30
rubidium	Rb	37	85.467_8	ytterbium	Yb	70	173.0_4
ruthenium	Ru	44	101.0_7	yttrium	Y	39	88.9059
rutherfordium	Rf	104	(261)	zinc	Zn	30	65.3_7
samarium	Sm	62	150.4	zirconium	Zr	40	91.22
scandium	Sc	21	44.9559				

PERIODIC TABLE

IA	IIA	IIIB	IVB	VB	VIB	VIIB	VIII			IB	IIB	IIIA	IVA	VA	VIA	VIIA	NOBLE GASES
1 H 1.0079†																1 H 1.0079†	2 He 4.00260
3 Li 6.941†	4 Be 9.01218											5 B 10.81	6 C 12.011	7 N 14.0067	8 O 15.9994†	9 F 18.99840	10 Ne 20.179†
11 Na 22.98977	12 Mg 24.305											13 Al 26.98154	14 Si 28.0855	15 P 30.97376	16 S 32.06	17 Cl 35.453	18 Ar 39.948†
19 K 39.0983	20 Ca 40.08	21 Sc 44.9559	22 Ti 47.90†	23 V 50.9414†	24 Cr 51.996	25 Mn 54.9380	26 Fe 55.847†	27 Co 58.9332	28 Ni 58.71†	29 Cu 63.546†	30 Zn 65.38	31 Ga 69.72	32 Ge 72.59†	33 As 74.9216	34 Se 78.96†	35 Br 79.904	36 Kr 83.80
37 Rb 85.4678†	38 Sr 87.62	39 Y 88.9059	40 Zr 91.22	41 Nb 92.9064	42 Mo 95.94†	43 Tc 98.9062	44 Ru 101.07†	45 Rh 102.9055	46 Pd 106.4	47 Ag 107.868	48 Cd 112.41	49 In 114.82	50 Sn 118.69†	51 Sb 121.75†	52 Te 127.60†	53 I 126.9045	54 Xe 131.30
55 Cs 132.9054	56 Ba 137.33	57 *La 138.9055†	72 Hf 178.49†	73 Ta 180.9479†	74 W 183.85†	75 Re 186.2	76 Os 190.2	77 Ir 192.22†	78 Pt 195.09†	79 Au 196.9665	80 Hg 200.59†	81 Tl 204.37†	82 Pb 207.2	83 Bi 208.9804	84 Po (210)	85 At (210)	86 Rn (222)
87 Fr (223)	88 Ra 226.0254	89 †Ac (227)	104 § (260)	105 § (260)													

* Lanthanoid Series

58 Ce 140.12	59 Pr 140.9077	60 Nd 144.24†	61 Pm (147)	62 Sm 150.4	63 Eu 151.96	64 Gd 157.25†	65 Tb 158.9254	66 Dy 162.50†	67 Ho 164.9304	68 Er 167.26†	69 Tm 168.9342	70 Yb 173.04†	71 Lu 174.97

† Actinoid Series

90 Th 232.0381	91 Pa 231.0359	92 U 238.029	93 Np 237.0482	94 Pu (244)	95 Am (243)	96 Cm (247)	97 Bk (247)	98 Cf (251)	99 Es (254)	100 Fm (257)	101 Md (258)	102 No (255)	103 Lr (256)

§The International Union for Pure and Applied Chemistry has not adopted official names or symbols for these elements.

†These weights are considered reliable to ±3 in the last place. Other weights are reliable to ±1 in the last place.

Atomic weights corrected to conform to the 1971 values of the Commission on Atomic Weights.

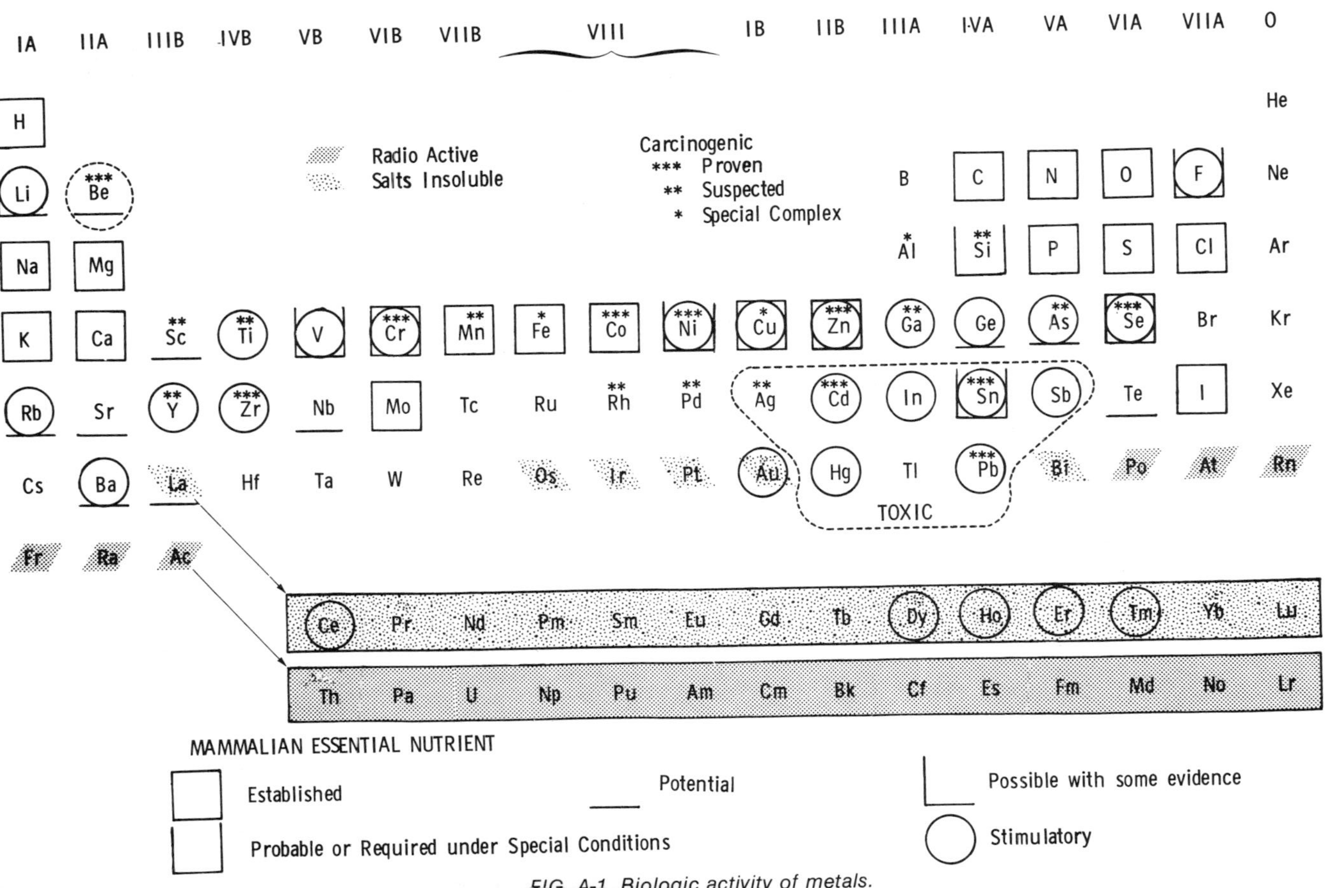

FIG. A-1 Biologic activity of metals.

GLOSSARY

aboral: opposite to the mouth or caudal

acidosis: acidemia or a lowered blood bicarbonate level

actinide: a radioactive element resembling actinium, the lightest member of the series

active transport: movement of material across a membrane with the use of chemical energy (ATP) against a concentration gradient

adenoma: a benign epithelial tumor

aerosol: a colloid in which the continuous phase is a gas

agglutination: the clumping of cells or particles; the precipitation of a complex as antigen–antibody

alkali metal: any of subgroup IA metals

alkaline earth metals: subgroup IIA metals

allergy: a hypersensitive state acquired by previous exposure to an allergen

allosteric effect: a change brought about to a site on a molecule other than the one directly affected

alluvial: material left or affected by running water

aluminum dextran: an aluminum derivative of a soluble starch

alveolus: a microscopic sac

Alzheimer's disease: presenile dementia

amphibole: a metal silicate such as asbestos

amphoteric: capable of reacting in two ways, i.e., acidic or basic

anabolism: constructive metabolism

anaphylaxis: a state of exaggerated allergic reaction

anionic: a state of carrying a negative net charge

antimetabolite: an inhibitor which competes with a metabolite for an active site on a macromolecule or membrane

anuresis: retention of urine in the bladder

apnea: transient cessation of the breathing impulse that follows forced breathing *apoferritin:* ferritin with no iron bound to it

arachoid cavity: a cavity with a network of spaces

argyria: silver poisoning

argyrol: a registered name for a silver protein complex

aromatic polynuclear hydrocarbon: a steroid or other multi-ring aromatic carbon compound

asbestos: a fibrous magnesium and calcium silicate which usually contains other metals and a variety of organic compounds

asbestosis: a form of lung disease caused by asbestos fiber inhalation

astringent: an agent which causes contraction of blood vessels

ataxia: failure of muscular coordination, irregularity of muscular action

atomic absorption: absorption of radiant energy by atoms and subsequent emission in spectral lines

bactericidal: destructive to bacteria

baritosis: pneumoconiosis due to the inhalation of dusts of barium salts or barites

basement membrane: delicate transparent layer underlying the epithelium of mucous membranes and secreting glands

beryllosis: beryllium poisoning

bidentate: two electron donor atoms in a chelate

biliary obstruction: blockage of the bile ducts

bilirubin: a bile pigment

biologic ligand: N-, O-, S-containing groups of biologic macromolecules, which bind metal ions

biotransformation: a change in a molecule brought about through metabolism

blood dyscrasias: abnormal composition of the blood

Bowen's disease or precancerous dermatitis: skin disease marked by the formation of a pinkish papule or tubercle covered by a thickened horn layer

bradycardia: abnormal slowness of heart beat or slowing of pulse rate

bronchiole: one of the small, final branches of the respiratory ducts

bronchiolitis: cirrhosis of the lung due to induration of the walls of bronchioles

cancer: cell growth or tumor which invades and metastasizes other tissues to the detriment of the individual

carbonyl: the CO radical or component of a compound, i.e., nickel carbonyl is $Ni(CO)_4$

carcinogen: a cancer-producing substance

carcinogenicity: the capability to produce cancer

carcinoma: one of the two broad catagories of cancer, a malignant growth of cells of epithelial origin

catabolic: processes on metabolism which degrade compounds

catechol: ortho dihydroxybenzene

cathartic: an agent which causes fast bowel elimination

cationic: containing a net positive charge

cell mediated immunity: any of the cell–cell interactions which protect the body from foreign cells

cellular eosinophilia: a condition of the cell when it could be readily stained with eosin

ceruloplasmin: a glycoprotein which transports copper in the blood

charge redistribution: distribution of electrical charges in macromolecules due to shifting of electrons

chelate: a compound in which the metal becomes incorporated into a ring structure

choreal movements: ceaseless, jerky movements

chromosomal aberration: abnormality in the number or the constitution of chromosomes which is expressed phenotypically

chronic toxicity: toxicity derived from repeated application of the material over many days

chrysiasis: deposition of gold in the tissues

chrysolite: magnesium iron silicate

chyme: the semifluid digesta in the gastrointestinal tract

cilia: minute finger-like projections of the cell, usually from a single area of the surface

cocarcinogen: an agent which enhances the carcinogenicity of a carcinogen

collagen: the supportive protein of connective tissue, skin, and skeletal tissues

colloid: finely dispersed matter in a dispersion medium

colloidoclastic shock: a breakup of the physical equilibrium of the colloids of the body producing an anaphylactoid crisis

coma: unconscious state

conjunctivitis: inflammation of the conjunctiva, the delicate membrane which lines the eyelids

coordination number: number of monovalent groups or radicals that could combine with a polyvalent ion through a coordinate bond

coprecipitation: the agglutination of an insoluble material by the precipitation of another complex

cornification: conversion of epithelium to heavy stratified squamous epithelium

corpus striatum: subcortical mass of gray and white matter in front of the thalamus in each cerebral hemisphere

crypt: a compartment formed by a blind tube or pit leading to an open surface

cumulative toxicity: toxicity resulting from accumulation of repeated applications of the toxic substances

cyanosis: blueness of the skin, e.g., cardiac malformation causing insufficient oxygenation of the blood

cytochrome: any of the electron-transfer hemoproteins; i.e., cytochrome *c*

decarboxylase: an enzyme which catalyzes the removal of the carboxyl group from a compound

delayed toxicity: the toxicity in which the symptoms are separated from the application by a significant time; a latent carcinogen may result in cancer 6–12 months later in mice, or 10–20 years later in humans

demulcent: a material which decreases irritation in a tissue

demyelinate: destroy or remove the myelin sheath of nerves

denaturation: the destruction of the natural structure of a protein; active proteins, such as enzymes, are inactivated by denaturation

desferrioxamine: a potent iron-chelating agent

desmosome: an intercellular attachment site

desquamation: skin shedding

detoxication: a reduction in toxicity

digesta: partially digested food; i.e., chyme

digital tremors: involuntary trembling of fingers

dipole distance: small distance which separates two electrically or magnetically charged particles of opposite sign

distal toxicity: toxicity associated with effects observed a long time following the administration of a toxic compound

dithiozon: diphenylthiocarbozone-chelating agent for heavy metals

diuresis: increased secretion of urine

diuretic: an agent which increases urine secretion

DTPA: diethylenetriaminepentaacetic acid, a metal-chelating compound

dysarthria: imperfect speech due to damaged nerves

dysphagia: difficulty in swallowing

dyspnea: difficult or labored breathing

eccrine sweat glands: ordinary exocrine sweat glands

EDDHA: ethylenediamine-*bis*-(1-hydroxy phenyl)-acetic acid, a metal-chelating compound

EDTA: a general chelating agent, *e*thylene*d*iamine*t*etra*a*cetic acid

effusion: a pouring forth

electronegativity: the ability to gain electrons

electron mobility: movement or shifting of electrons from a molecule

electron polarizability: ability of electron to cause polarization by being absorbed or removed from a molecule

electron orbitals: the electron energy levels assumed by electrons associated with an atomic nucleus

electron-transport system: the transference of electrons from one compound to another of increasing redox potential; in the mitochondria of mammals it is used to reduce oxygen in the formation of H_2O

electrophoretic mobility: the movement of an ion through a liquid in an electric field

electrophilic: having a strong attraction for electrons

electropositivity: the ability to lose electrons

electrovalent: chemical combination between two atoms, involving transfer of electrons *element:* an atom, one of the 110+ distinct kinds of matter

embolism: the blockage of a tube, i.e., an artery or vein, by an unusual body

emesis: vomiting

emetic: an agent that induces vomiting

encephalopathy: any degenerative disease of the brain

endocytosis: the engulfment of material with vacuolization of the cell membrane

endoplasmic reticulum: an ultramicroscopic system of membrane-bound cavities ramified throughout the cytoplasm of most cells

enteral: pertaining to the small intestine

enterohepatic system: the liver–small-intestine interchange of material absorbed by the ileum, carried to the liver and discharged into the intestine with the bile

enzyme: one of the many protein catalysts produced by each cell

erythrocyte: red blood cell

essential nutrient: an element or compound which can not be synthesized at a rate commensurate to the needs of the organism

exogenous: outside the cell or organism

exudative diathesis: diathesis marked by thickening of the lingual mucous membrane; seborrhea of the scalp, severe itching, and glandular enlargement, etc.

fascia: fibrous tissue associated with internal organs

ferritin: a protein–ion complex which contains 23% iron by weight

ferroxamine: a simple iron-chelating compound

fibrosarcoma: usually a malignant tumor, a sarcoma containing or derived from primitive connective tissue

fibroma: a tumor composed of connective tissue

fibrosis: formation (or degeneration) of fibrous tissue

filariasis: a disease state due to threadworm infection

follicle: a pouch-like cavity

foreign-body carcinogen: a nonspecific carcinogen which is dependent upon size, shape, and surface characteristics for its activity

gavage: forced feeding into the stomach

genotype: the genetic composition of the individual

germfree: having no demonstrable microorganisms

gingivitis: inflammation of the mucus membranes of the gums

gliosis: disease condition associated with the presence of gliomas or with the development of neuroglia tissue

glomerular filtration: the filtration of liquid in the kidney uriniferous tubules

gluteal muscle: the major muscles of the buttocks

glycocalyx: the glycoprotein–polysaccharide extrusion of cells

glycoprotein: proteins which contain appreciable quantities of carbohydrate

goblet cell: an epithelial cell containing a globule of mucin

granulation: the formation of small masses of tissues

granuloma (pl.-ata): a tumor-like mass of granulation tissue *granulomatous peritonitis:* inflammation of the peritoneum by exudation of granular material into the peritoneum

groups: vertical columns on the periodic chart

haptens: a simple substance which can unite with a protein to form an antigen

hard acid: acid fragment of a compound which holds its valence electrons firmly

hard base: base fragment of a compound which holds its electrons firmly

HEDTA: hydroxyethylenediaminetetraacetic acid, a metal-chelating compound

hematinic: an agent which improves the quality of the blood

hemochromatosis: excess deposition of iron in tissues

hemolytic anemia: anemia due to active destruction of red blood cells within the circulation

hemolytic crisis: crisis condition marked by hemolysis

hemosiderosis: abnormal storage of iron in tissues without tissue damage

histamine: a compound which causes dilation of blood vessels

histogenesis: the development of tissues from undifferentiated cells of the germ layer of the embryo

homeostatis: the tendency of the organism to maintain the normal internal environment

hormesis: the act of excitation

hormetic: stimulatory

hormetin: a stimulant

hormology: the knowledge of excitation, the study of stimulation

horny layer: the layer of cells in the epidermis which become hardened with fatty degeneration

HSAB: hard and soft acids and bases; a theory of combining properties promulgated by Pearson (1963)

humoral immunity: that part of the defense which encompasses antibody production

hydrocephalus: a condition in which fluid accumulates in the head to cause enlargement, mental deterioration, and convulsions

hydroxamate: compounds derived from the substitution of a hydrogen from the amino group of hydroxylamine (NH_2OH)

hyperemia: excess of blood in any part of the body

hypermagnesemia: an excess of magnesium in the blood

hypercalcemia: an excess of calcium in the blood

hyperplasia: an abnormal increase in number and/or size of cells in a tissue

hypophosphite: a salt of hypophosphorous acid, H_3PO_3

hypotension: abnormally low blood pressure

"idiopathic" hemachromatosis: metabolic disorder in the male with accumulation of large amounts of iron in body tissues

inherent toxicity: toxicity due to the internal character of the material

imino group: $RC{=}NH$

infarction: the formation of an infarct, the necrosis of tissue due to local deficiency of blood supply

interdigitation: an interlocking due to finger-like processes

interpolate: the extension of information in accordance with known information

interstitial: between cells or tissues

interstitial pneumonitis: localized acute inflammation of the lung without gross toxemia

interstitial fluid: the plasma-like fluid which bathes cells

intradermally: within the dermis

intraperitoneal: within the peritoneum

intrathecal: within the sheath (of the spinal cord)

intrinsic toxicity: inherent toxicity

inunction: application (with friction) to the skin

ionic radius: effective radius of a charged ion in H_2O

ionization potential: energy or work required (expressed in electron volts) to remove a given electron from its atomic orbit

iron dextran: a complex of iron with dextran, a soluble glucose polysaccharide

itai-itai: neurologic disease associated with women exposed to mercury intoxication, first observed in Japan

jaundice: yellow color due to excess of bilirubin in tissues

kala-azar: a parasitic infection from *Leishmania donorani* which is often fatal

keratin: an insoluble protein which is the main component of hair, nails, and horns

lamina propria: middle fibrous layer, e.g., of the tympanic membrane

lanthanides, lanthanoid: any of the series of elements with characteristics similar to lanthanum, the lightest of the series

latent carcinogenesis: carcinogenesis which develops an appreciable time following administration of the agent

LD_1, LD_{50}, LD_{100} : lethal dose for 1, 50, or 100% of the subjects to which the agent was administered; $LD_1 = LD_M$

leishmaniasis: infection of *Leishmania*

leukemia: a persistive excess of white blood cells

leukocyte: any of the white blood cells or amoeboid colorless tissue cells

ligand: an organic molecule which donates electrons to form coordinate covalent bonds with metal ions

liposarcoma: a malignant tumor derived from primitive lipoblastic cells

lymphosarcoma: a malignant growth of lymphoid tissue other than Hodgkin's disease

lysosome: a small cytoplasmic sac, or subcellular organelle, which contains hydrolytic enzymes

macromolecule: a polymeric compound with high molecular weight, usually protein, nucleic acid, carbohydrate, or a mixture of these with each other and/or lipid

macrophage: an amoeboid colorless cell which is a normal constituent of most tissues

malignant tumor: an invasive, metasticizing, and progressive tumor which would normally continue to grow to the detriment of the host

malignant neoplasm: a malignant tumor

manic psychosis: a disease characterized by emotional overreactions

mesothelioma: a tumor formed from the epithelium which covers serous membranes

metabolize: the sum of anabolism and catabolism, the change in molecules brought about by living organisms

metallothionein: protein isolated from liver capable of binding toxic metals such as mercury, cadmium, etc.

metastasize: to spread from one tissue to another by cell transfer, the main feature of malignant tumors

methemoglobinemia: presence of methemoglobin in the blood

microcalyx: the mucin extruded from a cell

micrognathia: unusual or undue smallness of the lower jaw with recession of the chin

microvilli: the finger-like protrusions from a cell lining the villus

milieu: the environment

millennia: thousands of years

Minamata disease: mercury poisoning characterized by severe neurologic disorder

MLD: minimum lethal dose

mole: the weight in grams equivalent to the formula mass

molar: the content as expressed by moles per liter

molal: the expression of content by moles per kilogram

mucosa: a mucous membrane

multidentate: more than 2^e donor atoms in a chelate

myoblast: an embryonic cell which develops into a muscle cell

myocardia: heart muscle

myocarditis: inflammation of the myocardium or muscular walls of the heart

myosarcoma: a malignant tumor derived from muscle tissue

nanogram: 10^{-9} gram, one-billionth of a gram

necrosis: cell death in a localized area

neoplasm: a new, abnormal growth

nephrotoxic: harmful to the kidney

neuraminic acid: sialic acid, a complex sugar

neuronal perikarya: the protoplasmic mass of a neuron

neutron activation: the production of unstable isotopes by neutron bombardment of atoms *neutron flux:* the density of neutrons in a neutron beam

nodular granulation: node filled with granular material

nonspecific carcinogen: surface carcinogen, a carcinogen which depends more upon physical size, shape, and polish than chemical character for its activity

occlusion: the hiding of a chemical by absorption or other means

olation: hydrolytic salt formation

oligodynamic: active in minute quantities; the oligodynamic action of heavy metals refers to the bactericidal activity of minute concentrations of heavy metal ions

oncogenic: carcinogenic

organelle: any discrete functioning part of a cell

organogenesis: the development of organs

osteomalacia: softening of the bones, which then become flexible, leading to deformities

oxolation: addition of OH groups to products of hydrolysis of metal salts

oxy acid: acids such as perchloric and sulfuric, containing oxygen, hydrogen, and another electronegative atom

oxy salts: a salt of an oxy acid

palliative: to ease without curing

pallor: paleness

papillary: resembling or pertaining to a nipple

papilloma: an epithelial benign tumor

parameter: a measured variable

parenteral: other than alimentary tract

peptidase: an enzyme which catalyzes peptide hydrolysis

percutaneous: through the skin

period: horizontal lines in the periodic chart

peritoneal cavity: the cavity formed within the peritoneum, fluid—the fluid within the peritoneal cavity

peritoneum: the serous membrane around the viscera

peroxidase: an enzyme which catalyzes the oxidation of substrate in the presence of hydrogen peroxide

Peyer's patches: lymphoid aggregates in the wall of the intestine

phagocytosis: the envelopment of solids by the cell membrane of a cell

pharyngeal: pertaining to the pharynx

pharmacotoxic: relationship of toxicity to a drug

phosphor: a phosphorescent compound or material

phosphorylase: an enzyme which catalyzes phosphorylysis

phylogenetic: pertaining to the development of the species or race

physiologic homeostasis: maintenance of the equilibrium of the organism by physiologic means

picogram: 10^{-12} gram, micromicrogram

pinocytosis: drinking through evagination of the cell membrane followed by vacuolization of the fluid

placental membrane: the semipermeable membrane which separates fetal from
maternal blood

pleural effusion: the presence of fluid in the pleural space

pneumoconiosis: chronic fibrous reaction in the lungs to the inhalation of dust

pollutant: any obnoxious material which renders the environment unclean

polydipsia: excessive thirst

polyencephalomalacia: morbid softness or softening of the brain

polymeric: compounds with only repetitive units in their structure

polycythemia: an increase in red cells in the body

polydactyly: excess digits on hands or feet

portal vein or blood: the vein or blood going from the intestine and spleen to the
liver

precancerous neoplasm: a conditioned tissue which will develop a tumor

primary carcinogen: a material which can induce cancer without undergoing
change

primary site: the place of first contact

primary target tissue: the most sensitive tissue

prophylaxis: preventive treatment

prostigmin: tradename for a preparation of neostigmine, a cholinergic drug

proximal toxicity: toxicity associated with immediate effect following the administration of the toxic compound

pulmonary emphysema: unnatural distention and rupture of the air vesicles of the
lungs

pulmonary fibrosis: formation of fibrous tissue or fibroid degeneration of the lungs

purgative: causing bowel evacuation

purulent: containing pus

quantum number: an index of the energy of a given electron level within an atom

quaternary ammonium hydroxide: ammonium hydroxide in which all four hydrogen atoms have been replaced by alkyl radicals

quaternary structure: the spatial relationships among separate subunits of a
macromolecule

radiocolloid: an especially small colloidal particle formed from a few molecules or
ions which can be detected by only radioactive tracers

radioactive metal: a metal which gives off quantities of radiation which are significant from a health viewpoint

radionuclide: a radioactive species of atom or isotope of an atom

recondite toxicity: hidden toxicity which would go unnoticed were it not for special
conditions

retrobulbar neuritis: inflammation in that portion of the optic nerve which is
posterior to the eyeball

sclerosis: hardening of tissue from inflammation and diseases of the interstitial substance

silicosis: disease of lungs caused by dust inhalation

soft acid: acid fragment of a compound not capable of holding its valence electrons firmly

soft base: base fragment of a compound not capable of holding its valence electrons firmly

spermatogenesis: process of formation of spermatozoa

spindle-cell fibrosarcoma: a sarcoma derived from fibroblasts in which the cells maintain their spindle shape

spirocheticidal: kills significant numbers of spirochetes

stomatitis: inflammation of the oral mucosa

stratum corneum: the outer strata of dead cells of the epidermis

stratum germanitivum: the innermost layer of reproducing cells of the epidermis

stratum granulosum: the granular layer of cells in the epidermis

stratum lucidum: the clear layers of cells beneath the stratum corneum

stratum mucosum: the thin layer of mucosus cells

styptic: astringent, or hemostatic

sublingual: beneath the tongue

sulfhydryl group: —SH as in R—SH

suprapharyngeal: structures associated with the upper part of the pharnyx

superoxide dismutase: copper-containing protein formerly known as cytocupreine, hepatocupreine, cerebrocupreine, and erythrocupreine

synergism: the working together to produce a greater action. Antonym of antagonism

[T]: molar or molal concentration of a toxin or toxicant

tachycardia: excessive rapidity in the action of the heart

teratogenesis: the production of physical fetal defects

teratology: the study of physical fetal anomalies

teratoma: a tumor containing congenitally derived fetal remains

terminal web: the clear area underlying the microvilli in a villus epithelial cell

tetrodotoxin: a toxin from the ovaries of the globe fish

thiol compounds: contain RSH and undergo the reactions

$$2\,RSH \rightarrow RSSR + 2H$$
$$RSH \rightarrow RS^- + H^+$$

thiomalate: ion or salt of mercaptosuccinic acid

thrombus: blood clot or clump of cells which blocks a vessel

tonoclonic convulsions: spasm or convulsion of the muscles

toxicant: a harmful agent other than the true protein toxins

toxicity: poisonousness. Different types are:

  acute: critical symptoms develop quickly

chronic: action from small quantities for a long time

cumulative: effect due to gradual accumulation of toxicant above a critical level

delayed: adaptation of organism delays the onset of symptoms

distal: effects observed long after administration of toxicant

latent: equivalent to distal

proximal: effects shown immediately upon administration

recondite: subtle effects following prolonged exposure to threshold amounts of relatively harmless compounds

toxin: a protein poison

toxicosis: the morbid condition induced by a poison

transcription: the copying of a message

transferrin: a serum β-globulin which binds and transports iron

transfollicular: movement across follicles

transit time: the time required to move from one designation to another

transition metal: metals with variable charges resulting from having electrons in varying orbits, i.e., iron

translation: the process of changing the information base

transuranium: those elements with atomic weights greater than uranium

trypanosome: a parasitic protozoan of the genus *Trypanosoma*

tubular reabsorption: the recovery of the filtrate by kidney tissues

tumor: a new growth of cells to give an unusual mass

unidentate ligand: chelate formation involving one ligand from each molecule

urethritis: inflammation of the urethra

urticaria: nettle, rash, or hives

vascular granulomata: tumor or neoplasm made up of granulation tissue and blood vessels

ventricular arrhythmia: variation from the normal rhythm of the ventricle

ventricular fibrillation: irregular heart beats due to incoordinate contraction of individual muscle fibers of heart muscle

zymogen activation: a pre-enzyme is converted into an active form

REFERENCES

Abdulla, M., Arnesjo, B., and Ihse, I. 1974. Metylering av kvicksilver vid tarmsjukdomar, *Svenska Lakartidningen 71:*810–811.

Aberg, B., Ekman, L., Falk, R., Persson, G., and Snihs, J-O. 1969. Metabolism of methyl mercury (^{203}Hg) compounds in man, *Arch. Environ. Health 19:*478.

Abruden-Bodea, I. B., and Bodea, C. 1973. Studies on substrate oxidation in rat liver nucleus: Comparison of nuclear and mitochondrial respiration, in: *Abstracts of the 9th International Congress of Biochemistry,* Stockholm, p. 239.

Acherman, H. 1948. *Preliminary Studies on the Toxicity of Thorium,* U.S.A.E.C. Report U-R-13, University of Rochester Atomic Energy Project.

Adam, M., and Kuhn, K. 1968. Investigation on the reaction of metals with collagen in vivo: 1. Comparison of reaction of gold thiosulfate with collagen *in vivo* and *in vitro, Eur. J. Biochem. 3:*407–410.

Addink, N. W. H. 1960. Subnormal levels of carbonic anhydrase in blood of carcinoma patients, *Nature 186:*253.

Addink, N. W. H., and Frank, L. J. P. 1959. Remarks apropos of analysis of trace elements in human tissues, *Cancer 12:*544–551.

Adelstein, S. A., and Vallee, B. L. 1961. Copper metabolism in man. *N. Engl. J. Med. 265:*892–897.

Agnew, W. F., and Cheng, J. T. 1971. Protein binding of tellurium 127 by maternal and fetal tissues of rat, *Toxicol. Appl. Pharmacol. 20:*346–356.

Agnew, W. F., Fauvre, F. M., and Pudenz, R. H. 1968. Tellurium hydrocephalus: distribution of tellurium 127 between maternal, fetal and neonatal tissues of the rat. *Exp. Neurol. 21:*120–132.

Aikawa, J. A. 1963. *The Role of Magnesium in Biologic Processes,* Thomas, Springfield.

Aikawa, J. A. 1971. *The Relationship of Magnesium to Disease in Domestic Animals and in Humans,* Thomas, Springfield.

Aiyar, A. S., and Sreenivasan, A. 1961. Effect of vanadium administration on coenzyme Q metabolism in rats. *Proc. Soc. Exp. Biol. Med. 107:*914–918.

Albert, R. A., Klevin, P., Fresco, J., Harley, J., Harris, W., and Eisenbud, M. 1955. Industrial hygiene and medical survey of a thorium refinery, *Arch. Ind. Health 11:*234–245.

Albert, R. E. 1966. *Thorium: Its Industrial Hygiene Aspects,* Academic Press, New York.

Aldrich, R. A. 1958. Acute iron toxicity, in: *Iron in Clinical Medicine* (R. O. Wallerstein and S. R. Mettler, eds.), University of California Press, Berkeley, pp. 99–104.

Aldridge, W. N., Barnes, J. M., and Denz, F. A. 1949. Experimental beryllium poisoning, *Brit. J. Exp. Pathol. 30:*375–405.

Aldridge, W. N., and Thomas, M. 1966. The inhibition of phosphoglucomutase by beryllium, *Biochem. J. 98:*100–104.

Alexander, C. S. 1969. Cobalt and the heart, *Ann. Intern. Med. 70:*411.

Alfrey, A. C., LeGendre, G. R., and Kaehny, W. D. 1976. The dialysis encephalopathy syndrome. Possible aluminum intoxication. *New Engl. J. Med. 294:*184–188.

Allaway, W. H., Kubota, J., Losee, F., and Roth, M. 1968. Selenium, molybdenum, and vanadium, *Arch. Environ. Health 16:*342–348.

Al-Rashid, R. A. 1971. Syndromes of iron overload, *Clin. Toxicol. 4:*571–578.

Altman, P. L., and Dittmer, D. S. 1968. *Metabolism,* Federation of American Societies for Experimental Biology, Bethesda, Md.

Altman, P. L., and Dittmer, D. S. 1973. *Biology Data Book,* 2nd ed., vol. 2, Federation of American Societies for Experimental Biology, Bethesda, Md.

Ambard, L., and Beujard, E. 1904. Causes de hypertension arterielle, *Arch. Gen. Med. 1:*520–524.

Amdur, M. L. 1947. Tellurium: Accidental exposure and treatment with BAL in oil. *Occup. Med. 3:*386–391.

Amdur, M. L. 1958. Tellurium oxide an animal study in acute toxicity. *Arch. Ind. Health 17:*665.

Ammerman, C. B., Flick, K. R., Hansard, S. L., and Miller, S. H. 1973. *Toxicity of Certain Minerals to Domestic Animals: A Review.* National Feed Ingredients Association, Des Moines, Iowa.

Andreoli, V. M. 1968. Sex differences in some toxic effects of lithium carbonate, *Experientia 24:*4155–4156.

Angino, E. E., Cannon, H. L., Hambidge, K. M., Voors, A. W., and Mertz, W. 1972, in: *Asilomar Workshop on the Geochemical Environment in Relation to Health and Disease, 1972. Geochemistry and the Environment,* Vol. I, National Academy of Sciences, Washington, pp. 36–42.

Anke, M., Hennig, A., Schneider, H. J., Ludke, H., Von Cagern, W., and Schlegal, H. 1970. The interrelations between cadmium, zinc, copper and iron in metabolisms of hens, ruminants and man. in: *Trace Element Metabolism in Animals* (C. F. Mills, ed.), Livingstone, Edinburgh, p. 317.

Anspaugh, L. R., and Robison, W. L. 1971. Trace elements in biology and medicine. in: *Progress in Atomic Medicine,* Vol. 4. *Recent Advances in Nuclear Medicine,* (J. H. Lawrence, ed.), Grune and Stratton, New York, p. 63.

Antonova, M. V., Pristupa, V. A., and Papov, B. P. 1968. Effect of small quantities of trace elements in food ration on the immunobiological reactivity of the body, *Gig. Sanit. 33:*39–42.

Aoki, F. Y., and Ruedy, J. 1971. Severe lithium intoxication: Management without dialysis and report of a possible teratogenic effect of lithium. *Can. Med. Ass. J. 105:*847.

Arbuthnott, J. P. 1970. Staphylococcal αtoxin. in: *Microbial toxins,* Vol. 3 (T. C. Montie, S. Kadis, and S. J. Ajl, eds.) Academic Press, New York.

Arena, J. M. 1970. *Poisoning,* 2nd ed., Thomas, Springfield, p. 369.

Arena, J. M. 1974. *Poisoning: Toxicology, Symptoms, Treatments,* 3rd ed., American Lecture Series, No. 903, American Lectures in Living Chemistry, Thomas, Springfield.

Arvela, P., and Karki, N. T. 1970. The effect of cerium on drug metabolizing activity in rat liver. *Acta Pharmacol. Toxicol. (Suppl.) (KBH) 28:*36.

Athanassiadis, Y. C. 1969a. *Air Pollution Aspects of Cadmium and its Compounds,* Litton Systems, Inc., National Technological Information Service, U. S. Dept. of Commerce.

Athanassiadis, Y. C. 1969*b*. *Air Pollution Aspects of Vanadium and Its Compounds,* Litton Systems, Inc., Rept. PB-188-093, Bethesda.

Bach, I., Gali, T., Savely, C., Cos, J., and Vdvardy, A. 1967. Effect of rubidium on aldosterone and corticosterone production in rats. *Endocrinology: 81:*913–14.

Backstrom, J., Hammarstrom, L., and Nelson, A. 1967. Distribution of zirconium and niobium in mice, *Acta Radiobiologica Ther. Phys. Biol. 6:*122–128.

Bacon, C. 1974. Death from accidental potassium poisoning in childhood, *Br. Med. J. 1:*389–390.

Baetjer, A. M. 1956. Relation of chromium to health, in: *Chromium,* Vol. I, *Chemistry of chromium and its compounds* (M. J. Udy, ed.), American Chemical Society Monograph, No. 132, pp. 76–104.

Baetjer, A. M., Damron, C., and Budacz, V. 1959. Distribution and retention of chromium in men and animals, *A.M.A. Arch. Ind. Health 20:*136–150.

Baggiolini, M., de Duve, C., Masson, P. L., and Heremans, J. F. 1970. Association of lactoferrin with specific granules in rabbit heterophil leukocytes, *J. Exp. Med. 131:*559.

Bagirov, S. N. 1968. Effect of cerium salts on the neurochemical shifts in the sympathetic nervous system, *Tr. Obv. Fiziol. Azerb. 1:*17–21.

Bailey, G. H., Davidson, P. B., and Bunting, C. H. 1925. The effect of germanium dioxide on the rabbit, *J. A. M. A. 84:*1722–24.

Bailey, R. R., Eastwood, J. B., and Clarkson, E. M. 1971. Effect of aluminum hydroxide on Ca, P and Al balances and the plasma parathyroid hormone in patients with chronic renal failure, *Clin. Sci. 41:*5–6.

Bair, W. J., and Thompson, R. C. 1974. Plutonium: Biochemical research, *Science 183:*715–722.

Baker, D. R., Schrader, W. H., and Hitchock, C. R. 1964. Small bowel ulceration apparently associated with thiazide and potassium therapy, *J. A. M. A. 190:*586–590.

Ball, R. A., Van Gelder, G., and Green, J. W. 1970. Neoplastic sequelae following subcutaneous implantation of mice with rare earth metals, *Proc. Soc. Exp. Biol. Med. 135:*462.

Ballotta, F. 1931. Richerche sperimentali sulla azione del vanadio, *Med. Lavoro 22(6):*250, in: *Biol. Abstr. 9:*346 (1935).

Ballou, J. E., and Hess, J. O. 1972. Biliary plutonium excretion in the rat. *Health Phys. 72:*369–372.

Ballou, J. E., and Thompson, R. C. 1958. Metabolism of Cs^{137} in the rat: Comparison of acute and chronic experiments, *Health Phys. 1:*85–89.

Ballou, J. E., Bair, W. J., Case, A. C., and Thompson, R. C. 1962. Studies with neptunium in the rat, *Health Phys. 8:*685–692.

Banis, R. J., Ponds, W. G., Walker, E. F., and O'Connor, J. R. 1969. Dietary Cd, Fe, and Zn interactions in growing rats, *Proc. Soc. Exp. Biol. Med. 130:*802–806.

Barber, R. S., Braude, R. and Mitchell, K. G. 1955. High copper mineral mixture for fattening Pigs, *Chem. and Ind.,* p. 601.

Barber, R. S., Braude, R., Mitchell, K. G. and Rook, J. A. F. 1956. Further studies on antibiotic and copper supplements for fattening pigs, *Proc. Nutr. Soc. 15:*IX–X.

Barclay, R. K., Peacock, W., and Karnofsky, D. A. 1953, Distribution and excretion of radioactive thallium in chick embryo, rat and man, *J. Pharmacol. Exp. Ther. 107:*178.

Barltrop, D., Barret, A. J., and Dingle, J. T. 1971. Subcellular distribution of lead in the rat, *J. Lab. Clin. Med. 77:*705–712.

Barnes, C. D., and Eltheringron, L. G. 1964. *Drug Dosage in Laboratory Animals,* University of Califiornia Press, Berkeley.

Barnes, J. M., and Stoner, H. B. 1959. The toxicology of tin compounds, *Pharmacol. Rev. 11:*211.

Baroni, C., Van Esch, G. J., and Saffioti, V. 1963. Carcinogenesis tests of two inorganic arsenicals, *Arch. Environ. Health 7:*668–674.

Barr, D. H., and Harris, J. W. 1973. Growth of *P*-388 leukemia as an ascites tumor in zinc deficient mice, *Proc. Soc. Exp. Biol. Med. 144:*284–287.

Barranco, V. P. 1972. Eczematous dermatitis caused by internal exposure to copper, *Arch. Dermatol. 106:*386–387.

Barranco, V. P., and Soloman, H. 1972. Eczematous dermatitis from nickel, *J. A. M. A. 220:*1244.

Barry, P. S. I., and Mossman, D. B. 1970. Lead concentrations in human tissues, *Brit. J. Ind. Med. 27:*339–351.

Battigelli, M. C. 1960. Mercury toxicity from industrial exposure. A critical review of literature, Parts I and II, *J. Occup. Med. 2:*337–394.

Bauer, D. J. 1958. Chemotherapeutic activity of copper and other metals, *Br. J. Exp. Pathol. 39:*480.

Bauer, G. C. H., Carlsson, A., and Lindquist, B. 1956. A comparative study of the metabolism of ^{140}Ba and ^{45}Ca in rats, *Biochem. J. 63:*535–542.

Bauer, G. C. H., Carlsson, A., and Lindquist, A. 1957. Metabolism of barium in man, *Acta Orthoped. Scand. 26:*241–254.

Baumann, E. J., Zizner, N., Oshrey, E., and Seidlin, S. M. 1949. Behavior of thyroid towards elements of the 7th periodic group: Rhenium, *Proc. Soc. Exp. Biol. 75:*502–504.

Bazell, R. J. 1971. Lead poisoning: Zoo animals may be the first victims, *Science 173:*130–131.

Beaser, S. B., Segel, A., and Vandam, L. 1942. The anticoagulant effect in rabbits and man of the intravenous injection of salts of the rare earths, *J. Clin. Invest. 24:*447–449.

Beasley, T. M., Palmer, H. E., and Nelp, W. B. 1966. Distribution and excretion of technetium in humans, *Health Phys. 12:*1425.

Beaver, D. L., and Burr, R. E. 1963. Bismuth inclusions in human kidney, *Arch. Pathol. 76:*89.

Becker, W. M., and Hoekstra, W. G. 1971. The intestinal absorption of zinc, in: *Intestinal Absorption of Metal Ions, Trace Elements and Radionuclides* (S. C. Skoryna and D. Waldron-Edward, eds.), Pergamon Press, Oxford, pp. 229–256.

Bell, M. C., and Sneed, N. N. 1970. Metabolism of tungsten by sheep and swine, in: *Trace Element Metabolism in Animals* (C. F. Mills, ed.), E. and S. Livingston, London.

Bencosme, S. A., Stone, R. S., Latta, H., and Madden, S. C. 1960. Acute tubular and glomerular lesions in rat kidney after uranium injury, *Arch. Pathol. 69:*470–476.

Benoy, C. J., Hooper, P. A., and Schneider, R. 1971. The toxicity of tin in canned fruit juices and solid food, *Food Cosmet. Toxicol. 9:*645–646.

Berg. L. R. 1963. Evidence of vanadium toxicity resulting from the use of certain commercial phosphorous supplements in chick rations, *Poult. Sci. 42:*766–769.

Berg, L. R., Bearse, G. E., and Merill, L. H. 1963. Vanadium toxicity in laying hens, *Poult. Sci. 42:*1407–1411.

Berke, H. L. 1968. The metabolism of rare earths: I. The distribution and excretion of intravenous europium in the rat, *Health Phys. 15:*301–312.

Berke, H. L. 1969. *Metabolism of Rare Earths Following Inhalation. Pathologic and Biochemical Response in the Lungs and Other Organs,* U. S. A. E. C. Rep. COD-1630-10.

Berke, H. L., Wilson, G. H., and Berke, E. S. 1968. Changes in size distribution of blood lymphocytes following inhalation of radio europium, *Int. J. Radiat. Biol. 14:*561–565.

Berlyne, G. M., Pest, D., Ben-Ari, J., Weinbarger, J., Stern, M., Gilmore, G. R., and Levine, R. 1970. Hyperaluminaemia from aluminum resins in renal failure, *Lancet 2:*494.

Berlyne, G. M., Ben Ari, J., Knopf, E., Yagil, R., Weinberger, G., and Danovitch, G. M. 1972. Aluminum toxicity in rats, *Lancet 1:*564–567.

Bernheim, F., and Bernheim, M. L. C. 1939. The effect of titanium on the oxidation of sulfhydryl groups by various tissues, *J. Biol. Chem. 127:*695–701.

Bessarabova, R. V. 1955. Voprosy gigieny truda, professional, *Patol. i Toksikol. V. Prom. Sverdlovsr. Ovlasti,* Sverdlovsk.

Bhattacharya, G. 1964. Influence of Li^+ on glucose metabolism in rats and rabbits, *Biochem. Biophys. Acta 93:*644–646.

Bingham, E., Barkley, W., Zerwas, M., Stemmer, K., and Taylor, P. 1972. Responses of alveolar macrophages to metals. I. Inhalation of lead and nickel, *Arch. Environ. Health 25:*406–414.

Bishop, C. W., and Surgenor, D. M. (eds.) 1964. *The Red Blood Cell: A Comprehensive Treatise,* Academic Press, New York, p. 110.

Bjegovic, M., and Randic, M. 1971. Effect of lithium ions on the release of acetyl choline from the cerebral cortex, *Nature 230:*587.

Block, W. D., Buchanan, M., and Freyburg, R. H. 1942. Metabolism, toxicity, and manner of action of gold compounds used in the treatment of arthritis: IV. Studies of the absorption, distribution, and excretion of gold following the intramuscular injection of gold thioglucose and gold calcium thiomalate, *J. Pharm. Exp. Ther. 76:*355–358.

Bloom, G., and Swensson, A. 1958. The reaction of thrombocytes to intravenously injected suspensions of submicroscopic particles, *Acta Med. Scand. 162:*423.

Bobyleva, V. R. 1968. Roll of indium, gallium, molybdenum, rubidium, and zirconium in preventing dental caries, in rats, *Stomatologiya 47(5):*19–22.

Boecker, B. B. 1963. Thorium distribution and excretion studies. III.: Single inhalation exposure to the citrate complex and multiple exposures to the chloride, *Am. Ind. Hyg. Assoc. J. 24:*155–160.

Boecker, B. B., Thomas, R. G., and Scott, J. K. 1963. Thorium distribution and excretion studies. II. General pattern following inhalation and the effect of inhaled dose, *Health Phys. 9:*165–170.

Bollet, A. J., and Shuster, A. 1960. Metabolism of mucopolysaccharides in connective tissue: Synthesis of glucosamine phosphate. *J. Clin. Invest. 39:*114–118.

Bonnell, J. A. 1955. Emphysoma and proteinuria in men casting cadmium alloys, *Br. J. Ind. Med. 12:*181–191.

Borsook, H., Fischer, E. H., and Keighley, G. 1957. Factors affecting protein synthesis *in vitro* in rabbit reticulocytes, *J. Biol. Chem. 229:*1059–1070.

Bortels, H. 1936. Die Bedeutung von Molybdan, Vanadium, Wolfram stickstaff bindende und andere Mikroorganisms. Cited by Browning, E. 1969. *Toxicity of Industrial Metals,* Butterworth, London.

Bothwell, T. H., and Finch, C. A. 1962. *Iron Metabolism,* Little, Brown and Co., Boston.

Boulos, B. M., Carnow, B., Naik, N., Bederkx, J. P., Jr., Kauffman, R. F., and Azarnoff, D. L. 1973. Placental transfer of lithium and environmental toxicants and their effects on the newborn, *Fed. Proc. 32 (3):*745.

Boutwell, R. K. 1963. A carcinogenicity evaluation of potassium arsenite and arsanilic acid, *J. Agri. Food Chem. 11:*381–385.

Bowler, K., and Duncan, D. J. 1970. The effect of copper on membrane enzymes, *Biochem. Biophys. Acta 196:*116–119.

Boyd, E. M., and Abel, M. 1966. Acute toxicity of barium sulfate administered intragastrically, *Can. Med. Assoc. J. 94:*849.

Boyd, E. M., and Shanas, M. N. 1961. The acute oral toxicity of potassium chloride, *Arch. Int. Pharmacodyn. Ther. 133:*275–281.

Boyd, E. M., and Shanas, M. N. 1963. The acute oral toxicity of sodium chloride, *Arch. Int. Pharmacodyn. Ther. 144:*86–95.

Boyd, G. E. 1959. Technetium and promethium, *J. Chem. Ed. 36:*3–14.

Boyland, E., Dukes, C. E., Grover, P. L., and Mitchley, B. C. V. 1962. The induction of renal tumors by feeding lead acetate to rats, *Br. J. Cancer 16:*283–288.

Bradley, W. R., and Frederick, W. G. 1940. The toxicity of antimony. *Ind. Med., Ind. Hygiene Sect. 10:*15–22, 170–174.

Braman, R. S., and Foreback, C. C. 1973. Methylated forms of arsenic in the environment, *Science 182:*1247–1249.

Branham, S. E. 1929. The effects of certain chemical compounds upon the course of gas production by bakers yeast, *J. Bact. 18:*247–264.

Braun, H. A., and Lusky, L. M. 1959. The protective action of disodium catechol disulfonate in experimental vanadium poisoning, *Toxicol. Appl. Pharmacol. 1:*38–41.

Bremner, I. 1974*a*. Copper and zinc proteins in ruminant liver, in: *Trace Elements Metabolism in Animals,* Vol II (W. G. Hoekstra, J. W. Suttie, H. E. Ganther, W. F. Mertz, eds.), University Park Press, Baltimore, pp. 489–492.

Bremner, I. 1974*b*. Heavy metal toxicities, *Quart Rev. Biophys. 7:*75–124.

Brieger, H., Semisch, C. W., Stasney, J., and Piatneck, D. A. 1954. Industrial antimony poisoning, *Ind. Med. Surg. 23:*521.

Bronner, F. 1964. Dynamics and function of calcium, in: *Mineral Metabolism,* Vol. II (C. L. Comar and F. Brommer, eds.), Part A, Academic Press, New York, pp. 342–444.

Brooksbank, W. A., and Leddicote, G. W. 1953. Ion-exchange separation of trace impurities, *J. Phys. Chem. 57:*819–824.

Brown, J. R., and Kulkarni, M. V. 1967. A review of toxicity and metabolism of mercury and its compounds, *Med. Ser. J. Can. 23(5):*786.

Brown, M. A., Thom, J. V., Orth, G. L., and Juarez, J. 1964. Food poisoning involving zinc contamination, *Arch. Environ. Health 8:*657–660.

Brownell, G. L., Ellet, W. H., Ojemann, R. G., and Sweet, W. H. 1960. The use of short-lived isotopes in the perfusion therapy of isolated organs, in: *Radioaktive Isotop in Klinick und Forschung,* Vol. IV, Urban and Schwarzenburg, Munich, pp. 126–141.

Browning, E. 1969. *Toxicity of Industrial Metals,* 2nd Ed., Butterworth and Co., London.

Bruce, D. W., Hietbrink, B. E., and DuBois, K. P. 1963. The acute mammalian toxicity of rare earth nitrates and oxides, *Toxicol. Appl. Pharmacol. 5:*750–759.

Brucer, M., Andrews, G. A., and Bruner, H. D. 1953. A study of [72]gallium, *Radiology 61:*534–540.

de Bruin, A. 1971. Certain biological effects of lead upon the animal organism, *Arch. Environ. Health 23:*249–264.

Brunot, F. R. 1933. The toxicity of osmium tetroxide (osmic acid), *J. Ind. Hyg. 15:*136.

Bryan, S. E., Lambert, C., Hardy, K. J., and Simons, S. 1974. Intranuclear localization of mercury *in vivo, Science 183:*486–832.

Buchan, R. F. 1947. Industrial selenosis, *Occup. Med. 3:*439–456.

Buchanan, W. D. 1962. *Toxicity of Arsenic Compunds,* Elsevier Monographs on Toxic Agents (E. Browning, ed.), Elsevier Publishing Co., Amsterdam, London, and New York.

Buldakov, L. A., and Burov, N. I. 1967. Behavior of radio cerium in sheep of different ages, *Radiobiologiya 7:*881.

Bullen, J. J., and Rogers, H. J. 1969. Bacterial iron metabolism and immunity to *Pasteurella spetica* and *Escherichia coli, Nature 224:*380.

Bunn, C. R., and Matrone, G. 1966. *In vivo* interactions of cadmium, copper, zinc, and iron in the mouse and rat, *J. Nutr. 90:*395.

Bunyan, J., Edwin, F. E., and Green, J. 1958. Protective effects of trace elements other than selenium against dietary necrotic liver dengeration, *Nature 181:*1801.

Burn, J. H. 1932. The physiological action of aluminum, *Analyst 57:*428.

Burr, R. F., Gotto A. M., and Beaver, D. L. 1965. Isolation and analysis of renal bismuth inclusion, *Toxicol. Appl. Pharmacol. 7:*588.

Burykina, L. N. 1962. The metabolism of radioactive ruthenium in the organism of experimental animals, in: *The Toxicology of Radioactive Substances, Strontium, Cesium, Ruthenium, and Radon,* Vol. 1 (A. A. Letavet and E. B. Kuryandskaya, eds.), Pergamon, New York, p. 60.

Butler, L. C., and Daniel, J. M. 1973. Copper metabolism in young women fed two levels of copper and two protein sources, *Am. J. Clin. Nutr. 26:*744–749.

Butt, E. M. 1960. Trace Metals in Health and Disease, Air Pollution Medical Conference, San Francisco, California.

Butt, E. M., Nussabaum, R. E., Gilmour, T. C., and DiDio, S. L. 1961. Trace metal patterns in disease states, *Am. J. Pathol. 30:*479–499.

Buttner, W. 1963. Action of trace elements on the metabolism of fluoride, *J. Dent. Res. 42:*453–458.

Byron, W. R., Bierbower, G. W., Brower, J. B., and Hansen, W. H. 1967. Pathologic changes in rats and dogs from two-year feeding of sodium arsenite and sodium arsenate, *Toxicol. Appl. Pharmacol. 10:*132–136.

Cade, J. F. J. 1949. Lithium salts in the treatment of psychotic excitement, *Med. J. Austr. 2:*349–352.

Calloway, D. H., and McMullen, J. J. 1966. Fecal excretion of iron and tin by men fed stored canned foods, *Am. J. Chem. Nutr. 18:*1–5.

Campbell, I. R., Cass, J. S., Cholak, J., and Kehoe, R. A. 1957. Aluminum in the environment of man, in view of its hygienic status, *Arch. Instr. Health 15:*359–448.

Campbell, J. K., and Mills, C. F. 1974. Effects of dietary cadmium and zinc on rats maintained on diets low in copper, *Proc. Nutr. Soc. 33:*15A–17A.

Campo, R. D., Tourtellote, C. D., and Ledrick, J. W. 1967. Selenium[75], an autoradiographic study of its disposition in cartilage and bone, *Proc. Soc. Exp. Biol. Med. 125:*512–515.

Capilna, S., Ghizari, E., Ababei, L., and Badescu, A. 1963. Effect of molybdenum and tungsten ions on intermediate metabolism of glutamine in rat liver and brain, *Nature 200:*470.

Caplan, R. M., and Curtis, A. C. 1961. Xanthoma of the skin, *J. A. M. A. 176:*859–864.

Caravaggi, C., Clarck, F. L., and Jackson, A. R. B. 1970. Experimental acute toxicity of orally administered sodium selenite in lambs, *Res. Vet. Sci. 11:*501–502.

Cardona, E., Lessler, M. A., and Brierly, G. P. 1971. Mitochondrial oxidative phosphorylation—interaction of lead with inorganic phosphate, *Proc. Soc. Exp. Biol. Med. 136:*300.

Carlton, W. W., and Kelley, W. A. 1967. Tellurium toxicosis, *Toxicol. Appl. Pharmacol. 11:*203.

Carrol, R. E. 1966. The relationship of Cd in the air to cardiovascular disease death rates, *J. A. M. A. 198:*177–180.

Carroll, K. G., and Tullis, J. L. 1968. Observations on the presence of titanium and zinc in human leukocytes, *Nature 217:*1172–1173.

Carroll, K. G., Spinelli, F. R., and Goyer, R. A. 1970. Electron probe micoanalyzer localization of lead in kidney tissue of poisoned rats, *Nature 227:*1076.

Castagnow, R., Paolette, C., and Larcebau, S. 1957. Absorption and distribution of barium after oral and intravenous administration in the rats, *C. R. Acad. Sci. 24:*2994.

Castronovo, F. P., and Wagner, H. N. 1971. Factors affecting the toxicity of the element indium, *Br. J. Exp. Pathol. 52:*543–559.

Caujolle, F., Caujolle, D., and Magna, H. 1963. Evolution de la localization du germanium chez la souris apres ingestion de tetraethylgermane, *C. R. Acad. Sci. Paris 257:*1563.

Cavanaugh, D. J., Harris, J., and Hearon, J. Z. 1955. Enzyme inhibition by complexing of substrates: Inhibition of tyrosinase by titanium compounds. *J. Amer. Chem. Soc. 77:*1531–1538.

Center for Disease Control. 1975. Acute copper poisoning, *Morb. Mort. Week. Rept.* (March) *15:*99.

Ceresa, C. 1947. Intossicazione alimentare acuta da Rama, *Med. Lavoro 38:*144–147.

Cerwenka, E. A., Jr., and Cooper, W. C. 1961. Toxicology of tellurium and selenium and their compounds, *Arch. Ind. Health 3:*189–194.

Chahovitch, X. 1955. The action of zinc on the growth of experimental tumors incited by carcinogens, *Gl. Srpske. Akad. Nauka Od Med. Nauka 215:*143–146.

Chance, B., and Estabrook, R. W. 1966. *Hemes and Hemoproteins,* Academic Press, New York.

Chandra, S. V., and Srivastava, S. P. 1970. Experimental production of early brain lesions in rats by parenteral administration of MgCl$_2$, *Acta Pharmacol. Toxicol. 28:*177.

Chang, L. W., Ware, R. A., and Desnoyers, P. A. 1973. A histochemical study on enzyme changes in the kidney, liver, and brain, after chronic mercury intoxication in the rat, *Food Cosmet. Toxicol. 11:*283–286.

Chavez, J. F., and Jafee, W. G. 1967. Nivel toxico de selenio en dietas para ratas, *Archs. Lat. Am. Nutr. 17:*69–76.

Cheftel, H. 1967. Tin in Food. Joint FAO/WHO Food Standard Program: Fourth meeting of the Codex Committee on Food Additives, September, 1967, Joint FAO/WHO Food Standards Branch, (Codex Alimentarius) FAO, Rome.

Chevallier, J., and Butow, R. A. 1971. Calcium binding to the sarcoplasmic reticulum of rabbit skeletal muscle, *Biochemistry 10:*2733.

Chiquoine, A. D., and Suntzeff, V. 1965. Sensitivity of mammals to cadmium necrosis in the testis, *J. Reprod. Fertil. 10:*455–460.

Chisolm, J. J., Jr. 1964. Disturbances in the biosynthesis of heme in lead intoxications, *J. Pediatr. 64:*174.

Chisolm, J. J., Jr. 1971. Lead poisoning, *Sci. Amer. 224:*15.

Choie, D. D., and Richter, G. W. 1972*a*. Cell proliferation in rat kidney produced by lead acetate and effects of uninephrectomy on the proliferation, *Amer. J. Pathol. 66:*265.

Choie, D. D., and Richter, G. W. 1972*b*. Lead poisoning: Rapid formation of intranuclear inclusions, *Science 117:*1195.

Choie, D. D., and Richter, G. W. 1973. Stimulation of DNA synthesis in rat kidney by repeated administration of lead, *Proc. Soc. Exp. Biol. Med. 142:*446–449.

Christensen, H. E., and Luginbyhl, T. T. 1974. The toxic substances list. USDA-NIOSH, Rockville, Md.

Christie, H., Mackay, R. J., and Fischer, A. M. 1963. Pulmonary effects of inhalation of aluminum by rats, *Am. Ind. Hyg. Assoc. J. 24:*47–56.

Chu, R. C., and Cox, D. H. 1972. Zinc, iron, copper, calcium, cytochrome oxidase, and phospholipid in rats of lactating mothers fed excess zinc, *Nutr. Rep. Int. 5:*61–66.

Chuttani, H. K., Gupta, P. S., Gulati, S., and Gupta, D. N. 1965. Acute copper sulfate poisoning, *Am. J. Med. 39:*849–854.

Chvapil, M., Ryan, J. N., and Zukoski, C. F. 1972. The effect of zinc and other metals on stability of lysosomes, *Proc. Soc. Exp. Biol. Med. 140:*642–644.

Clarkson, E. M., Luck, V. A., Hynson, W. V., Bailey, R. R., Eastwood, J. B., Woodhead, J. S., Clements, V. R., O'Riordan, J. L. H., and DeWardener, H. E. 1972. The effect of aluminum hydroxide on calcium, phosphorus and aluminum balances, the serum parathyroid hormone concentration and the aluminum content of bone in patients with chronic renal failure, *Clin. Sci. 43:*519–531.

Clarkson, T. W. 1965. Toxicological aspects of mercury, *Ann. Occup. Hyg. 8:*73–81.

Clarkson, T. W. 1972. Recent advances in the toxicology of mercury with emphasis on alkylmercurials, *CRC Critical Reviews in Toxicology 1 (2):*203.

Clarkson, T. W., and Megos, L. 1966. Studies on the binding of mercury in tissue homogenates, *Biochem. J. 99:*62–70.

Clarkson, T. W., Small, H., and Norseth, T. 1971. The effect of thiol containing resin on the gastrointestinal absorption and fecal excretion of methyl mercury compounds in experimental animals, *Fed. Proc. 30(3):*543.

Clary, J. J., Hopper, C. R., and Stokinger, H. E. 1972. Altered adrenal function as an inducer of latent chronic beryllium disease, *Toxicol. Appl. Pharmacol. 23:*365–375.

Clayton, C. C., and Baumann, C. A. 1949. Diet and azo dye: Effect of diet during a period when the dye is not fed, *Cancer Res. 9:*575.

Cochran, K. W., Doull, J., Mazur, M., and Du Bois, K. P. 1950. Acute toxicity of zirconium, columbium, strontium, lanthanum, cesium, tantalium, and yttrium. *Arch Ind. Hyg. Occup. Med 1:*637–650.

Cockburn, A., Barraco, R. A., Reyman, T. A., and Peck, W. A. 1975. Autopsy of an egyptian mummy, *Science 187:*1155–1160.

Cogburn, L. A., Parkhurst, C. R., and Thaxton, P. 1973. Relationship of mercury to reproductive efficiency and mating behavior in Japanese quail, Presented at Annual Meeting, Association of Southern Agricultural Workers, Atlanta.

Cohn, Z. A., and Fedorko, M. E. 1969. The formation and fate of lysosomes, in: *Lysosomes in Biology and Pathology,* Vol. 1 (J. T. Dingle and H. B. Fell, eds.), North Holland Pub. Co., London, p. 43.

Comar, C. L., and Bronner, F. (eds.). 1969. Calcium physiology, in: *Mineral Metabolism,* Vol. III, Academic Press, New York.

Comar, D., and Chevallier, F. 1967. Concentration of vanadium in the rat and its influence on cholesterol synthesis by neutron activation and isotopic balance, *Bull. Soc. Chim. Biol. 49:*1357–1364.

Comar, C. L., and Wasserman, R. H. 1964. Strontium, in: *Mineral Metabolism,* Vol. II, "Elements, Part A." (C. L. Comar and F. Bronner, eds.), Academic Press, New York, pp. 523–572.

Cook, J. A., Hoffman, E. O., and DiLuzio, N. R. 1975. Influence of lead and cadmium on the susceptibility of rats to bacterial challenge, *Proc. Soc. Exp. Biol. Med. 150:*741–747.

Cooper, W. C. 1967. Selenium toxicity in man, in: *Selenium in Biomedicine* (O. H. Muth, J. E. Oldfield, and P. H. Weswig, eds.), AVI Publishing Co., Westport, Conn., pp. 185–189.

Cooper, W. C. 1971. Toxicology of tellurium and its compounds, in: *Tellurium* (W. C. Cooper, ed.), Van Nostrand Reinhold Co., pp. 313–320.

Corcoran, A. C., Taylor, R. D., and Page, I. H. 1949. Lithium poisoning from the use of salt substitutes, *J. A. M. A. 139:*685–688.

Cortel, R., and Richards, R. E. 1942. Gold tolerance. *J. Pharmacol. Exp. Ther. 76:*17–24.

de Cotton, M., and Logan, M. E. 1966. Effects of antimony on the cardiovascular system and intestinal smooth muscle, *J. Pharmacol. 151:*7–11.

Cotzias, G. C. 1958. Manganese in health and disease. *Physiol. Rev. 38:*503.

Cotzias, G. C. 1964. Manganese, in: *Mineral Metabolism, An Advanced Treatise, Part B* (C. L. Comar and F. Bronner, eds.), p. 403, Academic Press, New York.

Cotzias, G. C., Borg, D. C., and Selleck, B. 1961. Virtual absence of turnover in cadmium metabolism: [109]Cd studies in the mouse, *Am. J. Physiol. 201:*927.

Coumbis, J. T. 1975. Environment (news report), *17* (7), page 21.

Cousins, R. J., Barber, A. K., and Trout, J. R. 1973. Cadmium toxicity in growing swine, *J. Nutr. 103:*964–972.

Cowan, G. A. B. 1947. Unusual case of poisoning by zinc sulphate, *Br. Med. J. 1:*451–452.

Cox, D. H., and Harris, D. L. 1960. Effect of excess dietary zinc on iron and copper in the rat, *J. Nutr. 70*:514–520.

Cox, D. H., and Harris, D. L. 1962. Reduction of liver xanthine oxidase activity and iron storage proteins in rats fed excess zinc, *J. Nutr. 78*:415.

Cramer, C. F., and Copp, D. H. 1959. Progress and rate of absorption of radio strontium through intestinal tracts of rats, *Proc. Soc. Exp. Biol. Med. 102*:514–517.

Crapper, D. R., Krishnan, S. S., and Dalton, A. J. 1973. Brain aluminum distribution in Alzheimer's disease and experimental neurofibrillary degeneration, *Science 108*:511–513.

Cravioto, H., Agnew, W. F., and Carregal, J. A. 1970. The distribution of tellurium in the nervous system of rat: An ultrastructural study, *J. Neuropath. Exp. Neurol. 29*:158–162.

Cremer, J. E., and Aldridge, W. N. 1964. Toxicological and biochemical studies on some trialkyl germanium compounds, *Br. J. Ind. Med. 21*:214.

Cross, C. E., Ibrahim, A. B., Ahmed, M., and Mustafa, M. G. 1970. Effect of Cd ion on respiration and ATPase activity of the pulmonary alveolar macrophages: A model for the study of environmental interference with pulmonary cell function, *Environ. Res. 3*:512.

Crowley, J. F., Hamilton, J. G., and Scott, K. G. 1949. Metabolism of carrier free Be[7] in the rat, *J. Biol. Chem. 177*:975.

Cummins, L. M., and Martin, J. L. 1961. Are selenocystine and seleomethionine synthesized *in vivo* from sodium selenite in mammals?, *Biochemistry 6*:532–538.

Cunningham, G. N., Wise, M. G., and Barrick, F. R. 1966. Effect of high dietary levels of manganese on the performance and blood constituents of calves, *J. Anim. Sci. 25*:532.

Curran, G. L., Azarnoff, D. L., and Bolinger, R. E. 1959. Inhibition of cholesterol synthesis from labelled acetate, *J. Clin. Invest. 38*:1251–1255.

Currie, A. M. 1947. The role of arsenic in carcinogenesis, *Br. Med. Bull. 4*:402–410.

Curry, A. S., Grech, J. L., Spiteri, L., and Vassallo, L. 1969. Death from thallium poisoning, *J. Europ. Toxicology 2*:260–270.

Cushney, A. R. 1941. *Pharmacology and Therapeutics,* Churchill, London.

Cuthbertson, D. P. 1973. Mineral and trace element requirements for normal growth and development, in: *Nutrition and Technology of Foods for Growing Humans* (J. C. Somogyi, ed.), Karger, Basel, pp. 65–91.

Czerwinski, A. W., and Gin, H. E. 1964. Bismuth neophrotoxicity, *Am. J. Med. 37*:969–971.

Daigneault, E. R. 1963. The distribution of intravenously administered YCl$_3$ (carrier free) in the rhesus monkey, *Toxicol. Appl. Pharmacol. 5*:331.

Daniel, E. P., and Lillie, R. D. 1938. Experimental vanadium poisoning in the white rat, *Publ. Health Rep. (Washington) 53*:765–770.

Davies, M., Lloyd, J. B., and Beck, F. 1971. The effect of trypan blue suramin and aurothiomalate on the breakdown of ^{125}I labeled albumin within rat liver lysosomes, *Biochem. J. 121*:21.

Davis, G. K. 1950. The influence of copper on the metabolism of phosporus and molybdenum. in: *Symposium on Copper Metabolism,* (W. D. McElroy and B. Glass, eds.), Johns Hopkins Press, Baltimore, pp. 216–229.

Davis, J. M., and W. E. Fann. 1971. Lithium, *Annu. Rev. Pharmacol. 11*:285–302.

Davis, P. S., Luke, C. G., and Deller, D. J. 1967. Gastric iron binding protein in iron chelation by gastric juice. *Nature 214*:1126.

Davies, T. A. L., and Harding, H. E. 1949. Manganese pneumonites, *Br. J. Ind. Med. 6*:82–89.

Debons, A. F., Krimsky, I., From, A., and Gloutier, R. J. 1970. Gold thioglucose induction of obesity: Significance of focal gold deposits in the hypothalamus, *Am. J. Physiol. 219(5)*:1403–1408.

Decker, L. E., Byerrum, R. U., Decker, C. F., Hoppert, C. F., and Langan, R. F. 1958.

Chronic toxicity studies: 1. Cadmium administered in drinking water to rats, *Arch. Ind. Health 18:*228–234.

Decker, W. J., Goldsmith, W. A., Mills, R. C., and Banez, R. J. 1972. Systemic absorption of copper after oral administration of radioactive copper sulfate emetic in rats, *Toxicol. Appl. Pharmacol. 21:*331–334.

De Feudis, F. V. 1973. Actions of lithium on cerebral carbohydrate metabolism, *Rev. Res. Commun. Chem. Path. Pharmacol. 5:*789–796.

De Feudis, F. V., and Delgado, J. M. R. 1970. Effects of lithium on the amino acid content of mouse brain *in vivo. Nature (London) 225:*747–748.

De Feudis, F. V., and Mitchell, J. F. 1970. Continuous collection of ^{3}H-γ-aminobutyric acid from the cerebral cortex of the rat, *J. Physiol. (London) 210:*83.

DeGrazia, J. A. 1971. The intestinal absorption of calcium, in: *Intestinal Absorption of Metal Ions and Trace Elements and Radionuclides* (S. C. Skoryna and D. Waldon-Edward, eds.), Pergamon Press, Oxford and New York, pp. 151–172.

Deiss, A., Lee, G. R., and Cartright, G. E. 1970. Hemolytic anemia in Wilson's disease, *Ann. Intern. Med. 73:*413–418.

DeLarrad, J., and Lazarini, H. J. 1954. *Arch. Maladies Prof. 15:*282–288.

Del Costillo, J., and Engback, L. 1954. The nature of the neuromuscular block produced by magnesium, *J. Physiol. (London) 124:*370–376.

Deobald, H. J., and Elvehjem, C. A. 1958. The effect of feeding high amounts of soluble iron and aluminum salts. *Am. J. Physiol. 111:*118.

DeRenzo, E. C. 1962. Molybdenum, in: *Mineral Metabolism: An Advanced Treatise,* Vol. II, Part B, (C. L. Comar and F. Bronner, eds.), Academic Press, New York, pp. 483–495.

Derr, R. F., Aaker, H., Alexander, C. S., and Nagasau, H. T. 1970. Synergism between cobalt and ethanol on rat growth rate, *J. Nutr. 100:*521–524.

Desselberger, U., and Wegener, H. H. 1971. Alcohol–cobalt-, and combined alcohol–cobalt-intoxication in guinea pigs, *Beitr. Pathol. 142:*150–176.

Dick, A. T. 1956*a*. Molybdenum in animal nutrition, *Soil Sci. 81:*229–234.

Dick, A. T. 1956*b*. Molybdenum and copper interrelationships in animal nutrition, in: *Inorganic Nitrogen Metabolism* (W. D. McElroy and B. Glass, eds.), Johns Hopkins Press, Baltimore, p. 445.

Diding, N., Ottoson, J. C., and Schou, M. 1969. Lithium in psychiatry, *Acta Psychiat. Neurol. Scand. 45 (Suppl.) 207:*1.

Diengott, D. 1964. Hypokalemia in barium poisoning, *Lancet 2:*343.

Diez-Ewald, M., Weintraub, L. R., and Crosby, W. H. 1968. Interrelationship of iron and manganese metabolism, *Proc. Soc. Exp. Biol. Med. 129:*448–451.

Di Luzio, N. 1972. Lead poisoning and endotoxins, *Chem. Eng. News 50(9):*56.

Dimond, E. G., Caravaca, J., and Benchimol, A. 1963. Vanadium: Excretion, toxicity, and lipid effect in man, *Am. J. Clin. Nutr. 12:*49.

Dinkel, C. A., Minyard, J. A., and Ray, D. E. 1963. Effects of season of breeding on reproductive and weaning performance of beef cattle grazing seleniferous range, *J. Anim. Sci. 22:*1043–1048.

Dixon, N. E., Gazzola, C., Blakeley, R. L., and Zernor, B. 1975. Jackbean urease a metalloenzyme—A simple biological role for nickel, *J. Am. Chem. Soc. 97:*4131.

Doak, R. L., Schmidtke, R. P., Wallach, J. D., Davis, L. E., and Niemeyer, K. H. 1965. Thallium intoxication: A specific antidote, supportive therapy and clinical evaluation. *Vet. Med./Small Anim. Clinic 60:*1277.

Dodds, E. C., Noble, R. L., Kinderkmecht, H., and Williams, P. C. 1937. Prolongation of action of the pituitary antidiuretic substance and of histamine by metallic salts, *Lancet 2:*309.

Doherty, P. C., Barlow, R. M., and Angus, K. W. 1969. Spongy changes in the brains of sheep poisoned by excess dietary copper, *Res. Vet. Sci. 10:*303–304.

Doisy, E. A., Jr. 1972. Micronutrient controls on biosynthesis of clotting proteins and cholesterol, in: *Trace Substances in Environmental Health, VI* (D. Hemphill, ed.), Proceedings of the University of Missouri's 6th Annual Conference on Trace Substances in Environmental Health, p. 193.

Douglas, W. W. 1964. The effects of alkaline earth and other divalent cations on adrenal medullary secretion, *J. Physiol. 175:*231–234.

Douglas, W. W., and Rubin, R. P. 1964. Stimulant action of barium on the adrenal medulla, *Nature 203:*305.

Dousa, T., and Hechter, O. 1970. The effect of sodium chloride and lithium chloride on vasopression sensitive adenyl cyclase, *Life Sci. 9:*765.

Downs, W. L., Scot, J. K., Maynard, E. A., and Hodge, H. C. 1959*a*. Studies on the toxicity of thorium nitrate, University of Rochester Atomic Energy Research Project, *USAEC Res. Dev. Report, U.R. 561,* pp. 1–35.

Downs, W. L., Scot, J. K., and Steadman, L. T. 1959*b*. The toxicity of indium, *USAEC Res. Dev. Report, U.R. 558.*

Downs, W. L., Scott, J. K., Steadman, L. T., and Maynard, E. A. 1960. Acute and sub-acute toxicity studies of thallium compounds, *Am. Ind. Hyg. Assoc. J. 21:*399–406.

Downs, W. L., Scott, J. K., Yuile, C. L., Caruso, F. S., and Wong, L. C. K. 1965. The toxicity of niobium salts, *Am. Ind. Hyg. Assoc. J. 26:*237.

Doyle, J. J., Pfander, W. H., Grebing, S. E., and Pierce, J. O., II. 1974. Cadmium absorption and cadmium tissue levels in growing lambs, *J. Nutr. 104:*160–166.

Dru, D., Agnew, W. F., and Greene, E. 1972. Effects of tellurium ingestion on learning capacity of the rat, *Psychopharmacologia 24:*508–515.

Duckett, S. 1972. Teratogenesis caused by tellurium, *Ann. N.Y. Acad. Sci. 192:*220–226.

Dudley, H. C. 1936. Toxicology of selenium. II. The urinary excretion of selenium, *Am. J. Hyg. 23:*181–186.

Dudley, H. C. 1953. Pharmacological studies of radiogermanium, *Arch. Ind. Hyg. 8:*528–533.

Dudley, H. C., and Levine, M. D. 1949. Gallium toxicity, *J. Pharmacol. 95:*487–493.

Dudley, H. C., and Marrer, H. H. 1952. Gallium metabolism—Deposition in and clearance from bone. *J. Pharm. Exp. Ther. 106:*129–134.

Dudley, H. C., Henrey, K. E., and Lindsay, B. F. 1950*a*. Studies of the toxic action of gallium, *J. Pharmacol. Exp. Ther. 98:*409–412.

Dudley, H. C., Munn, J. I., and Henrey, K. E. 1950*b*. Studies of the metabolism of gallium. II, *J. Pharm. Exp. Ther. 98:*105–110.

Duncan, G. C. 1953. Effect of zinc on cytochrome oxidase activity, *Proc. Soc. Exp. Biol. Med. 82:*625–628.

Duncan, J. R., Dreosti, I. E., and Albrecht, C. F. 1974. Zinc intake and growth of a transplanted hepatoma induced by 3′-methyl-4-dimethyl-aminoazobenzene in rats, *J. Natl. Cancer Inst. 53:*277–278.

Dunphy, B. 1967. Acute occupational cadmium poisoning: A critical review of the literature, *J. Occup. Med. 9:*22–26.

Durbin, P. W. 1960. Metabolic characteristics within a chemical family, *Health. Phys. 2:*225.

Durbin, P. W. 1972. Plutonium in man: A new look at the old data, in: *Radiobiology of Plutonium* (B. J. Stober and W. S. S. Jee, eds., pp. 469–481, The J. W. Press, Salt Lake City.

Durbin, P. W., Scott, K. G., and Hamilton, J. G. 1957. Biological behavior of specific radioisotopes of metals, *Univ. Calif. Pub. Pharmacol. 3:*1–9.

Durocher, N. L. 1969. Air pollution aspects of beryllium and its compounds, Litton Systems Inc., Maryland, National Technical Information Service Bulletin.

Dyckerhoff, H., and Grunewald, O. 1943. Uber den Reaktion mechanismus der Hemmung der Blutgerinnung durch eininge seltene Erden und durch Heparin, *Biochim. Z. 314*:124–138.

Dymond, A. M., Kaechele, L. E., Jurist, J. M., and Crandall, P. H. 1970. Brain tissue reactions to chronically implanted metals, *J. Neurosurg. 33*:574.

Eagle, H., Germuth, F. G., Jr., Magnuson, H. J., and Fleischman, R. 1947. The protective action of BAL in experimental antimony poisoning, *J. Pharmacol. 89*:196–204.

Earl, A. E., Diener, R. M., and Shoffstall, D. H. 1966. Effect of various potassium salts on the gastrointestinal trace of monkeys, *Toxicol. Appl. Pharmacol. 8*:339.

Eckhardt, A. 1909. Beitrag zur Frage der Zinnvergiftungen, *Z. Unters. Nahr u. Genussmittel 18*:193–196.

Edwards, C. L., and Hayes, R. L. 1970. Scanning malignant neoplasms with gallium 67, *J. A. M. A. 212*:1182–1190.

Edwards, C. L., Nelson, B., and Hayes, R. L. 1970. Localization of gallium in human tumors, *Clin. Res. 18*:89–93.

Eichner, D., and Opitz, K. 1974. Lithium tissue content in untreated animals (German), *Histochem. 42*:295–300.

Einer-Jensen, N., Larsson, S., Nielsen, R. A., and Secher, N. J. 1970. Pregnancy in exercised and food-restricted gold thioglucose obese mice, *Acta Physiol. Scand. 79(1)*:59–63.

Ekman, L. 1961. Distribution and excretion of radio cesium in goats, pigs, and hens, *Acta. Vet. Scand. 2(4)*:10.

Ekman, L., and Aberg, G. 1961. Excretion of niobium[95], yttrium[91], cerium[144], and promethium[147] in goats, *Res. Vet. Sci. 2*:100.

Elliot, K. A. C., and Van Gelder, N. M. 1970. The state of Factor I in rat brain: The effects of metabolic conditions and drugs. *J. Physiol. (London) 153*:423.

Ellis, W. G., and Huston, J. E. 1968. [144]Ce–[144]Pr as a particulate digesta flow marker in ruminants, *J. Nutr. 95*:67–68.

de Estable-ping, R. F., de Estable-ping, J. F., and Romero, C. 1971. Nuclear changes in glial cells after aluminum hydroxide injection, *Virchows Arch. Abt. B. Zell Pathol. 8*:267–273.

Epstein, F. H. 1963. Nephropathy of hypercalcemia, in: *Diseases of the Kidney* (M. B. Straws and L. G. Walt, eds.), Little, Brown and Company, Boston.

Evans, G. W. 1973. Copper homeostasis in the mammalian system, *Physiol. Rev. 53*:535–570.

Evodkimoff, V., and Wagner, H. N. 1972. Hepatic phagocytosis as a mechanism for increasing heavy metal toxicity, *J. Retinculoendothel. Soc. 11*:148–153.

Ewer, T. K. 1951. Rickets in sheep. *Br. J. Nutr. 5*:300–306.

Fahim, M. S., Webb, M., Hilderbrand, D. C., and Russell, R. L. 1972. Effect of lead acetate on male reproduction, *Fed. Proc. 31*:272.

Fahim, M. S., Der, R., Fahim, Z., and Harman, J., 1975. Effect of zinc on reproduction and metabolism, *J. Clin. Pharmacol.* (abstr.), in press.

Fairhill, L. T., and Hyslop, F. 1947. The toxicology of antimony, *U. S. Public Health Serv. Rep., Suppl. No. 195.*

Fairhill, L. T., and Hyslop, F. 1957. Physiological action of antimony and its compounds in animals and man—the toxicology of antimony, *U.S. Public Health Serv. Rep., Suppl. No. 195.*

Fairhill, L. T., and Neal, P. A. 1943, Industrial manganese poisoning, *Natl. Inst. Health Bull. No. 182.*

Fairhill, L. T., Dunn, R. D., Sharpless, N. E., and Prichard, E. A. 1945. Toxicity of molybdenum. *U. S. Public Health Serv. Bull. 293.*

Faulkner-Hudson, T. G. 1964. Vanadium: Toxicology and biological significance, *Elsevier Monographs on Toxic Agents*, Elsevier Publishing Co., Amsterdam and New York.

Fawcett, D. W., and Gens, J. P. 1943. Magnesium poisoning following an enema of epsom salt, *J. A. M. A. 123:*1028–1030.

Fay, M., Andersch, M. A., and Behrman, V. C. 1942. The biochemistry of strontium, *J. Biol. Chem. 114:*383.

Fed. Register. 1974. Selenium, *39:*1355–1358.

Feldman, S. L., and Cousins, R. J. 1973. Inhibition of 1-25 dihydrocholecalciferol synthesis by cadmium *in vitro, Fed. Pro. 32:*918 (abstr.).

Ferm, V. H. 1970. Protective effect of ferric dextran on the embryopathic action of indium, *Experientia 26:*633–635.

Ferm, V. H., and Carpenter, S. J. 1967. Teratogenic effect of cadmium and its inhibition by zinc, *Nature (London) 216:*1123.

Ferm, V. H., and Carpenter, S. J. 1968. The relationship of cadmium and zinc in experimental mammalian teratogenesis, *Lab. Invest. 18:*429–434.

Ferm, V. H., and Carpenter, S. J. 1970. Teratogenic and embryopathic effects of indium, gallium, and germanium, *Toxicol. Appl. Pharmacol. 16:*166–171.

Finkell, J. A., and A. M. Brues. 1960. *Argonne National Laboratories Summaries of Current Work* (Oct., 1960), p. 42.

Fleshman, D. G., Silva, A. J., and Shore, B. 1971. The metabolism of tantalum in the rat, *Health Phys. 21:*385.

Flick, D. F., Kraybill, H. F., and Dimitroff, J. M. 1971. Toxic effects of cadium: A review, *Environ. Res. 4:*71–85.

Follis, R. H., Jr. 1943. Histological effects in rats resulting from adding rubidium or cesium to a diet deficient in potassium, *Am. J. Physiol. 138:*246–251.

Follis, R. H. 1955. Bone changes resulting from parenteral strontium adminstration, *Fed. Proc. 14:*403.

Forbes, G. B. 1963. Sodium, in: *Mineral Metabolism,* Vol. II, Part B (C. L. Comar and F. Bronner, eds.), Academic Press, New York.

Forbes, G. B., and Reina, J. C. 1972. Effect of age on gastrointestinal absorption of Fe, Sr, and Pb in the rat, *J. Nutr. 102:*647–652.

Forbes, R. M. 1967. in: *New Methods of Nutritional Biochemistry with Application and Interpretations* (A. A. Alabanese, ed.), Academic Press, New York, p. 339.

Forbes, R. M., and Mitchel, H. H. 1957. Accumulation of dietary boron and strontium in young and adult albino rats, *Arch. Ind. Health 16:*489–492.

Forth, W., and Rummel, W. 1971. Absorption of iron and chemically related metals *in vitro* and *in vivo:* Specificity of the iron binding system in the mucosa of the jejunum, in: *Intestinal Absorption of Metal Ions, Trace Elements and Radionuclides* (S. C. Skoryna and D. Waldron-Edward, eds.), Pergamon Press, Oxford and New York.

Forth, W. and Rummel, W. 1973. Iron absorption, *Physiol. Rev. 53:*724–793.

Fox, M. R. S., and Fry, B. E., Jr. 1970. Cadmium toxicity decreased by dietary ascorbic acid supplements, *Science 169:*989.

Fox, M. R. S., Fry, B. E., Harland, B. F., Schertel, M. E., and Weeks, C. E. 1971. Effect of ascorbic acid on cadmium toxicity in the young coturnix, *J. Nutr. 101:*1295–1305.

Franke, K. W., and Moxon, A. L. 1936. Comparison of minimum fatal doses of selenium, tellurium, arsenic, and vanadium, *J. Pharm. Exp. Therap. 58:*454–459.

Franke, K. W., and Moxon, A. L. 1937. Toxicity of orally ingested arsenic, selenium, tellurium, vanadium, and molybdenum. *J. Pharm. Exp. Therap. 61:*89–102.

Fredrick, W. G., and Bradley, W. R. 1946. Report of the 8th Annual Meeting of the American Industrial Hygiene Association, *Quarterly Suppl. 8.*

Freeland, J. H., and Cousins, R. J. 1973. Effect of dietary cadmium on anemia and iron absorption, *Fed. Proc. 32(3):*924.

Freyberg, R. H. 1960. Gold therapy for rheumatoid arthritis, in: *Arthrites* (J. L. Hodander, ed.), Lea and Febiger, Philadelphia.

Friberg, L. 1957. Proteinuria in chronic cadmium poisoning after comparatively short exposure to cadmium dust, *AMA Arch. Ind. Health 16:*30–35.

Friberg, L., and Vostal, J. 1971. Mercury in the environment: A toxicological and epidermiological approach, *Pub. EPA,* Office of Air Programs, Res. Triangle Park, North Carolina.

Friberg, L., and Vostal, J. 1972. *Mercury in the Environment: An Epidemiological and Toxicological Appraisal,* Chemical Rubber Co. Press, Cleveland.

Friberg, L., Piscator, M., and Nordberg, G. 1971. *Cadmium in the Environment,* Chemical Rubber Co. Press, Cleveland.

Frost, D. V. 1967. Arsenicals in biology: Retrospect and prospect, *Fed. Proc. 26:*194.

Frost, D. V. 1970. Tolerances for arsenic and selenium, a psychodynamic problem, *World Review Pest Control 9:*6.

Frost, D. V. 1972. The two faces of selenium—Can selenophobia be cured?, *CRC Crit. Rev. Toxicol. 1 (4):*467.

Frost, D. V., and Lish, P. M. 1975. Selenium in biology, *Ann. Rev. Pharmacol. 15:*259–284.

Frost, D. V., Overly, L. C., and Spruth, H. C. 1955. Studies with arsanilic acid and related compounds, *J. Agric. Food Chem. 3:*235–242.

Gabbedy, B. J. 1970. Toxicity in sheep associated with the prophylactic use of selenium, *Aust. Vet. J. 46:*223–226.

Gabbedy, R. J., and Dickson J. 1969. Acute selenium poisoning in lambs, *Aust. Vet. J. 45(10):*470–472.

Gabbiani, G., Selye, H., and Tuchweber, B. 1962. Prevention of indium intoxication by ferric dextran, *Br. J. Pharmacol. 19:*508–514.

Gabbiani, G., Jacquin, M. L., and Richard, R. M. 1966. Soft tissue calcification induced by rare earth metals and its prevention by sodium pyrophosphate, *Br. J. Pharmacol. 27:*1–19.

Gammil, J. C., Wheeler, B., Carothers, E. L., and Hahn, P. F. 1950. Distribution of radioactive silver colloids in tissues or rodents, *Prod. Soc. Exp. Biol. (N.Y.) 74:*691–694.

Ganther, H. E. 1965. The fats of selenium in animals, *World Rev. Nutr. Dietet. 5:*893–899.

Ganther, H. E. 1968. Selenotrisulfides. Formation by the reaction of thiols with selenious acid. *Biochemistry 1:*2898–2901.

Ganther, H. E. 1971. Reduction of the selenotrisulfide derivative of glutathione to a persulfide analog by glutothione reductase, *Biochemistry 10:*4089–4098.

Ganther, H. E., Gouldie, C., Sundae, M. L., Kopecky, M. J., P. Wagner, Oh, S.-H., and Hoekstra, W. G. 1972. Selenium: Relation to decreased toxicity of methyl mercury added to diets containing tuna, *Science 175:*1122–1124.

Garber, G., and Wei, E. 1973. Adaptation to the toxic effects of lead. *Am. Ind. Hygiene Assoc. J. 33(11):*756–760.

Gardiner, M. R., and Nairn, M. E. 1971. Effect of cobalt and selenium in clover disease of ewes, *Aust. Vet. J. 45:*215–222.

Gardiner, M. R., and Nicol, H. 1971. Cobalt–selenium interactions in the nutrition of rat, *Aust. J. Exp. Biol. Med. Sci. 49:*291–296.

Garner, R. J., Jones, H. G., and Ekman, L. 1960. Fission products and the dairy cow: The fate of orally administered cerium[144], *J. Agric. Sci. 55:*107.

Garrett, N. E., Garrett, R. J. B., and Archdeacon, J. W. 1972. Placental transmission of mercury to the fetal rat, *Toxicol. Appl. Pharmacol. 22:*649–654.

Gehrcke, E., Lau, E., and Meinhart, O. 1939. The rare earths and the nervous system, *Z. Ges. Naturw. 5:*106–107.

Gesiler, A., Wraae, O., and Olesen, O. V. 1972. Adenyl cyclase activity in kidney of rats with lithium-induced polyuria, *Acta Pharmacol. Toxicol. 31:*203.

Gettler, A. O., and Weiss, L. 1943. Clinical toxicology of thallium, *Am. J. Clin. Pathol. 13:*422–432.

Gilman, J. P. W. 1966. Muscle tumorigenesis, in: *Proceedings of the Sixth Canadian Cancer Research Conference,* Honey Harbor, Ontario, Oxford, Pergamon Press, pp. 209–223.

Gilman, J. P. W., and Ruckerbauer, C. M. 1962. Metal carcinogenesis, *Cancer Res. 22:*152–158.

Gipp, W. F., Pond, W. G., Kallfelz, F. A., Tasker, J. B., Van Campen, D. R., Krook, L., and Visek, W. J. 1974. Effect of dietary copper, iron, and ascorbic acid levels on hematology, blood and tissue copper, iron and zinc concentrations, and ^{64}Cu and ^{59}Fe metabolism in young pigs, *J. Nutr. 104:*532–541.

Gisselbrecht, H., Baufle, G. H., and Duvernoy, J. 1957. Action inhibitrice des sels bismuth et d'alumine sur l'asborption intestinale du glucose, *Ann. Scient. Univ. Basançon. Med. 44:*29–33.

Gleason, M. N., Gosselin R. E., Hodge, H. C., and Smith, R. P. 1969. *Clinical Toxicology of Commercial Products,* 3rd ed, Williams and Wilkins, Baltimore, Sect. II, p. 131.

Glendening, B. L., Schrenk, W. G., and Parrish, D. B. 1956. Effect of rubidium in purified diet rats. *J. Nutr. 60:*563–569.

Glover, J. R. 1970. Selenium and its industrial toxicology, *Ind. Med. Surg. 39:*50–54.

Goldblatt, P. J., Liebermann, N. Q., and Witschi, H. 1973. Beryllium-induced ultrastructural changes in intact and regenerating liver, *Arch. Environ. Health. 26:*48–52.

Goodman, L. S., and Gilman, S. 1965. *Pharmacological Basis of Therapeutics,* MacMillan, New York.

Goodman, L. S., and Gilman, S. 1970. *The Pharamacological Basis of Therapeutics,* Mac-Millan and Company, London.

Gordynya, R. I. 1969. Effect of a ration containing a nickel salt additive on carbohydrate metabolism in experimental animals. *Vopr. Rastion Pitan. No. 5:*167–170 (Russian).

Gorodiskii, V. I., Veselaya, I. V., and Rostovtseva, D. N. 1956. Cu, Zn, Cd, and Ni content of muscles and tumors, *Vopr. Med.*

Gould, B. S. 1936. Effects of thorium, zirconium, titanium, and cerium on enzyme action, *Proc. Soc. Exp. Biol. Med. 34:*381–385.

Goyer, R. A. 1968. The renal tubule in lead poisoning: 1. Mitochondrial swelling and amino aciduria, *Lab. Invest. 19:*71–77.

Goyer, R. A. 1971*a*. Lead and the kidney, *Curr. Top. Pathol. 55:*147.

Goyer, R. A. 1971*b*. Lead toxicity: A problem in environmental pathology, *Am. J. Pathol. 64:*167–179.

Goyer, R. A., Leonard, D. L., Moore, J. F., Rhyme, B., and Krigman, M. R. 1970. Lead dosage and role of the intranuclear inclusion body, *Arch. Environ. Health. 30:*705–711.

Grabowski, C. T. 1966. Teratogenic effects of calcium salts on chick embryoes, *J. Embryol. Exp. Morph. 15:*113–120.

Graca, J. G., Davison, F. C., and Feavel, J. B. 1962. Comparative toxicity of stable rare earth compounds, *Arch. Environ. Health. 5:*437.

Gralla, E. J., and McIllhenny, H. M. 1972. Studies in pregnant rats, rabbits, and monkeys with lithium carbonate, *Toxicol. Appl. Pharmacol. 21:*428–433.

Grant, N. C., and Root, W. S. 1952. Fundamental stimulus for erythropoiesis, *Physiol. Rev. 32:*449–498.

Grant, W. M., and Kern, H. L. 1956. in: *Rare Earths in Biochemical and Medical Research* (G. C. Kyker and G. R. Anderson, eds.), Oakridge Institute of Nuclear Studies, Vol. 12, pp. 346–355.

Granta, A., Barbaro, M., and Maturo, L. 1970. Toxicological aspects of the effect of cadmium on human blood cells, *Arch. Mol. Prof. Med. Traver. Secur. Soc. 31:*357–363.

Grant-Frost, D. R., and Underwood, E. J. 1958. Zinc toxicity in the rat and its relation with copper, *Aust. J. Exp. Biol. 36:*339–346.

Grasso. P. R. 1969. The role of dietary silver in the production of liver necrosis in vitamin-E-deficient rats, *Exp. Mol. Pathol. 11(2):*186.

Gray, F. G., and Daniel, L. J. 1954. Some effects of excess molybdenum on nutrition of rats, *J. Nutri. 53:*43–47.

Greaves, M. W., and Skillen, A. W. 1970. Effects of long-continued ingestion of zinc sulphate in patients with venous leg ulceration, *Lancet 2:*889.

Greenfield, I., Zuger, M., Bleak, R. M., and Bakal, S. F. 1950. Lithium chloride intoxication, *N.Y. State J. Med. 50:*459–460.

Grice, H. C., Goodman, T., Munn, I. C., Wiberg, G. S., and Morrison, A. B. 1969*a*. Myocardial toxicity of cobalt in rat, *Ann. N.Y. Acad. Sci. 156:*189–194.

Grice, H. C., Munro, I. C., Wieberg, G. S., and Heggtreit, H. A. 1969*b*. The pathology of experimentally induced cobalt cardiomyopathies. A comparison with beer drinkers, *Cardiomyopathy Clin. Toxicol. 2:*273–286.

Griggs, R. C. 1964. Lead poisoning: Hematologic aspects, *Progr. Hematol. 4:*117.

de Groot, A. P. 1973. Subacute toxicity of inorganic tin as influenced by dietary levels of iron and copper, *Food Cosmet. Toxicol. 11:*955–962.

de Groot, A. P., Feron, V. J., and Til, H. P. 1973. Short-term toxicity studies on some salts and oxides of tin in rat, *Food Cosmet. Toxicol. 11:*19–29.

Grunert, R. R., Myer, J. H., and Phillips, R. H. 1950. The sodium and potassium requirements of the rat for growth, *J. Nutr. 42:*609–618.

Gunn, S. A., and Gould, T. C. 1967. Specificity of response in relation to cadmium, zinc and selenium, in: *Selenium in Biomedicine* (O. H. Muth, J. E. Oldfield, and P. H. Weswig, eds.), AVI Publishing Co., Connecticut, pp. 395–413.

Gunn, S. A., Gould, T. C., and Anderson, W. A. D. 1968*a*. Specificity in protection against lethality and testicular toxicity from cadmium, *Proc. Soc. Exp. Biol. Med. 128:*591.

Gunn, S. A., Gould, T. C., and Anderson, W. A. D. 1968*b*. Mechanisms of zinc, cysteine, and selenium protection against cadmium-induced vascular injury to mouse testis, *J. Reprod. Fert. 15:*65–69.

Gustavson, C. R., Garcia, J., Hankins, W. G., and Rusiniak, K. W. 1974. Coyote predation control by aversive conditioning, *Science 184:*581–583.

Guthrie, J. 1964. Histological effects of intratesticular injections of cadmium chloride in domestic fowl, *Br. J. Cancer 18:*225.

Hackett, C. 1962. Stimulative effects of aluminum on plant growth. *Nature (London) 195:*471–472.

Haddon, W., Jr. 1958. Lower nephron necrosis of a heavy metal type in rats given an inorganic tin preparation intraperitoneally. N.Y. State Dept. Health, *Ann. Rept. Div. Lab. Res.,* p. 41.

Haddow, A., and Horning, E. S. 1960. On the carcinogenicity of an iron–dextran complex, *J. Natl. Cancer Inst. 24:*109.

Haddow, A., Roe, F. J. C., Dukes, C. E., and Mitchley, B. C. V. 1964. Cadmium neoplasia: Sarcomata at the site of injection of cadmium sulphate in rats and mice, *Br. J. Cancer 18:*667.

Hadjimarkos, D. M. 1967*a*. Effect of selenium and type of diet on the pattern of food and water intake in the rats, *Experientia 23:*930–932.

Hadjimarkos, D. M. 1967*b*. Selenium and fluoride interaction in relation to dental caries, *Arch. Environ. Health 14:*881–886.

Hadjimarkos, D. M. 1969. Selenium: a caries-enhancing trace element, *Caries Res. 3:*14–17.

Hadjimarkos, D. M. 1973. Selenium in relation to dental caries, *Food Cosmet. Toxicol. 11:*1083–1095.

Haley, T. J. 1965. Pharmacology and toxicology of the rare earth elements, *J. Pharm. Sci. 54:*663–670.

Haley, T. J., and Cartwright, P. D. 1968. Pharmacology and toxicology of potassium perrhenate and rhenium chloride, *J. Pharm. Sci. 57:*321.

Haley, T. J., Raymond, K., Komesu, N., and Upham, H. C. 1961. Toxicology and pharmacological effects of gadolinum and samarium chlorides, *Br. J. Pharm. 17:*526–537.

Haley, T. J., Komesu, N., Mavis, L., Cawthorme, J., and Upham, H. C. 1962*a*. Pharmacology and toxicology of scandium chloride, *J. Pharm. Sci. 51:*1043–1049.

Haley, T. J., Komesu, N., and Raymond, K. 1962*b*. The pharmacology and toxicology of niobium chloride, *Toxicol. Appl. Pharmacol. 4:*385–392.

Haley, T. J., Raymond, K., Komesu, N., and Upham, H. C. 1962*c*. The toxicologic and pharmacologic effects of hafnium salts, *Toxicol. Appl. Pharmacol. 4:*238–246.

Haley, T. J., Komesu, N., Flescher, A. M., Mairs, L., Cawthorne, J., and Upham, H. C. 1963. Pharmacology and toxicology of terbium, thulium, and ytterbium chlorides, *Toxicol. Appl. Pharmacol. 5:*427–436.

Haley, T. J., Komesu, N., Efros, M., Koste, L., and Upham, H. C. 1964*a*. Pharmacology and toxicology of lutetium chloride, *J. Pharm. Sci. 53:*1186.

Haley, T. J., Komesu, N., Efros, M., Koste, L., and Upham, H. C. 1964*b*. Pharmacology and toxicology of Pr and Nd chlorides, *Toxicol. Appl. Pharmacol. 6:*614.

Haley, T. J., Komesu, N., Calvin, G., Koste, L., and Upham, H. C. 1965. Pharmacology and toxicology of europium chloride, *J. Pharm. Sci. 54:*643–653.

Haley, T. J., Koste, L., Komesu, N., Efros, M., and Upham, H. C. 1966. Pharmacology and toxicology of dysprosium, holmium, and erbium chlorides, *Toxicol. Appl. Pharmacol. 8:*37–43.

Hall, H. A. 1951. Preliminary observation on the toxicity of elemental selenium, *Arch. Ind. Health 4:*458.

Hall, J. L., and Smith, E. B. 1968. Cobalt heart disease—An electron microscopic and histochemical study in the rabbit, *Arch. Pathol. 86:*403.

Hall, R. H., Scott, J. K., Laskin, S., Skroud, C. A., and Stokinger, H. E. 1950. Acute toxicity of beryllium III, *AMA Arch. Ind. Hyg. Occup. Med. 2:*25–31.

Hall, R. H., Stroud, C. A., Scott, J. K., Root, R. E., Steadman, L. T., and Stokinger, H. E. 1951. Acute toxicity of inhaled thorium compounds, University of Rochester Atomic Energy Project, *USAEC Rept. U.R. 190,* pp. 1–27.

Halme, E. 1961. The carcinomatous effect of zinc in drinking water, *Vitalstoffe. Zivilisationskrawkh. 6:*59–61.

Halverson, A. W., Phifer, J. H., and Monty, K. G. 1960. A mechanism for the copper-molybdenum interrelationship, *J. Nutr. 71:*95.

Halverson, A. W., Tsay, D. T., Triebwasser, K. C., and Whitehead, E. I. 1970. Development of hemolytic anemia in rats fed selenites, *Toxicol. Appl. Pharmacol. 17:*151–160.

Hamilton, E. I. 1972. The concentration of uranium in man and his diet, *Health Phys. 22:*149–153.

Hamilton, J. G. 1947. The metabolism of fission products and the heaviest elements, *Radiology 49*:325–340.

Hamilton, J. G. 1949. Metabolism of radioactive elements created by nuclear fission, *New Engl. J. Med. 240*:863–865.

Hammond, P. B. 1969. Lead poisoning: An old problem with a new dimension, in: *Essays in Toxicology*, Vol. I (F. R. Blood, ed.), Academic Press, New York, pp. 115–155.

Hanlon, L. W., Romaine, M., Gilroy, F. J., and Dietrick, J. C. 1949. Lithium chloride as substitute for NaCl in the diet. *J. A. M. A. 139*:688–692.

Hapke, H. J. 1971. The effect of calcium salts in cattle, *Dtsch. Tierarztl. Wschr. 78*:628.

Hardy, H. L. 1965. Beryllium poisoning: Lessons in control of man-made disease, *New Eng. J. Med. 273 (22)*:1188.

Harkinen, M., and Kormano, M. 1970. Acute cadmium-induced changes in the energy metabolism of rat testis, *J. Reprod. Fertil. 21*:221.

Harr, J. R., and Muth, O. H. 1972. Selenium poisoning in domestic animals and its relationship to man, *Clin. Toxicol. 5*:175–184.

Harr, J. R., Bone, J. F., Tinsley, I. J., Weswig, P. H., and Yamamoto, R. 1967. Selenium toxicity in rats. II. Histopathology, in: *Selenium in Biomedicine* (O. H. Muth, J. E. Oldfield, and P. H. Weswig, eds.), AVI Publishing Co., Connecticut, pp. 153–178.

Harr, J. R., Exon, J. H., Whanger, P. D., and Weswig, P. H. 1972. Effect of dietary selenium on *N*-2-fluorenyl-acetamide (FAA): Induced cancer in vitamin-E-supplemented selenium-depleted rats, *Clin. Toxicol. 5*:187–194.

Harris, S. B., Wilson, J. G., and Printz, R. J. 1972. Embryotoxicity of methylmercury chloride in golden hamsters, *Teratology 6*:139–142.

Harrison, G. E., Raymond, W. H. A., and Tretheway, H. C. 1955. The metabolism of strontium in man, *Clin. Sci. 14*:681–693.

Harrold, G. C., Meek, S. F., Whitman, N., and McCord, C. P. 1943. The physiologic properties of indium and its compounds. *J. Ind. Hyg. 25*:233–237.

Hart, H. E., Greenburg, G., Levin, R., Spencer, H., Stern, K. G., and Lazlo, A. D. 1955. Metabolism of lanthanum and yttrium chelates, *J. Lab. Clin. Med. 46*:182–192.

Hart, M., Smith, C. F., Yancey, S. T., and Adamson, R. H. 1971. Toxicity and anti-tumor activity of gallium nitrate and related metal salts, *J. Natl. Cancer Inst. 47*:1121–1127.

Hartman, R. H., Matrone, G., and Wise, G. H. 1955. Effect of a high dietary manganese on hemoglobin formation, *J. Nutr. 57*:429–432.

Hassan, A. 1938. Quantitative study of excretion of antimony, *J. Egypt. Med. Assoc. 21*:126–130.

Hatem, S. 1958. Cancers de cobalt et complexe histamine sels de cobalt, *C. R. Acad. Sci. 247*:1681–1684.

Hatem-champy, S. 1961. Cancers du nickel et complexe nickel-imidazole, *C. R. Acad. Sci. 253*:2791–2792.

Hathcock, J. N., Hill, C. H., and Matrone, G. 1964. Vanadium toxicity and distribution in chicks and rats, *J. Nutr. 82*:106–110.

Hayes, A. D., and Rothstein, A. 1962. The metabolism of inhaled mercury vapors in the rat studied by isotope techniques, *J. Pharm. Exp. Ther. 138*:1–5.

Hayes, R. L., Nelson, B., and Swartzendruber, D. C. 1970. Gallium localization in rat and mouse tumors, *Science 167*:389–390.

Heath, J. C. 1956. The production of malignant tumors by cobalt in the rat, *Br. J. Cancer 10*:668–673.

Heath, J. C., and Daniel, M. R. 1964a. The production of malignant tumors by cadmium in the rat, *Br. J. Cancer 18:*124–130.

Heath, J. C., and Daniel, M. R. 1964b. The production of malignant tumors by nickel in the rat, *Br. J. Cancer 18:*261–264.

Heath, J. C., and Webb, M. 1967. Content and intracellular distribution of the inducing metal in the primary rhabdomyosarcomata induced in the rat by cobalt, nickel, and cadmium, *Br. J. Cancer 21:*768–774.

Heggtveit, H. A., Grice, H. C., and Wiberg, G. S. 1970. Cobalt cardiomyopathy: Experimental basis for the human lesion, *Pathol. Microbiol. 35:*110–113.

Heine, W. 1944. Observations and experimental investigation of manganese pneumonia, *Z. Hyg. Infektionskrankheiten 125:*1–10.

Hemphill, F. E., Kaeberle, M. L., and Buck, W. B. 1971. Lead suppression of mouse resistance to *Salmonella typhimurium, Science 172:*1031–1032.

Henschler, D., and Kirschner U. 1969. The adsorption and toxicity of selenium sulfide, *Arch. Toxicol. 24:*341–344.

Herdson, P. B., Garvin, P. J., and Jennings, R. B. 1964. Fine structural changes produced in rat liver by partial starvation, *Am. J. Pathol. 45:*157–164.

Hermann, M. M., and Bensch, K. G. 1964. Histologic changes caused by acute and chronic thallium intoxication in rats, *Fed. Proc. 23:*199.

van Heynigen, W. E. 1970. General characteristics. in: *Microbial toxins,* Vol. 1 (S. J. Ajl, S. Kadis, and T. C. Montie, eds.) Academic Press, New York.

van Heynigen, W. E., and Mellanby, J. 1971. Tetanus toxins, in: *Microbial toxins,* Vol 2A (S. Kadis, S. J. Ajl, and T. C. Montie, eds.) Academic Press, New York, pp. 69–108.

Heyroth, F. F. 1947. Thallium: A review and summary of medical literature, U.S. Pub. Health Serv. Pub., *Health Rept. Suppl. No. 197.*

Hilderbrand, D., Olds, M., Der, R., and Fahim, M. S. 1972. Effect of lead acetate on reproduction, *Am. J. Obstet. Gynecol. 115:*1058–1065.

Hill, C. H. 1974. Reversal of selenium toxicity in chicks by mercury, copper, and cadmium, *J. Nutr. 104:*593–598.

Hill, C. H., and Matrone, G. 1970. Chemical parameters in the study of *in vivo* and *in vitro* interactions of transition elements, *Fed. Proc. 29:*1474–1475.

Hobel, M., Maroske, D., Wegener, K., and Eichler, O. 1972. Über die toxische Wirkung von $CoCl_2$, Co (Co EDTA), oder Na_2 [Co-EDTA] enthaltender Aerosole auf die Ratte und die Verteilung von [Co EDTA] in Organ des Meerschweinschens, *Arch. Int. Pharmacodyn. Ther. 198:*213–222.

Hodge, H. C. 1956. Mechanism of uranium poisoning, *Arch. Ind. Health 14:*43–47.

Hodge, H. C., Maynard, E. A., and Leach, L. J. 1960. USAEC Res. Div. Rept., *U.R. 562.*

Hoey, M. J. 1966. The effects of metallic salts on the histology and functioning of the rat tests, *J. Reprod. Fertil. 12:*461–471.

Hofvander, Y. 1968. Hematological investigations in Ethiopia, *Acta Med. Scand. Suppl. 494:*1–74.

Holdsworth, E. S., and D'A Welling, D. 1971. The effect of bile salts on the absorption of calcium and other cations, in: *Intestional Absorption of Metal Ions and Trace Elements and radionuclides* (S. C. Skoryna and D. Waldon-Edward, eds.), Pergamon Press, Oxford and New York, pp. 339–358.

Hollins, I. G. 1969. The metabolism of tellurium in rats, *Health Phys. 17:*497.

Holmberg, R. E., and Ferm, V. H. 1969. Interrelationships of selenium, cadmium, and arsenic in mammalian teratogenesis, *Arch. Environ. Health 18:*873–877.

Hood, R. D. 1972. Effects of sodium arsenite on fetal development, *Bull. Environ. Contam. Toxicol. 7:*216–220.

Hood, R. D., and Bishop. S. L. 1972. Teratogenic effects of arsenate in mice, *Arch Environ. Health 24:*62–68.

Hood, S. L., and Comar, C. L. 1953, Metabolism of [137]cesium in rats and farm animals, *Arch. Biochem. Biophys. 45:*423–433.

Hood, S. L., and Comar, C. L. 1956. Tissue distribution and placental transfer of yttrium[91] in rats, in: *Rare Earths in Biochemical and Medical Research* Oakridge Institute of Nuclear Studies, pp. 280–300.

Hopkins, L. L., Jr., and Mohr, H. E. 1971. The biological essentiality of vanadium, in: *Newer Trace Elements in Nutrition* (W. Mertz and W. E. Cornatzer, eds.), Marcel Decker, New York, pp. 195–201.

Hopkins, L. L. Jr., and Schwarz, K. 1964. Chromium (III) binding to serum proteins, especially siderophilin, *Biochim. Biophys. Acta 90:*484–491.

Hoppe, J. O., Marcelli, G. M. A., and Tainter, M. L. 1955. A review of the toxicity of iron compounds, *Am. J. Med. Sci. 230:*558–571.

Horak, E., and Sunderman, F. W., Jr. 1973. Fecal nickel excretion by healthy adults, *Clin. Chem. 19:*429–430.

Horiuchi, K., Horiguchi, S., and Suekane, M. 1959. Studies on industrial lead poisoning, *Osaka C. Med. J. 5:*41.

Horovitz, C. T. 1975. Scandium: Biological significance and toxicology, in: *Scandium - Its Occurrence, Chemistry, Physics, Metallurgy, Biology and Technology* (C. T. Horovitz, K. A. Gschneidner, Jr., G. A. Melson, D. H. Youngblood,, and H. H. Schock, eds.), Academic Press, New York, pp. 513–546.

Howe, M., McGee, J., and Lengemann, F. W. 1972. Transfer of inorganic mercury to milk of goats, *Nature 237:*316–318.

Hsu, F. S., Brook, L., Shively, J. N., Duncan, J. R. and Pond, W. G. 1973. Lead inclusion bodies in osteoclasts, *Science 181:*447–448.

Heuper, W. C. 1958*a*. Pulmonary lesions in guinea pigs and rats exposed to prolonged inhalation of metallic nickel, *Arch. Pathol. 65:*600–604.

Hueper, W. C. 1958*b* Experimental studies in metal carcinogenesis IX. Pulmony lesions in guinea pigs and rats exposed to prolonged inhalation of powdered nickel, *Arch. Pathol. 64:*605–607.

Heuper, W. C., and Payne, W. W. 1959. Experimental studies in metal carcinogenesis. Experimental cancers in rats produced by chromium compounds and their significance to industry and public health, *Am Ind. Hyg. Assn. J. 20:*272–280.

Hueper, W. C., and Payne, W. W. 1962. Experimental studies in metal carcinogenesis, *Arch. Environ. Health 5:*445–462.

Hueper, W. C., Zuefle, J. H., Link, A. M., and Johnson, M. G. 1952. Experimental studies in metal cancerigenesis: II. Experimental uranium cancers in rats, *J. Natl. Cancer Inst. 13:*291–301.

Huggins, C. B., and Froehlich, J. P. 1966. High concentration of injected titanium dioxide in rat abdominal lymph nodes, *J. Exp. Med. 124:*1099.

Hughes, W. L. 1957. A physicochemical rationale for the biological activity of mercury and its compounds, *Am. N.Y. Acad. Sci. 65:*454–464.

Hunter, D. 1964. *The Diseases of Occupation,* English Universities Press, London, p. 474.

Hunter, D., and Russel, D. S. 1954. Focal cerebral and cerebellar atrophy in a human subject due to organic mercury compounds, *J. Neurol. Neurosurg. Psychiat. 17:*235.

Hunter, D., Milton, R., and Perry, K. 1945. Asthma caused by complex salts of platinum, *Br. J. Ind. Med. 2:*92–96.

Hursh, J. B., and Gates, A. A. 1950. Body radium content of individuals with no known occupational exposure, *Nucleonics 7:*46–49.

Hursh, J. B., and Suomela, J. 1968. Absorption of ^{212}Pb from the gastrointestinal tract of man, U.S. A. E. C. UCRL. 18140, pp. 217–233.

Husain, S. L. 1969. Oral zinc sulphate in leg ulcers, *Lancet 1:*1069–1071.

Huston, J. E., and Ellis, W. G. 1968. Evaluation of certain properties of radio cerium as an indigestible marker, *J. Agri. Food Chem. 16:*225–230.

Hutcheson, D. P., Gray, D. H., Venugopal, B., and Luckey, T. D. 1975*a*. Nutritional safety of heavy metals in mice, *J. Nutr. 105:*670–675.

Hutcheson, D. P., Gray, D. H., Venugopal, B., Luckey, T. D. 1975*b*. Safety of heavy metals as nutritional markers, *Environ. Qual. Saf., Suppl. 1:*74–80.

Irving, J. T. 1973. *Calcium and Phosphorous Metabolism,* Academic Press, New York and London.

Ishinishi, N., and Morishige, T. 1969. A study on the distribution of ^{95}Zr–^{95}Nb administered subcutaneously to rats: Comparison between young and old rats, *J. Radiat. Res. 10:*101.

Iverson, L. L., and Neal, M. J. 1968. The uptake of [^{3}H] GABA by slices of rat cerebral cortex, *J. Neurochem. 15:*1141.

Iverson, F., Downie, R. H., Paul, C., and Trenholm, H. L. 1973. Methyl mercury-acute toxicity, tissue distribution and decay profiles in the guinea pig, *Toxicol. Appl. Pharmacol. 24:*545–554.

Iwata, H., Kamato, H. O., and Ohswa, Y. 1973. Effect of selenium on methyl mercury poisoning, *Res. Commun. Chem. Pathol. Pharmacol. 53:*673–680.

Jackson, D. E. 1912. The pulmonary action of vanadium together with a study of the peripheral reactions to the metal, *Pharmacol. Exp. Ther. 4:*1–10.

Jackson, M. J., and Smyth, D. H. 1971. Intestinal absorption of sodium and potassium, in: *Intestinal Absorption of Metal Ions: Trace Elements and Radionuclides* (S. C. Skoryna and D. Waldron-Edward, eds.), Pergamon Press, Oxford and New York, pp. 137–150.

Jacobs, R. M., Fox, M. R. S., Fry, B. E., and Harland, B. F. 1974. Effect of a two-day exposure of dietary cadmium on the concentration of elements in the duodenal tissue of Japanese quail, in: *Trace Element Metabolism in Animals* (W. G. Hoekstra, J. W. Suttie, H. E. Ganther, and W. Mertz, eds.), pp. 684–686, University Park Press, Baltimore.

Jandl, J. H., and Simmons, R. L. 1957. The aggutination and sensitization of red cells by metallic atoms: Interactions between multivalent metals and the red cell membrane, *Br. J. Haematol. 3:*19.

Janoff, A. 1970. Inhibition of human granulocyte elastase, *Biochem. Pharmacol. 19:*626–628.

Jefferson, J. M., and Aukland, P. 1974. Enteric-coated potassium chloride—a continuing hazard, *Br. Med. J. 1:*456.

Jensen, L. S. 1975. Precipitation of selenium deficiency by high dietary levels of copper and zinc, *Proc. Soc. Exp. Biol. Med. 149:*113–116.

Jernigan, E. L. 1973. *Lead Poisoning in Man and the Environment,* MSS Information Corporation, New York.

Jet, R., Jr., Pierce, J. O., and Stemmer, K. L. 1968. Toxicity of alloys of ferrochromium. III. Transport of chromium (III) by rat serum protein studied by immunoelectrophoretic analysis and autoradiography, *Arch. Environ. Health 17:*29–34.

Johansen, K., and Ulrich, K. 1969. Preliminary studies of the possible teratogenic effect of lithium, in: *Lithium in Psychiatry, Acta Psychiat. Neurol. Scand. 45 (Suppl. 207):* 91–98.

Johnson, A. D., and Sigman, M. B. 1971. Early actions of cadmium in the rat and domestic fowl testis—IV. Autoradiographic location of [115]cadmium, *J. Reprod. Fertil. 24:*115.

Johnson, A. D., Gomes, W. R., and Van Demark, N. L. 1970. Early actions of Cd in the rat and domestic fowl testes, *J. Reprod. Fertil. 21:*383.

Johnson, F. A., and Stonehill, R. B. 1961. Chemical pneumonites from inhalation of zinc chloride, *Dis. Chest 40:*619–624.

Johnson, F. N. (ed). 1975. *Lithium Research and Therapy,* Academic Press, New York.

Johnson, J. L., Cohen, H. J., and Rajagopal, K. V. 1973. Molybdenum enzymes in the rat; Effect of tungsten administration on sulfite oxidase and xanthine oxidase, *Fed. Proc. 32:*923.

Johnson, R. F., and Ziemer, P. L. 1971. The deposition and retention of inhaled [152–154]europium oxide in the rat, *Health Phys. 20:*187–193.

Johnstone, R. M. 1963. Sulfhydryl agents: Arsenicals, in: *Metabolic Inhibitors,* Vol. II (R. M. Hochster and J. H. Quastel, eds.), Academic Press, New York.

Jonek, J., Jonderko, G., and Pacholek, A. 1965. Histochemische Nierenuntersuchungen nach chronishcher Manganvergiftung, *Arch. Gerwebepath. Gewerbehyg. 21:*347.

Jones, J. H. 1938. The metabolism of calcium and phosphorus as influenced by the addition to the diet of salts of metals which form insoluble phosphates, *Am. J. Physiol. 124:*230–237.

Jones, R. H., Williams, R. L., and Jones, A. M. 1971. Effects of heavy metals on the immune response: Preliminary findings for cadmium in rats, *Proc. Soc. Exp. Biol. Med. 137:*1231.

Jowsey, J., Rowland, R. E., and Marshall, J. H. 1958. The deposition of rare earths in bone, *Radiat. Res. 8:*490–497.

Jung, E. G., and Trachsel, B. 1970. Molekularbiologische Untersuchungen zur Arsencarcinogenese, *Arch. Klin. Exp. Derm. 237:*819–826.

Kahn, B., Straub, C. P., and Robbins, P. J. 1969. Retention of radio strontium, calcium, and phosphorus by infants, *Pediatrics (Suppl.) II 43(4):*651–664.

Kaluk, C. 1971. Study of the nature of metal binding sites and estimate of the distance between the metal binding sites in transferrin using trivalent lanthanide ions as fluorescent probes, *Biochemistry 10:*2838–2843.

Kanisawa, M., and Schroeder, H. A. 1967. Life term studies on the effects of arsenic, germanium, tin, and vanadium on spontaneous tumors in mice, *Cancer Res. 27:*1192–1195.

Kanisawa, M., and Schroeder, H. A. 1969. Renal arterial changes in hypersensitive rats given cadmium in drinking water, *Exp. Mol. Pathol. 10:*81.

Kao, R. L. C., and R. M. Forbes, 1973. Effects of lead on heme-synthesising enzymes and urinary aminolevulinic acid in the rat, *Proc. Soc. Exp. Biol. Med. 143:*234–247.

Karasek, F., Poupa, O., and Jelinek V. 1948, *Minerva Med. 39(11):*28.

Karasek, F., Poupa, O., and Jelinek, V. 1949, *Chem. Abstr. 43:*6291.

Kasprzak, K. S., and Sunderman, F. W., Jr. 1969. Metabolism of nickel carbonyl, *Toxicol. Appl. Pharmacol. 15:*295–301.

Katz, J., Kornberg, H. A., and Parker, H. M. 1955. Absorption of plutonium fed chronically to rats, *Am. J. Roentgenol. 73:*303.

Kaufman, M. 1913. Über ein neues Entfettungsmittal: Kolloidales Palladiumhydroxydul. *Munch. Med. Wschr. 60:*525–529.

Kawin, B., Copp, D. H., and Hamilton, J. G. 1950. Metabolism of certain fission products and plutonium, University of California Radiation Lab. Rept. *UCRL 812,* pp. 1–9.

Kay, M. 1976. Personal communication.

Kazantzis, G. 1966. Chronic mercury poisoning—clinical aspects, *Ann. Occup. Hyg. (London) 8:*65–72.

Kazantzis, G., and Hanbury, W. J. 1966. Induction of sarcoma in the rat by cadmium sulfide and by cadmium oxide, *Br. J. Cancer 20:*190.

Kehoe, R. A. 1963. Industrial lead poisoning, in *Hygiene and Toxicology,* 2nd ed., Vol. II (D. W. Fasset and D. D. Irish, eds.), pp. 941–985, John Wiley, New York.

Kehoe, R. A., Cholak, J., and Story, R. V. 1940. A spectrochemical study of the normal ranges of concentration of certain trace metals in biological materials, *J. Nutr. 19:*579–589.

Kent, N. L., and McCance, C. A. 1941*a*. Arsenic, germanium, tin, and vanadium in mice: Effects on growth, survival and tissue levels, *J. Nutr. 92:*245–251.

Kent, N. L., and McCance, C. A. 1941*b*. The absorption and excretion of "minor" elements by man: I. Silver, gold, lithium, boron, and vanadium, *Biochem. J. 35:*837–841.

Khera, K. S. 1973. Reproductive capability of male rats and mice treated with methyl mercury, *Toxicol. Appl. Pharamacol. 24:*167–177.

Kinard, F. W., and Aull, J. C. 1945. Distribution of tungsten in the rat following ingestion of tungsten compounds, *J. Pharm. Exp. Ther. 83:*53–55.

Kinard, F. W., and Van de Erve, J. 1941. Toxicity of orally ingested tungsten compounds in the rat, *J. Pharmacol. Exp. Ther. 72:*196–201.

King, B. J. 1971. Maximum daily intake of lead without excessive body lead burden in children, *Am. J. Dis. Child. 122:*337–340.

Kiryushkin, V. I., Doshchenko, V. N., Podgorodetskaya, V. N., and Tokarskaya, Z. B. 1963. Clinical manifestations following a single exposure to [137]cesium, *Fed. Proc. (Trans. Suppl.) 23:*T1321.

Klauder, D. S., Murthy, L., and Petering, H. G. 1972. Effect of dietary intake of lead acetate on copper metabolism in male rats, in: *Trace Substances in Environmental Health,* Vol. VI (D. D. Hemphill, ed.), University of Missouri Press, Columbia, pp. 131–136.

Klechkovsky, V. M. 1957. On the behavior of radioactive fission products in soil, their absorption by plants and their accumulation in crops, Academy of Sciences, USSR, *Transl. A.E.C.-TR. 2867.*

Klein, R., Herman, S. P., Brubaker, P. E., Lucier, G. W., and Krigman, M. R. 1972. A model of acute methyl mercury intoxication in rats, *Arch. Pathol. 93:*408–414.

Klemperer, F. W., and Liddey, R. E. 1952. The fate of beryllium compounds in the rat, *Arch. Biochem. Biophys. 41:*148.

Knorr, D. W. 1975. Resorption, perorale Toxizität und in lebensmitteln maximal zulassige Konzentration von Zinn, *Lebenson-Wiss. Technol. 8:*51–56.

Knorr, D., Schaller, S., Schon, C., and Herrmann, W. 1974. Evaluation of corrosion in lacquered and unlacquered tin cans containing tomato puree, *Int. J. Technol. Fruit-Veg. Proc. 19:*5–75.

Kobayasha, J., Nakahasa, H., and Hasegawa, T. 1971. Accumulation of cadmium in mice fed cadmium-polluted rice, *Nippon Eiseigaku Zasshi 26:*401–407.

Koelsch, F. 1959. Bleivergiftung und Zahn Ausfall, *Zbl. Anheitsmed. 9:*114.

Kojima, K., and Fujita, M. 1973. Summary of recent studies in Japan on methyl mercury poisoning, *Toxicology 1:*43–62.

Koller, L. D., Exon, J. H., and Roan, J. G. 1976. Humoral antibody response in mice after single exposure to lead or cadmium, *Proc. Soc. Exp. Biol. Med. 151:*339–342.

Kolomiitseva, M. G., and Stoliar, V. I. 1969. Effect of trace elements in the diet on the activity of some tissue enzymes in experimental animals, *Vopr. Pit. 2:*31–33.

Kones, R. J. 1975. *Glucose, Insulin, Potassium and the Heart,* Futural Pub. Co., Mount Kisco.

Koontz, A. R., and Kimberly, R. C. 1953. Tissue reactions of tantalum gauze and stainless steel gauze: An experimental comparison, *Ann. Surg. 137*:833–840.

Kortus, J. 1967. The carbohydrate metabolism accompanying intoxication by aluminum salts in the rat, *Experientia 23*:912–915.

Kosower, N. S., Vanderhoff, G. A., and London, I. M. 1967. The regeneration of reduced glutathione in normal and glucose-6-phosphate dehydrogenase deficient human red blood cells, *Blood 29*:313–319.

Kotb, A. R., and Luckey, T. D. 1972. Markers in nutrition, *Nutr. Abstr. Rev. 42(3)*:813–845.

Koutensky, J., Koutensky, E. M., Sykora, J., and Mertl, F. 1971 Influence of sodium diethyldithiocarbamate on the toxicity and distribution of copper in mice, *Eur. J. Pharm. 14*:389–392.

Koval, Y. F., and Kondrashev, V. M. 1967. Acceleration of the elimination of Y^{91}, La^{140}, Ce^{141}, Pr^{143}, Nd^{147}, and $plutonium^{147}$ from the body, *Med. Radiol. 12*:84–85.

Kovar, E. W., and Luckey, T. D. 1975. Hormology, stress, and growth promotants, *Abstracts of the Xth International Congress on Nutrition,* Tokyo, p. 147.

Krasnow, N. 1972. Effects of lanthanum and gadolinum ions on cardiac sarcoplasmic reticulum in dogs, *Biochim. Biophys. Acta 282*:187–194.

Krehl, W. A. 1972. Mercury: The slipper metal, *Nutr. Today 7 (6)*:4.

Krishmamachari, K. A. V. R., 1974. Some aspects of copper metabolism in pellagra, *Am. J. Clin. Nutr. 27*:108–111.

Kulcher, G. V., and Reynolds, W. J. 1942. Bismuth hepatitis a survey of 121 cases, *J. A. M. A. 120*:343.

Kulieva, T. K. 1971. Effect of vanadium on leukocyte phagocytic activites in guinea pig, *Izv. Akad. Nauk. Turkm. SSR Ser. Biol. Nauk (3)*:70–71.

Kulikova, V. G. 1966. Effect of six, age, castration, sex hormones and adrenalectomy on the behavior of Cs^{137} and Ce^{144} in the rat organism, *Tr. Inst. Biol. Mol. Skin Filial. Akad. Nauk. USSR 46*:111.

Kwantes, W. 1966. in: *Safety of Canned Foods,* Royal Soc. Health, Cox and Wyman, Ltd., London, p. 143.

Kyker, G. C. 1962. Rare earths, in: *Mineral metabolism,* Vol. II, Part B, "The Elements," (C. L. Comar and F. Bronner, eds.), Academic Press, New York and London, pp. 499–541.

Kyker, G. C., and Anderson, E. B. (eds.). 1956. in: *Rare earths in Biochemical and Medical Research,* A conference sponsored by Medical Division, Oak Ridge Institute of Nuclear Studies, 1955, *USAEC Rept. ORINS-12.*

LaBella, F. S., Dular, R., Lemon, O., Vivian, S., and Queen, G. 1973. Prolactin secretion is specifically inhibited by nickel, *Nature 245*:330–332.

LaManna, C., and Sakaguchi, G. 1971. Botulinal toxins and the problem of nomenclature of simple toxins, *Bacteriol. Rev. 35*:242–249.

Lamola, A. A., and Yamane, T. 1974. Zinc protoporphyria in the erthrocytes of patients with lead intoxication and iron deficiency anemia, *Science 186*:936–938.

Lampert, P., Garro, F., and Pentschew, A. 1967. Lead encephalopathy in suckling rats, in: *Brain Edema* (I. Klatzo and F. Seitelberger, eds.), Springer Verlag, New York, pp. 207–222.

Lampert, P., Garro, F., and Pentschew, A. 1970. Tellurium neuropathy, *Acta Neuropath. 15*:308–317.

Landolt, R. R., Berk, H. W., and Russel, H. T. 1972. Studies on the toxicity of rhodium chloride in rats and rabbits, *Toxicol. Appl. Pharmacol. 21*:589.

Larson, S. E., and Piscator, M. 1971. Effects of cadmium on skeletal tissue in normal and calcium deficient rats, *Israel J. Med. Sci. 7*:495–498.

Lau, T. J., Hackett, R. L., and Sunderman, F. W., Jr. 1972. The carcinogenicity of intravenous nickel carbonyl in rats, *Cancer Res. 32:*2253–2258.

Lawson, J. J. 1961. Toxicity of titanium tetrachloride, *J. Occup. Med. 3:*7–12.

Lazar, G. 1973. The reticuloendothelial-blocking effect of rare earth metals in rats, *J. Reticuloendothel. Soc. 13:*231–237.

Lazarus, J. H., and Bennie, E. H. 1972. Effect of lithium on thyroid function in man, *Acta Endocr. Copenh. 70:*277.

Lazarus, S. S., Goldner, M. G., and Volk, B. W. 1953. Selective destruction of pancreatic alpha cells by cobaltous chloride in the dog, *Metabolism 2:*513–516.

Laznitski, A., and Szorenyi, E. 1934. The influence of different cations on the growth of yeast cells, *Biochem. J. 28(2):*1678.

Lebedeva, G. A. 1967. Concerning morphological changes in the blood vessels of the gastrointestinal tract in injury with radioactive cerium, *Biol. Abstr. 49:*1481 (16704).

Leddicote, G. W., and Tipton, I. H. Personal communication. 1958. Cited by G. C. Kyker. 1962. Rare earths, in: *Mineral Metabolism,* Vol. II, Part B: "Rare Earths," Academic Press, New York.

Lee, A. M., and Fraumeni, J. F. 1969. Arsenic and respiratory cancer in man: An occupational study, *J. Natl. Cancer Inst. 42:*1045–1052.

Lee, D., and Matrone, G. 1969. Iron and copper effects on ceruloplasmin activity of rats with zinc-induced copper deficiency, *Proc. Soc. Exp. Biol. Med. 130:*1190–1193.

Lee, G. R. Nacht, S., Lukens, J. N., and Cartwright, G. E. 1968. Iron metabolism in copper deficient swine, *J. Clin. Invest. 47:*2058–2069.

Lee, W. S. S. 1972. Distribution and toxicity of [239]plutonium in bone, *Health Phys. 22:*583–595.

Lehman, K. B. 1910. Important technical and hygienic gases and vapors. XIV. Castors or zinc fever, *Arch. Hyg. 72:*538–542.

Lehman, K. B., and Herget, L. 1927. Studien über die hygienischen Eigenschaften des Titanoxyds und des Titanweiss, *Chemiker-Ztg. 82:*793–796.

Lehninger, A. L., and Carafoli, I. E. 1971. The interaction of La^{3+} with mitochondria in relation to respiration coupled Ca^{2+} transport, *Arch. Biochem. Biophys. 143:*506.

Lenihan, J. M. A., Loutit, J. F., and Martin, J. H. 1967. *Strontium Metabolism,* Academic Press, London and New York.

Lessler, M. A., and Walters, M. I. 1973. Erythrocyte osmotic fragility in the presence of lead or mercury, *Proc. Soc. Exp. Biol. Med. 142:*548–553.

Levander, O. A., and Argrett, L. C. 1969. Effects of arsenic, mercury, thallium, and lead on selenium metabolism in rats, *Toxicol. Appl. Pharmacol. 14:*308–314.

Levander, O. A., and Morris, V. C. 1970. Interactions of methionine, vitamin E, and antioxidants in selenium toxicity in the rat, *J. Nutr. 100:*1111–1118.

Levander, O. A., Young, M. L., and Meeks, S. A. 1970. Studies on the binding of selenium by liver homogenates from rats fed diets containing, *Toxicol. App. Pharmacol. 16:*79–85.

Levander, O. A., Morris. V. C., Higgs, D. J., and Feretti, R. J. 1975. Lead poisoning in vitamin E deficient rats, *J. Nutr. 105:*1481–1485.

Levene, G. M. 1971. Platinum sensitivity, *Br. J. Derm. 85:*590.

Levina, E. N., and Rabachevsky, E. G. 1955. Changes in the lung tissue with intratracheal injection of manganese, *Gig. i Sanit. 1:*25–27.

Levina, E. W., and Chekunova, M. P. 1965. Toxicity of antimony halides, *Fed. Proc. Trans. Suppl. T.* 7608.

Lewis, A. H. 1943. The teart pastures of Somerset I and II, *J. Agri. Sci. 33:*52–68.

Lewis, P. K., Hoekstra, W. G., and Grummer, R. H. 1957. Restricted calcium feeding versus zinc supplementation for the control of parakeratosis in swine, *J. Anim. Sci. 16:*578–588.

Lewis, R. C., Jr., McKee, F. S., and Longwell, B. B. 1944. The effect of excessive dietary Na and K on the carbohydrate metabolism in rats, *J. Nutr. 27*:11–17.

Libenston, L. 1945. Toxicity and mode of action of gold salts, *Exp. Med. Surg. 3*:146–152.

Lie, R., Thomas R. G., and Scot, J. K. 1960. The distribution and excretion of thallium[204] in the rat with suggested MPC's and a bioassay procedure, *Health Phys. 2.* 334–340.

Lin, J. H., and Duffy, J. L. 1970. Cobalt-induced myocardial lesions in rats, *Lab. Invest. 23*:158.

Lindenbaum, A., and Rosenthal, M. W. 1972. Deposition patterns and toxicity of plutonium and americium in the liver, *Health Phys. 22*:597–605.

Linder, M. C., and H. N. Munro. 1973. Iron and copper metabolism during development, *Enzymes 15*:111–138.

Lipkan, G. N. 1970. The effect of unithiol on the enzymatic activity of myosin and myosinoid proteins and the content of sulfhydryl groups in the presence of metals reacting with thiols, *Farmakol. Toksikol. Respub. Mezhvedom, S. B. 5*:191.

Lipsitz, P. J., and English, I. C. 1967. Hypermagnesemia in the newborn infant, *Pediatrics 40*:856–862.

Little, J. A., and Sunica, R. 1958. Cobalt-induced goiter with cardiomegaly and congestive failure, *J. Pediatr. 52*:284–288.

Lloyd, L. E., Rutherford, B. E., and Crampton, E. W. 1955. Titanium oxide and chromium oxide as index materials for determining apparent digestibility, *J. Nutr. 56*:265–270.

Loeser, D., and Konweiser, A. L. 1929. A study of the toxicity of strontium and comparison with other cations employed in therapeutics, *J. Lab. Clin. Med. 15*:35.

Losee, F. L., and Adkins, B. L. 1969. A study of the mineral environment of caries-resistant navy recruits, *Caries Res. 3*:23–27.

Lucis, O. J., Lucis, R., and Shaikh, Z. A. 1972. Cadmium and zinc in pregnancy and lactation, *Arch. Environ. Health 25*:14–19.

Luckey, T. D. 1975*a*. Introduction to heavy metal toxicology, safety and hormology, *Environ. Qual. Saf. Suppl. 1*:1–3.

Luckey, T. D. 1975*b*. Hormology with inorganic compounds, *Environ. Qual. Saf. Suppl. 1*:81–103.

Luckey, T. D., and Venugopal, B. 1977*a*. *Metal Toxicity in Mammals,* Vol. 1, "Physiologic and Chemical Basis for Metal Toxicity," Plenum Press, New York.

Luckey, T. D., and Venugopal, B. 1977*b*. pT, a new classification system for toxic compounds, *J. Toxicol. Environ. Health 2*:633–638.

Luckey, T. D., Kotb, A. R., Vogt, J. R., and Hutcheson, D. P. 1975. Rat feasibility studies of multiple nutrient markers, *J. Nutr. 105*:660–669.

Luckey, T. D., Venugopal, B., Hutcheson, D. P., and Gray, D. H. 1977. Lanthanide marker evidence for two compartments in the human alimentary tract, *Nutr. Rep. Internat. 16*:339–347.

Lund, A. 1956. The distribution of thallium in the organisms and its elimination, *Acta Pharmacol. Toxicol. 12*:251–259.

Lydtin, H. 1965. Über barium vergiftung, *Münch. Med. Wschr. 107*:1045.

MacDonald, N. W., Nusbaum, R. E., Alexander, G. V., Ezmirnan, F., Spain, P., and Rounds, D. E. 1952. The skeletal deposition of yttrium, *J. Biol. Chem. 195*:837–842.

MacIntyre, I. 1967. Magnesium metabolism, in: *Advances in Internal Medicine,* Vol. 13 (W. Dock and I. Snapper, eds.), Yearbook Medical Publishers, Inc., Chicago, pp. 143–154.

MacIntyre, I., Boss, S., and Troughton, V. A. 1963. Parathyroid hormone and magnesium homeostasis, *Nature 198*:1058–1060.

Mackay, K. M., and Mackay, R. A. 1969. *Introduction to Modern Inorganic Chemistry,* Interext Books, London.

Macleod, R. A., and Snell, E. E. 1950. The relation of ion antagonism to the inorganic nutrition of lactic acid bacteria, *J. Bacteriol. 59:*783–792.

Maenza, R. M., Pradhan, A. M., Sunderman, F. W., Jr. 1971. Rapid induction of sarcomas in rats by a combination of nickel sulfide and 3,4 benzpyrene, *Cancer Res. 31:*2067–2071.

Magee, A. C., and Matrone, G. 1960. Studies on growth, copper metabolism, and iron metabolism of rats fed high levels of zinc, *J. Nutr. 72:*233–242.

Magnusson, G. 1962. Acute toxic effects of yttrium, cerium, terbium, holmium, and ytterbium on the liver, *Acta Pharmacol. Toxicol. 20 (Suppl):*53–74.

Magos, L. 1968. Uptake of mercury by the brain, *Br. J. Ind. Med. 25:*315.

Mahaffey, K. R., Goyer, R. A., and Haseman, J. K. 1973. Dose-response to lead ingestion in rats fed low dietary calcium, *J. Lab. Clin. Med. 83:*92–100.

Mahlum, D. D. 1967*a*. Influence of hepato carcinogen feeding on the retention of cerium[144] by the liver, *Toxicol. Appl. Pharmacol. 11:*585.

Mahlum, D. D. 1967*b*. Neptunium[237] toxicity in rat. II. Intracellular distribution of neptunium and serium[144] rat liver, *Toxicol. Appl. Pharmacol. 11:*264.

Mahlum, D. D., and Sikov, M. R. 1968. Distribution of [144]cerium in the fetal and newborn rats, *Health Phys. 14:*127–129.

Malcolm, D. 1972. Potential carcinogenic effect of cadmium in animals, *Ann. Occup. Hyg. 15:*33–36.

Maletsky, B., and Blachly, P. H. 1970. The use of lithium in psychiatry. *C R Clin. Lab. Science 2:*279–345.

Marchetti, F., Cruz, B., and Lafuma, J. 1967. Rechèrches experimentales sur le metabolisme des terres rares, *Minerva Fisiconucleare 14:*123–125.

Maresh, F., Lustock, M. J., and Cohen, P. P. 1940. Physiological studies on rhenium compounds, *Proc. Soc. Exp. Biol. Med. 45:*574–576.

Mascitelli-Coriandoli, E., and Citterio, F. 1959*a*. Effects of vanadium upon liver coenzyme A in rats, *Nature 183:*1527.

Mascitelli-Coriandoli, E., and Citerrio, F. 1959*b*. Intracellular thiotic acid and coenzyme A following vanadium treatment, *Nature 184 (Suppl. 21):*1641–1643.

Mason, K. E., and Young, J. O. 1967. Effectiveness of selenium and zinc in protecting against cadmium-induced injury of rat testis, in: *Symposium: Selenium in Biomedicine* (O. H. Muth, ed.), AVI Publishing Co., Connecticut, pp. 383–394.

Massman, W. 1956. Experimentelle Untersuchungen über die biologische Wirkung von Vandadinverbindungen, *Arch. Toxicol. 16:*182–186.

Matrone, G., Hartman, R. H., and Clawson, A. J. 1959. Studies of a manganese iron antagonism in the nutrition of rabbits and baby pigs, *J. Nutr. 67:*309.

Matten, K. E., Schutter, K., Van Senden, K. G., and Spruit, D. 1969. Nickel sensitization and detergents, *Acta Derm. Vener. Stockholm 49:*10–13.

Mattis, P. A. 1950. Toxicological studies of certain thorium salts, Western Reserve University School of Medicine, Atomic Energy Project, *USAEC Rept. N.Y.O. 1607.*

Mayer, S. W., and Morton, M. E. 1956. Preparation and distribution of [90]yttrium fluoride, in: *Rare Earths in Biochemical Medical Research,* A conference sponsored by Medical Division, Oak Ridge Institute of Nuclear Studies, 1955, pp. 83–87.

Mazturzo, A. 1951. Sangue periferico e mielogramma nella intossicazione sperimentalle da osmio. *Folia Med., Napoli 34:*27.

McClinton, D. T., and Schubert, J. 1948. Zirconium and thorium toxicity, *J. Pharmacol. 94:*1–5.

McConnell, K. P. 1963. Metabolism of selenium in the mammalian organism, *J. Agri. Food Chem. 11:*385–391.

McConnell, K. P., and Carpenter, D. M. 1971. Interrelationships between selenium and specific trace elements, *Proc. Soc. Exp. Biol. Med. 137:*996–1101.

McCord, C. P., Meek, S. F., Harrold, G. C., and Heussner, C. E. 1942. The physiologic properties of indium and its compounds, *J. Ind. Hyg. 24:*243–254.

McLaughlin, A. I., Milton, R. G., and Perry, K. M. A. 1946. Toxic manifestations of osmium tetroxide, *Br. J. Ind. Med. 3:*183.

McLaughlin, A. I., Kazantzis, G., King, E., Teare, D., Porter, R. J., and Owen, R. 1962. Pulmonary fibrosis and encephalopathy associated with the inhalation of aluminum dust, *Br. J. Ind. Med. 19:*253–63.

Mealey, J. 1957. Turnover of carrier free Zr in man, *Nature 179:*673–674.

Meek, F. F., Harrold, C. C., and McCord, C. P. 1943. The physiological properties of palladium, *Ind. Med. Surg. 12:*447–451.

Meekes, M. J., Landolt, R. R., Kessler, W. V., and Born, G. S. 1971. Effect of vanadium on metabolism of glucose in the rat, *J. Pharm. Sci. 60:*482.

Mela, L. 1969*a*. Inhibition and activation of Ca transport in mitochondria—Effect of lanthanides and local anesthetic drugs, *Biochemistry 8:*2481–2486.

Mela, L. 1969*b*. Reactions of lanthanides with mitochondrial membranes, *Ann. N.Y. Acad. Sci. 147:*824–828.

Meltzer, H. L., and Liebermann, K. W. 1971. Chronic ingestion of rubidium without toxicity: Implications for human therapy, *Experientia 27:*672–674.

Meltzer, H. L., Taylor, R. M., and Platman, S. R. 1969. Rubidium: A potential modifier of effect and behavior, *Nature 223:*321–322.

Mena, I., Martin, O., Fuenzalida, S., and Cotzias, G. C. 1967. Chronic manganese poisoning: Clinical picture and turnover, *Neurology 17:*128–138.

Meneely, G. R. 1973. Toxic effects of dietary sodium chloride and protective effects of potassium, in: *Toxicants Naturally Occurring in Food* (F. M. Strong *et al.,* eds.), National Academy of Sciences, pp. 26–42.

Meneely, G. R., and Ball, C. O. T. 1958. Experimental epidemiology of chronic sodium chloride toxicity and protective effect of potassium chloride, *Am. J. Med. 25:*713–718.

Meneely, G. R., Ball, C. O. T., and Youmans, B. 1957. Chronic sodium chloride toxicity—The protective effect of added potassium chloride, *Ann. Intern. Med. 47:*263.

Merali, Z., Kacew, S., and Singhla, R. L. 1975. Response of hepatic carbohydrate and cyclic AMP metabolism, *Can. J. Physiol. Pharmacol. 53:*174–184.

Mertz, D. P., Koschnick, R., and Wilk, G. 1970. Renale Ausscheidungsbedingungen von Nickel beim Menschen, *Z. Klin. Chem. Klin. Biochem. 3:*387–390.

Mertz, W. 1967. Biological role for chromium, *Fed. Proc. 26:*168–172.

Mertz, W. 1969. Chromium: Occurrence and function in biological systems, *Physiol Rev. 49:*163–239.

Mertz, W. 1975. Effect and metabolism of glucose tolerance factor, *Nutr. Rev. 33:*129–135.

Mertz, W., and Roginsky, E. E. 1975. Some biological properties of chromium (Cr) nicotinic acid (NA) complexes, *Fed. Proc. 34:*922.

Metz, E. N., and Sagone, A. L., Jr. 1972. The effect of copper on erythrocyte hexose monophosphate shunt pathway, *J. Lab. Clin. Med. 80:*405–413.

Miller, C. A., and Levine, E. M. 1974. Effects of aluminum salts on cultured neuroblastoma cells, *J. Neurochem. 22:*751–758.

Miller, C. O. 1967. Diesel smoke depression by fuel additive treatment, Society of Automotive Engineers. *S. A. E. 670093.*

Miller, J. K., and Byrne, W. F. 1970. Comparison of scandium[46] and cerium[144] as non-absorbed reference materials in studies with cattle, *J. Nutr. 100:*1287–1295.

Miller, J. K., Perry, S. C., Chandler, P. T., and Cragle, R. G. 1967. Evaluation of radio cerium as a non-absorbed reference material for determining gastrointestinal sites of nutrient absorption and excretion in cattle. *J. Dairy Sci. 50:*355–361.

Miller, J. K., Moss, B. R., and Byrne, W. F. 1971. Distribution of cerium in the digestive tract of calf according to time after dosing, *J. Dairy Sci. 54:*497.

Miller, R. F., and Engel, R. W. 1960. Interrelations of copper, molybdenum, and sulfate sulfur in nutrition, *Fed. Proc. 19:*666.

Miller, W. J. 1969. Absorption, tissue distribution, endogenous excretion and homeostatic control of zinc in ruminants, *Am. J. Clin. Nutrition 22:*1323.

Miller, W. J. 1971. Cadmium nutrition and metabolism in ruminants: Relationship to concentrations in tissues and products, *Feedstuffs,* p. 43.

Miller, W. J., Lampp, B., Powell, C. W., Salotti, C. A., and Blackman, D. M. 1967. Influence of a high level of dietary cadmium on cadmium content in milk, excretion and cow performance, *J. Dairy Sci. 50:*1404–1408.

Miller, W. J., Blackmore, D. M., Gentry, R. P., and Pate, F. M. 1969. Effect of dietary cadmium on tissue distribution of cadmium following a single oral dose in young goats, *J. Dairy Sci. 52:*2029–2035.

Miller, W. J., Blackmore, D. M., Gentry, R. P., and Pate, F. M. 1970. Effects of high but non-toxic levels of zinc in practical diets on ^{65}Zn and Zn metabolism in Holstein calves, *J. Nutr. 100:*893–902.

Mills, C. F., and Dalgarno, A. C. 1972. Copper and zinc status of ewes and lambs increased dietary concentrations of cadmium, *Nature 239:*171–173.

Milner, J. E. 1963. Effect of ingested arsenic on methylcholanthrene induced skin tumors in mice, *Arch. Environ. Health 18:*7–11.

Ming-Hsin, H., Shaoh-chi, C., Yu-Siu, P., and Kuo-Yuei, Y. 1958. Mechanism and treatment of cardiac arrhythmias in tartar emetic intoxication, *Chin. Med. J. 76:*103–110.

Mital, V. P., Wahal, D. K., and Bansal, O. P. 1966. A study of erythrocyte glutathione in acute copper sulfate poisoning, *Indian J. Pathol. Bact. 9:*156–162.

Mitchel, W. G. 1953. Antagonism of toxicity of vanadium by ethylene diamine tetraacetic acid in mice, *Proc. Soc. Exp. Biol. Med. 83:*346–348.

Mitchel, W. G., and Floyd, E. P. 1954. Ascorbic acid and ethylene diaminetetracetate as antidotes in experimental vanadium poisoning, *Proc. Soc. Exp. Biol. Med. 85:*206–209.

Moeschlin, S. 1965. *Poisoning: Diagnosis and Treatment,* Grune and Stratton, New York.

Mogilevskaya, O. Y. 1956. Action of aerosols of titanium and its oxide, *Gig. i Sanit. 21(3):*20–23, in: *Chem. Abstr. 50(1):*9990.

Moinuddin, J. F., and Lee, H. W. 1960. Alimentary, blood, and other changes due to feeding MnSo$_4$, MgSO$_4$, and Na$_2$SO$_4$, *Am. J. Physiol. 199:*77–83.

Molfino, F. 1938. Contributi sperimentalle allo studio dell' intoxicazione professionale da vanadio, *Rass. Med. Industr. 9:*362. *Abstr. J. Industr. Hyg. Toxicol. 21:*96. (1939).

Montie, T. C., and Ajl, S. J. 1970. Nature and synthesis of murine toxins of *Pasteurella pestes,* in: *Microbial toxins,* Vol. 3 (T. C. Montie, S. Kadis, and S. J. Ahl, eds.), Academic Press, New York.

Monty, K. J., and Click, E. M. 1961. A mechanism for the copper–molybdenum interrelationship. III. Rejection by the rat of molybdate containing diets. *J. Nutr. 75:*303.

Moore, C. V., and Dubach, R. 1962. Iron, in: *Mineral Metabolism: An Advanced Treatise,* Vol. II, Part B (C. L. Comar and F. Bronner, eds.), Academic Press, New York, p. 288.

Moore, W., Hysell, D., Crocker, D., and Stara, J. 1975. Biological fate of a single administration of 191 pT in rats following different routes of exposure, *Environ. Res. 9:*152–158.

Morgan, B. N., Thomas, R. G., and McCellan, R. O. 1970. Influence of chemical state of cerium[144] on its metabolism following inhalation by mice, *Am. Ind. Hyg. Assoc. J. 31:*479.

Morishige, T. 1968. Experimental studies on the effect of Sr[95]–Nb[95] on albino rats: Distribution of Zr–Nb among the organs—microautoradiography of lungs on intratracheally incubated albino rats, *Jap. J. Hyg. 23:*404.

Morris, V. C., and Levander O. A. 1970. Selenium content of foods, *J. Nutr. 100:*1383–1388.

Moskalev, Y. I. 1972. [239]Pu: Problems of its biologic effects, *Health Phys. 22:*723–729.

Mountain, J. T., Delker, L. L., and Stokinger, H. E. 1953. Studies in vanadium toxicity, *Arch. Ind. Hyg. Occup. Med. 8:*406–411.

Moxon, A. L, and Rhian, M. 1943, Selenium poisoning, *Physiol. Rev. 23:*305.

Mraz, F. R., Leloir, M., Pinajian, J. J., and Patrick, H. 1957. Influence of potassium and sodium on uptake and retention of [134]cesium in rats, *Arch. Biochem. Biophys. 66:*177–184.

Muhler, J. C., Stooky, G. K., and Wagner, M. J. 1959. Effect of Sr[89] on fluoride retention in the rat, *Proc. Soc. Exp. Biol. Med. 102:*644–647.

Muhler, J. C., Stookey, G. K., and Beck, C. W. 1970. Preparation and properties of indium heptafluorozirconate, *J. Dent. Res. 49:*529–531.

Murphy, J. V. 1970. Intoxication following ingestion of elemental zinc, *J. A. M. A. 212:*2119.

Murthy, L., Klevan, L. M., and Petering, H. G. 1974. Interrelationships of zinc and copper nurtriture in the rat, *J. Nutr. 104:*1458–1465.

Muth, O. H., Oldfield, J. E., and Weswig, P. H. (eds.). 1967. *Selenium in Biomedicine,* AVI Publishing Co., Connecticut.

Myers, E. D. 1972. Prophylactic lithium in recurrent affective disorders, *Lancet 1:*1287.

Nason, A. 1958. The metabolic role of vanadium and molybdenum in plants and animals, in: *Trace Elements* (C. A. Lamb, O. G. Bentley, and J. M. Beattie, eds.), Academic Press, New York, pp. 269–296.

Nath, I., Sood, S. K., and Nayak, N. C. 1972. Experimental siderosis and liver injury in the rhesus monkey, *J. Pathol. 106:*103–111.

National Academy of Sciences. 1972. Committee on Animal Nutrition. *Food Chem. News 14:*24.

National Academy of Sciences. 1974. Chromium, *Nat. Acad. Sci.,* Washington, D.C.

National Academy of Sciences. 1975. Committee on medical and biologic effects of environmental pollutants. Nickel, *Nat. Acad. Sci.,* Washington, D.C.

National Research Council. 1971. Selenium in nutrition, *Nat. Acad. Sci.,* Washington, D.C.

Neathery, M. W., Miller, W. J., Gentry, R. P., Blackman, D. M., and Stake, P. E. 1973. Methyl mercury and cadmium metabolism in lactating Jersey cows, *J. Dairy Sci. 56:*307.

Nelson, A. A., Fitzhugh, O. G., and Calvery, H. O. 1943. Liver tumors following cirrhosis by selenium in rats, *Cancer Res. 3:*365–366.

Nelson, S., Chen, C., Tsai, A., and Dyer, I. A. 1973. Effect of chelating agents on chromium absorption in rats, *J. Nutr. 103:*1182–1186.

Neubaur, O. 1946. Arsenical cancer: A review, *Br. J. Cancer 1:*192–196.

Newberne, P. M., Glaser, O., Griedman, L., and Stillings, B. E. 1972. Chronic exposure of rats to methyl mercury in fish protein, *Nature 237:*40–42.

Nicaud, P., Lafitte, A., and Gros, A. 1942. Les troubles de l'intoxication chronique par le cadmium, *Arch Mal. Prof. Med. Tran. 4:*192–198.

Nielsen, F. H. 1971. Studies on the essentiality of nickel, in: *Newer Trace Elements in Nutrition* (W. Mertz and W. E. Cornatzer, eds.), Marcel Dekker, Inc., New York, pp. 215–253.

Nielsen, F. H. 1975. Arsenic essentiality, *Fed. Proc. 34(3):*923.

Nielsen, F. H., and Ollerich, D. A. 1972. Studies on a vanadium deficiency in chicks, *Fed. Proc. 32:*929.

Nielsen, F. H., Myron, D. R., Givand, S. H., Zimmerman, T. J., and Ollerich, D. A. 1975. Nickel deficiency in rats, *J. Nutr. 105:*1620–1630.

Nishizumi, M. 1972. Electron microscope study of Cd nephrotoxicity in the rat, *Arch Environ. Health 24:*215.

Noddack, I., and Noddack, W. 1940. Die Häufigkeiten der Schwermetalle in Meerestiern, *Ark. Zoo. 32:*1–7.

Nomiyama, K., and Foulkes, E. C. 1968. Some effects of uranyl acetate on proximal tubular function in rabbit kidney, *Toxicol. Appl. Pharmacol. 13:*89–98.

Nomiyama, K., Sato, C., and Yamamoto, A. 1973. Early signs of cadmium intoxication in rabbits, *Toxicol. Appl. Pharmacol. 24:*624–635.

Nomoto, S., McNeely, M. D., and Sunderman, F. W., Jr. 1971. Isolation of a nickel macroglobulin from rabbit serum, *Biochemistry 10:*1647–1651.

Norris, W. P., Lisco, H., and Brues, A. M. 1956. The radiotoxicity of cerium and yttrium, in: *Rare Earths in Biochemical and Medical Research,* A conference sponsored by Medical Division, Oak Ridge Institute of Nuclear Studies, 1955, pp. 116–123.

Norseth, T. 1968. Intracellular distribution of mercury in rat liver after a single injection of mercuric chloride, *Biochem. Pharmacol. 17:*581.

Norseth, T., and Clarkson, T. W. 1970. Biotransformation of methyl mercury salts in the rat studied by specific determination of inorganic mercury, *Biochem. Pharmacol. 19:*2775–2783.

Noshkin, V. E. 1972. Ecological aspects of plutonium dissemination in aquatic environments, *Health Phys. 22:*537 (PIAEAS Intl. Atom. Ener. Agency) (NIL).

Novey, H. S., and Martel, S. H. 1969. Asthma, arsenic, and cancer, *J. Allergy 44:*315–319.

Noyes, R., Jr. 1969. Lithium carbonate: A review, *Dis. Nerv. Syst. 30:*318–321.

Oberleas, D., Muhrer, M. E., and O'Dell, B. L. 1966. Dietary metal-complexing agents and zinc availability in the rat, *J. Nutr. 90:*56.

Ochoa-Solano, A., and Gittler, C. 1968. Incorporation of ^{75}Se-selenomethionine and ^{35}S-methionine into chicken egg white proteins, *J. Nutr. 94:*243–248.

O'Dell, B. L. 1969. Effect of dietary components upon zinc availability: A review with original data, *Am. J. Clin. Nutr. 22:*1315.

O'Dell, B. L., and Campbell, B. J. 1971. Trace elements: Metabolism and metabolic function, in: *Comprehensive Biochemistry,* Vol. 21, (M. Florkin and E. H. Stotz, eds.), Elsevier Publishing Co., Amsterdam, pp. 179–266.

O'Dell, G. D., Miller, W. J., King, W. A., Moore, S. L., and Blackman, D. M. 1970a. Nickel toxicity in the young bovine, *J. Nutr. 100:*1447.

O'Dell, G. D., Miller, W. J., Moore, S. L., and King, W. A. 1970b. Effect of nickel as the chloride and the carbonate on palatability of cattle feed, *J. Dairy Sci. 53:*1266–1269.

O'Dell, G. D., Miller, W. J., Moore, S. L., King, W. A., Ellers, J. C., and Jurecek, H. 1971. Effect of dietary nickel level on excretion and nickel content of tissues in male calves, *J. Anim. Sci. 32:*769.

Oehme, F. W. 1972. Mechanisms of heavy metal toxicities, *Clin. Toxicol. 5(2):*151–167.

Oelschläger, W., and Menke, K. H. 1969. Selenium content of plants, animals, and other materials, *Z. Ernähr Wiss. 9:*216–222.

O'Gara, R. W., and Brown, J. M. 1968. Comparison of carcinogenic actions of subcutaneous implants of iron and aluminum in rodents, *J. Natl. Cancer Inst. 38(6):*947–952.

Ogata, E., Rasmussen, H., and Gruden, N. 1971. Basis of action of hormones on calcium absorption, in: *The Intestinal Absorption of Metal Ions and Trace Elements and Radionuclides* (S. C. Skoryna and D. Waldon-Edward, eds.), Pergamon Press, Oxford and New York, pp. 359–372.

Ogg, C. S., Pearson, J. D., and Veall, N. 1968. A method for measuring the gastro-intestinal absorption of Ca^{47} using Sc^{47} as an inert marker, *Clin. Sci. 34:*327–332.

Ohta, Y. 1971. Studies on the absorption and elution of mercury on human hair, *Jap. J. Ind. Health 11:*9.

O'Kelley, R. E., and Fontenout, J. P. 1973. Effects of feeding different magnesium levels to drylot-fed gestating beef cows, *J. Anim. Sci. 36:*994–1000.

Olsen, K. B., Heggan, G., Edwards, C. F., and Gorham, L. W. 1954. Trace element content of cancerous and non-cancerous liver tissue, *Science 199:*772.

Ondreicka, R., and Ginter, E. 1966. Glycide metabolism in rats during chronic and acute aluminum trichloride intoxication, *Biologia, Bratislava 21:*27–32.

Ondreicka, R., Ginter, E., and Kortus, J. 1966. Chronic toxicity of aluminum in rats and mice and its effects on phosphorus metabolism, *Br. J. Ind. Med. 23:*305–312.

Ondreicka, R., Kortus, J., and Ginter, E. 1971. Aluminum: Its absorption distribution and effects on phosphorus metabolism, in: *Intestinal Absorption of Metal Ions, Trace Elements and Radionuclides,* (S. C. Skoryna and D. Waldron-Edward, eds.), Pergamon Press, Oxford and New York, pp. 293–306.

Oppenheimer, B. S., Oppenheimer, E. T., Danishofsky, I., and Stout, A. P. 1956. Carcinogenic effect of metal in rodents, *Cancer Res. 16:*439–441.

Ordonez, J. V., Carrillo, J. A., Miranda, M., and Gale, J. L. 1966. Estudio epidemiologico de una enfermedad considerada como encefalitis en la region de los altose de Guatemala Bol, *Ofic. Sani. Panamerican 60:*510–515.

Orestano, G. 1933. Azione farmacologica del tidsolfato d'oro e di sodio, *Arch. Int. Pharmacol. Ther. 44:*259–262.

Orten, J. M., and Bucciero, M. C. 1948. The effect of cysteine, histidine, and methionine on the production of polycuthemia by cobalt, *J. Biol. Chem. 176:*961–968.

Osswald, H., and K. Goerttler. 1971. Arsenic-induced leucoses in mice after diaplacental and postnatal application, *Verh. Deut. Ges. Pathol. 55:*289–293.

Ott, E. A., Smith, W. H., Harrington, R. B., and Beeson, W. M. 1966*a*. Zinc toxicity in ruminants. I. Effects of high levels of dietary zinc on gains, feed consumption and feed efficiency of lambs, *J. Anim. Sci. 25:*414–418.

Ott, E. A., Smith, W. H., Harrington, R. B., and Beeson, W. M. 1966*b*. Zinc toxicity in ruminants. II. Effect of high levels of dietary zinc on gains feed consumption and feed efficiency of beef cattle, *J. Anim. Sci. 25:*419–423.

Ovcharenko, E. P. 1972. An experimental evaluation of the effects of transuranic elements on reproductive ability, *Health Phys. 22:*641.

Palmer, I. S., Gunsalus, R. P., Halverson A. W., and Olson, O. E. 1970. Trimethylselenonium ion as a general excretory product in rat urine, *Biochem. Biophys. Acta 208:*260–266.

Palmer, R. F., and Queen, F. B. 1958. Normal abundance of radium in cadavers from the Pacific north coast, *Am. J. Roengenol. Ra. Ther. Nuc. Med. 79:*521–529.

Paola, J. A. 1964. The potentiation of lymphosarcomas in the mouse by manganous chloride, *Fed. Proc. 23:*393.

Pappenheimer, A. M., and Maechling, E. H. 1934. Inclusions in renal epithelial cells following the use of certain bismuth preparations, *Am. J. Pathol. 10:*577.

Parizek, J. 1957. The destructive effect of cadmium ion on esticular tissue and its prevention by zinc, *J. Endocrinol. 15:*56–63.

Parizek, J., Benes, D., Ostadalova, I., Babicky, A., Benes, T., and Lenar, T. 1969. Metabolic interrelations of trace elements. The effect of some organic and inorganic compounds of selenium on the metabolism of cadmium and mercury in the rat, *Physiol. Bohemostov. 18:*95–103.

Parrot, J. L., Hebert, R., Saindelle, A., and Ruff, F. 1969. Platinum and platinosis: Allergy and histamine release due to platinum salts, *Arch. Environ. Health 19:*685.

Passow, H., Rothstein, A., and Clarkson, T. W. 1961. The general pharmacology of heavy metals, *Pharmacol. Rev. 13:*185–224.

Patrick, S. J. 1948. Some effects of the administration of thorium nitrate to mice, *Can. J. Res. 26(E):*303–304.

Patt, E. L. 1976. The Effects of Dietary Lithium Levels on Tissue Lithium Content, Growth Rate and Reproduction in the Rat, M.S. Thesis, University of Missouri, Columbia.

Paulson, G. D., Broderick, G. A., Baumann, C. A., and Pope, A. I. 1968. Effect of feeding sheep selenium fortified trace mineralized salt: Effect of tocopherol, *J. Anim. Sci. 27:*195–202.

Pearson, J. D. 1966. Use of Cr^{51}-labelled hemoglobin and Sc^{47} as inert fecal markers, *Int. J. Appl. Radiat. Isotop. 17:*13–16.

Pecher, C., and Pecher, J. 1941. Radiocalcium and radiostrontium metabolism in pregnant mice, *Proc. Soc. Exp. Biol. Med. N.Y. 46:*91–93.

Pederson, L. A., and Libby, W. F. 1972. Unseparated rare earth–cobalt oxides as auto exhaust catalysts, *Science 176:*1355–1356.

Peisach, J., Aisen, P., and Blumberg, W. E. (eds.). 1966. *The Biochemistry of copper,* Proceedings, September 8–10, 1965, Academic Press, New York.

Peoples, S. A. 1964. Arsenic toxicity in cattle, *Ann. N.Y. Acad. Sci. 111:*644.

Perry, H. M., and Schroeder, H. A. 1955. Concentration of trace metals in urine of treated and untreated hypersensitive patients compared with normal subjects, *J. Lab. Clin. Med. 45:*936–940.

Perry, H. M., Jr., Perry, E. F., and Purifoy, J. E. 1971. Antinatriuretic effect of intramuscular cadmium in rats, *Proc. Soc. Exp. Biol. Med. 136:*1240.

Persellin, R. H., and Ziff, M. 1966. The effect of gold salt on lysosomal enzymes of peritoneal macrophage, *Arthritis Rheum. 9:*57–65.

Peters, E. E. 1942. Bismuth stomatitis and albuminuria, *Amer. J. Syph. 26:*84.

Pham-Huu-chanh. 1964. Comparative action of sodium chromate, molybdate, tungstate, and metavanadate on the enzymatic action of tyrosinase, *Aggressologie 5:*3179–3182.

Pham-Huu-chanh. 1965. The comparative toxicity of sodium chromate molybdate, tungstate, and metavanadate, *Arch. Int. Pharmacodyn. 154:*243.

Phatak, S. S., and Patwardhan, V. N. 1950. Toxicity of nickel, *Ind. J. Sci. Ind. Res. 9B (3):*70–76.

Phatak, S. S., and Patwardhan, V. N. 1952. Toxicity of nickel, accumulation of nickel in rats fed on nickel containing diets, and its elimination, *Ind. J. Sci. Res. 11 B(5):*173–176.

Phillips, C. S. G., and Williams, R. J. P. 1965. *Inorganic Chemistry,* Vol. 1, "Principles and Non-Metals," The Oxford University Press, Oxford.

Pinto, S. S., and Bennet, B. M. 1963. Effect of arsenic trioxide exposure on mortality, *Arch. Environ. Health 7:*583–591.

Piscator, M. 1966. Proteinuria in chronic cadmium poisoning: IV. Gel filtration and ion exchange chromatography of urinary proteins from cadmium workers, *Arch. Environ. Health 12:*345–360.

Piscator, M., and Axelsson, B. 1970. Serum proteins and kidney function after exposure to cadmium, *Arch. Environ. Health 21:*604–609.

Platonow. N., and Abbey, H. K. 1968. Toxicity of vanadium in calves, *Vet. Rec. 82:*292–298.

Plenge, P., Mellerup, E. T., and Rafaelsen, O. J. 1969. Effect of lithium on carbohydrate metabolism, *Lancet 2:*1012.

Plume, C. 1971. Influence du débit du perfusion sur la toxicite aigue du gluconate de calcium, *Arch. Int. Pharmacodyn. Ther. 191:*44.

Pollack, S. 1968. Iron absorption, in: *Handbook of Physiology* (C. H. Code and W. Heidel, eds.), Amer. Physiol. Soc., Washington, D. C., p. 1553.

Pollack, S., Georg, J. N., Reba, R. C., Kaufman, R. N., and Crosby, W. H. 1965. The absorption of non-ferrous metals in iron deficiency, *J. Clin Invest.* *44:*1470–1473.

Pond, W. G., Chapman, P., and Walker, E. 1966. Influence of dietary zinc, corn oil, and cadmium on certain blood components, weight gain, and parakeratosis in young pigs, *J. Anim. Sci. 25:*122–127.

Porter, H. 1971. Neonatalhepatic mitochondrocuprein, *Biochem. Biophys. Acta 229:*143–154.

Potter, G. D., McIntyre, D. R., and Valtuone, G. M. 1971. Fate and implications of [203]lead in a dairy cow and calf, *Health Phys. 20:*650–653.

Potter, S. D., and Matrone, G. 1973. Effect of selenite on the toxicity and retention of dietary methyl mercury and mercuric chlorides, *Fed. Proc. 32:*929 (abstr.).

Potter, S. D., and Matrone, G. 1974. Effect of selenite on the toxicity of dietary methyl mercury and mercuric chloride in rat, *J. Nutr. 104:*638–647.

Potts, A. N., Simon, F. P., Tobias, J. M., Postal, S., Swift, M. N., Patt, H. M., and Gerard, R. W. 1950. Distribution and fate of cadmium in the animal body, *AMA Arch. Ind. Hyg. Occup. Med. 2:*175–180.

Potts, C. L. 1965. Cadmium proteinuria: The health of battery workers exposed to cadmium oxide dust, *Ann. Occup. Hyg. 8:*55–60.

Powell, G. W., Miller, W. J., Morton, J. D., and Clifton, C. M. 1964. Influence of dietary cadmium level and supplemental zinc on cadmium toxicity in the bovine, *J. Nutr. 84:*205.

Pragay, D. A. 1962. Musuclar dystrophy in chicks caused by dietary aluminum hydroxide gel, *Fed. Proc. 21:*388.

Prasad, A. S. (ed.). 1966. *Zinc Metabolism,* Charles C. Thomas, Springfield, Illinois.

Prick, J. J. G., Sillievis Smitt, W. G., and Muller, L. 1955. *Thallium Poisoning,* Elsevier Publishing Co., Amsterdam.

Prior, J. T., Cronk, G. P., and Ziegler, D. B. 1960. Pathological changes associated with the inhalation of sodium zirconium lactate, *Arch. Environ. Health 1960(1):*297–301.

Proescher, F., Seil, H. A., and Stilliams, A. W. 1917. A contribution to the action of vanadium with particular reference to syphillis, *Am. J. Syph. 1:*347–358.

Pulido, P., Kagi, J. H. R., and Vallee, B. L. 1966. Isolation and some properties of human metallothioneine, *Biochemistry 5:*1768–1777.

Rabinowitz, M. B., Wetherill, G. W., and Kopple, J. D. 1973. Lead metabolism in the normal human: Stable isotope studies, *Science 182:*725–727.

Rakimova, M. J. 1968. Phagocytic activity of human PMN cells: Decrease in lead intoxication, *Gig. Tr. Prof. Zabol. 12:*39.

Ramasastry, B. V. 1966. Fission products: retention and elimination of radionuclide pair Ru^{103}–Rh^{103} by rats, *Toxicol. Appl. Pharmacol. 9:*413.

Ramasastry, B. V., Owens, L. K., and Ball, C. O. T. 1964. Differences in the distribution of Zr^{95}–Nb^{95} in the rat, *Nature 201:*410.

Ramsey, T. A., Mendels, J., Stokes, J. W., and Fitzgerald, R. G. 1972. Lithium carbonate and kidney function: A failure in renal concentrating ability, *J. A. M. A. 219:*1446.

Randall, R. E., Osheroff, R. J., Bakerman, S., and Setter, J. G. 1972. Bismuth nephrotoxicity, *Ann. Intern. Med. 77:*481.

Raskova, H., and Masek, K. 1971. Microbial toxins, in: *Pharmacology and Toxicology of Naturally Occurring Toxins,* Vol. 1, Part 1 (O. B. Henriques, ed.), Pergamon Press, Elmsford, N.Y.

Read, W. O. 1949. Effects of germanium dioxide on the oxygen uptake of rat tissue, Thesis, University of Missouri.

Reece, W. O., Talbot, R. B., and Swenson, M. J. 1967. Effects of inhaled yttrium oxide on blood lactic acid, erythrocyte volume, and histologic features of lungs in exercised dogs, *Am. J. Vet. Res. 28:*979.

Reed, D., Crawley, J., Faro, S. N., Pieper, S. J., and Kurland, L. T. 1962. Thallotoxicosis, *J. A. M. A. 183:*516–522.

Reeves, A. L. 1965. Absorption of beryllium from the gastrointestinal tract, *Arch. Environ. Health 11:*209.

Reines, L., and Leonard, C. S. 1932. Role of NH_2, OH, and AS—A groups in parasitoxic action of arsphenamine derivatives, *Proc. Soc. Exp. Biol. Med. 29:*946.

Reissman, K. R., and Coleman, T. J. 1955. Acute intestinal intoxication of iron, II, *Blood 10:*46.

Relman, A. S. 1957. The physiological behavior of rubidium and cesium in relation to that of potassium, *Yale J. Biol. Med. 29:*248–262.

Relman, A. S., Lambie, A. T., Burrows, B., and Roy, A. M. 1957. The displacement of potassium by rubidium and cesium in the living animal, *J. Clin. Invest. 36:*1249–1253.

Report of an international committee. 1969. Maximum allowable concentration of mercury compounds, *Arch. Environ. Health 19:*891.

Resen, J. C., Cohen, N., and Wrenn, M. E. 1972. Short term metabolism of [241]americium in baboon, *Health Phys. 22:*621–626.

Rhyne, B. C., and R. A. Goyer. 1971. Cytochrome content of kidney mitochondria in experimental lead poisoning, *Exp. Mol. Pathol. 14:*386–391.

Ribas-Ozonas, B., Guelbenzu, M. D., and Ruiz, A. S. 1971. Inhibitory effect of cadmium on alkaline phosphatase of kidney and prostrate of guinea pig, in: *Trace Element Metabolism in Animals* (C. F. Mills, ed.), E. And E. Livingston, Edinburgh, pp. 173–176.

Richmond, H. G. 1959. Induction of sarcoma in the rat by iron–dextran complex, *Br. Med. J. 1959(1):*947.

Rifkin, R. J. 1965. *In vitro* inhibition of Na^+ and Mg^{2+} ATPases by mono, di, and trivalent cations, *Proc. Soc. Exp. Biol. Med. 120:*802–804.

Rigby, P. G. 1971. The utilization of iron, *Clin. Tox. 4:*559–569.

Ritchie, H. D., Leucke, R. W., Baltzer, B. V., Miller, E. R., Ullrey, D. E., and Hoefer, J. A. 1963. Copper and zinc interrelationships in the pig, *J. Nutr. 79:*117–130.

Riviere, M. R., Chouroulinskov, I., and Guerin, M. 1960. The production of tumors by means of intratesticular injection of zinc chloride in the rat, *Bull. Assoc. Fr. Cancer 47:*55–60.

Rizzo, A. M., and Furst, A. 1972. Mercury teratogenesis in the rat, *Proc. West Pharmac. Soc. 15:*52–54.

Roberts, A. E. 1951. Platinosis, *Arch. Ind. Hyg. 4:*549.

Roberts, T. M., Hutchinson, T. C., Paciga, J., Chattopadhyay, A., Jerris, R. E., VanLoon, J., and Parkinson, D. K. 1974. Lead contamination around secondary smelters: Estimation of dispersal and accumulation by humans, *Science 186:*1120–1122.

Robertson, D. S. E. 1970. Selenium—a possible teratogen, *Lancet 1:*518–519.

Robinson, F. R., Schaffner, F., and Trachtenberg, E. 1965. Ultrastructure of the lungs of dogs exposed to beryllium-containing dusts, *Arch. Ind. Health 17:*65–72.

Robson, A. O., and Jelliffe, A. J. 1963. Medical arsenic poisoning and lung cancer, *Br. Med. J. 2:*207–209.

Rochow, E. G., and Sindler, B. M. 1950. Dimethyl germanium oxide: Toxicity and effect on blood composition, *J. Am. Chem. Soc. 72:*1218–1220.

Roe, F. J. C. 1964. Cadmium neoplasia testicular atrophy and Leydig cell hyperplasia and neoplasia in rats and mice following the subcutaneous injection of cadmium salts, *Br. J. Cancer 18:*674–683.

Roe, F. J. C. 1971. Cadmium pollution and itai-itai disease, *Lancet 1971:*382.

Roe, F. J. C., and Carter, R. L. 1969. Chromium carcinogenesis calcium chromate as a potent carcinogen for the subcutaneous tissues of the rat, *Br. J. Cancer 23:*172–176.

Roe, F. J. C., Boyland, E., and Millican, K. 1965. Effects of oral administration of two tin compounds over prolonged periods, *Food Cosmet. Toxicol. 3:*277–280.

Romney, E. M., Mork, H. M., and Larsen, K. H. 1970. Persistence of plutonium in soil, plants, and small mammals, *Health Phys. 19:*487–494.

Rosenberg, B., Van Camp, L., Trosko, J. E., and Masour, V. H. 1969. Platinum compounds— New class of potent anti-tumor agents, *Nature 222:*385–386.

Rosenfeld, G. 1954. Metabolism of germanium, *Arch. Biochem. Biophys. 48:*84.

Rosenfeld, G., and Wallace, E. D. 1953. Studies of acute and chronic toxicity of germanium, *Arch. Ind. Hyg. 8:*466.

Rosenfeld, I., and Beath, O. A. 1947. Congenital malformations on eyes of sheep, *J. Agri. Res. 75:*93–95.

Rosenfeld, I., and Beath, O. A. 1954. Effect of selenium on reproduction in rats, *Proc. Soc. Exp. Biol. Med. 87:*295–299.

Rosenfeld, I., and Beath, O. A. 1964. *Selenium,* Academic Press, New York.

Rosenweig, N. S., and Hendrix, T. R. 1967. The differential effects of sodium, lithium, and potassium on the absorption of glucose, xylose, and water from rat jejunum *in vivo, Johns Hopkins Med. J. 121:*412–420.

Roshchin, I. V. 1967. Toxicology of vanadium compounds used in modern industry, *Gig. i Sanit. 32:*26.

Rosoff, B. 1963. Cesium^{137} metabolism in man, *Radiat. Res. 19:*643–654.

Roth, F. 1958. Bronchial cancer in vintners exposed to arsenic, *Virchow Arch. Path. Anat. 331:*119–137.

Rubin, L., Slepyan, A. H., Weber, L. F., and Neuhauser, I. 1956. Granuloma of the axilla caused by deodorants, *J. A. M. A. 162:*953–955.

Rundo, J. 1964. A survey of the metabolism of cesium in man, *Br. J. Rad. 37:*108–114.

Rusiecki, W., and Brzezinski, J. 1966. Influence of sodium selenate on acute thallium poisoning, *Acta Polon. Pharm. 23:*75.

Ryan, R., McNeil, J. S., Flamenbaum, W., and Nagle, R. 1973. Uranyl nitrate induced acute renal failure in the rat: Effect of varying doses and saline loading, *Proc. Soc. Exp. Biol. Med. 143:*289–296.

Rygh, O. 1949. Rechérches sur les oligo-elements. III. Sur l'importance du strontium, du vanadium, du varyum, du thallium, et du zinc dans le scorbut, *Bull. Soc. Chim. Biol. 31:*1408–1412.

Sadasivan, V. 1951*a*. Studies on the biochemistry of zinc. I. Effect of feeding zinc on the livers and bones of rats, *Biochem. J. 48:*527–532.

Sadasivan, V. 1951*b*. II. Effect of intake of zinc on the metabolism of rats maintained on a stock diet, *Biochem. J. 49:*186–191.

Sadasivan, P. 1952. Further investigations on the influence of zinc on metabolism, *Biochem. J. 52:*452–460.

Saffiotti, U., Cefis, F., and Kolb, L. H. 1968. A new method for the experimental induction of bronchogenic carcinoma, *Cancer Res. 28:*104.

Salerno, P. R., and Mattis, P. A. 1951. Absorption and distribution of thorium nitrate in the rat, *J. Pharmacol. Exp. Ther. 101:*31 (abstr.).

Saltzer, E. I., and Wilson, J. W. 1968. Allergic contact dermatitis due to copper, *Arch. Derm. 98:*375–376.

Salvidio, E., Pannacciulli, I., and Tizianello, A. 1963. Glucose-6-phosphate and 6-phosphogluconic dehydrogenase activities in the red blood cells of several animal species, *Nature (London) 200:*372–373.

Sanders, C. L. 1972. Deposition patterns and toxicity of transuranium elements in the lung, *Health Phys. 22:*607–615.

Sasser, L. B., Kienholz, E. W., and Ward, G. M. 1969. Interaction of rubidium and potassium in chick diets, *Poultry Sci. 48:*114–118.

Satterlee, H. S. 1960. The arsenic poisoning epidemic of 1900—Its relation to lung cancer in 1960—An exercise in retrospective epidemiology, *New Engl. J. Med. 263:*676.

Sauerhoff, M. W., and Michaelson, I. A. 1973. Hyperactivity and brain catecholamines in lead-exposed developing rats, *Science 182:*1022–1024.

Schade, S. G., Felsher, B. F., Bernier, G. M., and Conrad, M. E. 1970. Interrelationship of cobalt and iron absorption, *J. Lab. Clin. Med. 75:*435–441.

Scharpf, L. G., Hill, I. D., Wright, P. L., Plank, J. B., Keplinger, M. L., and Calandra, J. C. 1972. Effect of nitroloacetate on toxicity, teratogenicity, and distribution of cadmium, *Nature 239:*231–233.

Schepers, G. W. 1955*a*. The biological action of particulate tungsten metal, *Arch. Ind. Health 12:*134–136.

Schepers, G. W. 1955*b*. Biological action of tungsten carbide and cobalt, *Arch. Ind. Health. 12:*137–139.

Schepers, G. W. 1955*c*. Biological action of tantalum oxide. *Arch. Ind. Health 12:*121.

Schepers, G. W., Durkan, T. M., Delahant, A. B., and Creedon, F. T. 1957. The biological action of inhaled beryllium sulfate, *Arch. Ind. Health 15:*32.

Schienberg, I. H., and Sternlieb, I. 1960. Copper metabolism, *Pharmacol. Rev. 12:*355–381.

Schienberg, I. H., and Sternlieb, I. 1964. Wilson's disease, *Ann. Rev. Med. 16:*119.

Schlaepfer, W. W. 1971. Sequential study of endothelial changes in acute cadmium intoxication, *Lab. Invest. 25:*556–561.

Schlicker, S. A., and Cox, D. H. 1968. Maternal dietary zinc, and development. Zinc, iron, and copper content of the rat fetus, *J. Nutr. 95:*287.

Schmidt, C. L. A., and Hoagland, D. R. 1912. The determination of aluminum in feces, *J. Biol. Chem. 11:*387–91.

Schnahl, D., and Steinhoff, D. 1960. Experimental carcinogenesis in rats with colloidal silver and gold sols, *Z. Krebsforsch. 63:*586.

Schoffeniels, E. 1966. The activity of L-glutamic acid dehydrogenase. *Arch. Int. Physiol. Biochem. 74:*665–669.

Schou, M. 1957. Biology and pharmacology of the lithium ion, *Pharmacol. Rev. 9:*17.

Schou, M. 1958. Lithium studies, *Acta Pharmakol. (KBH) 15:*69.

Schou, M., and Amdisen, A. 1970. Lithium in pregnancy (letter), *Lancet 1:*1391.

Schrauzer, G. M., and Ishmael, D. 1974. Effects of selenium and of arsenic on the genesis of spontaneous mammary tumors in inbred C_3H mice, *Am. Clin. Lab. Sci. 4:*441–447.

Schroeder, H. A. 1965. Cadmium as a factor in hypertension, *J. Chron. Dis. 18:*647.

Schroeder, H. A. 1966. Municipal drinking water and cardiovascular death rates, *J. A. M. A. 195:*81.

Schroeder, H. A. 1967. Effects of selenate, selenite, and tellurite on the growth and early survival of mice and rats, *J. Nutr. 92:*334–338.

Schroeder, H. A. 1968. The role of chromium in mammalian nutrition, *Am. J. Clin. Nutr. 21:*230–240.

Schroeder, H. A. 1973. Recondite toxicity of trace elements, in: *Essays in Toxicology,* Vol. 4 (W. J. Hayes, Jr., ed.), Academic Press, New York.

Schroeder, H. A., and Balassa, J. J. 1961. Abnormal trace metals in man: Cadmium, *J. Chron. Dis. 14:*236–248.

Schroeder, H. A., and Balassa, J. J. 1965. Abnormal trace metals in man: Niobium, *J. Chron. Dis. 18:*229.

Schroeder, H. A., and Balassa, J. J. 1966*a*. Abnormal trace metals in man: Arsenic, *J. Chron. Dis. 19:*85.

Schroeder, H. A., and Balassa, J. J. 1966*b*. Abnormal trace metals in man: Zirconium, *J. Chron. Dis. 19:*573–586.

Schroeder, H. A., and Balassa, J. J. 1967*a*. Abnormal trace metals in man: Germanium, *J. Chron. Dis. 20:*211.

Schroeder, H. A., and Balassa, J. J. 1967*b*. Arsenic, germanium, tin and vanadium in mice: Effects on growth, survival and tissue levels, *J. Nutr. 92:*245–251.

Schroeder, H. A., and Mitchner, M. 1971*a*. Scandium chromium, gallium, yttrium, rhodium, palladium, and indium in mice: Effects on growth and life span, *J. Nutr. 101:*1431–1438.

Schroder, H. A., and Mitchner, M. 1971*b*. Toxic effects of trace elements on the reproduction of mice and rats, *Arch. Environ. Health 23:*102–106.

Shcroeder, H. A., and Mitchner, M. 1971*c*. Selenium and tellurium in rats: Effects on growth, survival and tumors, *J. Nutr. 101:*1531–1537.

Schroeder, H. A., and Mitchner, M. 1972. Selenium and tellurium in mice, *Arch. Environ. Health 24:*66–71.

Schroeder, H. A., and Nason, A. P. 1971. Trace element analysis in clinical chemistry, *Clin. Chem. 17:*461–474.

Schroeder, H. A., and Nason, A. P. 1974. Interactions of trace metals in rat tissues: Cadmium and nickel with zinc, chromium, copper and manganese, *J. Nutr. 104:*167–178.

Schroeder, H. A., Balassa, J. J., and Tipton, I. N. 1962*a*. Abnormal trace metals in man: Chromium, *J. Chron. Dis. 15:*941–964.

Schroeder, H. A., Balassa, J. J., and Tipton, I. H. 1962*b*. Abnormal trace metals in man: Nickel, *J. Chron. Dis. 15:*51.

Schroeder, H. A., Vinton, W. H., Jr., and Balassa, J. J. 1963*a*. Effect of chromium, cadmium, and other trace metals on the growth and survival of mice, *J. Nutr. 80:*39–47.

Schroeder, H. A., Vinton, W. H., Jr., and Balassa, J. J. 1963*b*. Effects of chromium, cadmium, and lead on the growth and survival of rats, *J. Nutr. 80:*43–54.

Schroeder, H. A., Balassa, J. J., and Tipton, I. H. 1963*c*. Abnormal trace metals in man: Titanium, *J. Chron. Dis. 16:*55–69.

Schroeder, H. A., Balassa, J. J., and Tipton, I. H. 1963*d*. Abnormal trace elements in man: Vanadium, *J. Chron. Dis. 16:*1047.

Schroeder, H. A., Balassa, J. J., and Tipton, I. H. 1964*a*. Abnormal trace metals in man: Tin, *J. Chron. Dis. 17:*483–502.

Schroeder, H. A., Balassa, J. J., and Vinton, W. H., Jr. 1964*b*. Chromium, lead, cadmium, nickel, and titanium in mice: Effect of mortality, tumors, and tissue levels, *J. Nutr. 83:*239–250.

Schroeder, H. A., Balassa, J. J., and Vinton, W. H. 1965. Chromium, cadmium and lead in rats. Effects on life span, tumors, and tissue levels, *J. Nutr. 86:*51–66.

Schroeder, H. A., Balassa, J. J., and Tipton, I. H. 1966*a*. Manganese: A study in homeostasis, *J. Chron. Dis. 19:*545.

Schroeder, H. A., Nason, A. P., Tipton, I. H., and Balassa, J. J. 1966*b*. Essential trace metals in man: Copper, *J. Chron. Dis. 19:*1007–1034.

Schroeder, H. A., Nason, A. P., and Tipton, I. H. 1967*a*. Essential trace metals in man: Cobalt, *J. Chron. Dis. 20:*869–890.

Schroeder, H. A., Nason, A. P., Tipton, I. H., and Balassa, J. J. 1967*b*. Essential trace metals in man: Zinc—Relation to environmental cadmium, *J. Chron. Dis. 20:*179–210.

Schroeder, H. A., Buckman, J., and Balassa, J. J. 1967*c*. Abnormal trace elements in man: Tellurium, *J. Chron. Dis. 20:*147–161.

Schroeder, H. A., Kanisawa, M., Frost, D. V., and Mitchner, M. 1968*a*. Germanium, tin, and arsenic in rats: Effects on life spans, tumors, and tissue levels, *J. Nutr. 96:*37–45.

Schroeder, H. A., Kanisawa, M., Frost, D. V., and Mitchener, M. 1968*b*. Germanium, tin, and arsenic in rats: Effects on growth, survival, pathological lesions, and life span, *J. Nutr. 96:*37.

Schroeder, H. A., Mitchner, M., Balassa, J. J., Kanisawa, M. and Nason, A. P. 1968c. Zirconium, niobrium, antimony, and fluorine in mice. Effects on growth, survival, and tissue levels, *J. Nutr. 95:*95–101.

Schroeder, H. A., Nason, A. P., and Tipton, I. H. 1969. Essential metals in man—Magnesium. *J. Chron. Dis. 21:*815–841.

Schroeder, H. A., Balassa, J. J., and Tipton, I. H. 1970a. Essential trace metals in man: Molybdenum, *J. Chron. Dis. 23:*481–499.

Schroeder, H. A., Frost, D. V., and Balassa, J. J. 1970b. Essential trace metals in man— Selenium, *J. Chron. Dis. 23:*227–243.

Schroeder, H. A., Mitchner, M., and Nason, A. P. 1970c. Zirconium, niobium, antimony, vanadium, and lead in rats: Life term studies, *J. Nutr. 100:*59–68.

Schroeder, H. A., Tipton, I. H., and Nason, A. P. 1972. Trace metals in man. Strontium and Barium, *J. Chron. Dis. 25:*491–517.

Schubert, J. 1949. An experimental study of the effect of zirconium and sodium citrate treatment on the metabolism of plutonium and radioyttrium, *J. Lab. Clin. Med. 34:*313–317.

J. Schubert. 1958. Beryllium and berylliosis, *Sci. Am. 199 (2):*27–33.

Schultz, H. 1888. Über Hefegifte, *Arch. Ges. Physiol, Vide Pelvgers Arch. 42:*517–54.

Schultz, S. G. 1974. Principles of electrophysiology and their application to epithelial tissue, in: *Physiology,* Vol. 4 (E. D. Jacobson and L. L. Shanbour, eds.), University Park Press, Baltimore.

Schwartz, L., Roginski, E. E., and Foltz, C. M. 1959. Ineffectiveness of Mo, Os, and Co in dietary necrotic liver degeneration, *Nature 183:*472.

Schwartz, L., Tulipan, L., and Peck, S. 1947. *A Textbook of Occupational Diseases of the Skin,* Lea and Febiger, Philadelphia.

Schwarz, K. 1971. Essentiality of tin, in: *Newer Trace Elements in Nutrition* (W. Mertz and W. E. Cornatzer, eds.), Marcel Decker, Inc., New York, p. 315.

Schwarz, K. 1974. New essential trace elements (Sn, V, F, and Si): Progress report and outlook, in: *Trace Elements Metabolism in Animals* (W. G. Hoekstra, J. W. Suttie, H. E. Ganther, and W. Mertz, eds.), University Park Press, Baltimore, pp. 355–380.

Schwarz, K., and Milne, D. B. 1971. Growth effects of vanadium in the rat, *Science 174:*426–428.

Schwarz, K., Milne, D. B., and Vineyard, E. 1970. Growth effects of tin compounds in rats maintained in a trace element controlled environment, *Biochem. Biophys. Res. Commun. 40:*22–24.

Schweitzer, G. K. 1956. The radio colloidal properties of the rare earth elements, *US AEC Rept. ORINS-12,* pp. 31–38.

Schweitzer, G. K., and Jackson, W. M. 1952a. Radiocolloids, *J. Chem. Educ. 29:*513–518.

Schweitzer, G. K., and Jackson, W. M. 1952b. Studies in low-concentration chemistry: I. The radiocolloidal properties of lanthanum-140, *J. Amer. Chem. Soc. 74:*4178–4182.

Schweitzer, G. K., and Scot, H. E. 1955. Low-concentration chemistry. X. Further observations on yttrium, *J. Am. Chem. Soc. 77:*2753–2757.

Scot, G. H., and Canaga, B. L. 1939. Cesium in the mammalian retina, *Proc. Soc. Exp. Biol. Med. 40:*275–276.

Scot, M. L. 1962. Selenium, in: *Mineral Metabolism,* Vol. II, Part B (P. Comar and F. Bronner, eds.), Academic Press, New York, p. 543.

Scot, M. L. 1967. Selenium deficiency in chicks and poults, in: *Selenium in Biomedicine* (O. H. Muth, J. E. Oldfield, and P. H. Weswig, eds.), AVI Publishing Co., Connecticut, pp. 231–238.

Scott, K. G., Hamilton, J. G., and Wallace, P. C. 1951. Deposition of carrier-free vanadium in the rat following intravenous administration. University of California Radiation Laboratory, *Rept. UCRL-1318.*

Seelig, M. 1972. Relationships of copper and molybdenum to iron metabolism, *Am. J. Clin. Nutr. 25:*1022–1037.

Sel'tser, V. K. 1967. Effect of incorporated radioisotopes of ^{90}Sr, ^{137}Cs and ^{144}Ce on cardiac activity in rats, *Radiobiologiya 7:*302–306.

Selye, H. 1962. *Calciphylaxis,* University of Chicago Press, Chicago.

Selye, H., Tuchweber, B., and Gabbiani, G. 1962. Prevention by ferric dextran of the topical calcification induced by direct calcifiers, *Medna. Exp. 7:*181–184.

Selye, H., Tuchweber, B., and Bertog, L. 1966. Effect of lead on the susceptibility of rats to bacterial endotoxins, *J. Bact. 91:*884–886.

Semenov, D. I., Moskalev, Y. I., and Buldakov, L. A. 1967. Behavior in the organism of Ca, Sr, Y, Ce, Zr, Nb, and Ru when they are inhaled by rats, *Tr. Inst. Biol. Ural. Fil. Aka. Nauk, USSR. 46:*49, Biol. Abstr., 1967, No. 89235.

Seramivof, N., and Tuchweber, B. 1965. Histamine liberators and calcification, *Arch. Int. Pharmacodyn. Ther. 154:*251–254.

Sernka, T. J. 1974. Gastrointestinal mucosal metabolism, in: *Gastrointestinal Physiology,* Vol. 4 (E. D. Jacobson and L. L. Shanbour, eds.), University Park Press, Baltimore.

Setchel, B. P., and Waites, G. M. 1970. Changes with the permeability of testicular capillaries and of the blood testes barrier after injection of $CdCl_2$ in the rat, *J. Endocrin. 47:*81.

Settlemire, C. T., and Matrone, G. 1967*a*. *In vivo* interference of zinc with ferritin iron in the rat, *J. Nutr. 92:*153–158.

Settlemire, C. T., and Matrone, G. 1967*b*. *In vivo* effect of zinc on iron turnover in rats and life span of the erythrocyte, *J. Nutr. 92:*159–164.

Shacklette, H. T. 1970. Mercury content of plants, in: *Mercury in the Environment,* Geological Survey Professional Paper 713, U.S. Geological Survey, Dept. of the Interior.

Shaikh, Z. A., and Lucis, O. J. 1969. Distribution and binding of ^{109}Cd and ^{65}Zn in experimental animals, *Proc. Can. Fed. Br. Soc. 12:*101–102.

Shaikh, Z. A., and Lucis, O. J. 1971. Isolation of cadmium binding proteins, *Experientia 27:*1024–1025.

Shaikh, Z. A., and Lucis, O. J. 1972. Cadmium and zinc binding in mammalian liver and kidney, *Arch. Environ. Health 24:*418–422.

Shamberger, R. J. 1970. Relationship of selenium to cancer. I. Inhibitory effect of selenium on carcinogenesis, *J. Natl. Cancer Inst. 44:*931–936.

Shamberger, R. J. 1971. Is selenium a teratogen?, *Lancet 2:*1137.

Shamberger, R. J. 1974. Antioxidants and cancer. III. Selenium and other antioxidants decrease carcinogen-induced chromosome linkage, in: *Trace Element Metabolism in Animals III* (W. G. Hoekstra, J. W. Suttie, H. E. Ganther, and W. Mertz, eds.), University Park Press, Baltimore, pp. 593–597.

Shamberger, R. J., and Frost, D. V. 1969. Possible protective effect of selenium against human cancer, *Can. Med. Ass. J. 100:*682.

Shamberger, R. J., and Willis C. E. 1971. Selenium distribution and human cancer mortality, in: *CRC Critical Reviews in Clinical Laboratory Sciences,* Vol. 3 CRC Publishing Company, Cleveland, pp. 211–221.

Shaver, C. G., and Riddell, A. R. 1947. Lung changes associated with the manufacture of alumina abrasives, *J. Ind. Hyg. Toxicol. 29:*145–147.

Sheard, C. 1955. Contact dermatitis from platinum and related metals, *AMA Arch. Derm. Syph. 71:*357–364.

Sheehan, R. C., and Frenkel, E. P. 1972. The control of Fe absorption by the gastrointestinal mucosal cell, *J. Clin. Invest. 51:*226–231.

Sheldon, J. H., and Ramage, H. 1931. CLXXIII. A spectrographic analysis of human tissues, *Biochem. J. 25:*1608.

Shellabarger, C. J. 1956. Studies on the thyroidal accumulation of rhenium in rat, *Endocrinology 58:*13.

Shelly, W. B., and Hauley, H. J. 1957. Experimental evidence for allergic basis for granuloma formation in man, *Nature 180:*1060–1061.

Shelly, W. B., and Hauley, H. J. 1958. The allergic origin of zirconium deodorant ganulomas, *Br. J. Derm. 70:*75–78.

Shelly, W. B., Hurley, J. J., Maycock, R. L. Close, H. P., and Catheart, H. C. 1958. Intradermal tests with metals and their inorganic elements in sarcoidosis and anthrosilicosis, *J. Invest. Dermatol. 31:*301–303.

Shopsin, B. 1970. Effects of lithium on thryoid function, *Dis. Nerv. Syst. 31:*237–244.

Shrader, D. A. 1972. The physiology and toxicology of lithium, *J. Kansas Med. Soc. 73:*24–30.

Shrier, H. R., Lane, W. T., and Schrack, W. D. 1974. Acute copper poisoning, *CDC Morbid. and Mort. Week. Rept.,* March, p. 99.

Shysh, A., Noujaim, A. A., and Riedel, B. E. 1969. Studies on the uptake and distribution of cerium[141] in the rat, *Can. J. Pharm. Sci. 4:*23.

Sikov, M. R., Thomas, J. M., and Mahlum, D. D. 1969. Comparison of passage of a tracer through the gastrointestinal tract of neonatal and adult rats, *Growth 33:*57–68.

Silverberg, D. S., Kidd, E. G., Shnitka, T. K., and Ulan, R. A. 1970. Gold nephropathy: A clinical pathologic study, *Arthritis Rheum. 13(6):*812–825.

Simkins, K. L., and Pensack, J. M. 1970. Effect of gold thioglucose on survival, food consumption, and body weight of broilers, *Poultry Sci. 49:*1341.

Simmons, D. J., and Failla, P. 1966. Strontium retention in mice treated with thyroid hormone, *Health Phys. 12:*1249–1259.

Singer, I., and Rotenburg, D. 1973. Mechanisms of lithium action, *New Engl. J. Med. 289:*254–260.

Singer, I., Rotenberg, D., and Duschett, J. B. 1972. Lithium induced nephrogenic diabetes insipidus—*in vivo* and *in vitro* studies, *J. Clin. Invest. 51:*1081–1091.

Six, K. M., and Goyer, R. A. 1972. The influence of iron deficiency on tissue content and toxicity of ingested lead in the rat, *J. Lab. Clin. Med. 79:*128–136.

Smith, B. S. W., Field, A. C., and Suttle, N. F. 1968. Effect of intake of copper molybdenum and sulfate on copper metabolism in the sheep. III. Studies with radioactive copper in male castrated sheep, *J. Comp. Pathol. 78:*449–461.

Smith, G. A., and Scot, J. L. 1957. Metabolism of indium by subcutaneous injection, *A.E.C. Res. and Develop. Rept., U.R. 507.*

Smith, G. A., Thomas, R. G., Black, D., and Scot, J. L. 1947. Metabolism of [114]indium administered to rat by intratracheal interbation, *A.E.C. Res. and Develop. Rept., U.R. 500.*

Smith, J. C., and Hackley, B. 1968. Distribution and excretion of nickel administered intravenously to rats, *J. Nutr. 95:*541–545.

Smith, M. I., Franke, K. W. and Westfall, B. B. 1936. The selenium problem in relation to public health. A preliminary survey to determine the possibility of selenium intoxication in the rural population living on seleniferous soil. *Publ. Health Rept. 51,* Washington, D.C., p. 1496.

Smith, R. J., and Cantrera, J. F. 1972. The effect of cobalt on erythropoietin and kininogen levels in rat plasma, *Proc. Soc. Exp. Biol. Med. 141:*895–897.

Smith, S. E., and Larsen, E. J. 1946. Zinc toxicity in rats, antagonistic effect of copper and liver, *J. Biol. Chem. 163:*29–32.

Snegireff, L. S., and Lombard, O. M. 1951. Arsenic and cancer—Observations in the metallurgical industry, *Ind. Hyg. Occup. Med. 4:*199–205.

Snowdon, C. T., and Sanderson, B. A. 1974. Lead pica produced in rats, *Science 183:*92–94.

Snyder, F., Cress, E. A., and Kyker, G. C. 1959. Lipid responses to intravenous rare earths in rats, *J. Lipid Res. 1:*125–131.

Sollman, T. 1948. *A Manual of Pharmacology,* 5th ed., W. B. Saunders Co., Philadelphia.

Sollman, T. 1953. *A Manual of Pharmacology and Its Application to Therapeutics and Toxicology,* 8th ed., W. B. Saunders Company, Philadelphia.

Sollman, T., and Seifter, J. 1942. Intravenous injection of soluble bismuth compounds, *J. Pharmacol. Exp. Ther. 74:*134.

Sollman, T., Cole, H. N., and Henderson, K. 1938. Clincal excretion of bismuth. VI. The autopsy distribution of Bi and patient after clinical Bi treatment, *Am. J. Syph. 22:*555.

Sommerville, J., and Davies, B. 1962. Effect of vanadium on serum cholesterol, *Am. Heart J. 64:*54–56.

Soremark, R. 1967. Vanadium in some biological specimens, *J. Nutr. 92:*183.

Sowden, E. M., and Stitch, S. R. 1957. Trace elements in human tissue: 2. Estimation of the concentrations of stable strontium and barium in human bone, *Biochem. J. 67:*104.

Spence, J. T. 1965. The biological function of molybdenum, *Z. Naturwiss. Med. Grundlagenforsch. 3(3):*267–283.

Spencer, I. O. B. 1951. Ferrous sulphate poisoning in children, *Br. Med. J. 2:*1112–1117.

Spencer-Lazlo, H., Samachson J., and Hardy, E. P. 1963. Strontium balances in man, *Clin. Sci. 24:*405–408.

Spiegel, C. J., LaFrance, L., and Ashworth, B. J. 1953. Blood and urine changes in experimental beryllium poisoning, *Arch. Ind. Hyg. Occup. Med. 7:*319–324.

Spink, D. M. 1961. Less common metals. Reactive metals zirconium, hafnium, and titanium, *Ind. Engr. Chem. 53:*97–104.

Spiridonova, V. S., and Suvorov, S. V. 1965. Influence of trace metals on the activity of cholinesterase in experimental organisms, *Mikr. Sred. Ih. Znachenie-M:*28–60.

Spirtes, M. A., and Gary, R. 1973. Rubidium levels in the blood and various parts of the brain monkey ingesting rubidium chloride, *Fed. Proc. 32(3):*732.

Spyker, J. M., and Smithberg, M. 1972. Effects of methylmercury on prenatal development in mice, *Teratology 5:*181–190.

Stara, J. F., Nelson, N. S., Krieger, H. L., and Kahn, B. 1971. Gastrointestinal absorption and tissue retention of radioruthenium, in: *Intestinal Absorption of Metal Ions, Trace Elements and Radionuclides* (S. C. Skoryna and D. Waldron-Edward, eds.), Pergamon Press, Oxford, p. 307.

Stavinoha, W. B., Emerson, G. A., and Nash, J. B. 1959. The effects of some sulfur compounds on thallotoxicosis in mice, *Toxicol. Appl. Pharmacol. 1:*638–646.

Sterne, T. L., Whitaker, C., and Webb, C. H. 1955. Fatal cases of bismuth intoxication, *J. L. State Med. Soc. 107:*332–334.

Sterner, J. H., and Eisenbud, M. 1951. Epidemiology of beryllium intoxication, *Arch. Ind. Hyg. Occup. Med. 4:*123–128.

Stewart, C. P., and Stolman, T. (eds.). 1960. *Toxicology—Mechanism and Analytical Methods,* Academic Press, New York and London.

Stich, S. R. 1957. Trace elements in human tissue. 1. A semiquantitiative spectrographic survey, *Biochem. J. 67:*97–103.

Stocks, P. 1960. On the relation between atmospheric pollution in urban and rural localities and mortality from cancer, bronchitis, and pneumonia with particular reference to 3,4-benzopyrene, beryllium, molybdenum, vanadium, and arsenic, *Br. J. Cancer 14:*397–410.

Stokinger, H. E. 1963. Metals excluding lead, in: *Industrial Hygiene and Toxicology,* Vol. II (D. W. Fasset and D. D. Irish, eds.), Interscience.

Stokinger, H. E. (eds.) 1966. *Beryllium: Its Industrial Hygiene Aspects,* Academic Press, New York.

Stokinger, H. E., and Coffin, D. L. 1968. Biological effects of air pollutants, in: *Air Pollution,* Academic Press, New York.

Stokinger, H. E., Stroud, C. A., and Root, R. E. 1951. Anemia in acute experimental beryllium poisoning, *J. Lab. Clin. Med. 38:*173.

Stokinger, H. E., Altman, R. I., and Salmon, K. 1953. The effect of various pathological conditions on *in vivo* hemoglobin synthesis: I. Hemoglobin synthesis in beryllium-induced anemia as studied with a ^{14}C-acetate, *Biochim. Biophys. Acta 12:*439.

Stoll, E. 1972. Medical portraits—Goya and Van Gogh, *The Sciences 12(4):*16.

Stolman, T., and Stewart, C. P. 1960. Metallic poisons, in: *Toxicology—Mechanism and Analytical Methods,* Vol. I (C. P. Stewart and T. Stolman, eds.), Academic Press, New York and London.

Stone, O. J. 1969. The effect of arsenic on inflammation: Infection and carcinogenesis, *Tex. St. J. Med. 65(10):*40–43.

Strasia, C. A. 1971. Vanadium: Essentiality and toxicity in the laboratory rat, Purdue University, *Dissertation Abst. 32:*646-B, 60 pp.

Sukai, K. 1972. Effect of methyl mercury on rat spermatogenesis, *Kumamoto Med. J. 25:*94–100.

Sullivan, D. J., McDonald, T. P., and Bell, M. C. 1969. Acute radio toxicity of ^{144}Ce–^{144}Pr after intravenous administration to sheep, *Cornell Vet. 59:*236–248.

Sullivan, R. J. 1969. Air pollution aspects of manganese and it compounds, Litton Systems, Inc., *U.S. Dept. of Commerce Pub. No. PB 188079.*

Sunderman, F. W., Jr. 1967. Nickel carbonyl inhibition of cortisone inductin of hepatic tryptophan pymolase, *Cancer Res. 27:*1595–1599.

Sunderman, F. W., Jr. 1972. Metal carcinogenesis in experimental animals, *Food Cosmet. Toxicol. 9:*105–120.

Sunderman, F. W., Jr., and Selin, C. E. 1968. The metabolism of nickel carbonyl, *Toxicol. Appl. Pharmacol. 12:*207–218.

Sunderman, F. W., and Sunderman, F. W., Jr. 1961. Nickel poisoning indication of nickel as a pulmonary carcinogen in tobacco smoke, *Am. J. Clin. Pathol. 35:*203–209.

Sunderman, F. W., Donnelly, A. J., West, B., and Kincaid, J. F. 1959. Nickel poisoning. II–IX. Carcinogenesis in rats exposed to nickel carbonyl, *AMA Arch. Ind. Health 20:*36–41.

Suttle, N. F., and Mills, C. F. 1966. Studies of the toxicity of copper to pigs: 1. Effect of oral supplements of zinc and iron salts on the development of copper toxicosis: 2. Effect of protein source and other dietary components on the response to high and moderate intakes of copper, *Br. J. Nutr. 20:*135–149.

Sutton, W. R., and Nelson, V. E. 1937. Studies on zinc, *Proc. Soc. Exp. Biol. Med. 36:*211–214.

Swartz, H. A., Christian, J. E., and Andrews, F. N. 1960. Distribution of sulfur[35] and gold[198] labelled gold thioglucose in mice, *Am. J. Physiol. 199:*67–72.

Syed, I. B., and Hosain, F. 1972. Determination of LD_{50} of barium chloride and allied agents, *Toxicol. Appl. Pharmacol. 22:*150.

Symeonides, P. P., Paschaloglou, C., and Papageorigiou, S. 1973. An allergic reaction after internal fixation of a fracture using a Vitallium plate, *J. Allergy Clin. Immunol. 51:*251–252.

Szabo, K. T. 1970. Teratogenic effect of lithium carbonate in the fetal rat, *Nature 225:*73.

Szabo, K. T., Hawk, A. M., and Henry, M. 1970. The teratogenic effect of lithium carbonate upon the palate of random bred mice, *Toxicol. Appl. Pharmacol. 17:*274 (abstr.).

Szmigielski, B., and Litwin, J. 1964. The histochemical demonstration of zinc in blood granulocytes. The new test in diagnosis of neoplastic diseases, *Cancer 17:*1381–1384.

Tabershaw, I. R. (ed.). 1972. *The Toxicology of Beryllium,* USDHEW Publ. Serv., Washington, D.C. (2173:23:16386).

Takada, K., Fujita, M., and Suzuki, M. 1970. Effects of carrier on the retention, excretion, and distribution of ^{144}Ce in the rat, *J. Radiat. Res. 11:*24.

Talbot, R. B., Davison, F. G., Green, J. W., Reece, W. O., and Vangelder, G. 1965. Effects of subcutaneous implantation of rare earth metals, *USAEC Report 1170.*

Tan, T. N., Weston, R. H., Warner, A. C. I., and Hogan, J. P. 1968–1969. Annual Report, Division of Animal Physiology, Commonwealth Scientific and Industrial Research Organization, Sydney, Australia.

Tank, G. and Storvick, C. A. 1960. Effect of naturally occurring selenium and vanadium on dental caries, *J. Dent. Res. 39:*473–479.

Tannebaum, A. 1951. *Toxicology of Uranium,* McGraw-Hill, New York.

Tapp, E. 1966. Beryllium-induced sarcomas of the rabbit tibia, *Br. J. Cancer. 20:*779.

Taylor, D. M. 1972. Interactions between transuranium elements and the components of cells and tissues, *Health Phys. 22:*575–581.

Taylor, D. M., Bligh, P. H., and Duggan, M. H. 1962. The absorption of calcium stronium, barium, and radium from the gastrointestinal tract of rat, *Biochem. J. 83:*25–29.

Taylor, T. J., Rieders, F., and Kocsis, J. J. 1973. Role of Hg^{2+} and methyl mercury on lipid peroxidation, *Fed. Proc. 32(3):*261.

Tedeschi, R. E., and Sunderman, F. W. 1957. Nickel poisoning. V. The metabolism of nickel under normal conditions and after exposure to nickel carbonyl, *AMA Arch. Ind. Health 16:*486–488.

Tepper, L. B. 1971. Beryllium, *CRC Crit. Rev. Toxicol. 1(3):*235–260.

Tepper. L. D., Hardy, H. L., and Chamberlin, R. I. 1961. in: *Toxicity of Beryllium Compounds* (E. Browning, ed.), Elsevier Pub. Co., Amsterdam and New York.

Thakur, M. L., Gunsaekera, S., and Merrick, M. V. 1973. Indium radionuclide helps detect cancers, *Chem. Engr. News 14:*12.

Thomas, J. A., and Thiery, J. P. 1953. Production élective de liposarcoma chez des lapins par les oligoelements zinc et cobalt, *C. R. Acad. Sci. 236:*1387–1389.

Thomas, J. W., and Moss, S. 1951. The effect of orally administered molybdenum on growth, spermatogenesis, and testis histology of young dairy bulls, *J. Dairy Sci. 34:*929–936.

Thomas, M., and Aldridge, W. N. 1966. The inhibition of enzymes by beryllium, *Biochem. J. 98:*94–99.

Thomas, W. C., Jr., and Howard, J. E. 1962. Disturbances in calcium metabolism, in: *Mineral Metabolism,* Vol. II. Part A (C. L. Comar and F. Bronner, eds.), Academic Press, New York and London.

Thompson, J. N., Erdody, P., and Smith, D. C. 1975. Selenium content of food consumed by Canadians, *J. Nutr. 105:*274–277.

Thompson, R. C., and Bair, W. J. 1972. Toxicology of plutonium, *Health Phys. 22:*533–539.

Thompson, R. C., and Hollis, O. L. 1958, Irradiation of the gastrointestinal tract of the rat by ingested ruthenium-106, *Am. J. Physiol. 194:*308–312.

Thompson, R. H., and Todd, J. R. 1970. Chronic copper poisoning in sheep—Biochemical studies of the hemolytic process, in: *Trace Element Metabolism in Animals* (C. F. Mills, ed.), E. and S. Livingston, London.

Thurston, H., Gilmore, G. R., and Swales, J. E. 1972. Aluminum retention and toxicity in renal failure, *Lancet 1:*881–883.

Thyresson, N. 1951. Experimental investigation of thallium poisoning in the rat, *Acta Derm.-vener. 31:*3–10.

Tipton, I. H., and Cook, M. J. 1963. Trace elements in human tissue: Adult subjects from the United States, *Health Phys. 9:*103.

Todd, J. R. 1969. Chronic copper toxicity of ruminants, *Proc. Nutr. Soc. 28:*189–198.

Todd, J. R., and Thompson, R. H. 1963. Studies on chronic copper poisoning. II. Biochemical studies on blood of sheep during hemolytic crisis, *Br. Vet. J. 119:*161–173.

Todd, J. R., and Thompson, R. H. 1965. Studies on chronic copper poisoning: Biochemistry of the toxic syndrome in the calf, *Br. Vet. J. 121:*90–96.

Trapmann, H., 1959. New catalytic processes involving metal ions especially the rare earths: Their effect on cell processes, Parts 1. and 2, *Arzneimittelforsch.* 9:341–346, 403–410.

Trautner, E. M., and Morris, R. 1955. The excretion and retention of ingested lithium and its effect on the ionic balance of man, *Med. J. Aust. 1:*277.

Trautner, E. M., Pennycuik, P. R., Morris, J. H., Gershon, S., and Shankly, K. H. 1958. The effects of prolonged sub-toxic lithium ingestion on pregnancy in rats, *Aust. J. Exp. Biol. Med. Sci. 36:*305.

Trentini, G. P., De Gaetani, C. F., and Saviano, M. S. 1969. Modificazioni istomorfologiche ed istoenzimatiche del corticosurrene di ratto in carenza e in trattamento con zinco, *Boll. Soc. Ital. Biol. Sper. 45:*607–610.

Truhaut, R. 1959. Rechèrches sur la toxicologie du thallium, Instut. National Securité pour la Prevention des Accidents du Travail, Paris.

Tscherkes, L. A., Aptekar, S. G., and Volgarev, M. N. 1961. Hepatic tumors induced by selenium, *Byull. Eksperimental noi Biologii i Meditsiny 53:*78–82 (Russian).

Tscherkes, L. A., Volgarev, M. N., and Aptekar, S. G. 1963. Selenium caused tumors, *Acta Un. Int. Cancer 19:*632–633.

Tsevetkova, R. P. 1970. Influence of cadmium compounds on the generative function, *Gig. Tr. Prof. Gabol. 14:*31–33.

Tu, A. T. 1977. *Venoms. Chemistry and Molecular Biology,* John Wiley, New York.

Tu, A. T., and Bjarnason, B. 1976. Personal Communication.

Tuchweber, B., and Savoie, L. 1968. Rare earth metals and soft tissue calcification, *Proc. Soc. Exp. Biol. Med. 128:*473.

Tupper, R., Watts, R. W. E., and Normall, A. 1955. The incorporation of zinc in mammary tumors and some other tissues of mice after injections of the isotope, *Biochem. J. 59:*264–268.

Underhill, F. P., and Peterman, P. I. 1929. Studies on the metabolism of aluminum: 3. Absorption and excretion of aluminum in normal man, *Am. J. Physiol. 90:*40–51.

Underhill, F. P., Peterman, P. I., and Steel, S. L. 1929. Studies on the metabolism of aluminum: 4. The fate of intravenously injected aluminum, *Am. J. Physiol. 90:*52–61.

Underwood, E. J. 1971. *Trace Elements in Human and Animal Nutrition,* Academic Press, New York and London, p. 416–424.

Underwood, E. J. 1975. Cobalt, *Nutr. Rev. 33:*65–69.

Usov, G. P. 1969. Effect of selenium on basic nervous processes in sheep, *Nater. Respub. Knof. Probl. "Microelem, Med. Zhivotnod,"* 1968.

Vacher, J. 1972. Immunological responses of guinea pigs to beryllium salts, *J. Med. Microbiol.* 5:91–108.

Vacher, J., and Stoner, H. B. 1968a. Transport of beryllium in rat blood, *Biochem. Pharmacol. 17:*93–99.

Vacher, J., and Stoner, H. B. 1968b. The removal of injected beryllium from the blood of rat: The role of reticuloendothelial system, *Br. J. Exp. Pathol. 49:*315.

Vacher, J., Deraedt, R., and Benzoni, I. 1973. Compared effects of two beryllium salts (soluble and insoluble): Toxicity and blockade of reticuloendothelial systems, *Toxicol. Appl. Pharmacol. 24:*497–506.

Vaidya, S. G., Chaudhri, M. A., and Morrison, R. 1970. Localization of gallium in malignant neoplasms, *Lancet 2:*911–914.

Valberg, L. S. 1971. Cobalt absorption, in: *Intestinal Absorption of Metal Ions, Trace Elements and Radionuclides* (S. C. Skoryna and D. Waldron-Edward, eds.), Pergamon Press, Oxford, pp. 257–263.

Vallee, B. L. 1962. Zinc, in: *Mineral Metabolism,* Vol. II, Part B (C. L. Comar and F. Bronner, eds.), Academic Press, New York, p. 443.

Vallee, B. L., Ulmer, D. D., and Wacker, W. E. C. 1960. Arsenic: Toxicology and biochemistry, *Arch. Ind. Health 21:*132.

Van Campen, D. R. 1971. Absorption of copper from the gastrointestinal tract, in: *Intestinal Absorption of Metal Ions, Trace Elements and Radionuclides* (S. C. Skoryna and D. Waldron-Edward, eds.), Pergamon Press, Oxford, p. 211.

Van Campen, D. R., and Scaife, P. V. 1967. Zinc interference with copper absorption in rats, *J. Nutr. 91:*473–476.

Van Cleave, C. D., and Kaylor, C. T. 1955. Distribution, retention, and elimination of [7]Be in the rat after intratracheal injection, *Arch. Ind. Health 11:*375.

Vandi, M. 1969. Cadmium content in cigarettes, *Lancet 2* (7634):132.

Van Heyningen, W., and Gladstone, G. P. 1953. The neurotoxin of *Shigella shigae:* 3. The effect of iron on production of the toxin, *Br. J. Exp. Pathol. 34:*221–229.

Van Kien, K. L., and Thai tuong, V. 1955. Etude de l'elimination de l'eau minerale et organique, *C.R. Soc. Biol. (Paris) 149:*2196.

Van Niekerk, R. N. 1937. Einige pharmakologische Untersuchungen mit Salzen von reinem Zirkonium und reinem Hafnium, *Naunyn Schmiedeberg's Arch. Exp. Pathol. Pharmakol. 184:*686–693.

Van Ordstrand, D., and Deodhar, S. 1974. Cited in: Beryllium, *Environment 16(3):*35–36.

Venugopal, B., and Luckey, T. D. 1975. Toxicology of non-radioactive heavy metals and their salts, *Environ. Qual. Saf. Suppl. 1:*4–73.

Venugopal, B., Hutcheson, D. P., Gray, D. H., Luckey, T. D. 1974. Rare earth metal oxides as nutritional markers in humans, *Fed. Proc. 33:*703 (abstr.).

Verity, M. A., and Reith, A. 1967. Effect of mercurial compounds on the structure-linked latency of lysosomal hydrolases, *Biochem. J. 105:*685.

Vernetti Blina, L. 1928. Richerche clinica e sperimentale sull assido di Titanio, *Rif. Med. 47:*1516–1518.

Vigiliani, E. C. 1969. The biopathology of cadmium, *Am. Ind. Hyg. Ass. J. 30:*329–340.

Vignoli, L., Poursines, Y., Oliver, H., and Merland, R. 1946. Contribution a study of experimental intoxication by indium, *Arch. Mal. Prof. 7:*356–360.

Vikbladh, I. 1950. Studies on zinc in blood, *Scand. J. Clin. Invest. 2:*143–148.

Visek, W. J., Whitney, I. B., Kuhn, V. S. G., and Comar, C. L. 1963. Metabolism of chromium[51] by animals as influenced by chemical state, *Proc. Soc. Exp. Biol. Med. 84:*610–613.

Voegtlin, C. 1925. The pharmacology of arsphenamine and related arsenicals, *Physiol. Rev. 5:*63–94.

Voegtlin, C., and Hodge, H. C. (eds.) 1953. *Pharmacology and Toxicology of Uranium Compounds,* McGraw Hill, New York.

Volf, V. 1971. Intestinal absorption of strontium, in: *Intestinal Absorption of Metal Ions, Trace Elements and Radio Nuclides* (S. C. Skoryna and D. Waldron-Edward, eds.), Pergamon Press, Oxford, pp. 277–293.

Volgarev, M. N., and Tscherkes, L. A. 1967. Further studies in tissue changes associated with sodium selenate, in: *Selenium in Biomedicine* (O. H. Muth, J. E. Oldfield, and P. H. Westwig, eds.), AVI Publishing Co., Connecticut, pp. 179–184.

Von Oettingen, W. F. 1935. Manganese: Its distribution, pharmacology, and health hazards, *Physiol. Rev. 15:*175.

Vorwald, A. J. 1959. Experimental pulmonary cancer in monkeys, Progress Report, U.S. Public Health Service Grant C-2507 (C4) SEOH.

Vorwald, A. J., Reeves, A. L., and Urban, E. C. J. 1966. Experimental beryllium toxicology, in: *Beryllium: Its Industrial Hygiene Aspects* (H. C. Stokinger, ed.), Academic Press, New York.

Wacker, W. E. C., and Vallee, B. L. 1959. Nucleic acids and metals. Chromium, manganese, nickel, iron and other metals in ribonucleic acid from diverse biological sources, *J. Biol. Chem. 234:*3257–3262.

Wacker, W. E. C., and Vallee, B. L. 1964. Magnesium, in: *Mineral Metabolism,* Vol. II, Part A, "Elements" (C. L. Comar and F. Bronner, eds.), Academic Press, New York and London, pp. 483–521.

Waechter, R. V., Henning, H., and Frey, G. 1969. Lethal hemorrhagic diathesis after gold therapy, *Med. Klin. 64(22):*1046–1050.

Wahlstrom, R. C., and Olson, O. E. 1959. The effect of selenium on reproduction in swine, *J. Anim. Sci. 18:*141–145.

Waldron, H. A. 1966. Anaemia of lead poisoning: A review, *Br. J. Ind. Med. 23:*83.

Walker-Smith, J. A., and Blomfield, J. 1973. Wilson's disease or chronic copper poisoning, *Arch. Dis. Child. 48:*476–479.

Wallace, W. 1947. Methemoglobinemia in an infant as a result of administration of bismuth nitrate, *J. A. M. A. 133:*1280–1283.

Wallach, S., Bellavia, J. V., Reizenstein, D. L., and Gamponia, P. J. 1967. Tissue distribution and transport of electroytes Mg^{28} and Ca^{47} in hypermagnesemia, *Metabolism 16:*451–464.

Walters, M., and Roe, F. J. C. 1965. A study of the effects of tin and zinc administered orally to mice over a prolonged period, *Food Cosmet. Toxicol. 3:*271–276.

Waltschewa, W., Slatewa, M., and Michailow, I. 1972. Adverse effects of chronic nickel sulfate poisoning in male rats, *Exp. Pathol. 6:*116–121.

Warburton, S., Udler, W., Ewert, R. M., and Haynes, W. S. 1962. Outbreak of foodborne illness attributed to tin, *Pub. Health Rep. 77:*798–800.

Warren, C. O., Schubmehl, Q. D., and Wood, I. R. 1944. Studies on the mechanims of cobalt polycythemia, *Am. J. Physiol. 142:*173–176.

Wassarman, M., and Mihail, G. 1964. Significant indicators for the early detection of manganism in miners working in manganese mines, *Acta Medicinae Legalis et Socialis 17:*61–65.

Wasserman, R. H. 1962. Metabolic behavior of ^{137}Cs in lactating goats, *Int. Rad. Biol. 4:*299–310.

Waxman, H. S., and Rabinowitz, M. 1966. Control of reticulocyte polyribosome content and hemoglobin synthesis by heme, *Biochem. Biophys. Acta 129:*369–379.

Weaver, J. C., Koslainsek, V. M., and Richards, P. D. N. 1956. Cobalt tumor of thyroid gland, *Calif. Med. 85:*110–112.

Weber, C. W., and Reid, B. L. 1969*a*. Effect of dietary cadmium on mice, *Toxicol. Appl. Pharmacol. 14:*420.

Weber, G. B., and Reid, B. L. 1969*b*. Nickel toxicity in young growing mice, *J. Anim. Sci. 28:*620–623.

Weiss, G. B., and Goodman, F. R. 1969. Effects of lanthanum on contraction, calcium distribution, and Ca movement in intestinal smooth muscle, *J. Pharm. Exp. Ther. 169:*46.

Weitz, A., and Ober, E. E. 1965. Physiological distribution of antimony after administration of Sb^{124} labeled tartar emetic. *Bull. Wld. Hlth. Org. 33:*137.

Welsh, S. O., and Soares, J. H., Jr. 1973. Serum transaminase levels and the interacting effect of selenium and vitamin E in mercury toxicity, *Fed. Proc. 32:*261 (abstr.).

Wenzel, W. J., Thomas, R. G., and McCellen, R. O. 1969. Effect of stable yttrium concentration on the distribution and excretion of inhaled radio yttrium in the rat, *Am. Ind. Hyg. 30:*630.

Westernhagen, B. V. 1970. Histochemical demonstrable changes of metabolism in the inner ear of guinea pig after chronic arsenic poisoning, *Arch. Ohr. Nas.-Kehlkheilk 197:*7–13.

Whanger, P. D. 1973. Effect of dietary cadmium on intracellular distribution of hepatic iron in rats, *Res. Commun. Chem. Pathol. Pharmacol. 5:*733–740.

Whitten, C. F., and Brough, A. J. 1971. The pathophysiology of acute iron poisoning, *Clin. Toxicol. 4(4):*585–597.

Wiberg, G. S. 1968. The effect of cobalt ions on energy metabolism in the rat, *Can. J. Biochem. 46:*549.

Wiberg, G. S., Munro, I. C., and Grice, H. C. 1967. Studies on cobalt toxicity, *Toxicol. Appl. Pharmacol. 10:*395.

Wiberg, G. S., Munro, I. C., Meranger, J. C., Morrison, A. B., Grice, H. C., and Heggtreit, H. A. 1969. Factors affecting cardiotoxicity of cobalt, *Clin. Toxicol. 2:*257–272.

Widdowson, E. M., Chan, H., and Milner, R. D. G. 1972. Accumulation of Cu, Zn, Mn, Cr, and Co in the human liver before birth, *Biol. Neonate 20:*300–367.

Wiederanders, R. E., Evans, G. W., and Wasdahl, A. 1968. Acute and chronic copper poisoning in the rat, *Lancet 88:*286–288.

Wieland, T., and Wieland, O. 1972. The toxic peptides of Amanta species. in: *Microbial Toxins,* Vol. 8 (S. Kadis, A. Ciegler, and S. J. Ajl, eds.), pp. 249–280, Academic Press, New York.

Wilbanks, G. D., Bressler, B., Peete, C. H., Jr., Cherny, W. B., and London, W. L. 1970. Toxic effects of lithium carbonate in a mother and newborn infant, *J. A. M. A. 213:*865.

Wilkinson, D. R., and Palmer, W. 1975. Lead in teeth as a function of age, *Am. Lab.* (Mar.) 67–70.

Williams, P. A., and Peacocke, A. R. 1967. The binding of Ca^{2+} and Y^{3+} to a glycoprotein from bovine cortical bone, *Biochem. J. 105:*1177.

Willoughby, R. A., MacDonald, E., McSherry, B. J., and Brown, G. 1972. The interaction of toxic amounts of lead and zinc fed to young growing horses, *Vet. Rec. 91:*382.

Wilson, H. B., Stokinger, H. E., and Sylvester, G. E. 1953. Acute toxicity of carnotite ore dust, *Arch. Ind. Hyg. 7:*301–309.

Wilson, S. J., Health, H. E., Nelson, P. L., and Ens, G. G. 1958. Blood coagulation in acute iron intoxication, *Blood 13:*483–491.

Witschi, H. P. 1968. Inhibition of DNA synthesis in regenerating liver by beryllium, *Lab. Invest. 19:*67–69.

Witschi, H. P. 1970. Effects of beryllium on DNA-synthesizing enzymes in regenerating rat liver, *Biochem. J. 120:*623–628.

Witschi, H. P. 1971. Interference of beryllium with enzyme induction in liver, *Toxicol. Appl. Pharmacol. 20:*565.

Witschi, H. P. 1972. A comparative study of *in vivo* RNA and protein synthesis in rat liver and lung, *Cancer Res. 32:*1686–1694.

Witzleben, C. L., and Chaffey, N. J. 1966. Acute ferrous sulfate poisoning, *Arch. Pathol. 82:*454.

Worker, N. A., and Migicovsky, B. B. 1961. Effects of vitamin D on the utilization of zinc, *J. Nutr. 75:*222–226.

Worowski, K. 1968. The hypercoagulability in mercury chloride intoxicated dogs, *Thromb. Diath. Haemorrh. 19:*236.

Worwood, M., and Jacobs, A. 1971. Absorption of ⁵⁹Fe in the rat; Iron binding substances in the soluble fraction of intestinal mucosa, *Life Sci. 10(1):*1363–1373.

Wraae, O., Geisler, A., and Olesen, O. V. 1972. Relation between vasopression stimulation of renal adenyl cyclase and Li-induced polyuria, *Acta Pharmacol. Toxicol. 31:*314–317.

Wright, T. L., Hoffman, L. H., and Davies, J. 1970. Lithium teratogenicity, *Lancet 2:*876.

Wright, T. L., Hoffman, L. H., and Davies, J. 1971. Teratogenic effects of lithium in rats, *Teratology 4:*151.

Wright, W.R. 1968. Metabolic Interrelationship between Vanadium and Chromium, Thesis, North Carolina State University, Raleigh.

Wustenberg, P. W. 1972. Magnesium metabolism from the nephrologic viewpoint, *Z. Urol. Nephrol. 65:*241–257.

Yang, M. G., Wang, J. H. C., Garcia, J. D., Post, E., and Lei, K. Y. 1973. Mammary transfer of ^{203}Hg from mothers to brains of nursing rats, *Proc. Soc. Exp. Biol. Med. 142:*723–726.

Yoshikawa, H. 1970. Preventive effect of pretreatment with low dose of metals on the acute toxicity of metals in mice, *Ind. Health 8:*184–190.

Zanni, A. C. 1965. Contribution to the pharmacology of europium, *Rev. Fac. Farm. Bioquim. (Univ. of Sao Paulo) 3:*199–240.

Zanni, A. C., and Ramos, A. O. 1965. Electrocardiographic changes caused by europium in rats, *Folia. Clin. Biol. 34:*115–118.

Zeya, H. I., and Spitznagel, J. K. 1968. Arginine-rich proteins of polymorphonuclear leukocyte lysosomes: Antimicrobial specificity and biochemical heterogeneity, *J. Exp. Med. 127:*927.

Zollinger, H. W. 1953. Kidney adenomas and carcinomas in rats caused by chronic lead poisoning and their relationship to corresponding human neoplasma, *Virchows Arch. Path. Anat. 323:*694–710.

Zorlein, V., Langendorf, V., Hannover, R., Schulte, S., and Siebert, G. 1969. Rubidum metabolism in the nuclei of rat liver cells, *Hoppe-Seylers Z. Physiol. Chem. 350:*1683–1685.

INDEX